Heat-Transfer Calculations

Heat-Transfer Calculations

Myer Kutz
Editor

McGraw-Hill
New York Chicago San Francisco Lisbon London Madrid
Mexico City Milan New Delhi San Juan Seoul
Singapore Sydney Toronto

CIP Data is on file with the Library of Congress

Copyright © 2006 by The McGraw-Hill Companies, Inc. All rights reserved. Printed in the United States of America. Except as permitted under the United States Copyright Act of 1976, no part of this publication may be reproduced or distributed in any form or by any means, or stored in a data base or retrieval system, without the prior written permission of the publisher.

1 2 3 4 5 6 7 8 9 0   DOC/DOC   0 1 0 9 8 7 6 5

ISBN 0-07-141041-4

The sponsoring editor for this book was Kenneth P. McCombs, the editing supervisor was Stephen M. Smith, and the production supervisor was Pamela A. Pelton. It was set in Century Schoolbook by TechBooks. The art director for the cover was Anthony Landi.

Printed and bound by RR Donnelley.

McGraw-Hill books are available at special quantity discounts to use as premiums and sales promotions, or for use in corporate training programs. For more information, please write to the Director of Special Sales, McGraw-Hill Professional, Two Penn Plaza, New York, NY 10121-2298. Or contact your local bookstore.

This book is printed on recycled, acid-free paper containing a minimum of 50% recycled, de-inked fiber.

Information contained in this work has been obtained by The McGraw-Hill Companies, Inc. ("McGraw-Hill") from sources believed to be reliable. However, neither McGraw-Hill nor its authors guarantee the accuracy or completeness of any information published herein and neither McGraw-Hill nor its authors shall be responsible for any errors, omissions, or damages arising out of use of this information. This work is published with the understanding that McGraw-Hill and its authors are supplying information but are not attempting to render engineering or other professional services. If such services are required, the assistance of an appropriate professional should be sought.

To fond memories of Doc Draper's Thermal Lab, and to Archie, Charlie, and the rest of the fine people I worked with there

Contents

Contributors xiii
Preface xvii

Part 1 Introductory Calculations   1.1

Chapter 1. Multiphase Films and Phase Change
Greg F. Naterer   1.3

Chapter 2. Industrial Heat-Transfer Calculations
Mohammad G. Rasul   2.1

Part 2 Steady-State Calculations   3.1

Chapter 3. Heat-Transfer and Temperature Results for a Moving Sheet Situated in a Moving Fluid
John P. Abraham and Ephraim M. Sparrow   3.3

Chapter 4. Solution for the Heat-Transfer Design of a Cooled Gas Turbine Airfoil
Ronald S. Bunker   4.1

Chapter 5. Steady-State Heat-Transfer Problem: Cooling Fin
David R. Dearth   5.1

Chapter 6. Cooling of a Fuel Cell
Jason M. Keith   6.1

Chapter 7. Turbogenerator Rotor Cooling Calculation
James T. McLeskey, Jr.   7.1

Chapter 8. Heat Transfer through a Double-Glazed Window
P. H. Oosthuizen and David Naylor   8.1

Part 3 Transient and Cyclic Calculations   9.1

Chapter 9. Use of Green's Function to Solve for Temperatures in a Bimaterial Slab Exposed to a Periodic Heat Flux Applied to Corrosion Detection
Larry W. Byrd   9.3

viii   Contents

Chapter 10. Lumped-Capacitance Model of a Tube Heated by a
Periodic Source with Application to a PDE Tube
   Larry W. Byrd, Fred Schauer, John L. Hoke, and Royce Bradley   10.1

Chapter 11. Calculation of Decoking Intervals for Direct-Fired Gas
and Liquid Cracking Heaters
   Alan Cross   11.1

Chapter 12. Transient Heat-Transfer Problem: Tape Pack Cooling
   David R. Dearth   12.1

Chapter 13. Calculation of the Transient Response of a Roof to
Diurnal Heat Load Variations
   Jack M. Kleinfeld   13.1

Chapter 14. Transient Heating of a Painted Vehicle Body Panel in an
Automobile Assembly Plant Paint Shop Oven
   Thomas M. Lawrence and Scott Adams   14.1

Chapter 15. Thermal System Transient Response
   James E. Marthinuss, Jr., and George Hall   15.1

Chapter 16. Thermal Response of Laminates to Cyclic Heat Input
from the Edge
   Wataru Nakayama   16.1

Chapter 17. A Simple Calculation Procedure for Transient Heat and
Mass Transfer with Phase Change: Moist Air Example
   Pedro J. Mago and S. A. Sherif   17.1

Chapter 18. A Calculation Procedure for First- and Second-Law
Efficiency Optimization of Refrigeration Cycles
   J. S. Tiedeman and S. A. Sherif   18.1

Chapter 19. Transient Analysis of Low Temperature, Low Energy
Carrier (LoTEC©)
   Francis C. Wessling   19.1

Chapter 20. Parameter Estimation of Low Temperature, Low Energy
Carrier (LoTEC©)
   Francis C. Wessling and James J. Swain   20.1

Part 4   Heat-Transfer-Coefficient Determination   21.1

Chapter 21. Calculation of Convective Heat-Transfer Coefficient
Using Semi-infinite Solid Assumption
   Srinath V. Ekkad   21.3

Chapter 22. Determination of Heat-Transfer Film Coefficients by the Wilson Analysis
Ronald J. Willey   22.1

Part 5 Tubes, Pipes, and Ducts   23.1

Chapter 23. Calculation of Local Inside-Wall Convective Heat-Transfer Parameters from Measurements of Local Outside-Wall Temperatures along an Electrically Heated Circular Tube
Afshin J. Ghajar and Jae-yong Kim   23.3

Chapter 24. Two-Phase Pressure Drop in Pipes
S. Dyer Harris   24.1

Chapter 25. Heat-Transfer Calculations for Predicting Solids Deposition in Pipeline Transportation of "Waxy" Crude Oils
Anil K. Mehrotra and Hamid O. Bidmus   25.1

Chapter 26. Heat Transfer in a Circular Duct
Nicholas Tiliakos   26.1

Part 6 Heat Exchangers   27.1

Chapter 27. Air Cooling of a High-Voltage Power Supply Using a Compact Heat Exchanger
Kirk D. Hagen   27.3

Chapter 28. Energy Recovery from an Industrial Clothes Dryer Using a Condensing Heat Exchanger
David Naylor and P. H. Oosthuizen   28.1

Chapter 29. Sizing of a Crossflow Compact Heat Exchanger
Dusan P. Sekulic   29.1

Chapter 30. Single-Phase Natural Circulation Loops: An Analysis Methodology with an Example (Numerical) Calculation for an Air-Cooled Heat Exchanger
Daniel M. Speyer   30.1

Chapter 31. Evaluation of Condensation Heat Transfer in a Vertical Tube Heat Exchanger
Karen Vierow   31.1

Chapter 32. Heat-Exchanger Design Using an Evolutionary Algorithm
Keith A. Woodbury   32.1

Part 7 Fluidized Beds — 33.1

Chapter 33. Fluidized-Bed Heat Transfer
Charles J. Coronella and Scott A. Cooper — 33.3

Part 8 Parameter and Boundary Estimation — 34.1

Chapter 34. Estimation of Parameters in Models
Ashley F. Emery — 34.3

Chapter 35. Upper Bounds of Heat Transfer from Boxes
Wataru Nakayama — 35.1

Chapter 36. Estimating Freezing Time of Foods
R. Paul Singh — 36.1

Part 9 Temperature Control — 37.1

Chapter 37. Precision Temperature Control Using a Thermoelectric Module
Marc Hodes — 37.3

Part 10 Thermal Analysis and Design — 38.1

Chapter 38. Thermal Analysis of a Large Telescope Mirror
Nathan E. Dalrymple — 38.3

Chapter 39. Thermal Design and Operation of a Portable PCM Cooler
R. Letan and G. Ziskind — 39.1

Chapter 40. A First-Order Thermal Analysis of Balloon-Borne Air-Cooled Electronics
Angela Minichiello — 40.1

Chapter 41. Thermal Analysis of Convectively Cooled Heat-Dissipating Components on Printed-Circuit Boards
Hyunjae Park — 41.1

Chapter 42. Design of a Fusion Bonding Process for Fabricating Thermoplastic-Matrix Composites
F. Yang and R. Pitchumani — 42.1

Part 11 Economic Optimization   43.1

Chapter 43. Economic Optimization of Heat-Transfer Systems
Thomas M. Adams   43.3

Chapter 44. Turkey Oven Design Problem
Jason M. Keith   44.1

Index follows Chapter 44

Contributors

John P. Abraham *School of Engineering, University of St. Thomas, St. Paul, Minnesota* (Chap. 3)

Scott Adams *Paint Engineering, Ford Motor Company, Vehicle Operations General Office, Allen Park, Michigan* (Chap. 14)

Thomas M. Adams *Mechanical Engineering Department, Rose-Hulman Institute of Technology, Terre Haute, Indiana* (Chap. 43)

Hamid O. Bidmus *Department of Chemical and Petroleum Engineering, University of Calgary, Calgary, Alberta, Canada* (Chap. 25)

Royce Bradley *Innovative Scientific Solutions, Inc., Wright-Patterson AFB, Ohio* (Chap. 10)

Ronald S. Bunker *Energy and Propulsion Technologies, General Electric Global Research Center, Niskayuna, New York* (Chap. 4)

Larry W. Byrd *AFRL/VASA, Wright-Patterson AFB, Ohio* (Chaps. 9, 10)

Scott A. Cooper *Chemical and Metallurgical Engineering Dept., University of Nevada, Reno, Nevada* (Chap. 33)

Charles J. Coronella *Chemical and Metallurgical Engineering Dept., University of Nevada, Reno, Nevada* (Chap. 33)

Alan Cross *Little Neck, New York* (Chap. 11)

Nathan E. Dalrymple *Air Force Research Laboratory, Space Vehicles Directorate, National Solar Observatory, Sunspot, New Mexico* (Chap. 38)

David R. Dearth *Applied Analysis & Technology, Huntington Beach, California* (Chaps. 5, 12)

Srinath V. Ekkad *Mechanical Engineering Department, Louisiana State University, Baton Rouge, Louisiana* (Chap. 21)

Ashley F. Emery *Mechanical Engineering Department, University of Washington, Seattle, Washington* (Chap. 34)

Afshin J. Ghajar *School of Mechanical and Aerospace Engineering, Oklahoma State University, Stillwater, Oklahoma* (Chap. 23)

Kirk D. Hagen *Manufacturing and Mechanical Engineering Technology Department, Weber State University, Ogden, Utah* (Chap. 27)

George Hall *Northrop Grumman Electronics Systems, Linthicum, Maryland* (Chap. 15)

S. Dyer Harris *Equipment Engineering Services, P.A., Wilmington, Delaware* (Chap. 24)

Marc Hodes *Bell Laboratories, Lucent Technologies, Murray Hill, New Jersey* (Chap. 37)

John L. Hoke *Innovative Scientific Solutions, Inc., Wright-Patterson AFB, Ohio* (Chap. 10)

Jason M. Keith *Department of Chemical Engineering, Michigan Technological University, Houghton, Michigan* (Chaps. 6, 44)

Jae-yong Kim *School of Mechanical and Aerospace Engineering, Oklahoma State University, Stillwater, Oklahoma* (Chap. 23)

Jack M. Kleinfeld *Kleinfeld Technical Services, Inc., Bronx, New York* (Chap. 13)

Thomas M. Lawrence *Department of Biological and Agricultural Engineering, Driftmier Engineering Center, University of Georgia, Athens, Georgia* (Chap. 14)

R. Letan *Heat Transfer Laboratory, Mechanical Engineering Department, Ben-Gurion University of the Negev, Beer-Sheva, Israel* (Chap. 39)

Pedro J. Mago *Department of Mechanical Engineering, Mississippi State University, Mississippi State, Mississippi* (Chap. 17)

James E. Marthinuss, Jr. *Northrop Grumman Electronics Systems, Linthicum, Maryland* (Chap. 15)

James T. McLeskey, Jr. *Department of Mechanical Engineering, Virginia Commonwealth University, Richmond, Virginia* (Chap. 7)

Anil K. Mehrotra *Department of Chemical and Petroleum Engineering, University of Calgary, Calgary, Alberta, Canada* (Chap. 25)

Angela Minichiello *Space Dynamics Laboratory, Utah State University, North Logan, Utah* (Chap. 40)

Wataru Nakayama *ThermTech International, Kanagawa, Japan* (Chaps. 16, 35)

Greg F. Naterer *University of Ontario Institute of Technology, Oshawa, Ontario, Canada* (Chap. 1)

David Naylor *Department of Mechanical and Industrial Engineering, Ryerson University, Toronto, Ontario, Canada* (Chaps. 8, 28)

P. H. Oosthuizen *Department of Mechanical Engineering, Queen's University, Kingston, Ontario, Canada* (Chaps. 8, 28)

Hyunjae Park *Thermofluid Science and Energy Research Center (TSERC), Department of Mechanical Engineering, Marquette University, Milwaukee, Wisconsin* (Chap. 41)

R. Pitchumani *Advanced Materials and Technologies Laboratory, Department of Mechanical Engineering, University of Connecticut, Storrs, Connecticut* (Chap. 42)

Mohammad G. Rasul *School of Advanced Technologies and Processes, Faculty of Engineering and Physical Systems, Central Queensland University, Rockhampton, Queensland, Australia* (Chap. 2)

Fred Schauer *Pulsed Detonation Research, AFRL/PRTS, Wright-Patterson AFB, Ohio* (Chap. 10)

Dusan P. Sekulic *Department of Mechanical Engineering, University of Kentucky, Lexington, Kentucky* (Chap. 29)

S. A. Sherif *Department of Mechanical and Aerospace Engineering, University of Florida, Gainesville, Florida* (Chaps. 17, 18)

R. Paul Singh *Department of Biological and Agricultural Engineering, University of California, Davis, California* (Chap. 36)

Ephraim M. Sparrow *Department of Mechanical Engineering, Univeristy of Minnesota, Minneapolis, Minnesota* (Chap. 3)

Daniel M. Speyer *Mechanical Engineering Department, The Cooper Union for the Advancement of Science and Art, New York, New York* (Chap. 30)

James J. Swain *Department of Industrial and Systems Engineering and Engineering Management, University of Alabama in Huntsville, Huntsville, Alabama* (Chap. 20)

J. S. Tiedeman *BKV Group, Minneapolis, Minnesota* (Chap. 18)

Nicholas Tiliakos *Technology Development Dept., ATK-GASL, Ronkonkoma, New York* (Chap. 26)

Karen Vierow *School of Nuclear Engineering, Purdue University, West Lafayette, Indiana* (Chap. 31)

Francis C. Wessling *Department of Mechanical and Aerospace Engineering, University of Alabama in Huntsville, Huntsville, Alabama* (Chaps. 19, 20)

Ronald J. Willey *Department of Chemical Engineering, Northeastern University, Boston, Massachusetts* (Chap. 22)

Keith A. Woodbury *Mechanical Engineering Department, University of Alabama, Tuscaloosa, Alabama* (Chap. 32)

F. Yang *Advanced Materials and Technologies Laboratory, Department of Mechanical Engineering, University of Connecticut, Storrs, Connecticut* (Chap. 42)

G. Ziskind *Heat Transfer Laboratory, Mechanical Engineering Department, Ben-Gurion University of the Negev, Beer-Sheva, Israel* (Chap. 39)

Preface

Several years ago, when Ken McCombs, my editor at McGraw-Hill, telephoned me out of the blue and asked if I would put together a handbook of heat-transfer calculations, I was, I must admit, dubious about such a project, even though I did have a leg up on it, so to speak. (The situation was unusual, moreover; my standard practice in developing handbooks is to conceive of the central topic myself and sell the idea to a publisher. But no matter.) As it happens, I had worked as a heat-transfer engineer on the Apollo project at the MIT Instrumentation Lab early in my career and later worked on thermal design issues at a Cambridge company that made rocket-borne telescopes that examined solar flares and a Manhattan company that made de-icing equipment. My first book was *Temperature Control*, which was published when I was in my late twenties. Later in my career I edited a number of heat-transfer volumes while working as an acquisitions editor at a major publisher. Nevertheless, my skepticism deepened the more I thought about the project. After all, I knew as well as anyone else that there was no shortage of heat-transfer books at all levels, from undergraduate to professional, that contained standard calculations, often with example problems. So the question I was faced with was this: How do I make this book different and, therefore, useful and needed?

The charge I eventually gave myself for creating this handbook was to find contributors who, as I wrote in e-mails soliciting potential contributions, "have, or wish to develop, any interesting, unusual, or even unique—but at the same time practical and instructive—calculations or procedures covering any aspect of heat transfer, whether involving a device, machinery, an assembly, or an environment, that you would consider submitting for the volume of Heat Transfer Calculations. I am looking not for standard calculations, but for new and novel techniques not available in undergraduate or graduate textbooks. An example might involve a micro-scale heat transfer application in the biomedical area. Contributions must be original and not previously published. Calculations not published in detail in journal articles or published in theses are acceptable."

I have to say that I did not arrive at this charge to myself and to potential contributors immediately. At the outset, in fact, I developed a proposed Table of Contents that looked like what you will find in many, many other heat-transfer volumes. But after perusing a number of them, I concluded that a different approach to assembling a handbook of calculations was warranted, to ensure that it comprised work that would complement and supplement the vast literature already available. What I wanted, I decided, were calculations that addressed real-life problems encountered in engineering practice, and I recognized that for the most part they would be far more complex than those typically found in heat-transfer books and would be, as a result, more difficult to categorize. But that would be a small price to pay, I reckoned, in exchange for the richness of the material that would be going into the handbook.

The result of my solicitations is a book of some 44 chapters, which I have divided into 11 parts. While the part titles may be, in some cases, a bit unorthodox for a typical heat-transfer handbook, the flow of the material will feel comfortable to practicing engineers. Each part has a brief introduction that sets the stage for what is in the part and helps provide continuity for the book as a whole. The emphasis is on the industrial problems that the calculations address, as well as on the nature of the calculations themselves. Readers will note a small degree of overlap among several parts, which adds to the flow.

The calculations in this handbook are valuable not only for what they say about how to solve the particular problems that contributors confronted, but also because they serve as prompts and guides to solving the different problems that readers have to contend with. The substantial collection of references in the book provides additional, and in some cases, vital help in this regard. I should also point out that the calculations are fresh; they were submitted in the latter half of 2004 and the first two months of 2005.

I would like to express my heartfelt appreciation to each and every one of the contributors. I know how difficult it is to find the time to contribute a chapter, which involves not only processing text and equations, but also preparing illustrations and lists of references, as well as dealing with copy editor queries and checking page proof. So it is no minor miracle when anyone makes the effort. Thanks also to the McGraw-Hill production team, which brings manuscripts to printed pages in a very short time. Ken McCombs continues to be encouraging, and I thank him for that. Finally, my wife Arlene, who keeps me healthy and happy, knows how I feel about her love and support. Still, I don't want to miss this opportunity to thank her again.

Myer Kutz

Part 1

Introductory Calculations

Heat-Transfer Calculations *opens with eight brief industrial heat-transfer calculations. While they are short, especially when compared to the calculations that make up the bulk of this handbook, these opening calculations are not trivial. And they deal with real life, which is the hallmark of the calculations throughout the book:*

- *A chemical processing technology that needs a thinner liquid film coating for better adhesive properties of a new product*
- *A new passive method of cooling an electronic assembly that uses a closed-loop thermosyphon*
- *During casting of an industrial metal component, a more efficient gating and risering system that minimizes the amount of metal poured to produce a good casting*
- *A furnace wall in a coal-fired power plant*
- *A reinforced-concrete smokestack that must be lined with a refractory on the inside*
- *The condenser in a large steam power plant*
- *The duct of an air conditioning plant*
- *A concentric tube heat exchanger with specified operating conditions*

Chapter 1

Multiphase Films and Phase Change

Greg F. Naterer
University of Ontario Institute of Technology
Oshawa, Ontario, Canada

Problem 1

A chemical processing technology needs a thinner liquid film coating for better adhesive properties of a new product. The film flows down under gravity along an inclined plate. Effective thermal control is needed to achieve a specified level of adhesive quality. This goal is accomplished by changing the wall temperature at a certain point along the plate, which affects the thermal boundary layer within the film. Before evaluating heat transfer based on the Nusselt number, the film velocity is needed to predict thermal convection within the film. Then the growth of the thermal boundary layer can be determined. The spatial profile of temperature remains constant, after the point where the edge of the thermal boundary layer reaches the edge of the film. This point is needed for sizing of the processing equipment. Thus, a thermal analysis is needed to find the film velocity, boundary-layer growth, and Nusselt number for evaluating the temperature within the film. This processing temperature affects the adhesive properties of the film.

Consider a liquid film (ethylene glycol at 300 K) during testing of the chemical process. The film slows steadily down along a flat plate in the x direction. The plate is inclined at an angle of θ with respect to the horizontal direction. The x position is measured along the direction of the plate. It is assumed that the film thickness δ remains nearly uniform. A reference position of $x = 0$ is defined at a point where the wall boundary conditions change abruptly. Before this point, the film, wall,

1.4 Introductory Calculations

and air temperatures are T_0. Then the surface temperature increases to T_n (downstream of $x = 0$) and a thermal boundary layer δ_T develops and grows in thickness within the film.

Part (a) Using a reduced form of the momentum equation in the liquid film, as well as appropriate interfacial and wall boundary conditions, derive the following velocity distribution in the film:

$$\frac{u}{U} = 2\left(\frac{y}{\delta}\right) - \left(\frac{y}{\delta}\right)^2$$

where U refers to the velocity at the film-air interface. Neglect changes of film velocity in the x direction along the plate.

Part (b) Consider a linear approximation of the temperature profile within the thermal boundary layer, specifically, $T = a + by$, where the coefficients a and b can be determined from the boundary conditions. Then perform an integral analysis by integrating the relevant energy equation to obtain the thermal boundary-layer thickness δ_T in terms of x, δ, U, and α. For $U = 1$ cm/s and a film thickness of 2 mm, estimate the x location where δ_T reaches the edge of the liquid film.

Part (c) Outline a solution procedure to find the Nusselt number, based on results in part (b).

Solution

Part (a) Start with the following mass and momentum equations in the film:

$$\frac{\partial u}{\partial x} + \frac{\partial v}{\partial y} = 0 \tag{1.1}$$

$$\rho u \frac{\partial u}{\partial x} + \rho v \frac{\partial u}{\partial y} = -\frac{\partial p}{\partial x} + \mu \frac{\partial^2 u}{\partial y^2} + \rho g (\sin \theta) \tag{1.2}$$

From the continuity equation, the derivative involving u vanishes since the terminal velocity is attained, and thus v equals zero (since $\partial v/\partial y = 0$ subject to $v = 0$ on the plate). As a result, both terms on the left side of Eq. (1.2), as well as the pressure gradient term (flat plate), disappear in the momentum equation. Also, we have the boundary condition with a zero velocity at the wall ($y = 0$) and an approximately zero velocity gradient at the air-water interface. The gradient condition means that $\partial u/\partial y = 0$ at $y = \delta$ as a result of matching shear stress values ($\mu \, \partial u/\partial y$) on both air and water sides of the interface and $\mu_a \ll \mu_w$.

Integrating the reduced momentum equation, subject to the boundary conditions, we obtain the result that

$$u(y) = \frac{g(\sin\theta)}{\nu}\left(\delta y - \frac{1}{2}y^2\right) \tag{1.3}$$

Substituting $y = \delta$ into Eq. (1.3) gives the following expression for the velocity at the air-water interface:

$$U = \frac{g(\sin\theta)}{\nu}\left(\frac{\delta^2}{2}\right) \tag{1.4}$$

Combining Eqs. (1.3) and (1.4), we have

$$\frac{u}{U} = 2\left(\frac{y}{\delta}\right) - \left(\frac{y}{\delta}\right)^2 \tag{1.5}$$

Part (b) Assuming steady, two-dimensional (2D), laminar, incompressible flow in the liquid film, with negligible stream-wise conduction and viscous dissipation, we have the following governing energy equation:

$$\frac{\partial}{\partial x}(uT) + \frac{\partial}{\partial y}(vT) = \alpha\frac{\partial^2 T}{\partial y^2} \tag{1.6}$$

where the second term vanishes as a result of the previous discussion ($v = 0$). Multiplying Eq. (1.6) by dy, integrating across the thermal boundary layer from 0 to δ_T, and invoking the Leibniz rule, we have

$$\frac{d}{dx}\int_0^{\delta_T}(uT)dy - \frac{\partial \delta_T}{\partial y}u_{\delta_T}T_{\delta_T} = \alpha\left.\frac{\partial T}{\partial y}\right|_{\delta_T} - \alpha\left.\frac{\partial T}{\partial y}\right|_0 \tag{1.7}$$

The first term on the right side of Eq. (1.7) vanishes since it will be assumed that the fluid above and near $y = \delta_T$ is isothermal at $T = T_0$. Although this assumption cannot be directly applied in the linear temperature profile, it nevertheless arises in more accurate temperature profiles (such as quadratic or cubic curve fits) and thus is adopted here.

Since the temperature distribution is assumed as $T = a + by$, we can now obtain the unknown coefficients and use the results in Eq. (1.7). After applying the boundary conditions of $T = T_n$ at $y = 0$ and $T = T_0$ at $y = \delta_T$, we obtain the following distribution:

$$T = T_n + \left(\frac{T_0 - T_n}{\delta_T}\right)y \tag{1.8}$$

1.6 Introductory Calculations

Substituting Eqs. (1.5) to (1.8) into Eq. (1.7) and dividing by $U(T_n - T_0) = U \wedge T$, we obtain

$$\frac{d}{dx} \int_0^{\delta_T} \left[2\left(\frac{y}{\delta}\right) - \left(\frac{y}{\delta}\right)^2 \right] \left[\frac{T_n}{\Delta T} - \left(\frac{y}{\delta_T}\right) \right] dy - \frac{d\delta_T}{dx} \left[2\left(\frac{\delta_T}{\delta}\right) - \left(\frac{\delta_T}{\delta}\right)^2 \right] \frac{T_n}{\Delta T}$$

$$= \frac{\alpha}{U\delta_T} \qquad (1.9)$$

Performing the integration, simplifying, and rearranging terms, we obtain

$$\frac{d}{dx}\left[\frac{2\delta_T^3}{9\delta} - \frac{3\delta_T^4}{48\delta^2} \right] = \frac{\alpha}{U} \qquad (1.10)$$

Integrating Eq. (1.10) subject to $\delta_T = 0$ at $x = 0$, we have

$$\frac{2}{9}\frac{\delta_T^3}{\delta} - \frac{3}{48}\frac{\delta_T^4}{\delta^2} = \frac{\alpha}{U}x \qquad (1.11)$$

Solving this equation for $\delta_T = \delta$ gives the location where the film and thermal boundary-layer edges coincide. At this location, $x \approx 0.16 U\delta^2/\alpha$. Substituting numerical values for the specified interfacial velocity, film thickness, and property values of ethylene glycol at 300 K, the x location becomes 6.8 cm.

Part (c) Finally, the local Nusselt number can be obtained by

$$\text{Nu}_x = \frac{hx}{k} = \frac{(q_w/\Delta T)x}{k} = \frac{k\Delta T x}{k\Delta T \delta_T} = \frac{x}{\delta_T} \qquad (1.12)$$

where δ_T is obtained from the solution of Eq. (1.11).

Problem 2

A new passive method of cooling an electronic assembly involves the use of a closed-loop thermosyphon, which absorbs heat along its boiling surface and rejects heat in the condensing section (see Fig. 1.1). The approach must be tested to determine whether it can be used as a retrofit option to augment thermal management of an existing system. As the heat flux increases and generates a higher rate of condensate return flow along the walls, the higher vapor velocity may produce appreciable effects of shear forces that entrain liquid from the return flow. This entrainment disrupts the thermosyphon operation, unless an adequate surface area of return flow is provided. This problem investigates how the required surface area is related to the condensation flow rate within a thermosyphon.

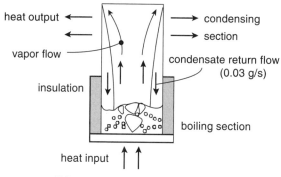

Figure 1.1 Schematic of a closed-loop thermosyphon.

Consider a cylindrical copper thermosyphon divided into a lower insulation section (3 cm) and an upper condensing section (3 cm). The thermosyphon is oriented vertically with the heater surface at 120°C along the bottom boundary. Saturated water is boiled at atmospheric pressure above the heater. The top surface (above the condensation section) is well insulated. What surface area is required to produce a total condensation flow rate of 0.03 g/s at steady state?

Solution

A thermosyphon of 6 cm length under steady-state operation is considered in this problem. The properties of saturated liquid water at 100°C are $\mu_l = 279 \times 10^{-6}$ kg/ms, $Pr_l = 1.76$, $h_{fg} = 2257$ kJ/kg, $c_{p,l} = 4217$ J/kg K, $\rho_l = 958$ kg/m³, and $\sigma = 0.059$ N/m. For saturated vapor at 100°C, the density is $\rho_v = 0.596$ kg/m³. For saturated liquid water at 82°C (initially assumed surface temperature of the condenser), the properties are $\rho_l = 971$ kg/m³, $c_{p,l} = 4.2$ kJ/kg K, $k_l = 0.671$ W/m K, and $\mu_l = 343 \times 10^{-6}$ kg/ms.

For nucleate boiling, the following correlation is adopted [1]:

$$q''_{s,b} = \mu_l h_{fg} \left[\frac{g(\rho_l - \rho_v)}{\sigma} \right]^{1/2} \left[\frac{c_{p,l}(T_{s,b} - T_{\text{sat}})}{c_{s,f} h_{fg} Pr_l^n} \right]^3 \quad (1.13)$$

Using appropriate coefficient values for $c_{s,f}$ and n based on a water-copper arrangement, we obtain

$$q''_{s,b} = 279 \times 10^{-6} \times 2{,}257{,}000 \left[\frac{9.8(958 - 0.596)}{0.059} \right]^{1/2}$$

$$\times \left[\frac{4217(120 - 100)}{0.013 \times 2{,}257{,}000 \times 1.76} \right]^3 \quad (1.14)$$

and so $q''_{s,b} = 1.09 \times 10^6$ W/m².

1.8 Introductory Calculations

Assuming a condensing surface temperature of $T_{s,c} = 82°C$ (returning and iterating afterward as required)

$$h'_{fg} = h_{fg} + 0.68 c_{p,l}(T_{\text{sat}} - T_{s,c}) = 2257 \times 10^3 + 0.68(4200)18$$

$$= 2.308 \times 10^3 \text{ kJ/kg} \tag{1.15}$$

Thus, the heat removed through the condensing section is

$$q_c = \dot{m} h'_{fg} = 3 \times 10^{-5}(2.308 \times 10^6) = 69.2 \text{ W} \tag{1.16}$$

Also, on the basis of conservation of energy for the entire system, the heat removed through the condenser must balance the heat added from the heat source to the boiling section:

$$q_c = q''_{s,b}\left(\frac{\pi D^2}{4}\right) \tag{1.17}$$

which requires a thermosyphon diameter of

$$D = \left[\frac{4(69.2)}{\pi(1.09 \times 10^6)}\right]^{1/2} = 0.009 \text{ m} = 9 \text{ mm} \tag{1.18}$$

For a cylindrical thermosyphon length of 6 cm, this diameter yields a total surface area of about 0.0018 m².

The initially assumed condensing surface temperature should be checked. For heat transfer in the condensing section, we obtain

$$q_c = 69.2 = \bar{h}_c \pi D L_c (T_{\text{sat}} - T_{s,c}) \tag{1.19}$$

where

$$\bar{h}_c = 0.943 \left[\frac{g \rho_l (\rho_l - \rho_v) k_l^3 h'_{fg}}{\mu_l (T_{\text{sat}} - T_{s,c}) L_c}\right]^{1/4} \tag{1.20}$$

Substituting numerical values yields

$$69.2 = 0.943 \left[\frac{9.8(971)(971 - 0.596)0.671^3(2.308 \times 10^6)}{343 \times 10^{-6}(0.03)}\right]^{1/4} \pi(0.009)0.03(100 - T_{s,c})^{3/4} \tag{1.21}$$

Solving this equation gives a surface temperature of 95.5°C.

The updated temperature value can be used to recalculate property values, leading to updated h'_{fg} and q_c values, and iterating until convergence is achieved. Also, the flow Reynolds number is

$$\mathrm{Re}_\delta = \frac{4\dot{m}}{\mu_l \pi D} = \frac{4(3 \times 10^{-5})}{343 \times 10^{-6}(\pi)0.009} = 12.4 \quad (1.22)$$

which supports the previous use of a laminar flow correlation since $\mathrm{Re}_\delta < 30$. The Reynolds number is smaller than the value corresponding to transition to laminar-wavy flow.

Problem 3

During casting of an industrial metal component, a more efficient gating and risering system is needed, in order to minimize the amount of metal poured to produce a good casting. Solidification of liquid metal within the mold during the casting process depends largely on the external rate of cooling. This problem considers a one-dimensional analysis to predict spatial and time-varying changes of temperature within a solidified material. These estimates of temperature profiles within the solid are needed to predict transition between columnar and equiaxed dendrite formations, as well as dendrite fragmentation. Mechanical properties of the solidified material are adversely affected if the solid temperature falls too rapidly and produces an unstable dendritic interface.

Consider a region of liquid tin, which is initially slightly above the fusion temperature ($T_f = 505$ K), that is suddenly cooled at a uniform rate through the wall at $x = 0$. Solidification begins and the solid-liquid interface moves outward from the wall. The thickness of the liquid region is much larger than the growing solid layer, so conduction-dominated heat transfer within a one-dimensional (1D) semi-infinite domain is assumed.

Part (a) Using a quasi-stationary approximation, estimate the temperature variation within the solid as a function of position x. Express your answer in terms of the wall cooling rate, time, thermophysical properties, and T_f.

Part (b) How long would it take for the temperature to drop to 500 K (or 99 percent of T_f) at $x = 4$ cm, when the wall cooling rate is 8 kW/m²?

Solution

Part (a) According to the 1D quasi-stationary approximation for heat transfer in the solid, we have

$$\frac{\partial^2 T_s}{\partial x^2} = 0 \quad (1.23)$$

1.10 Introductory Calculations

subject to $T_s(X, t) = T_f$ with $X(0) = 0$. The following interfacial heat balance at $x = X$ is obtained because no initial superheating occurs in the liquid region:

$$k_s \frac{\partial T_s}{\partial x} = q_w \qquad (1.24)$$

where q_w is a specified constant.

Integrating Eq. (1.23) and imposing the condition of Eq. (1.24), we obtain

$$\frac{dT_s}{dx} = C_1 = \frac{q_w}{k_s} \qquad (1.25)$$

Integrating again and matching the temperature with T_f at $x = X$, we have

$$T_s(X, t) = \frac{q_w}{k_s} X + C_2 = T_f \qquad (1.26)$$

This result yields the following result for temperature in the solid:

$$T_s = T_f - \frac{q_w}{k_s}(X - x) \qquad (1.27)$$

Substituting the differentiated temperature into Eq. (1.24), we obtain

$$k_s \left(\frac{q_w}{k_s} \right) = \rho L \frac{dX}{dt} \qquad (1.28)$$

The right side represents latent heat released at the interface, which balances the rate of heat conduction from the interface on the left side of the equation. Integrating the result subject to $X(0) = 0$, we have

$$X = \frac{q_w}{\rho L} t \qquad (1.29)$$

which can be substituted into Eq. (1.27) to give

$$T_s = T_f - \frac{q_w}{k_s} \left(\frac{q_w}{\rho L} t - x \right) \qquad (1.30)$$

This result is approximately valid for $t > 0$ since the transient derivative was neglected in Eq. (1.23).

Part (b) At a specified time t_s, the temperature reaches a specified fraction of T_f, denoted by aT_f, when

$$aT_f = T_f - \frac{q_w}{k_s} \left(\frac{q_w}{\rho L} t_s - x \right) \qquad (1.31)$$

This result can be rearranged in terms of the position x_1 as follows:

$$x = \frac{q_w}{\rho L} t_s - (1-a)\frac{k_s T_f}{q_w} \tag{1.32}$$

Using the thermophysical properties for tin, as well as the given numerical values in the problem statement, we obtain

$$0.04 = \frac{8000}{6940 \times 59{,}000} \times t_s - 0.01 \times \frac{32.9 \times (505)}{8000} \tag{1.33}$$

A time of 3116 s is required before the temperature at $x = 0.04$ m reaches 500 K, or 99 percent of the fusion temperature.

Reference

1. G. F. Naterer, *Heat Transfer in Single and Multiphase Systems*, CRC Press, Boca Raton, Fla., 2002.

Chapter 2

Industrial Heat-Transfer Calculations

Mohammad G. Rasul
School of Advanced Technologies and Processes
Faculty of Engineering and Physical Systems
Central Queensland University
Rockhampton, Queensland, Australia

Introduction

Heat transfer is an important aspect of thermodynamics and energy. It is fundamental to many engineering applications. There is a heat or energy transfer whenever there is a temperature difference. Engineers are often required to calculate the heat-transfer rate for the application of process technology. A high heat-transfer rate is required in many industrial processes. For example, electrical energy generation in a power plant requires a high heat transfer from the hot combustion gases to the water in the boiler tubes and drums. There are three different modes of heat transfer: conduction, convection, and radiation. In practice, two or more types of heat transfer occur simultaneously, but for ease of calculation and analysis, the modes may be separated initially and combined later on. Details of heat-transfer analysis, design, and operation can be found elsewhere in the literature [1–5]. In this chapter basic heat-transfer equations with several worked-out examples on heat-transfer calculations and thermal design of heat-exchange equipment, usually used in process industries, are presented.

2.2 Introductory Calculations

Basic Heat-Transfer Equations

The *heat-transfer rate* is the amount of heat that transfers per unit time (usually per second). If a hot metal bar has a surface temperature of t_2 on one side and t_1 on the other side, the basic heat-transfer rate due to *conduction* can be given by

$$\dot{Q} = UA \, \Delta t \qquad (2.1)$$

where
$\dot{Q}$ = heat-transfer rate in watts (W)
U = heat-transfer coefficient in W/m² K
A = surface area of the hot metal bar in m²
$\Delta t = t_2 - t_1$ = temperature difference in kelvins

If a hot wall at a temperature t_2 is exposed to a cool fluid at a temperature t_1 on one side, the *convective* heat-transfer rate can be given by

$$\dot{Q} = hA \, \Delta t \qquad (2.2)$$

where h is the convective heat-transfer coefficient in W/m² K. The convective heat-transfer coefficient is usually given a special symbol, h, to distinguish it from the overall heat-transfer coefficient U.

Because of the many factors that affect the convection heat-transfer coefficient, calculation of the coefficient is complex. However, dimensionless numbers are used to calculate h for both free convection and forced convection [1–5]. For forced convection, the formula expressing the relationship between the various dimensionless groups may generally be written in the following form:

$$\mathrm{Nu} = A \, \mathrm{Re}^a \, \mathrm{Pr}^b \qquad (2.3)$$

where $\mathrm{Nu} = hd/\lambda$, $\mathrm{Re} = \rho v d/\mu$, $\mathrm{Pr} = \mu C_p/\lambda$, and A, a, and b are constant for the particular type of flow [1–5]. Here, Nu is the Nusselt number, Re is the Reynolds number, and Pr is the Prandtl number; d is the diameter or characteristic dimension in meters, λ is the thermal conductivity in W/m K, ρ is the density in kg/m³, μ is the dynamic viscosity in pascals, and C_p is the specific heat capacity at constant pressure in kJ/kg K.

For composite materials in series as shown in Fig. 2.1, the overall heat-transfer coefficient U due to combined conduction and convection heat transfer is given by

$$\frac{1}{U} = \frac{1}{h_A} + \Sigma \frac{x}{\lambda} + \frac{1}{h_B} \qquad (2.4)$$

where h_A = heat-transfer coefficient of fluid film A
h_B = heat-transfer coefficient of fluid film B

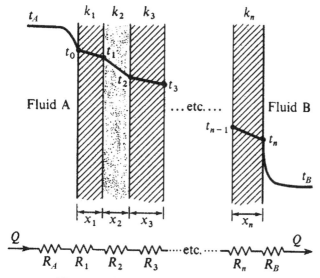

Figure 2.1 Heat transfer through a composite wall [5].

λ = thermal conductivity of materials
x = wall thickness

Thermal resistance R is the reciprocal of U for unit area; therefore

$$R_A = \frac{1}{h_A A}, \quad R_1 = \frac{x_1}{\lambda_1 A}, \ldots, R_n = \frac{x_n}{\lambda_n A}, \quad R_B = \frac{1}{h_B A}$$

The total resistance to heat flow *in series* is then given by

$$R_T = R_A + R_1 + R_2 + \cdots + R_n + R_B \quad \text{or} \quad R_T = \frac{1}{h_A A} + \Sigma \frac{x}{\lambda A} + \frac{1}{h_B A} \quad (2.5)$$

Similarly, the total resistance *in parallel* is given by

$$\frac{1}{R_T} = \frac{1}{R_A} + \frac{1}{R_1} + \frac{1}{R_2} + \cdots + \frac{1}{R_n} + \frac{1}{R_B} \quad (2.6)$$

The overall heat-transfer coefficient U is then given by $U = 1/R_T A$.

The total resistance to heat flow through a *cylinder*, which has cross-sectional inner radius of r_1 and outer radius of r_2, is given by

$$R_T = \frac{1}{h_o A_o} + \Sigma \frac{\ln(r_2/r_1)}{2\pi\lambda} + \frac{1}{h_i A_i} \quad (2.7)$$

2.4 Introductory Calculations

The total resistance to heat flow through a *sphere*, which has inner radius of r_1 and outer radius of r_2, is given by

$$R_T = \frac{1}{h_o A_o} + \Sigma \frac{r_2 - r_1}{4\pi \lambda r_1 r_2} + \frac{1}{h_i A_i} \tag{2.8}$$

where subscripts i and o represent inside and outside, respectively.

Any material which has a temperature above absolute zero radiates energy in the form of electromagnetic waves. The wavelength and frequency of these waves vary from one end to the other end over the spectral band of radiation. It was discovered experimentally by Stefan and verified mathematically by Boltzmann that the energy radiated is proportional to the fourth power of the absolute temperature. The Stefan-Boltzmann law of *radiation* is given by

$$\dot{Q} = \varepsilon \sigma A (t_2^4 - t_1^4) \tag{2.9}$$

where t_2 = absolute temperature of body in kelvins
t_1 = absolute temperature of surroundings in kelvins
σ = Stefan-Boltzmann constant (56.7×10^{-9} W/m^2 K^4)
ε = emissivity (dimensionless)

Equation (2.9) indicates that if the absolute temperature is doubled, the amount of energy radiated increases 16 times. However, over a restricted range of Celsius temperatures, little error is introduced by calculating an equivalent heat-transfer coefficient for radiation h_R defined by [1]

$$\dot{Q} = \varepsilon \sigma A (t_2^4 - t_1^4) = h_R A (t_2 - t_1) \tag{2.10}$$

Therefore, $h_R = [\varepsilon \sigma (t_2^4 - t_1^4)/(t_2 - t_1)]$ equivalent radiation heat-transfer coefficient.

Heat Exchanger

The equipment used to implement heat exchange between two flowing fluids that are at different temperatures and separated by a solid wall is called a *heat exchanger*. Heat exchangers may be found in applications such as space heating and air conditioning, power production, waste-heat recovery, and chemical processing. Different types of heat exchangers are used in process industries [2,7]:

Concentric tube
- Counterflow
- Parallel flow

Crossflow
- Single pass
- C_{max} (mixed), C_{min} (unmixed)
- C_{min} (mixed), C_{max} (unmixed)

Shell and tube
- Single-shell pass and double-tube pass
- Multipass and so on

There are two basic methods of heat-exchanger analysis: *logarithmic mean-temperature-difference* (LMTD) method and the *number-of-transfer-units* (NTU) method. Details on heat-exchanger analysis, design, and operation can be found in the book by Shah and Sekulic [2]. In LMTD method, the heat-transfer rate can be calculated by the following equation for a concentric tube heat exchanger as shown in Fig. 2.2:

$$Q = UA\Delta t_{lm} \quad \text{where} \quad \Delta t_{lm} = \frac{\Delta t_1 - \Delta t_2}{\ln(\Delta t_1/\Delta t_2)} \quad \text{and} \quad A = \pi \, dl \quad (2.11)$$

In the NTU method, effectiveness (ε) is calculated through NTU as follows:

$$\varepsilon = \frac{e^{-NTU(1-R)}}{1 - R.e^{-NTU(1-R)}} \quad (2.12)$$

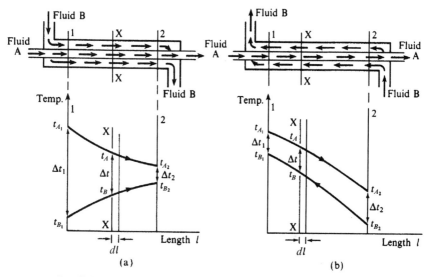

Figure 2.2 Parallel-flow (*a*) and counterflow (*b*) concentric heat exchangers and the temperature distribution with length [5].

2.6 Introductory Calculations

where NTU = UA/C_{min}, $R = C_{min}/C_{max}$, $\varepsilon = \dot{Q}/\dot{Q}_{max}$, and $\dot{Q}_{max} = C_{min}(t_{h,i} - t_{c,i})$. Then $C = \dot{m}C_p$ and C_p is evaluated at the average temperature of the fluid $t_{avg} = (t_i + t_o)/2$.

Worked-out Examples

The following assumptions have been made to solve the worked-out examples presented below:

- Negligible heat transfer between exchanger and surroundings
- Negligible kinetic-energy (KE) and potential-energy (PE) change
- Tube internal flow and thermal conditions fully developed
- Negligible thermal resistance of tube material and fouling effects
- All properties constant

Example 1

In a coal-fired power plant, a furnace wall consists of a 125-mm-wide refractory brick and a 125-mm-wide insulating firebrick separated by an airgap as shown in Fig. 2.3. The outside wall is covered with a 12 mm thickness of plaster. The inner surface of the wall is at 1100°C, and the room temperature is 25°C. The heat-transfer coefficient from the outside wall surface to the air in the room is 17 W/m² K, and the resistance to heat flow of the airgap is 0.16 K/W. The thermal conductivities of the refractor brick, the insulating firebrick, and the plaster are 1.6, 0.3, and 0.14 W/m K, respectively. Calculate

1. The rate of heat loss per unit area of wall surface.
2. The temperature at each interface throughout the wall.
3. The temperature at the outside surface of the wall.

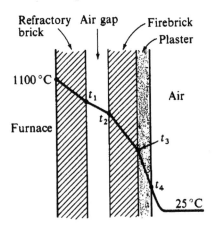

Figure 2.3 Composite wall for Example 1.

Solution

Known. Specified thermal conductivities and thickness of wall materials, the room temperature, and the wall surface temperature.

Analysis

1. Consider 1 m² of surface area of the wall in Fig. 2.3. Then, using equation for resistance, we have

$$\text{Resistance of refractory brick} = \frac{x}{\lambda A} = \frac{125}{1.6 \times 10^3}$$

$$= 0.0781 \text{ K/W}$$

Similarly

$$\text{Resistance of insulating firebrick} = \frac{125}{0.3 \times 10^3} = 0.417 \text{ K/W}$$

$$\text{Resistance of plaster} = \frac{12}{0.14 \times 10^3} = 0.0857 \text{ K/W}$$

$$\text{Resistance of air film on outside surface} = \frac{1}{h_o A} = \frac{1}{17} = 0.0588 \text{ K/W}$$

$$\text{Resistance of airgap} = 0.16 \text{ K/W}$$

Hence, the total resistance is $R_T = 0.0781 + 0.417 + 0.0857 + 0.0588 + 0.16 = 0.8$ K/W. Using the heat-transfer rate equation, we have $\dot{Q} = (t_A - t_B)/R_T = (1100 - 25)/0.8 = 1344$ W; thus the rate of heat loss per square meter of surface area is 1.344 kW.

2. Referring to Fig. 2.3, the interface temperatures are t_1, t_2, and t_3; the outside temperature is t_4. Using the thermal resistance calculated above, we have $\dot{Q} = 1344 = (1100 - t_1)/0.0781$; therefore $t_1 = 995°C$. Similarly, $\dot{Q} = 1344 = (t_1 - t_2)/0.16$; therefore $t_2 = 780°C$. Also, $\dot{Q} = 1344 = (t_2 - t_3)/0.417$; therefore $t_3 = 220°C$, and $\dot{Q} = 1344 = (t_3 - t_4)/0.0857$; therefore $t_4 = 104°C$.

3. The temperature t_4 can also be determined by considering the heat-transfer rate through the air film, $\dot{Q} = 1344 = (t_4 - 25)/0.0857$; therefore $t_4 = 104.1°C$. The temperature at the outside surface of the wall is 104.1°C.

Example 2

In a process industry, a reinforced-concrete smokestack must be lined with a refractory on the inside. The inner and outer diameters of the smokestack are 90 and 130 cm, respectively. The temperature of the

2.8 Introductory Calculations

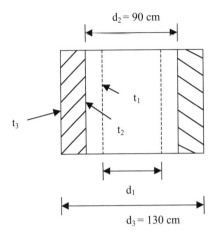

Figure 2.4 Figure for Example 2.

inner surface of the refractory lining is 425°C, and the temperature of the inner surface of the reinforced concrete should not exceed 225°C. The thermal conductivity λ of the lining material is 0.5 W/m K, and that of the reinforced concrete is 1.1 W/m K. Determine the thickness of the refractory lining and temperature of the outer surface of the smokestack under the conditions that the heat loss must not exceed 2000 W per meter height.

Solution

Known. Referring to Fig. 2.4, $r_2 = 45$ cm, $r_3 = 65$ cm, $t_1 = 425°$C, and $t_2 = 225°$C, and thermal conductivities are as specified.

Analysis. The smokestack can be considered as a case of heat transfer through a cylinder. For heat transfer across a cylinder, we have

$$\dot{Q} = \frac{2\pi(t_1 - t_2)}{(1/\lambda_1)\ln(r_2/r_1)} = \frac{2\pi(t_2 - t_3)}{(1/\lambda_2)\ln(r_3/r_2)}$$

Substituting values, we obtain

$$2000 = \frac{2\pi(425 - 225)}{(1/0.5)\ln(45/r_1)}$$

Therefore, $r_1 = 32.87$ cm or $d_1 = 65.74$ cm. Then the lining thickness is

$$\delta = \frac{d_2 - d_1}{2} = \frac{90 - 65.74}{2} = 12.13 \text{ cm}$$

Therefore
$$2000 = \frac{2\pi(225 - t_3)}{(1/1.1)\ln(65/45)} \quad \text{or} \quad t_3 = 118.6°C$$

The temperature of the outer surface of the smokestack is 118.6°C.

Example 3

In a large steam power plant, the *condenser* (where steam is condensed to liquid) can be assumed as a shell-and-tube heat exchanger. The *heat exchanger* consists of a single shell and 30,000 tubes, each executing two passes. The tubes are constructed with $d = 25$ mm and steam condenses on their outer surface with an associated convection coefficient of $h_o = 11,000$ W/m² K. The heat-transfer rate Q is 2×10^9 W, and this is accomplished by passing cooling water through the tubes at a rate of 3×10^4 kg/s (the flow rate per tube is therefore 1 kg/s). The water enters at 20°C, and the steam condenses at 50°C. What is the temperature of the cooling water emerging from the condenser? What is the required tube length per pass?

Solution

Known. Specifications of a shell-and-tube heat exchanger with specified heat flow rate and heat-transfer coefficient. Water inlet and outlet temperatures.

Properties. Assume cold-water average temperature = 27°C, from thermophysical properties of saturated water [3, app. A6], $C_p = 4179$ J/kg K, $\mu = 855 \times 10^{-6}$ N s/m², $\lambda = 0.613$ W/m K, and Pr = 5.83.

Analysis. The cooling-water outlet temperature can be determined from the following energy-balance equation:

$$\dot{Q} = \dot{m}_c C_{p,c}(t_{c,o} - t_{c,i}) \quad \text{or} \quad 2 \times 10^9 = 30,000 \times 4179(t_{c,o} - 20)$$

Therefore $t_{c,o} = 36°C$.

The heat-exchanger design calculation could be based on the LMTD or NTU method. Assuming the LMTD method for calculation, we have $\dot{Q} = UA \, \Delta t_{lm}$, where $A = N \times 2 \times \pi d l$ and

$$U = \frac{1}{(1/h_i) + (1/h_o)}$$

Here, h_i may be estimated from an internal flow correlation, with Reynolds number (Re) given as

$$\text{Re} = \frac{4\dot{m}}{\pi d \mu} = \frac{4 \times 1}{\pi (0.025) \, 855 \times 10^{-6}} = 59{,}567$$

Therefore, the flow is turbulent. Using dimensionless correlation [1,3], we have $\text{Nu} = 0.023 \text{Re}^{0.8} \text{Pr}^{0.4} = 0.023(59,567)^{0.8}(5.83)^{0.4} = 308$. Hence

$$h_i = \text{Nu}\left(\frac{\lambda}{d}\right) = 308 \times \frac{0.613}{0.025} = 7552 \text{ W/m}^2 \text{ K}$$

Therefore

$$U = \frac{1}{(1/7552) + (1/11,000)} = 4478 \text{ W/m}^2 \text{ K}$$

$$\Delta t_{lm} = \frac{(t_{h,i} - t_{c,o}) - (t_{h,o} - t_{c,i})}{\ln[(t_{h,i} - t_{c,o})/(t_{h,o} - t_{c,i})]} = \frac{(50 - 36) - (50 - 20)}{\ln(14/30)} = 21°\text{C}$$

$$l = \frac{\dot{Q}}{U(2N\pi d)\Delta t_{lm}} = \frac{2 \times 10^9}{(4478)(30,000 \times 2\pi \times 0.025) \times 21} = 4.51 \text{ m}$$

Example 4

In an air-conditioning plant, fluid at 20°C flows through a duct which has a wall thickness of 100 mm. The surrounding temperature is 10°C. The convective heat-transfer coefficient inside the duct is 12 W/m² K, and that of outside is 5 W/m² K. The thermal conductivity of the duct is 0.5 W/m K, and the external emissivity is 0.85. Assuming negligible internal radiation, determine the heat flow rate through the duct per square meter of duct surface and external wall surface temperature. Use an initial trial value of 40°C for external duct wall temperature.

Solution

Known. Specifications of an air-conditioning duct and specified operating conditions. Inside and outside heat-transfer coefficients.

Analysis

Trial 1. The external duct wall temperature $t = 40°\text{C} = 313$ K. The equivalent heat-transfer coefficient for external radiation is given by

$$h_R = \frac{\varepsilon\sigma A(t_2^4 - t_1^4)}{t_2 - t_1} = \frac{0.85 \times 56.7 \times 10^{-9}(313^4 - 283^4)}{40 - 10} = 5.115 \text{ W/m}^2 \text{ K}$$

Therefore, the combined convection-radiation heat-transfer coefficient outside is $h_o = 5.115 + 5 = 10.115$ W/m² K.

The overall heat-transfer coefficient U can be determined by

$$\frac{1}{U} = \frac{1}{h_i} + \Sigma\frac{x}{\lambda} + \frac{1}{h_o} = \frac{1}{12} + \frac{0.1}{0.5} + \frac{1}{10.115} = 0.382$$

Therefore, the overall heat-transfer coefficient $U = 2.616$ W/m² K and $\dot{Q} = UA\,\Delta t = 2.616 \times 190 = 497.13$ W.

The outside temperature difference is $\Delta t = 497.13/10.115 = 49.15°C$. Therefore, $t = 49.15 + 10 = 59.15°C$.

Trial 2. $t = 59.15°C = 332.15$ K. The equivalent heat-transfer coefficient for external radiation is now

$$h_R = \frac{\varepsilon\sigma A(t_2^4 - t_1^4)}{t_2 - t_1} = \frac{0.85 \times 56.7 \times 10^{-9}(332.15^4 - 283^4)}{49.15} = 5.645 \text{ W/m}^2 \text{ K}$$

The combined convection-radiation heat transfer coefficient outside is then $h_o = 5.645 + 5 = 10.645$ W/m² K. Therefore

$$\frac{1}{U} = \frac{1}{h_i} + \Sigma\frac{x}{\lambda} + \frac{1}{h_o} = \frac{1}{12} + \frac{0.1}{0.5} + \frac{1}{10.645} = 0.377$$

Therefore the overall heat-transfer coefficient $U = 2.65$ W/m² K and $Q = UA\,\Delta t = 2.65 \times 190 = 503.6$ W. The outside temperature difference is $\Delta t = 503.6/10.645 = 47.3°C$; therefore $t = 47.3 + 10 = 57.3°C$.

Further iterations could be made to refine the accuracy even more, but these are unnecessary as this result is only a small improvement. Therefore, the heat flow rate is 503.6 W and the wall temperature is 57.3°C.

Example 5

A concentric tube heat exchanger with an area of 50 m² operating under the following conditions:

Description	Heat capacity rate C_p, kW/K	Inlet temperature t_i, °C	Outlet temperature t_o, °C
Hot fluid	6	70	—
Cold fluid	3	30	60

1. What would be the outlet temperature of hot fluid?
2. Is the heat exchanger operating in counterflow or parallel flow or cannot be determined from the available information?
3. Determine the overall heat-transfer coefficient.
4. Calculate the effectiveness of the heat exchanger.
5. What would be the effectiveness of this heat exchanger if its length were greatly increased?

2.12 Introductory Calculations

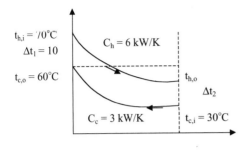

Figure 2.5 Counterflow concentric heat exchanger and its temperature distribution for Example 5.

Solution

Known. Specified operating conditions of heat exchanger including heat capacity rate.

Analysis

1. The outlet temperature of hot fluid can be determined using the energy-balance equation as follows: $\dot{Q} = C_c(t_{c,o} - t_{c,i}) = C_h(t_{h,i} - t_{h,o})$ or $3000(60 - 30) = 6000(70 - t_{h,o})$. Therefore, $t_{h,o} = 55°C$.

2. The heat exchanger must be operating in counterflow, since $t_{h,o} < t_{c,o}$. See Fig. 2.5 for temperature distribution.

3. The rate equation with $A = 50$ m² can be written as

$$\dot{Q} = C_c(t_{c,o} - t_{c,i}) = UA\,\Delta t_{lm}$$

$$\Delta t_{lm} = \frac{\Delta t_1 - \Delta t_2}{\ln(\Delta t_1/\Delta t_2)} = \frac{(70-60)-(55-30)}{\ln(10/25)} = 16.37°C$$

Therefore, $3000(60 - 30) = U \times 50 \times 16.37$, so $U = 109.96$ W/m² K.

4. The effectiveness with the cold fluid as the minimum fluid $C_c = C_{min}$ is

$$\varepsilon = \frac{\dot{Q}}{\dot{Q}_{max}} = \frac{C_c(t_{c,o} - t_{c,i})}{C_{min}(t_{h,i} - t_{c,i})} = \frac{60 - 30}{70 - 30} = 0.75$$

5. For a very long heat exchanger, the outlet of the minimum fluid $C_{min} = C_c$ will approach $t_{h,i}$:

$$\dot{Q} = C_{min}(t_{c,o} - t_{c,i}) = \dot{Q}_{max} \qquad \therefore \quad \varepsilon = \frac{\dot{Q}}{\dot{Q}_{max}} = 1$$

Example 6

You have been asked to design a heat exchanger that meets the following specifications:

Description	Mass flow rate m, kg/s	Inlet temperature t_i, °C	Outlet temperature t_o, °C
Hot fluid	28	90	65
Cold fluid	27	34	60

As in many real-world situations, the customer hasn't revealed or doesn't know the additional requirements that would allow you to proceed directly to a final design. At the outset, it is helpful to make a first-cut design based on simplifying assumptions, which may be evaluated later to determine what additional requirements and tradeoffs should be considered by the customer. Your submission must list and explain your assumptions. Evaluate your design by identifying what features and configurations could be explored with your customer in order to develop more complete specifications.

Solution

Known. Specified operating conditions of heat exchanger including mass flow rate.

Properties. The properties of water at mean temperature, determined from thermophysical properties of saturated water [3], are given in Table 2.1.

Analysis. Using the rate equation, we can determine the value of the UA product required to satisfy the design criteria. Sizing the heat exchanger involves determining the heat-transfer area A (tube diameter, length, and number) and associated overall convection coefficient U such that $U \times A$ satisfies the required UA product.

TABLE 2.1 Properties of Saturated Water at Mean Temperature

Description	Units	Hot-water mean temperature, 77.5°C	Cold-water mean temperature, 47°C
Density ρ	kg/m^3	972.8	989
Specific heat C_p	kJ/kg K	4.196	4.18
Viscosity μ	kg/m·s	3.65×10^{-04}	5.77×10^{-04}
Conductivity λ	W/m K	0.668	0.64
Prandtl number Pr	—	2.29	3.77

2.14 Introductory Calculations

Your design approach should have the following five steps:

1. *Calculate the UA product.* Select a configuration and calculate the required UA product.
2. *Estimate the area A.* Assume a range of area for overall coefficient, calculate the area, and consider suitable tube diameter(s).
3. *Estimate the overall convection coefficient U.* For selected tube diameter(s), use correlations to estimate hot- and cold-side convection coefficients and overall coefficient.
4. *Evaluate first-pass design.* Check whether the A and U values ($U \times A$) from steps 2 and 3 satisfy the required UA product; if not, then follow step 5.
5. *Repeat the analysis.* Iterate on different values for area parameters until a satisfactory match is made $U \times A = UA$.

To perform the design and analysis, the effectiveness-NTU method may be chosen.

1. *Calculate the required UA.* For initial design, select a concentric tube, counterflow heat exchanger, as shown in Fig. 2.6. Calculate UA using the following set of equations:

$$\varepsilon = \frac{e^{-NTU(1-R)}}{1 - R.e^{-NTU(1-R)}}$$

where $NTU = UA/C_{min}$, $R = C_{min}/C_{max}$, $\varepsilon = \dot{Q}/\dot{Q}_{max}$, and $\dot{Q}_{max} = C_{min}(t_{h,i} - t_{c,i})$. Then $C = \dot{m}C_p$, and C_p is evaluated at the average temperature of the fluid, $t_{avg} = (t_i + t_o)/2$.

Substituting values, we find $\varepsilon = 0.464$, $NTU = 0.8523$, $Q = 2.94 \times 10^6$ W, and $UA = 9.62 \times 10^4$ W/K.

2. *Estimate the area A.* The typical range of U for water-to-water exchangers is 850 to 1700 W/m² K [3]. With $UA = 9.62 \times 10^4$ W/K, the range for A is 57 to 113 m², where $A = \pi d_i l N$ and l and d are length and diameter of the tube, respectively. Considering different values of d_i with length $l = 10$ m to describe the heat exchanger, we have three different cases as given in Table 2.2.

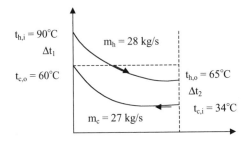

Figure 2.6 Counterflow concentric heat exchanger and its temperature distribution for Example 6.

TABLE 2.2 Result of Analysis for Three Different d_i Values

Case	d_i, mm	l, m	N	A, m^2
1	25	10	73–146	57–113
2	50	10	36–72	57–113 (answer)
3	75	10	24–48	57–113

3. *Estimate the overall coefficient U.* With inner (i.e., hot) and outer (i.e., cold) water in the concentric tube arrangement, the overall coefficient is given by

$$\frac{1}{U} = \frac{1}{h_i} + \frac{1}{h_o}$$

where h_i and h_o are estimated using dimensionless correlation assuming fully developed turbulent flow.

Coefficient, hot side, h_i: For flow in the inner side (hot side), the coefficient h_i can be determined using the Reynolds number formula $\text{Re}_{di} = 4\dot{m}_{h,i}/\pi d_i \mu_h$ and $\dot{m}_h = \dot{m}_{h,i} N$, and the correlation, $\text{Nu} = h_i d_i/\lambda = 0.037 \text{Re}^{0.8} \text{Pr}^{0.3}$, where properties are evaluated at average mean temperature $t_{h,m} = (t_{h,i} + t_{h,o})/2$.

Coefficient, cold side, h_o: For flow in the annular space $d_o - d_i$, the relations expressed above apply where the characteristic dimension is the hydraulic diameter given by $d_{h,o} = 4A_{c,o}/P_o$, $A_{c,o} = [\pi(d_o^2 - d_i^2)]/4$, and $P_o = \pi(d_o + d_i)$. To determine the outer diameter d_o, the inner and outer flow areas must be the same:

$$A_{c,i} = A_{c,o} \quad \text{or} \quad \frac{\pi d_i^2}{4} = \frac{\pi(d_o^2 - d_i^2)}{4}$$

The results of the analysis with $l = 10$ m are summarized in Table 2.3. For all these cases, the Reynolds numbers are above 10,000 and turbulent flow occurs.

4. *Evaluate first-pass design.* The required UA product value determined in step 1 is 9.62×10^4 W/K. The $U \times A$ values for cases 1 and 3 are, respectively, larger and smaller than the required value of 9.62×10^4 W/K. Therefore, case 2 with $N = 36$ and $d_i = 50$ mm could be considered as satisfying the heat-exchanger specifications in the first-pass design. A strategy can now be developed in step 5 to iterate the

TABLE 2.3 Results of h_i, h_o, U, and UA Product for Three Different d_i and N Values

Case	d_i, mm	N	A, m^2	h_i, W/m^2 K	h_o, W/m^2 K	U, W/m^2 K	$U \times A$, W/K
1	25	73	57	4795	4877	2418	1.39×10^5
2	50	36	57	2424	2465	1222	6.91×10^4
3	75	24	57	1616	1644	814	4.61×10^4

analysis on different values for d_i and N, as well as with different lengths, to identify a combination that will meet complete specifications.

References

1. R. Kinskey, *Thermodynamics: Advanced Applications*, McGraw-Hill, Sydney, Australia, 2001.
2. R. K. Shah and D. P. Sekulic, *Fundamentals of Heat Exchanger Design*, Wiley, Hoboken, N.J., 2003.
3. F. P. Incropera and D. P. DeWitt, *Fundamentals of Heat and Mass Transfer*, 5th ed., Wiley, New York, 2002.
4. S. C. Arora and S. Domkundwar, *A Course in Heat & Mass Transfer*, 3d ed., Dhanpat Rai & Sons, Delhi, India, 1987.
5. T. D. Eastop and A. McConkey, *Applied Thermodynamics for Engineering Technologist*, 5th ed., Prentice-Hall, UK, 1993.
6. M. J. Moran and H. W. Shapiro, *Fundamentals of Engineering Thermodynamics*, 5th ed., Wiley, Hoboken, N.J., 2004.

Part 2

Steady-State Calculations

The collection of calculations in this part deals with a variety of circumstances encountered in industrial settings. They are as follows:

- *The fabrication of sheet- and fiberlike materials*
- *High-temperature gas turbine components*
- *Hammer-bank coils on a high-speed dot-matrix printer*
- *Fuel cells*
- *Large turbogenerators*
- *Double-paned windows*

Contributors include academics, an engineer working in industry, and a consultant. All of the calculations are steady-state and mostly involve cooling using gases and liquids. Determination of heat-transfer coefficients, a subject encountered later in this handbook, is a key factor in several calculations. Thermal resistance plays an important role.

Chapter 3

Heat-Transfer and Temperature Results for a Moving Sheet Situated in a Moving Fluid

John P. Abraham
School of Engineering
University of St. Thomas
St. Paul, Minnesota

Ephraim M. Sparrow
Department of Mechanical Engineering
University of Minnesota
Minneapolis, Minnesota

Abstract

A method is developed for determining the heat transfer between a moving sheet that passes through a moving fluid environment. The method also enables the streamwise variation of the sheet temperature to be determined. The numerical information needed to apply the method is presented in tabular form. With this information, the desired results can be calculated by algebraic means.

Introduction

In the fabrication of sheet- and fiberlike materials, the material customarily is in motion during the manufacturing process. The processing may involve heat transfer between the material and an adjacent

3.4 Steady-State Calculations

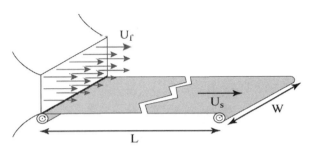

Figure 3.1 Processing station consisting of a moving sheet situated in a parallel gas flow.

fluid which may also be in motion. Typical materials include polymer sheets, paper, linoleum, roofing shingles, and fine-fiber matts. Illustrative diagrams showing possible processing configurations are presented in Figs. 3.1 and 3.2. In Fig. 3.1, an unfinished or partially finished material is unrolled and becomes a moving sheet. In order to complete the finishing operation, a fluid flows over the sheet either to heat or cool it before rollup. In Fig. 3.2, a polymeric sheet is seen emerging from a die. Air flowing through an opening in the die passes in parallel over the sheet to facilitate the heat-transfer process. Another alternative (not illustrated) is to eliminate the blown fluid flows of Figs. 3.1 and 3.2 and instead pass the moving sheet through a confined liquid bath.

A necessary prerequisite for the solution of the heat-transfer problem for such processing applications is the determination of local and average heat-transfer coefficients. The issue of heat transfer and the degree to which it is affected by the simultaneous motion of the participating media will be the topic of the present chapter. From an analysis of laminar fluid flow and convective heat transfer, a complete set

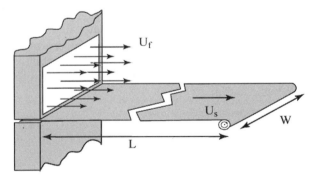

Figure 3.2 A moving sheet extruded or drawn through a die. A parallel fluid stream passes over the sheet.

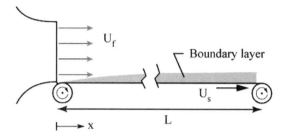

Figure 3.3 Side view showing the web-processing station of Fig. 3.1.

of heat-transfer results have been obtained. The results are parameterized by the ratio U_f/U_s and the Prandtl number. A design-oriented example will be worked out to provide explicit instructions about how to use the tabulated and graphed heat-transfer coefficients.

In practice, it is highly likely that the thermal interaction between the moving sheet and the moving fluid would give rise to a streamwise variation of the sheet temperature. An algebraic method will be presented for dealing with such variations.

The model on which the analysis will be based is conveyed by the diagrams shown in Figs. 3.3 and 3.4. These diagrams are, respectively, side views of the pictorial diagrams of Figs. 3.1 and 3.2. Each diagram shows dimensional nomenclature and coordinates. The problems will be treated on a two-dimensional basis, so that it will not be necessary to deal further with the z coordinate which is perpendicular to the plane of Figs. 3.3 and 3.4.

Similarity Solutions and Their Physical Meanings

The velocity problem, solved in Ref. 1, was based on a similarity solution of the conservation equations for mass and momentum. To understand

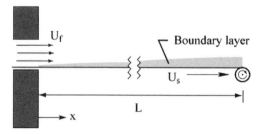

Figure 3.4 Side view showing the extrusion process of Fig. 3.2.

3.6 Steady-State Calculations

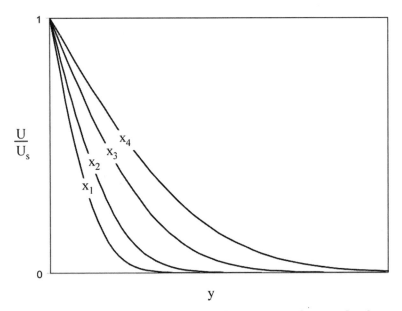

Figure 3.5 Velocity profiles in the boundary layer corresponding to a flat sheet moving through an otherwise quiescent fluid. The profiles correspond to four streamwise locations $x_1 < x_2 < x_3 < x_4$.

the nature of a similarity solution, the reader may refer to Fig. 3.5, which shows representative velocity profiles for the case of a moving surface situated in an otherwise stationary fluid and where u is the local streamwise velocity in the fluid and y is the coordinate perpendicular to the surface of the sheet. Each velocity profile corresponds to a streamwise location x. The nomenclature is ordered with respect to position so that $x_1 < x_2 < x_3 < x_4$.

If a more concise presentation of the velocity information were sought, the optimal end result would be a single curve representing all velocity profiles at all values of x. Such a simple presentation is, in fact, possible if the y coordinate of Fig. 3.5 is replaced by a composite coordinate which involves both x and y. Such a coordinate is normally referred to as a *similarity variable*. For the case of a moving surface in a stationary fluid, the similarity variable is

$$\xi = y\sqrt{\frac{U_s}{\nu x}} \qquad (3.1)$$

When the velocity profiles of Fig. 3.5 are replotted with ξ as the abscissa variable, they fall together as shown in Fig. 3.6.

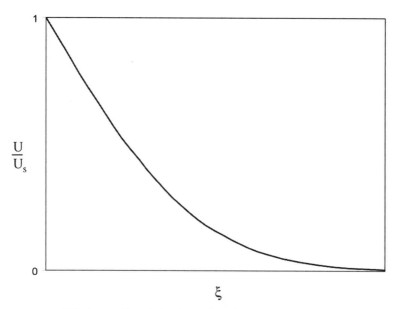

Figure 3.6 Velocity profiles of Fig. 3.5 replotted as a function of the similarity variable ξ of Eq. (3.2). All the profiles fall together on a single curve.

The temperature problem will be solved for parametric values of the ratio $\alpha = U_f/U_s$ in the range between 0 and ∞. The temperatures of the fluid and the surface, respectively T_f and T_s, will not enter the solution as such because they are removed by the introduction of a dimensionless temperature variable. The solutions will encompass both streamwise-uniform and streamwise-varying sheet temperatures. The fluids to be considered are gases (primarily air) and liquid water. The Prandtl numbers of these fluids vary between 0.7 and 10. The worked-out example will demonstrate how to deal with sheet surfaces whose temperatures vary in the streamwise direction.

Analysis

Analysis of the heat-transfer problem has, as its starting point, the conservation-of-energy principle. The equation expressing conservation of energy for a boundary layer is

$$\rho_f c_f \left(u \frac{\partial T}{\partial x} + v \frac{\partial T}{\partial y} \right) = k_f \frac{\partial^2 T}{\partial y^2} \tag{3.2}$$

To continue the analysis, it is necessary to make use of the existing solutions for the velocity components u and v. The expressions for u and

3.8 Steady-State Calculations

v are available from Eqs. (3.A6) and (3.A7). These expressions involve a special similarity variable which is appropriate for the situation in which both of the participating media are in motion. In Eqs. (3.A6) and (3.A7), the quantity U_{rel} appears. Its definition is

$$U_{\text{rel}} = |U_s - U_f| \tag{3.3}$$

To make Eq. (3.2) compatible with the expressions for u and v, it is necessary to transform the physical coordinates, x and y, to the similarity variable η, which is used for the velocity problem as shown in the appendix at the end of this chapter. Furthermore, for compactness, it is useful to define a dimensionless temperature Θ as

$$\Theta(x, y) = \frac{T(x, y) - T_f}{T_s - T_f} \tag{3.4}$$

so that

$$T(x, y) = (T_s - T_f)\Theta(x, y) + T_f \tag{3.5}$$

The fluid temperature T_f that appears in Eq. (3.4) will be a constant for all values of x. On the other hand, the temperature T_s of the moving sheet may vary with x as the sheet either cools down or heats up as a result of heat exchange with the fluid. To keep the analysis sufficiently general to accommodate variations of T_s with x, the temperature difference $(T_s - T_f)$ that appears in Eq. (3.5) will be assumed to have the form

$$T_s = T_f + ax^n \tag{3.6}$$

where a is a constant. The exponent n may be zero or any positive integer. The case $n = 0$ corresponds to a uniform sheet temperature.

We will now consider the quantity $\partial T/\partial x$ appearing in Eq. (3.2). This quantity can be evaluated by applying the operation $\partial/\partial x$ to Eq. (3.5), with the result

$$\frac{\partial T}{\partial x} = \frac{\partial\left[(T_s - T_f)\Theta + T_f\right]}{\partial x} = (T_s - T_f)\frac{\partial \Theta}{\partial x} + \Theta\frac{\partial(T_s - T_f)}{\partial x}$$

$$= (T_s - T_f)\frac{\partial \Theta}{\partial x} + \Theta n a x^{n-1} \tag{3.7}$$

where the derivative $\partial T_s/\partial x$ has been evaluated using Eq. (3.6). Similarly, the terms $\partial T/\partial y$ and $\partial^2 T/\partial y^2$ follow as

$$\frac{\partial T}{\partial y} = \frac{\partial\left[(T_s - T_f)\Theta + T_f\right]}{\partial y} = (T_s - T_f)\frac{\partial \Theta}{\partial y} \tag{3.8}$$

and

$$\frac{\partial^2 T}{\partial y^2} = \frac{\partial\left[(T_s - T_f)\dfrac{\partial \Theta}{\partial y}\right]}{\partial y} = (T_s - T_f)\frac{\partial^2 \Theta}{\partial y^2} \qquad (3.9)$$

Although Θ depends on both x and y, it is advantageous to tentatively assume that Θ is a function only of the similarity variable η, which is a combination of x and y. Then, with $\Theta = \Theta(\eta)$, it remains to evaluate the quantities $\partial\Theta/\partial x$, $\partial\Theta/\partial y$, and $\partial^2\Theta/\partial y^2$ in terms of derivatives involving the similarity variable η. For that evaluation, and with the definition of η taken from Eq. (3.A3), we obtain

$$\frac{\partial \eta}{\partial y} = \frac{\partial \left(y\sqrt{U_{rel}/\nu x}\right)}{\partial y} = -\frac{\eta}{2x} \qquad (3.10)$$

and

$$\frac{\partial \eta}{\partial y} = \frac{\partial \left(y\sqrt{U_{rel}/\nu x}\right)}{\partial y} = \sqrt{\frac{U_{rel}}{\nu x}} \qquad (3.11)$$

With these relationships, the quantities $\partial\Theta/\partial x$, $\partial\Theta/\partial y$, and $\partial^2\Theta/\partial y^2$ follow as

$$\frac{\partial \Theta}{\partial x} = \frac{d\Theta}{d\eta}\frac{\partial \eta}{\partial x} = -\frac{\eta}{2x}\frac{d\Theta}{d\eta} = -\frac{\eta}{2x}\Theta' \qquad (3.12)$$

$$\frac{\partial \Theta}{\partial y} = \frac{d\Theta}{d\eta}\frac{\partial \eta}{\partial y} = \sqrt{\frac{U_{rel}}{\nu x}}\frac{d\Theta}{d\eta} = \sqrt{\frac{U_{rel}}{\nu x}}\Theta' \qquad (3.13)$$

$$\frac{\partial^2 \Theta}{\partial y^2} = \frac{\partial\left[\sqrt{U_{rel}/\nu x}\,(d\Theta/d\eta)\right]}{\partial y} = \sqrt{\frac{U_{rel}}{\nu x}}\frac{\partial (d\Theta/d\eta)}{\partial y} = \sqrt{\frac{U_{rel}}{\nu x}}\frac{d^2\Theta}{d\eta^2}\frac{\partial \eta}{\partial y}$$

$$= \frac{U_{rel}}{\nu x}\frac{d^2\Theta}{d\eta^2} = \frac{U_{rel}\Theta''}{\nu x} \qquad (3.14)$$

When Eqs. (3.7) to (3.9), (3.12) and (3.13), (3.A6), and (3.A7) are substituted into the energy-conservation equation [Eq. (3.2)] and if the notation $\Theta' = d/d\eta$ is used, there results

$$nf'\Theta - \frac{1}{2}f\Theta' = \frac{1}{Pr}\Theta'' \qquad (3.15)$$

Everything in this equation is a function only of η. As a final specification of the temperature problem, it is necessary to state the boundary

3.10 Steady-State Calculations

conditions. In physical terms, they are

$$T = T_s \text{ at } y = 0 \quad \text{and} \quad T \to T_f \text{ as } y \to \infty \quad (3.16)$$

These conditions may be transformed in terms of the new variables Θ and η to yield

$$\Theta = 1 \text{ at } \eta = 0 \quad \text{and} \quad \Theta \to 0 \text{ as } \eta \to \infty \quad (3.17)$$

Numerical Solution Scheme

A suggested method of numerically solving the dimensionless energy equation [Eq. (3.15)], along with its boundary conditions [Eq. (3.17)] will now be presented. The relevant equations are

$$\Theta''(\eta) = \Pr[nf'(\eta)\Theta(\eta) - 1/2 f(\eta)\Theta'(\eta)] \quad (3.18)$$

$$\Theta(\eta + \Delta\eta) = \Theta(\eta) + \Theta'(\eta)\Delta\eta \quad (3.19)$$

$$\Theta'(\eta + \Delta\eta) = \Theta'(\eta) + \Theta''(\eta)\Delta\eta \quad (3.20)$$

to which Eqs. (3.A14) to (3.A17) along with Eq. (3.A10) must be added if the velocity problem has not already been solved. This scheme involves simultaneous solution of the velocity and temperature problems. For these equations, the quantities U_f/U_s, n, and $\Pr$ are prescribed parameters.

The solution of these equations, including Eqs. (3.A14) to (3.A17), starts by assembling the values of Θ, Θ', f, f', and f'' at $\eta = 0$. Among these, the known values are $\Theta = 1$, $f = 0$, $f' = U_s/U_{rel}$, and f'' from Table 3.A1. Since the value of Θ' at $\eta = 0$ must be known in order to initiate the numerical integration of Eqs. (3.18) to (3.20), it is necessary to make a tentative guess of the value of $\Theta'(0)$. Once this guess has been made, the known values of $\Pr$, n, f, f', Θ, and Θ' at $\eta = 0$ are introduced into Eq. (3.18) to yield $\Theta''(0)$. Then, the right-hand sides of Eqs. (3.19) and (3.20) are evaluated at $\eta = 0$, yielding the values of $\Theta(\Delta\eta)$ and $\Theta'(\Delta\eta)$. At the same time, the f equations, Eqs. (3.A14) to (3.A17), are used as described in the text following Table 3.A1 to advance f, f', f'', and f''' from $\eta = 0$ to $\eta = \Delta\eta$.

A similar procedure is followed to advance all the functions from $\eta = \Delta\eta$ to $\eta = 2\Delta\eta$. These calculations are repeated successively until Θ takes on a constant value. If the thus-obtained Θ value does not equal zero as required by Eq. (3.17), the tentatively estimated value of $\Theta'(0)$ must be revised. In this case, a new value of $\Theta'(0)$ is selected, and the foregoing computations are repeated. This process is continued until the behavior of Θ stipulated by Eq. (3.17) is achieved.

Extraction of Heat-Transfer Coefficients

For later application, heat-transfer coefficients are needed. These coefficients can be extracted from the numerical solutions that have just been described. The local heat-transfer coefficient at any location x is defined as

$$h_n = \left| \frac{q}{T_s - T_f} \right| = \frac{\left| k_f \frac{\partial T}{\partial y} \right|_{y=0}}{|T_s - T_f|} \qquad (3.21)$$

where the subscript n corresponds to the exponent n in the sheet surface temperature distribution of Eq. (3.6). The use of the absolute-value notation is motivated by the fact that heat may flow either from the moving surface to the fluid ($T_s > T_f$) or vice versa ($T_f > T_s$). The derivative $\partial T/\partial y$ can be transformed into the variables of the analysis by making use of Eqs. (3.8) and (3.13), with the end result

$$h_n = k_f \sqrt{\frac{U_{\text{rel}}}{\nu x}} \, |\Theta'_n(0)| \qquad (3.22)$$

A dimensionless form of the heat-transfer coefficient can be expressed in terms of the local Nusselt number Nu_x whose definition is

$$\text{Nu}_{x,n} = \frac{h_n x}{k_f} = \sqrt{\frac{U_{\text{rel}}}{\nu x}} \cdot x \cdot |\Theta'_n(0)| = \sqrt{\frac{U_{\text{rel}} x}{\nu}} |\Theta'_n(0)| = \sqrt{\text{Re}_x} \, |\Theta'_n(0)| \qquad (3.23)$$

where Re_x represents the local Reynolds number: $\text{Re}_x = U_{\text{rel}} x/\nu$. Values of $|\Theta'_n(0)|/\text{Pr}^{0.45}$ needed to evaluate h_n and $\text{Nu}_{x,n}$ for three Prandtl numbers (0.7, 5, and 10) are listed in Table 3.1 for $n = 0$ (uniform sheet temperature), 1, and 2. The factor $\text{Pr}^{0.45}$ is used to diminish the spread of the results with the Prandtl number and to improve the accuracy of the interpolation for Prandtl numbers other than those listed. Inspection of the table reveals that the values of $|\Theta'(0)|$ are greater at higher Prandtl numbers. This is because the thermal boundary layer becomes thinner as the Prandtl number increases, and, as a consequence, the temperature $\Theta_n(\eta)$ must decrease more rapidly between its terminal values of 1 and 0. Also noteworthy is the observation that the sensitivity of $|\Theta'_n(0)|$ to the velocity ratio α increases with increasing Prandtl number.

For the case of uniform sheet temperature ($n = 0$), the local Nusselt number can be integrated to obtain a surface-averaged heat-transfer

3.12 Steady-State Calculations

TABLE 3.1 Listing of $|\Theta'(0)|/Pr^{0.45}$ Values for $n = 0$ (Uniform Sheet Temperature), 1, and 2

	$n = 0$			$n = 1$			$n = 2$		
$\alpha = \dfrac{U_f}{U_s}$	0.7	5	10	0.7	5	10	0.7	5	10
0.00	0.410	0.559	0.596	0.943	1.164	1.224	1.316	1.573	1.647
0.10	0.453	0.594	0.631	1.014	1.233	1.294	1.407	1.662	1.739
0.20	0.500	0.635	0.673	1.097	1.313	1.376	1.510	1.769	1.848
0.30	0.554	0.685	0.723	1.194	1.410	1.476	1.633	1.897	1.980
0.40	0.618	0.746	0.786	1.311	1.532	1.599	1.786	2.057	2.144
0.50	0.696	0.825	0.866	1.459	1.685	1.758	1.980	2.261	2.354
0.60	0.800	0.931	0.974	1.657	1.893	1.972	2.238	2.537	2.639
0.70	0.946	1.085	1.132	1.942	2.197	2.285	2.612	2.941	3.055
0.80	1.186	1.342	1.395	2.412	2.705	2.809	3.235	3.616	3.752
0.90	1.715	1.915	1.987	3.459	3.846	3.986	4.625	5.134	5.321
1.11	1.700	1.931	1.911	3.373	3.688	3.812	4.484	4.913	5.078
1.25	1.164	1.299	1.286	2.288	2.478	2.556	3.032	3.294	3.400
1.43	0.917	1.004	0.994	1.786	1.912	1.966	2.356	2.535	2.612
1.67	0.766	0.818	0.809	1.471	1.552	1.592	1.933	2.054	2.110
2.00	0.656	0.683	0.676	1.245	1.291	1.317	1.626	1.701	1.741
2.50	0.572	0.576	0.569	1.068	1.082	1.098	1.384	1.419	1.445
3.33	0.504	0.487	0.480	0.920	0.905	0.910	1.182	1.179	1.191
5.00	0.445	0.410	0.401	0.792	0.747	0.741	1.005	0.962	0.960
10.0	0.392	0.343	0.328	0.675	0.599	0.580	0.842	1.241	0.737
∞	0.344	0.279	0.258	0.564	0.452	0.417	0.686	0.548	0.506

coefficient $\bar{h}_0$, defined as

$$\bar{h}_0 = \left|\frac{Q}{A(T_s - T_f)}\right| = \frac{1}{A|T_s - T_f|}\int_0^L |q|\,dA$$

$$= \frac{1}{WL|T_s - T_f|}\int_0^L h_0\,|T_s - T_f|\,W\,dx = \frac{1}{L}\int_0^L h_0\,dx \quad (3.24)$$

where W is the width of the sheet. Equation (3.24) shows that $\bar{h}_0$ can be obtained as the length average of h_0. The indicated integration can be carried out by substituting Eq. (3.22) into the Eq. (3.24), giving

$$\bar{h}_0 = \frac{1}{L}\int_0^L k_f\sqrt{\frac{U_{\text{rel}}}{\nu x}}\,|\Theta'_0(0)|\,dx = \frac{2k_f\,|\Theta'_0(0)|}{L}\sqrt{\frac{U_{\text{rel}}L}{\nu}} \quad (3.25)$$

or

$$\text{Nu}_{L,0} = \frac{\bar{h}_0 L}{k_f} = 2\,|\Theta'_0(0)|\sqrt{\frac{U_{\text{rel}}L}{\nu}} = 2\,|\Theta'_0(0)|\,\sqrt{\text{Re}_L} \quad (3.26)$$

For situations where the sheet temperature is not uniform ($n \ne 0$), the average heat-transfer coefficient *cannot* be obtained by averaging h.

Illustrative Example

The purpose of this worked-out example is to demonstrate how to deal with generalized streamwise temperature variations of the moving sheet.

Problem statement

A linoleum sheet moving at 0.5 ft/s is to be cooled by a parallel stream of air whose velocity is 5 ft/s. The temperature of the sheet emerging from a die is 160°F, while the air temperature is 80°F. The thickness of the sheet is 1/32 in. (1/384 ft), its width is 3 ft, and the distance between the die and the takeup roll is 5 ft. The thermophysical properties of the participating media are

Linoleum sheet

Density (ρ_s) = 22 lb_m/ft^3
Specific heat (c_s) = 0.21 Btu/lb_m-°F

Air

Density (ρ_f) = 0.0684 lb_m/ft^3
Specific heat (c_f) = 0.241 Btu/lb_m-°F
Thermal conductivity (k_f) = 0.0160 Btu/h-ft-°F
Kinematic viscosity (ν_f) = 0.690 ft^2/h
Prandtl number (Pr) = 0.7

Find

The temperature decrease of the sheet as it cools between its departure from the die and its arrival at the takeup roll.

Discussion

The solution strategy to be used to achieve the required result is as follows:

1. Obtain an *estimate* of the change in the temperature of the moving sheet between the initiation of cooling ($x = 0$) and the termination of cooling ($x = 5$ ft). This estimate will enable you to judge whether the issue of a streamwise temperature variation is worthy of further pursuit from the standpoint of engineering practice. In this regard, it should be noted that the maximum temperature difference between the moving sheet and the adjacent air is 80°F. If a 5 percent decrease in this

3.14 Steady-State Calculations

temperature difference is deemed to be allowable, then a temperature drop of 4°F along the length of the sheet would be regarded as acceptable.

2. If the estimated overall temperature decrease is judged to be significant, then a refined approach is warranted. The first step in that approach is to determine the variation of the sheet temperature T_s^I with x by making use of the local heat-transfer coefficients that correspond to uniform surface temperature. Here, the superscript I is used to denote the temperature variation obtained from the first iteration. This approach, while seemingly inconsistent, is appropriate as the starting point in the successive refinement of the solution.

3. The temperature $T_s^I(x)$ is then fitted with a polynomial of the form

$$T_s^I(x) \approx a_0 + a_1 x + a_2 x^2$$

This polynomial approximation to the sheet temperature will be required in the first step in the refinement process.

4. It is noted that the fitted curve $T_s^I(x)$ contains x^0, x^1, and x^2. This observation, taken together with Eq. (3.6), suggests that heat-transfer coefficient information is needed for the cases $n = 0$, 1, and 2.

5. Use the heat-transfer coefficients identified in paragraph 4 to determine the local heat flux $q(x)$ at the surface of the sheet.

6. The $q(x)$ information enables $T_s^{II}(x)$ to be determined.

7. Compare $T_s^I(x)$ with $T_s^{II}(x)$ and, if necessary, continue iterating until convergence.

Step 1. An upper-bound estimate of the streamwise decrease in temperature of the moving sheet can be made by utilizing a heat-transfer model in which the sheet temperature is constant during the process. This calculation requires the determination of the surface-averaged heat-transfer coefficient $\bar{h}_0$ from Eq. (3.25). The value of $\Theta_0'(0)$ can be found in Table 3.1 for the velocity ratio $U_f/U_s = 10$, Pr $= 0.7$, and $n = 0$ to be $0.392(0.7^{0.45}) = 0.334$. Next, the Reynolds number Re_L is calculated to be

$$\text{Re}_L = \frac{U_{\text{rel}} L}{\nu} = \frac{(4.5 \text{ ft/s}) 5 \text{ ft}}{0.690 \text{ ft}^2/\text{h}} \frac{3600 \text{ s}}{\text{h}} = 117{,}000 \qquad (3.27)$$

With this information, the heat-transfer coefficient emerges as

$$\bar{h}_0 = \frac{2 k_f |\Theta_0'(0)|}{L} \sqrt{\frac{U_{\text{rel}} L}{\nu}} = \frac{2 \left(0.0160 \text{ Btu/h-ft-°F}\right) 0.334}{5 \text{ ft}} \sqrt{117{,}000}$$

$$= 0.731 \frac{\text{Btu}}{\text{h-ft}^2\text{-°F}} \qquad (3.28)$$

An energy balance on the moving sheet can be used to connect the estimated heat loss due to convection with the change in temperature of the moving sheet. That energy balance is

$$\dot{m}_s c_s \Delta T_s = -\bar{h}_0 A (T_s - T_f) \tag{3.29}$$

where $\dot{m}_s$ is the mass flow rate of the sheet, c_s is the specific heat of the sheet, and ΔT_s is the temperature change experienced by the sheet during the cooling process. The convective area A includes both the upper and lower surfaces of the sheet. When the mass flow rate of the sheet and the convective area are expressed in terms of known quantities, Eq. (3.29) becomes

$$\rho_s W t U_s c_s \Delta T_s = -2\bar{h}_0 W L (T_s - T_f) \tag{3.30}$$

where W, t, and L represent, respectively, the width, thickness, and length of the sheet. Equation (3.30) can be solved for the sheet temperature change to give

$$\Delta T_s = -\frac{2\bar{h}_0 L (T_s - T_f)}{\rho_s t U_s c_s}$$

$$= -\frac{2(0.731 \text{ Btu/h-ft}^2\text{-}°\text{F})(5\text{ ft})(80°\text{F})(1\text{ h}/3600\text{ s})}{(22\text{ lb}_m/\text{ft}^3)\left(\frac{1}{384}\text{ft}\right)(0.5\text{ ft/s})(0.21\text{ Btu/lb}_m\text{-}°\text{F})}$$

$$= -27.0°\text{F} \tag{3.31}$$

A temperature decrease of this magnitude warrants a refined approach that accounts for temperature variations all along the sheet.

Step 2. The next step in the solution procedure is to determine T_s^I based on local heat-transfer coefficients that correspond to uniform surface temperature. The symbol T_s^I represents a preliminary temperature variation of the sheet which will be used as the basis of further refinements. An energy balance can be performed on the sheet to connect the T_s^I with the local rate of heat loss. This balance leads to

$$\frac{dT_s^I}{dx} = -\frac{2q(x)}{\rho_s t c_s U_s} = -\frac{2h_0(x)\left(T_s^I(x) - T_f\right)}{\rho_s t c_s U_s} \tag{3.32}$$

which can be rearranged and integrated along the sheet length to give

$$\int_{T_0}^{T_s} \frac{dT_s^I}{T_s^I(x) - T_f} = -\int_0^x \frac{2h_0(\lambda)d\lambda}{\rho_s t c_s U_s} = -\frac{2}{\rho_s t c_s U_s} \int_0^x h_0(\lambda)d\lambda \tag{3.33}$$

3.16 Steady-State Calculations

in which λ is a variable of integration. After Eq. (3.22) is substituted into the right hand side, and the indicated integrations are performed, there follows

$$\ln\left(\frac{T_s^{\rm I}(x) - T_f}{T_s^{\rm I}(0) - T_f}\right) = -\frac{2}{\rho_s t c_s U_s} \int_0^x k_f \sqrt{\frac{U_{\rm rel}}{\nu \lambda}} \left|\Theta_0'(0)\right| d\lambda$$

$$= -\frac{2 k_f \left|\Theta_0'(0)\right|}{\rho_s t c_s U_s} \sqrt{\frac{U_{\rm rel}}{\nu}} \int_0^x \sqrt{\frac{1}{\lambda}} d\lambda \qquad (3.34)$$

After integrating on the right-hand side and inserting the limits of integration, we obtain

$$\ln\left(\frac{T_s^{\rm I}(x) - T_f}{T_s^{\rm I}(0) - T_f}\right) = -\frac{4 k_f \left|\Theta_0'(0)\right|}{\rho_s t c_s U_s} \sqrt{\frac{U_{\rm rel} x}{\nu}} \qquad (3.35)$$

which can be solved explicitly for $T_s^{\rm I}(x)$ to give

$$T_s^{\rm I}(x) = \left(T_s^{\rm I}(0) - T_f\right) \exp\left[-\left(\frac{4 k_f \left|\Theta_0(0)\right|}{\rho_s t c_s U_s} \sqrt{\frac{U_{\rm rel} x}{\nu}}\right)\right] + T_f$$

$$= \left(T_s^{\rm I}(0) - T_f\right) e^{-\Omega} + T_f \qquad (3.36)$$

where Ω, which was used to simplify the expression of Eq. (3.36), is found to be

$$\Omega = \frac{4\,(0.0160\ {\rm Btu/h\text{-}ft\text{-}°F})\,0.334}{(22\ {\rm lb_m/ft^3})\left(\frac{1}{384}{\rm ft}\right)(0.21\ {\rm Btu/lb_m\text{-}°F})(1800\ {\rm ft/h})} \sqrt{\frac{4.5\ {\rm ft/s}\cdot x}{\frac{0.690}{3600}\ {\rm ft^2/s}}}$$

$$= 0.151\sqrt{x} \qquad (3.37)$$

Equations (3.36) and (3.37) can be combined to give

$$T_s^{\rm I}(x) - 80(°{\rm F}) = (80)e^{-0.151\sqrt{x}}(°{\rm F}) \approx 75.86 - 6.907x + 0.6641x^2(°{\rm F}) \qquad (3.38)$$

where the polynomial on the right-hand side is a least-squares fit used to approximate $T_s^{\rm I}$.

Step 3. A refined calculation can be carried out using the polynomial expression of Eq. (3.38). The first step in the refinement is the recognition that the first-law equation [Eq. (3.2)] is linear with the temperature T; that is, wherever T appears, it is to the power 1. It is also known that for a linear differential equation, solutions for that equation for different boundary conditions can be added together. This realization suggests that a suitable solution form for $T(x, y)$ in the presence of a quadratic variation of the sheet temperature is

$$T(x, y) - T_f = a_0 \Theta_0(\eta) + a_1 \Theta_1(\eta) x + a_2 \Theta_2(\eta) x^2 \tag{3.39}$$

where the Θ_n terms are functions of η alone. Note that at the sheet surface, where $\Theta_n(0) = 1$ for all n, Eq. (3.39) reduces to

$$T_s(x) - T_f = a_0 + a_1 x + a_2 x^2 \tag{3.40}$$

which is identical in form to Eq. (3.38). In addition, as $y \to \infty$, it follows that $\Theta_n(0) \to 0$ for all n. Furthermore, substitution of Eq. (3.39) into the first law [Eq. (3.2)] shows that the differential equation obeyed by $\Theta_n(\eta)$ is that of Eq. (3.15). In light of these facts, it follows that the solutions for $\Theta_n(\eta)$ have already been obtained and tabulated for $n = 0, 1,$ and 2 (Table 3.1). The heat flux at the surface of the moving sheet can be obtained from Eq. (3.39) by application of Fourier's law, which gives

$$q(x) = -k_f \left.\frac{\partial T}{\partial y}\right|_{y=0} = -k_f \left.\frac{\partial \left(a_0 \Theta_0(\eta) + a_1 \Theta_1(\eta) x + a_2 \Theta_2(\eta) x^2\right)}{\partial y}\right|_{y=0} \tag{3.41}$$

When the indicated differentiations are carried out, there follows

$$q(x) = k_f \sqrt{\frac{U_{\text{rel}}}{\nu x}} \left(a_0 |\Theta_0'(0)| + a_1 x |\Theta_1'(0)| + a_2 x^2 |\Theta_2'(0)| \right) \tag{3.42}$$

The numerical values of $|\Theta_0'(0)|, |\Theta_1'(0)|,$ and $|\Theta_2'(0)|$ were determined from Table 3.1 to be 0.334, 0.575, and 0.717, respectively. When the values of $a_0, a_1,$ and a_2 are taken from Eq. (3.38) and inserted into Eq. (3.42), the local heat flux emerges as

$$q(x) = k_f \sqrt{\frac{U_{\text{rel}}}{\nu x}} \left[25.3 - 3.97x + 0.476x^2 \right] \tag{3.43}$$

Step 4. Equation (3.43) can be used as input to the energy balance of Eq. (3.32) to give

$$\frac{dT_s^{\text{II}}}{dx} = -\frac{2q(x)}{\rho_s t c_s U_s} = -\frac{2 k_f \sqrt{\frac{U_{\text{rel}}}{\nu x}} (25.3 - 3.97x + 0.476x^2)}{\rho_s t c_s U_s} \tag{3.44}$$

3.18 Steady-State Calculations

or, on integration

$$\int_{T_s(0)}^{T_s(x)} dT_s^{II} = -\int_0^x \frac{2k_f \sqrt{\frac{U_{rel}}{\nu\lambda}}(25.3 - 3.97\lambda + 0.476\lambda^2)d\lambda}{\rho_s t c_s U_s} \quad (3.45)$$

where the variable of integration λ has been introduced. When the integration of Eq. (3.45) is carried out, we obtain

$$T_s^{II}(x) - T_s^{II}(0) = -\frac{2k_f\sqrt{U_{rel}/\nu}}{\rho_s t c_s U_s}(50.6x^{1/2} - 2.65x^{3/2} + 0.190x^{5/2}) \quad (3.46)$$

When the known quantities are substituted into Eq. (3.46), the second-iteration sheet temperature is found to be

$$T_s^{II}(x) = -0.226(50.6x^{1/2} - 2.65x^{3/2} + 0.190x^{5/2}) + 160 \quad (3.47)$$

A comparison of T_s^I with T_s^{II} will enable a determination of whether further iteration is required. That comparison, set forth graphically in Fig. 3.7, shows satisfactory agreement between T_s^I (dashed curve) and T_s^{II} (solid curve). The close agreement (within $1.7°F$) obviates the need for further revision. Equation (3.47) can be used to determine the overall temperature decrease experienced by the sheet throughout

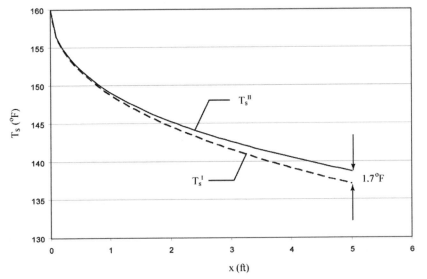

Figure 3.7 Comparison between the first- and second-iteration results T_s^I and T_s^{II}.

the entire cooling process. That decrease is found to be

$$T_s^{II}(L) = -0.226 \left(50.6 L^{1/2} - 2.65 L^{3/2} + 0.190 L^{5/2}\right) + 160 = 138.7°F \tag{3.48}$$

Concluding Remarks

A method has been developed for calculating the heat exchange between a moving fluid and the surface of a sheet which also moves. Both the fluid and the sheet move in the same direction. The heat transfer is initiated by a temperature difference between the fluid and the sheet which exists at their first point of contact. The initial temperature of the sheet may be set by upstream processes to which the sheet is subjected. Under normal conditions, the sheet temperature will vary in the streamwise direction as a result of thermal interaction with the moving fluid. Determination of the temperature variation of the sheet is another focus of the methodology developed here.

The information needed to determine both the heat-transfer rate and the sheet temperature variation is provided in tabular form. With this information, the desired results can be obtained by algebraic operations.

Nomenclature

a	Constant value used in Eq. (3.6)
a_0, a_1, a_2	Coefficients used in the polynomial fit of T_s^I
A	Surface area of sheet
c	Specific heat
f	Similarity function
h	Local heat-transfer coefficient
$\bar{h}$	Averaged heat-transfer coefficient
k	Thermal conductivity
L	Length of sheet
$\dot{m}$	Mass flow rate of moving sheet
n	Constant value used in Eq. (3.6)
Nu	Nusselt number
p	Pressure
Pr	Prandtl number
q	Local heat flux
Q	Rate of heat flow
Re	Reynolds number
t	Sheet thickness
T	Temperature

3.20 Steady-State Calculations

T_s^{I}	First-iteration sheet temperature
T_s^{II}	Second-iteration sheet temperature
u, v	x, y velocities
U	Velocity
W	Sheet width
x, y	Coordinates

Greek

η	Similarity variable based on the relative velocity
Θ	Dimensionless temperature
μ	Dynamic viscosity
ν	Kinematic viscosity
ξ	Similarity variable based on the sheet velocity
ρ	Density
ψ	Streamfunction
Ω	$= \dfrac{4k_f |\Theta'(0)|}{\rho_s t c_s U_s} \sqrt{\dfrac{U_{\mathrm{rel}} x}{\nu}}$

Subscripts

f	Fluid
rel	Relative
s	Surface
0, 1, 2	Corresponding to zeroth-, first-, and second-order terms in the temperature polynomial

Reference

1. John P. Abraham and Ephraim M. Sparrow, Friction Drag Resulting from the Simultaneous Imposed Motions of a Freestream and Its Bounding Surface, *International Journal of Heat and Fluid Flow* 26(2):289–295, 2005.

Appendix: Similarity Solutions for the Velocity Problem

The starting point of the analysis is the equations expressing conservation of mass and x momentum for laminar boundary-layer flow of a constant-property fluid. These are, respectively

$$\frac{\partial u}{\partial x} + \frac{\partial v}{\partial y} = 0 \tag{3.A1}$$

and

$$u\frac{\partial u}{\partial x} + v\frac{\partial u}{\partial y} = -\frac{1}{\rho}\frac{dp}{dx} + v\frac{\partial^2 u}{\partial y^2} \qquad (3.A2)$$

In these equations, u and v are the velocity components that respectively correspond to x and y, with p the pressure, and ρ and v, respectively, the density and the viscosity of the fluid. Although both the freestream flow and the bounding surface move, the streamwise pressure gradient dp/dx is essentially zero, as it is for the two limiting cases of a moving fluid in the presence of a stationary surface and a moving surface passing through a stationary fluid.

These equations may be transformed from the realm of partial differential equations to that of ordinary differential equations by using the similarity transformation from the x, y plane to the η plane. To begin the derivation of the similarity model, it is useful to define

$$\eta = y\sqrt{\frac{U_{\text{rel}}}{vx}} \qquad (3.A3)$$

and

$$\psi = \sqrt{U_{\text{rel}} vx}\, f(\eta) \qquad (3.A4)$$

where $U_{\text{rel}} = |U_f - U_s|$. These variables can be employed to transform the conservation equations, Eqs. (3.A1) and (3.A2), to similarity form

$$\frac{d^3 f}{d\eta^3} + \frac{1}{2}\frac{d^2 f}{d\eta^2}\frac{df}{d\eta} = f''' + \frac{1}{2}f'' f' = 0 \qquad (3.A5)$$

In the rightmost member of this equation, the notation $' = d/d\eta$ for compactness. The fact that this equation is an ordinary differential equation is testimony to the fact that the magnitude of the relative velocity is a suitable parameter for the similarity transformation. It still remains to demonstrate that the boundary conditions depend only on η and are independent of x. To this end, the velocity components are needed. They are expressible in terms of f and f' as

$$u = U_{\text{rel}} f' \qquad (3.A6)$$

$$v = \frac{1}{2}\sqrt{\frac{vU_{\text{rel}}}{x}}(\eta f' - f) \qquad (3.A7)$$

The physical boundary conditions to be applied to Eqs. (3.A6) and (3.A7) are

$$v = 0 \quad \text{and} \quad u = U_s \quad \text{at} \quad y = 0 \qquad (3.A8)$$

$$u \rightarrow U_f \quad \text{as} \quad y \rightarrow \infty \qquad (3.A9)$$

3.22 Steady-State Calculations

The end result of the application of the boundary conditions is

$$f(0) = 0 \qquad f'(0) = \frac{U_s}{U_{rel}} \qquad f'(\eta \to \infty) = \frac{U_f}{U_{rel}} \qquad (3.A10)$$

The transformed boundary conditions of Eq. (3.A10) are seen to be constants, independent of x. Therefore, the similarity transformation is complete.

From the standpoint of calculation, it is necessary to prescribe the values of f' at $\eta = 0$ and as $\eta \to \infty$. To this end, it is convenient to define

$$\alpha = \frac{U_f}{U_s} \qquad (3.A11)$$

so that

$$f'(0) = \frac{U_s}{U_{rel}} = \frac{1}{|1 - U_f/U_s|} = \frac{1}{|1 - \alpha|} \qquad (3.A12)$$

and

$$f'(\eta \to \infty) = \frac{U_f}{U_{rel}} = \frac{U_f/U_s}{|1 - U_f/U_s|} = \frac{\alpha}{|1 - \alpha|} \qquad (3.A13)$$

A review of Eqs. (3.A5), (3.A10), (3.A12), and (3.A13) indicates a complete definition of the *similarity-based, relative-velocity model*. To implement the numerical solutions, values of $\alpha = U_f/U_s$ were parametrically assigned from 0 (stationary fluid, moving surface) to ∞ (stationary surface, moving fluid). The values of $f''|_{\eta = 0}$ that correspond to the specific values of α are listed in Table 3.A1. The knowledge of $f''|_{\eta = 0}$, together with an algebraic algorithm, enables Eq. (3.A5) to be solved without difficulty.

Many algorithms can be used for solving Eq. (3.A5). In the present instance, where $f''|_{\eta = 0}$ is known, the equations which constitute the algorithm are

$$f'''(\eta) = -\frac{1}{2} \cdot f(\eta) \cdot f''(\eta) \qquad (3.A14)$$

$$f''(\eta + \Delta\eta) = f''(\eta) + \Delta\eta \cdot f'''(\eta) \qquad (3.A15)$$

$$f'(\eta + \Delta\eta) = f'(\eta) + \Delta\eta \cdot f''(\eta) \qquad (3.A16)$$

$$f(\eta + \Delta\eta) = f(\eta) + \Delta\eta \cdot f'(\eta) \qquad (3.A17)$$

The use of the algorithms set forth in Eqs. (3.A14) to (3.A17) will now be illustrated. The first step is to choose a velocity ratio as represented

TABLE 3.A1 Listing of $f''(\eta = 0)$ Values for the Similarity-Based, Relative-Velocity Model

$\alpha = U_f/U_s$	$f''(\eta = 0)$
0.000	−0.4439
0.1000	−0.4832
0.2000	−0.5279
0.3000	−0.5797
0.4000	−0.6421
0.5000	−0.7204
0.6000	−0.8238
0.7000	−0.9713
0.8000	−1.214
0.9000	−1.717
1.111	1.726
1.250	1.177
1.429	0.9257
1.667	0.7689
2.000	0.6573
2.500	0.5704
3.333	0.4991
5.000	0.4377
10.00	0.3829
∞	0.3319

by α. Then, by substituting the value of α into Eq. (3.A12), the value of $f'|_{\eta = 0}$ is determined. With this information, the values of f, f', and f'' at $\eta = 0$ are known, the latter from Table 3.A1. In addition, the value $f''' = 0$ at $\eta = 0$ follows directly from the differential equation [Eq. (3.A14)].

The numerical work is initiated by evaluating the right-hand sides of Eqs. (3.A14) to (3.A17) at $\eta = 0$. These calculations yield the numerical values of $f(\Delta\eta), f'(\Delta\eta), f''(\Delta\eta)$, and $f'''(\Delta\eta)$, which are, in turn, used as new inputs to the right-hand sides of Eqs. (3.A14) to (3.A17). The end result of this second step are the values of $f(2\Delta\eta), f'(2\Delta\eta), f''(2\Delta\eta)$, and $f'''(2\Delta\eta)$. These operations are continued repetitively until an η value is reached at which f' becomes a constant. For checking purposes, the value of that constant should be equal to that obtained from Eq. (3.A13) for the selected value of α.

Chapter 4

Solution for the Heat-Transfer Design of a Cooled Gas Turbine Airfoil

Ronald S. Bunker
Energy and Propulsion Technologies
General Electric Global Research Center
Niskayuna, New York

Introduction: Gas Turbine Cooling

The technology of cooling gas turbine components via internal convective flows of single-phase gases has developed over the years from simple smooth cooling passages to very complex geometries involving many differing surfaces, architectures, and fluid-surface interactions. The fundamental aim of this technology area is to obtain the highest overall cooling effectiveness with the lowest possible penalty on thermodynamic cycle performance. As a thermodynamic Brayton cycle, the efficiency of the gas turbine engine can be raised substantially by increasing the firing temperature of the turbine. Modern gas turbine systems are fired at temperatures far in excess of the material melting temperature limits. This is made possible by the aggressive cooling of the hot-gas-path (HGP) components using a portion of the compressor discharge air, as depicted in Fig. 4.1. The use of 20 to 30 percent of this compressed air to cool the high-pressure turbine (HPT) presents a severe penalty on the thermodynamic efficiency unless the firing temperature is sufficiently high for the gains to outweigh the losses. In all properly operating cooled turbine systems, the efficiency gain is significant enough to justify the added complexity and cost of the

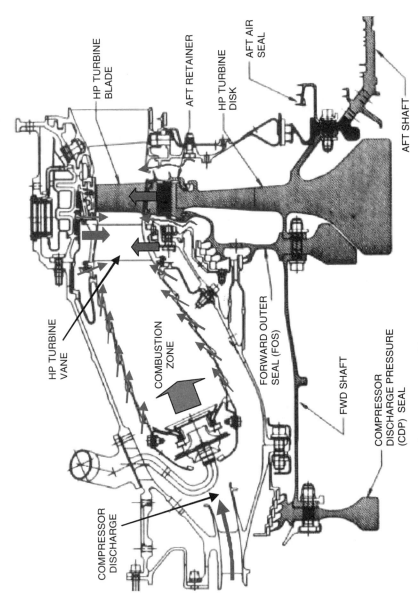

Figure 4.1 Cross section of high-pressure gas turbine.

cooling technologies employed. Actively or passively cooled regions in both aircraft engine and power-generating gas turbines include the stationary vanes or nozzles and the rotating blades or buckets of the HPT stages, the shrouds bounding the rotating blades, and the combustor liners and flame-holding segments. Aircraft engines also include cooling in the exhaust nozzles and in some low-pressure turbines (LPTs). All such engines additionally cool the interfaces and secondary flow regions around the immediate hot-gas path.

Cooling technology, as applied to gas turbine components such as the high-pressure vane and the blade, consists of four main elements: (1) internal convective cooling, (2) external surface film cooling, (3) materials selection, and (4) thermomechanical design. *Film cooling* is the practice of bleeding internal cooling flows onto the exterior skin of the components to provide a heat-flux-reducing cooling layer. Film cooling is intimately tied to the internal cooling technique used in that the local internal flow details will influence the flow characteristics of the film jets injected on the surface. Materials most commonly employed in cooled parts include high-temperature, high-strength nickel- or cobalt-based superalloys coated with yttria-stabilized zirconia oxide ceramics [thermal barrier coating (TBC)]. The protective ceramic coatings are today actively used to enhance the cooling capability of the internal convection mechanisms. The thermomechanical design of the components must marry these first three elements into a package that has acceptable thermal stresses, coating strains, oxidation limits, creep rupture properties, and aeromechanical response. Under the majority of practical system constraints, this means the highest achievable internal convective heat-transfer coefficients with the lowest achievable friction coefficient or pressure loss. In some circumstances, pressure loss is not a concern, such as when plenty of head is available and the highest available heat-transfer enhancements are sought, while in other applications pressure loss may be so restricted as to dictate a very limited means of heat-transfer enhancement.

In many respects, the evolution of gas turbine internal cooling technologies began in parallel with heat-exchanger and fluid-processing techniques, "simply" packaged into the constrained designs required of turbine airfoils (aerodynamics, mechanical strength, vibrational response, etc.). Turbine airfoils are, after all, merely highly specialized and complex heat exchangers that release the cold-side fluid in a controlled fashion to maximize work extraction. The enhancement of internal convective flow surfaces for the augmentation of heat transfer was initially improved around 1975 to 1980 through the introduction of rib rougheners or turbulators, and also pin-banks or pin-fins. Almost all highly cooled regions of the high-pressure turbine components involve the use of turbulent convective flows and heat transfer. Very few, if

4.4 Steady-State Calculations

any, cooling flows within the primary hot section are laminar or transitional. The enhancement of heat-transfer coefficients for turbine cooling makes full use of the turbulent flow nature by seeking to generate mixing mechanisms in the coolant flows that actively exchange cooler fluid for the heated fluid near the walls. These mechanisms include shear layers, boundary-layer disruption, and vortex generation. In a marked difference from conventional heat exchangers, few turbine-cooling means rely on an increase in cooling surface area, since the available surface area–volume ratios are very small. Surface area increases are beneficial, but are not the primary objective of enhancements. The use of various enhancement techniques typically results in at least 50 and as much as 300 percent increase in local heat-transfer coefficients over that associated with fully developed turbulent flow in a smooth duct.

Problem Statement

Estimate the minimum cooling flow requirements for a prescribed high-pressure turbine inlet guide vane (Fig. 4.2) meeting predefined system and component constraints, using basic engineering thermal-fluid relationships. Determine the optimal configuration of cooling flow and airfoil wall composition. The hot-gas (air) inlet total pressure and

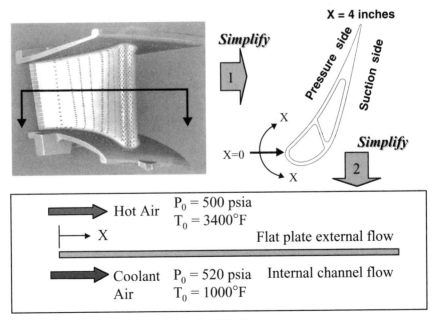

Figure 4.2 Problem geometry and flow simplification.

temperature are 500 psia (3.446 MPa) and 3400°F (1870°C), respectively. The air coolant supply total pressure and temperature are 520 psia (3.584 MPa) and 1000°F (538°C), respectively. The substrate metal of the vane may be assumed to have an isotropic thermal conductivity of 10 Btu/h/ft/°F (17.296 W/m/K), while the protective ceramic TBC thermal conductivity is 1 Btu/h/ft/°F (1.73 W/m/K). The airfoil chordal length is 4 in. (10.16 cm), and the airfoil is 3 in. (7.62 cm) in height. The wall thickness may be considered a constant value along the length and height of the airfoil. The external hot-gas Mach number over the airfoil is represented by a linear variation from 0.005 at $x = 0$ to 1.0 at $x = L$, where L is the total length. This distribution represents the average of the pressure and suction sides of the airfoil, as well as the endwalls, and suits the average nature of the present estimation. The airfoil is considered choked aerodynamically. The average surface roughness k_{rms} on the airfoil is 400 µin. (10 µm). The conditions are considered to be steady-state, representing a maximum power operating point.

Certain constraints must be placed on the solution as typically dictated by material properties or fluid pressure fields. The substrate metal is a Ni-based superalloy composition, but must not exceed a maximum temperature of 2000°F (1094°C) at any given location. The TBC thickness may not exceed 0.010 in. (0.254 mm) as a result of strain limits of bonding with the metal. Also, the TBC maximum temperature must not be more than 2300°F (1260°C) to avoid the infiltration of molten particulates from the hot gases. The overall temperature increase of the cooling air due to convective exchange with the airfoil interior surfaces must not exceed 150°F (66°C). We will utilize impingement jet arrays to cool the airfoil inside, and film cooling to reduce the external incident heat flux. The maximum impingement jet Reynolds number cannot exceed 50,000 without reducing coolant pressure to the point of allowing hot-gas ingestion inside the airfoil. The average film cooling effectiveness may be assumed to be 0.4 for the entire airfoil. Coolant temperature rise as the flow passes through the film holes is approximately 50°F (10.5°C).

Solution Outline

In the complete design of such a high-temperature component, many interdependent aspects of design must be considered for both the individual component and its integration into the engine system. The design system is an iterative method in which aerothermomechanical requirements and constraints must be balanced with material limitations and high-level system objectives. We will consider here only the thermal design portion of the total solution. In addition, component design may entail one of several levels of analysis, from preliminary

4.6 Steady-State Calculations

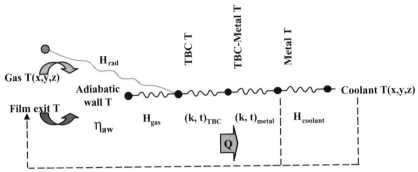

Figure 4.3 One-dimensional thermal resistance model for a cooled airfoil.

estimates, to detailed two-dimensional analyses, to complete three-dimensional computational predictions including the conjugate effects of the convective and radiative environments. Each level of analysis has its use as the design progresses from concept to reality. Preliminary design uses mostly bulk quantities and one-dimensional simplified equations to arrive at approximate yet meaningful estimates of temperatures and flow requirements. Two-dimensional design incorporates boundary-layer analyses, network flow and energy balances, and some thermal gradient estimates to refine the results for local temperature and flow predictions suitable for use in finite-element stress modeling. Three-dimensional design may use complete computational fluid dynamics and heat-transfer modeling of the internal and external flow fields to obtain the most detailed predictions of local thermal effects and flow losses. In the present solution, only the simplified preliminary design method will be used. Figure 4.2 shows the vane airfoil and endwalls reduced first to a constant cross section of the aerodynamic shape, and then again to a basic flat plate representing flow from the leading-edge stagnation point to the trailing edge. While the actual airfoil or endwall shape involves many complexities of accelerating and decelerating flows, secondary flows, and discrete film injection holes, a good estimate may still be obtained using fundamental flat-plate relations.

The solution may be best understood as a sequence of several steps leading to an overall model that is optimized for material thicknesses, cooling configuration, and cooling flow. Figure 4.3 shows the one-dimensional thermal model that applies to any location on the airfoil. These steps include

1. Estimation of the external heat-transfer coefficient distribution on the airfoil
2. Inclusion of airfoil surface roughness effects for the external heat-transfer coefficients

3. Calculation of the average adiabatic wall temperature due to film cooling
4. Estimation of the internal heat-transfer coefficients due to impingement cooling
5. Calculation of the required cooling flow rate
6. Optimization of the solution

The solution will be iterative to account for fluid property changes with temperature, both internal and external to the airfoil. Thermal radiation heat load incident on the airfoil will be neglected here, but should be included for any locations that are in direct view of the combustor flame. The required properties for air may be calculated by the following equations:

Density: $\rho = P/RT$, lb_m/ft^3

Dynamic viscosity: $\mu = 28.679\{[(T/491.6)^{1.5}]/(T+199)\}$, $lb_m/ft/h$

Thermal conductivity: $k = 1.065 \times 10^{-4}(T)^{0.79}$, Btu/h/ft/°R

Specific heat: $C_p = 2.58 \times 10^{-5} \cdot T + 0.225$, Btu/$lb_m$/°R

Prandtl number: $Pr = 0.7$

Temperature T is a static reference value in °R, and pressure P is a static value in psia. The local hot-gas static temperature is determined from the isentropic relationship between total temperature and local Mach number:

$$T_{total}/T_{static} = 1 + [(\gamma - 1)/2] \cdot M^2 \qquad \gamma(T) = \text{specific-heat ratio}$$

External Boundary Conditions

External heat-transfer coefficients

By utilizing a simple flat-plate model for the airfoil in which the start (leading edge) is at $X = 0$ in. and the end (trailing edge) is at $X = 4$ in., we can employ boundary-layer heat-transfer equations for the assumed conditions of constant freestream velocity and zero pressure gradient. The laminar and turbulent heat-transfer coefficients (HTCs) at some location x may then be expressed per Kays and Crawford [1] as

Laminar $\qquad h(x) = 0.332 \dfrac{k}{x} \cdot Re^{0.5} Pr^{1/3}$

Turbulent $\qquad h(x) = 0.0287 \dfrac{k}{x} \cdot Re^{0.8} Pr^{0.6}$

4.8 Steady-State Calculations

where the local Reynolds number (Re) is based on the distance as $Re = \rho V x / \mu$ and the fluid properties are evaluated at a simple average film temperature $T_{film} = 0.5 \cdot (T_{gas} + T_{wall})$. The film temperature is a function of local wall temperature, and since the local wall temperature will change with the external $h(x)$ distribution, the solution becomes iterative. Moreover, when the internal cooling is altered in a subsequent step, wall temperature will again be changed, leading to iteration throughout the solution. The gas temperature in this case is the static hot-gas value.

Calculation of the laminar and turbulent heat-transfer coefficients is not sufficient; we must also approximate the location of transition from laminar to turbulent flow, and model the transition zone. Typical transition of a simple boundary layer occurs between Reynolds numbers 200,000 and 500,000 [1]. We will model the external heat-transfer coefficients as laminar up to Re = 200,000, turbulent after Re = 500,000, and use the local average value of the two in the short transition region.

Film cooling

Film cooling of gas turbine components is the practice of bleeding the internal cooling air through discrete holes in the walls to form a layer of protective film on the external surface. Figure 4.4 shows the ideal case of a two-dimensional tangential slot of film cooling covering a surface. In this ideal case, the layer of film has the exit temperature of the coolant after having been heated by internal cooling and energy pickup through the slot. The gas "film" temperature driving external heat flux to the surface is then that of the coolant rather than that of the hot gas. As the film-cooling layer progresses downstream,

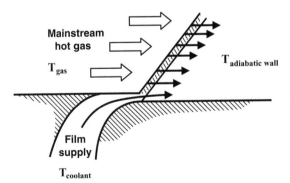

Figure 4.4 Ideal tangential slot film cooling (three-temperature problem).

diffusion, shear mixing, and turbulence act to mix the coolant with the hot gas and reduce the effective film temperature. Eventually, the two streams of fluid would be completely mixed, and, because of the overwhelming amount of hot gases, the film temperature would return to that of the hot gas. This is also true for the realistic case of film cooling via many discrete jets, where coolant injection and mixing are highly three-dimensional processes. In either case, the film cooling serves to reduce the driving potential for external convective heat flux. This three-temperature problem is characterized by the adiabatic film effectiveness defined as

$$\eta = \frac{T_{\text{gas}} - T_{\text{adiabatic wall}}}{T_{\text{gas}} - T_{\text{coolant exit}}}$$

where the adiabatic wall temperature is that which would exist in the absence of heat transfer. This adiabatic wall temperature is the reference mixed gas temperature that drives heat flux to the wall in the presence of film cooling. In the present model, the average adiabatic film effectiveness is 0.4. The local value of adiabatic wall temperature is then determined from the local gas static temperature (from the Mach number distribution) and the coolant exit temperature. The latter temperature is a function of the degree of internal cooling.

Surface roughness effect

In most gas turbines, a significant portion of the airfoil external surfaces become rough as a result of deposits, erosion, or corrosion. Roughness serves to elevate heat transfer coefficients drastically when the roughness elements become large enough to disrupt or protrude through the momentum thickness layer δ_2. Roughness is generally characterized using a so-called sandgrain roughness k_s that is related to the actual physical average height k_{rms} of the rough peaks [2]. Depending on the type of roughness, the relationship is in the range of $k_s = 2k_{\text{rms}}$ to $10k_{\text{rms}}$ [3]. For the present estimation purposes we shall use $k_s = 5k_{\text{rms}}$. A local roughness Reynolds number is then defined on the basis of k_s. A surface is said to be smooth if the roughness Re_{k_s} is less than 5, and fully rough when greater than 70, with a transition region between these values [2]. This roughness Re_{k_s} also employs the coefficient of friction as

$$\text{Re}_{k_s} = \frac{\rho V k_s}{\mu} \left(\frac{C_f}{2}\right)^{1/2}$$

4.10 Steady-State Calculations

In the specific case of a fully rough surface, the coefficient of friction is then related to the momentum thickness by

$$C_f = \frac{0.336}{[\ln(864\delta_2/k_s)]^2}$$

$$\delta_2 = 0.036 \cdot x \cdot \text{Re}^{-0.2}$$

Finally, the fully rough surface heat-transfer coefficient in an external boundary layer may be expressed with the model due to Dipprey and Sabersky [4] as

$$h_{\text{rgh}} = \frac{(k/x) \cdot \text{Re} \cdot \text{Pr} \cdot (C_f/2)}{0.9 + \left[(C_f/2)^{1/2}/\text{Re}_{k_s}^{-0.2} \cdot \text{Pr}^{-0.44}\right]}$$

We may now estimate the magnitude and form of the external heat-transfer coefficient distribution by assuming a constant value of external wall temperature, at least until a specific wall temperature distribution is determined using internal cooling. Knowing that our target maximum value is 2300°F (1260°C), we will use 2200°F (1205°C) for now as the surface temperature everywhere. We also know that the coolant is allowed to heat up by as much as 150°F (66°C) inside the airfoil, plus another 50°F (10.5°C) for the film hole through-wall heating. We shall therefore use a film-cooling exit temperature of 1200°F (650°C) as an average value. As an aside, it is generally best to fully utilize the allowable internal cooling before exhausting the air as film cooling. All calculations are performed in a spreadsheet format. Figure 4.5 shows the current intermediate estimates for Re and Re_{k_s} distributions, the smooth- and rough-surface heat-transfer coefficient distributions, and for reference, the Mach number. The locations of laminar and turbulent transition are reflected in the discontinuities present in the heat-transfer coefficients. The fully rough condition of $\text{Re}_{k_s} = 70$ occurs at $x/L \simeq 0.25$. The effect of roughness is seen to be nearly a doubling of the local heat-transfer coefficients. Overall, it is somewhat startling to note that heat-transfer coefficients may vary by an order of magnitude over the entire surface. We should note that the current estimate has not accounted for the airfoil leading-edge stagnation region heat transfer, which actually acts as a cylinder in crossflow with augmentation due to freestream turbulence and roughness. Hence the heat-transfer coefficients for $x/L < 0.1$ should be much higher, but the addition is not large compared to the rest of the airfoil. We should also mention that film injection affects local heat-transfer coefficients, sometimes significantly, but the complexity is too much to include in this simple estimation.

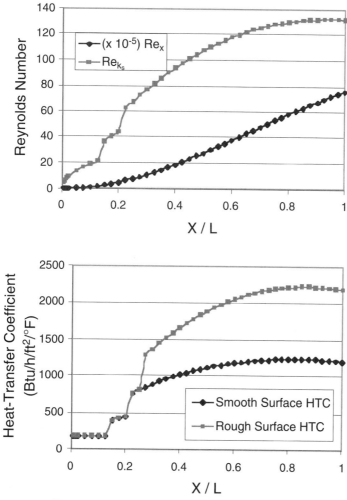

Figure 4.5 External flow Reynolds numbers and heat-transfer coefficients.

Internal Cooling

Impingement cooling

The use of impingement jets for the cooling of various regions of modern gas turbine engines is widespread, especially within the high-pressure turbine. Since the cooling effectiveness of impingement jets is very high, this method of cooling provides an efficient means of component heat load management given sufficient available pressure head and geometric space for implementation. Regular arrays of impingement jets are used within turbine airfoils and endwalls to provide relatively uniform and controlled cooling of fairly open internal surface regions. Such regular impingement arrays are generally directed against the target

4.12 Steady-State Calculations

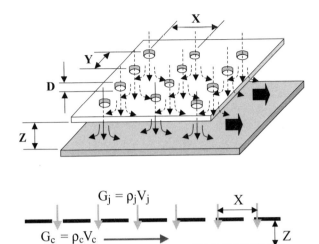

Figure 4.6 General jet array impingement geometry.

surfaces, as shown in Fig. 4.6, by the use of sheetmetal baffle plates, inserts, or covers which are fixed in position relative to the target surface. These arrangements allow for the design of a wide range of impingement geometries, including in-line, staggered, or arbitrary patterns of jets. There are several pertinent parameters in such jet array impingement arrangements. The primary fluid parameter is the jet Re_D, based on the orifice diameter and exit velocity. The main geometric parameters include the target spacing Z/D, and the jet-to-jet (interjet) spacings x/D and y/D. Also important in an array of jets is the relative crossflow strength ratio of postimpingement mass velocity to jet mass velocity G_c/G_j or $(\rho V)_c/(\rho V)_j$. This crossflow effect can in severe cases actually shut down the flow of coolant from subsequent jets. Fortunately, in the design of turbine inlet vanes, the crossflow ratio is relatively weak, usually around $G_c/G_j = 0.2$, as a result of the large number of distributed film holes over the surface that extract the air.

Jet array impingement correlations for surface-average heat-transfer coefficients exist in the literature for various conditions. The correlation of Bailey and Bunker [5] for square jet arrays is used here:

$$h_D = \frac{k}{D} \left\{ 47.1 - 5.5\frac{x}{D} + \frac{Z}{D}\left(7.3 - 2.3\frac{Z}{D}\right) \right.$$
$$+ Re_D \left(4 \times 10^{-3} - 1.3 \times 10^{-4} \cdot \frac{x}{D} - 1.5 \times 10^{-8} \cdot Re_D\right)$$
$$\left. + \frac{G_c}{G_j}\left(61.2 - 13.7\frac{x}{D} - 28\frac{Z}{D}\right) \right\}$$

This expression is valid for $1 < x/D < 15$, $1.5 < Z/D < 5$, and $Re_D < 100{,}000$. The jet Re_D is based on the supply cooling conditions. The coolant properties used in this equation are based on an internal film temperature, the average of the coolant supply temperature, and the local internal wall temperature.

Coolant temperature rise

The last element of this process is simply a check of the allowed temperature rise in the coolant. This is obtained by calculating the heat flux through the composite wall, iterating for closure on the wall temperatures, and summing the total heat flux that must be absorbed by the coolant. The total coolant flow rate m_c is merely the sum of the impingement jet flows as determined from the jet diameter D, jet Re_D, and the number of jets that fit into the airfoil wall chord and height space. The coolant temperature rise is then

$$\Delta T_c = \frac{\sum \left[h_D \cdot A \cdot (T_{\text{wall internal}} - T_{\text{coolant}}) \right]}{m_c \cdot C_p}$$

Optimization of Solution

Optimization of the current design based on average conditions means determining the minimum coolant usage possible while still satisfying the required constraints. The parameters that are available for variation include the impingement jet array geometry, jet Re_D, metal wall thickness, and TBC thickness. Actually, maximum thermal protection is obtained with the maximum TBC thickness of 0.010 in. (0.254 mm), so this parameter is set by common sense. Metal thickness must be set to a value that can be manufactured. Aside from this, a minimum metal thickness will lead to more effective internal heat flux; 0.050 in. (1.27 mm) is selected to meet both needs. In exploring the design space available, the constraints cannot all be exactly satisfied by a single solution. In addition, calculations show that the most limiting constraint is that of the 150°F (66°C) allowable coolant temperature rise. This means that the coolant impingement heat-transfer coefficient will be limited in order to maintain the temperature-rise limit. This solution is represented by the following design:

Metal thickness	0.050 in. (1.27 mm)
TBC thickness	0.010 in. (0.254 mm)
Jet diameter	0.031 in. (0.787 mm)
Jet Re_D	20,000
Jet x/D	8

4.14 Steady-State Calculations

Jet Z/D	3
Number of jets	195
Cooling HTC	404 Btu/h/ft^2/°F (2293 W/m^2/K)
Maximum metal T	1951°F (1066°C)
Maximum TBC T	2245°F (1230°C)
Average external T	2079°F (1138°C)
Bulk metal T	1782°F (973°C)
Coolant ΔT	149.5°F (65.6°C)
Coolant flow rate	0.195 lb$_m$/s (88.64 g/s)

The metal and TBC maximum temperatures are each about 50°F (10.5°C) below their allowable limits. The complete thermal distribution is shown in Fig. 4.7. The external heat-transfer coefficient distribution is essentially the same as that of Fig. 4.5, with roughness. Hot-gas static temperature decreases significantly as the Mach number increases. The effect of film cooling makes it clear why this cooling method is the best line of defense in design, serving to directly reduce the driving potential for heat flux to the surface. Low external heat-transfer coefficients near the airfoil leading edge are balanced by high thermal potentials, while high heat-transfer coefficients at the trailing edge encounter lower thermal potentials. Using this coolant flow rate for each of the airfoil sides, and approximating the same amount again for the two endwalls, the entire airfoil cooling flow is 0.78 lb$_m$/s (355 g/s). A turbine airfoil of this size has a count of about 44 in the engine, making the total vane segment cooling 34.32 lb$_m$/s (15.6 kg/s). For an engine

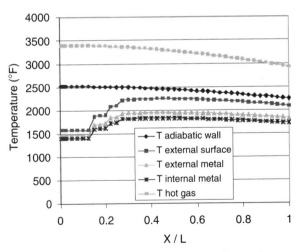

Figure 4.7 Optimized wall temperature distribution for constrained problem.

flow rate of approximately 300 lb$_m$/s (136.4 kg/s), the vane cooling flow represents 11.44 percent of the engine compressor airflow, which is very typical of high-performance gas turbine inlet guide vanes.

An additional note is appropriate here. Real turbine vane cooling designs make full use of available parameter space in a local manner rather than on an average basis. Film cooling will be varied in strength over the airfoil surface, and internal cooling will also be adjusted locally, to obtain a more uniform distribution of resulting material temperatures leading to lower in-plane stresses.

References

1. Kays, W. M., and Crawford, M. E., 1980, *Convective Heat and Mass Transfer*, 2d ed., McGraw-Hill.
2. Schlichting, H., 1979, *Boundary Layer Theory*, 7th ed., McGraw-Hill.
3. Bunker, R. S., 2003, *The Effect of Thermal Barrier Coating Roughness Magnitude on Heat Transfer with and without Flowpath Surface Steps*, Paper IMECE2003-41073, International Mechanical Engineering Conference, Washington, D.C.
4. Dipprey, R. B., and Sabersky, R. H., 1963, Heat and Mass Transfer in Smooth and Rough Tubes at Various Prandtl Numbers, *International Journal of Heat and Mass Transfer*, Vol. 6, pp. 329–353.
5. Bailey, J. C., and Bunker, R. S., 2002, *Local Heat Transfer and Flow Distributions for Impinging Jet Arrays of Both Sparse and Dense Extent*, GT-2002-30473, IGTI Turbo Expo, Amsterdam.

Chapter 5

Steady-State Heat-Transfer Problem: Cooling Fin

David R. Dearth
Applied Analysis & Technology
Huntington Beach, California

Introduction

In one of my early engineering consulting projects, finite-element analysis (FEA) techniques were utilized to investigate extending service life of hammer-bank coils on a high-speed dot-matrix printer. One aspect of this overall project was assigned to estimate reductions in coil temperatures when additional cooling fins are added. Figure 5.1 shows the original single-fin and two new dual-cooling-fin design concepts selected for investigation.

Before beginning development of the FEA idealizations, I recommend doing a few sample "warmup" problems with known textbook solutions. These test problems are extensively investigated prior to tackling the real problem. This sample problem was also a "sanity" check, hand solution created to ensure that I was using the correct units on the heat-transfer parameters *before* I invested time on the actual FEA mathematical idealizations.

The representative warmup problem selected for investigating this heat-transfer project was a simplified cooling fin idealized as a uniform rod or cylinder. A search through the engineering literature addressing heat-transfer problems located various portions of each individual

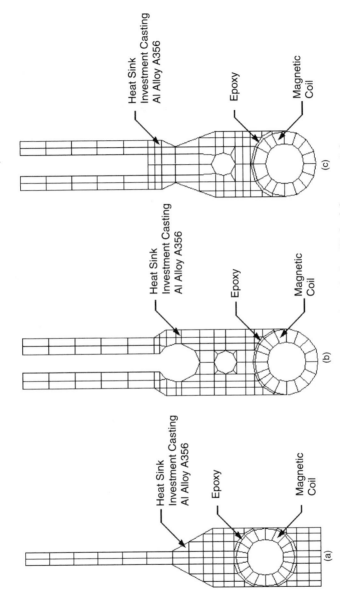

Figure 5.1 Design concepts for cooling fins: (*a*) original single fin; (*b*) dual-fin concept 1; (*c*) dual-fin concept 2.

aspect of these types of heat-transfer problems. However, none of the engineering references contained a complete approach to the problem from beginning to end. This sample fin problem contains all the features of any real-life problem as a comprehensive, step-by-step solution to the simplified version of the cooling fins with an FEA model to cross-check the results. Once we were confident of our methodology, techniques, and procedures from investigating this sample problem, we proceeded to address the real-life problem.

Sample Problem: Heat-Transfer Solution for a Simplified Cooling Fin

Figure 5.2 shows a sample problem of a cylindrical fin with heat source at the base temperature $T_b = 250°F$ and surrounding still air at $T_\infty = 72°F$. The fin is fabricated from a generic aluminum alloy with thermal conductivity $K = 130$ Btu/h-ft-°F.

The first step in the analysis is to determine an average heat-transfer film coefficient h_{avg} for free (natural) convection to the surrounding ambient air (fluid). For purposes of this sample problem $h_{avg} = 1.30$ Btu/h-ft²-°F. The arithmetic for estimating h_{avg} can be found in the detailed hand calculations that follow.

For our sample problem, it is desired to estimate the distribution of temperatures °F along the length of the fin and at the tip.

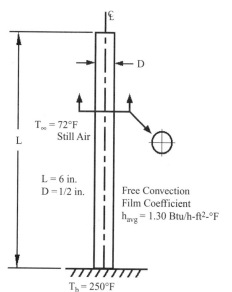

Figure 5.2 Sample heat-transfer "fin" model for steady-state temperatures.

5.4 Steady-State Calculations

To gain confidence in our solution, we will present two approaches: (1) hand solutions using conventional equations found in most engineering textbooks on heat transfer and (2) correlate results with a finite-element idealization.

Answers

An aluminum rod ($K = 130$ Btu/h-ft-°F) having a diameter of 0.500 in. and a length of 6 in. is attached to a surface where the base temperature is 250°F. The ambient air is 72°F, and the average free convective heat-transfer film coefficient along the rod length and end is $h_{avg} = 1.30$ Btu/h-ft^2-°F.

Determine the temperature distribution along the length of the cylindrical rod, accounting for the heat transfer at the endface.

Step 1: Compute average film coefficient from mean film surface temperature T_{film}

$$T_{film} = \frac{T_b + T_\infty}{2} \tag{5.1}$$

Substituting into Eq. (5.1), we obtain

$$T_{film} = \frac{250 + 72}{2} = 161°F$$

Table 5.1 lists properties of air at $T_{film} = 161°F$ using liner interpolation from data in table A-4 of Ref. 1.

Step 2: Compute Grashof number Gr and check for laminar or turbulent free convection

Let $T_{film} = T_{surface} = T_s$:

TABLE 5.1 Properties of Air at 161°F

Density ρ	=	0.0634 lb/ft^3
Specific heat C_p	=	0.2409 Btu/lb-°F
Dynamic viscosity μ	=	1.3787×10^{-5} lb/s-ft
Kinematic viscosity ν	=	21.83×10^{-5} ft^2/s
Thermal conductivity k	=	0.01713 Btu/h-ft-°F
Thermal diffusivity α	=	1.1263 ft^2/h
Prandtl number Pr	=	0.698

For gas (air),

$$\text{Gr} = \frac{g\beta(T_s - T_\infty)L^3}{v^2} \tag{5.2}$$

$$\beta = \frac{1}{T\ ^\circ R} = \frac{1}{161 + 460} = 1.610 \times 10^{-3}\ ^\circ R^{-1} \tag{5.3}$$

Let characteristic length $L = L_v = (6/12)$ ft. Substituting into Eq. (5.2), we obtain

$$\text{Gr} = \frac{(32.2\ \text{ft/s}^2)(1.610 \times 10^{-3}\ ^\circ R^{-1})(250-72)\ ^\circ R\ (6/12\ \text{ft})^3}{(21.83 \times 10^{-5}\ \text{ft}^2/\text{s})^2}$$

$$= 2.421 \times 10^7$$

Compute the Rayleigh number Ra:

$$\text{Ra} = \text{Gr Pr} = (2.421 \times 10^7)(0.698) = 1.690 \times 10^7$$

Since $\text{Ra} = \text{Gr Pr} < 10^9$, laminar flow conditions may be assumed.

Step 3: Compute average film heat-transfer coefficient $h_{\text{avg}} = \overline{h}$

First compute the Nusselt number:

$$\overline{\text{Nu}} = \frac{\overline{h}L}{k} \tag{5.4}$$

The Nusselt number can also be represented using empirical relations depending on the geometry of the free convective system:

$$\overline{\text{Nu}} = C(\text{Gr Pr})^a \tag{5.5}$$

Then

$$\frac{\overline{h}L}{k} = \overline{\text{Nu}} = C(\text{Gr Pr})^a \tag{5.6}$$

For a long cylinder $L/D > 10$ and laminar flow, select the following coefficients: $C = 0.59$ and $a = 1/4$, from table 7.1 in Ref. 2.

Substituting and solving for $\overline{h}$, we obtain

$$\overline{h} = \frac{0.01713\ \text{Btu/h-ft-}^\circ F}{6/12\ \text{ft}}(0.59)(1.690 \times 10^7)^{1/4}$$

$$= 1.30\ \text{Btu/h-ft}^2\text{-}^\circ F$$

5.6 Steady-State Calculations

Step 4: Solve for temperature distribution along the length of the fin

The solution to the differential equations for combined conduction-convection of the fin with heat loss at the endface can be found in Ref. 2:

$$\frac{T(x) - T_\infty}{T_b - T_\infty} = \frac{\cosh[m(L-x)] + (\overline{h}/mK)\sinh[m(L-x)]}{\cosh(mL) + (\overline{h}/mK)\sinh(mL)} \quad (5.7)$$

where

$$m = \sqrt{\frac{\overline{h}P}{KA}} \iff \sqrt{\frac{2\overline{h}}{Kr}} \quad \text{for a cylinder} \quad (5.8)$$

Substituting and solving Eq. (5.8), we obtain

$$m = \sqrt{\frac{2 \times (1.30 \text{ Btu/h-ft}^2\text{-}°\text{F})}{(130 \text{ Btu/h-ft-}°\text{F})[(0.50/2)/12 \text{ ft}]}} = 0.9798 \text{ ft}^{-1}$$

Also

$$\frac{\overline{h}}{mK} = \frac{1.30 \text{ Btu/h-ft}^2\text{-}°\text{F}}{(0.9798 \text{ ft}^{-1})(130 \text{ Btu/h-ft-}°\text{F})} = 0.010206$$

Substituting into Eq. (5.7), we obtain

$$\frac{T(x) - 72}{250 - 72} = \frac{\cosh[0.9798(^6\!/_{12} - x)] + (0.010206)\sinh[0.9798(^6\!/_{12} - x)]}{\cosh[0.9798(^6\!/_{12})] + (0.010206)\sinh[0.9798(^6\!/_{12})]} \quad (5.9)$$

Step 5: Evaluate $T(x)$ at $1/2$-in. intervals (0.50/12-ft intervals)

Table 5.2 lists estimates for external surface temperatures of the fin by using a spreadsheet to minimize roundoff in the arithmetic of Eq. (5.9). Figure 5.3 plots the data of Table 5.2.

Step 6: Verify hand calculations using FEA idealization

Table 5.2 estimates the steady-state fin tip temperature $T_{\text{fin tip}} = 229.9090°\text{F}$. Figure 5.4 summarizes results for an FEA model idealization in which Fig. 5.2 is processed using MSC/Nastran. Figure 5.4 shows an average tip temperature $T_{\text{fin tip, avg}} = 229.9119°\text{F}$. A comparison between the solutions gives a percentage difference equal to ± 0.00126 percent.

TABLE 5.2 Fin External Surface Temperatures

	x, in.	$T(x)$, °F
Base	0.00	250.0000
	0.50	246.7976
	1.00	243.8856
	1.50	241.2594
	2.00	238.9144
	2.50	236.8468
	3.00	235.0532
	3.50	233.5305
	4.00	232.2763
	4.50	231.2885
	5.00	230.5654
	5.50	230.1058
Tip	6.00	229.9090

Applying the Principles

With a high degree of confidence in defining and solving a simplified version of the transient heat-transfer problem, the final step is to apply the principles. Because of the complexity of the geometry, FEA methods are better suited for addressing solutions for the actual three fin design concepts shown in Fig. 5.1.

What Is Finite-Element Analysis, and How Does It Work?

Finite-element analysis methods were first introduced in 1943. Finite-element analysis uses the Ritz method of numerical analysis and

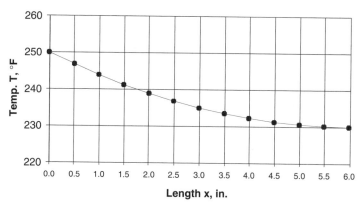

Figure 5.3 Fin external surface temperatures plot.

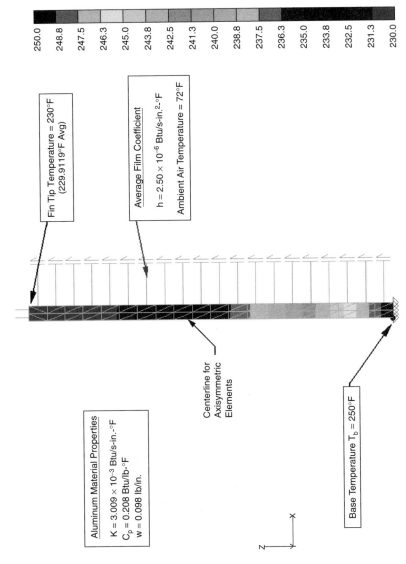

Figure 5.4 Summary of results of FEA solution: steady-state fin temperatures.

minimization of variational calculus to obtain approximate solutions to systems. By the early 1970s, FEA was limited to high-end companies involved in the aerospace, automotive, and power plant industries that could afford the cost of having mainframe computers. After the introduction of desktop personal computers, FEA was soon made available to the average user. FEA can be used to investigate new product designs and improve or refine existing designs.

The FEA approach is a mathematical idealization method that approximates physical systems by idealizing the geometry of a design using a system of grid points called *nodes*. These node points are connected together to create a "finite" number of regions or mesh. The regions that define the mesh are assigned properties such as the type of material and the type of mesh used. With the mesh properties assigned, the mesh regions are now called "elements." The types of mesh can be defined as rod elements, beam elements, plate elements, solid elements, and so on. Constraints or reactions are entered to simulate how the idealized geometry will react to the defined loading.

Analysis models can represent both linear and nonlinear systems. *Linear* models use constant parameters and assume that the material will return to its original shape after the loading is removed. *Nonlinear* models consist of stressing the material past its elastic capabilities, and permanent deformations will remain after the loading is removed. The definitions for the materials in these nonlinear models can be complex.

For heat transfer, FEA models can be idealized for both steady-state and transient solutions. The mesh elements are defined using the material conductivity or thermal dynamics to represent the system being analyzed. *Steady-state transfer* refers to materials with constant thermoproperties and where heat diffusion can be represented by linear equations. Thermal loads for solutions to heat-transfer analysis include temperatures, internal heat generation, and convection.

FEA Results: Comparison of Fin Design Concepts

With a high degree of confidence in defining and solving the sample heat-transfer problem, the final step is to apply the principles. Figure 5.5 shows summary results when the FEA model idealizations of fin designs shown in Fig. 5.1 are analyzed using MSC/Nastran. In Fig. 5.5 magnetic coils for each fin design concept are shown generating internal heat at the rate of 16 Btu/h per coil.

Table 5.3 lists a comparison of average coil temperatures for each fin design concept. Table 5.3 shows *dual-fin concept 1*, which produces the largest estimate of a 22 percent reduction in coil temperatures.

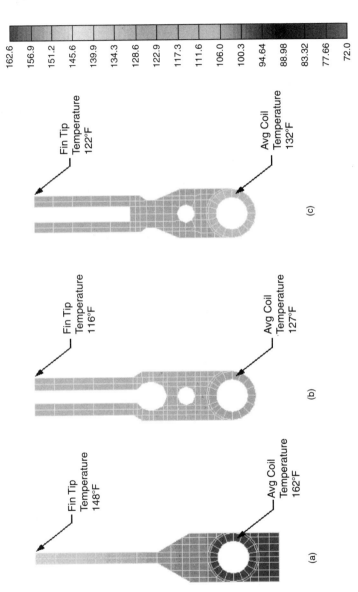

Figure 5.5 Summary of results of FEA solution: fin design concepts. (*a*) Original single fin; (*b*) dual-fin concept 1; (*c*) dual-fin concept 2.

TABLE 5.3 Summary of Average Coil Temperatures

	Original single-fin design	Dual-fin concept 1	Dual-fin concept 2
Average coil temperature	162°F	127°F	132°F
Percent reduction	N/A	−22%	−19%

Dual-fin concept 2 comes in at a close second at 19 percent reduction in coil temperatures.

As a result of this steady-state heat-transfer analysis, an approximate 20 percent reduction in coil operating temperatures could significantly extend the operational life of the magnetic coils. Therefore it is recommended that both dual-fin concepts be evaluated by performing actual physical testing before implementation of any production changes.

References

1. E. R. G. Eckert and Robert M. Drake, Jr., *Heat and Mass Transfer*, 2d ed., McGraw-Hill, 1959.
2. J. P. Holman, "Conduction-Convection Systems," in *Heat Transfer*, 5th ed., McGraw-Hill, 1981, chap. 2-9.
3. W. M. Rohsenow and J. P. Hartnett, *Handbook of Heat Transfer*, McGraw-Hill, 1973.
4. Donald R. Pitts and Leighton E. Sissom, *Heat Transfer*, 2d ed., Schaum's Outline Series, McGraw-Hill, 1977.

Chapter 6

Cooling of a Fuel Cell

Jason M. Keith
Department of Chemical Engineering
Michigan Technological University
Houghton, Michigan

Introduction

This problem applies fundamentals of heat transfer to the cooling of a fuel cell, which may have applications in laptop computers, cell phones, and the motor vehicles of the future. In this problem you will predict the electrical current generated within the fuel cell for a certain hydrogen mass flow rate, estimate the efficiency of the fuel cell in generating electricity, estimate the heat loss, and design a cooling system for the fuel cell.

Background: The Proton-Exchange Membrane Fuel Cell

Each cell within a fuel cell is basically an electrochemical device which converts hydrogen and oxygen gas into water and direct-current electricity. The by-product of the reaction is heat, which must be removed to prevent damage to the fuel cell components. A rough schematic of a single cell within a fuel cell is shown in Fig. 6.1. Hydrogen gas flows within channels etched in the bipolar plate (slanted lines in Fig. 6.1). The gas is transferred by convection to the surface of the gas diffusion layer (horizontal lines in Fig. 6.1), which it penetrates to reach the anode (area to left of center with vertical lines in Fig. 6.1). At the interface between the anode and electrolyte is a platinum catalyst, which oxidizes

6.2 Steady-State Calculations

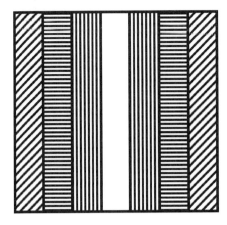

Figure 6.1 Schematic of one cell of a proton-exchange membrane fuel cell. The slanted lines represent the bipolar plates; the horizontal lines, the gas diffusion layer; the vertical lines, the electrodes (left block is the anode; right block is the cathode); and the open block, the electrolyte.

the hydrogen into electrons and protons:

$$2\,H_2 \rightarrow 4\,H^+ + 4\,e^- \quad (6.1)$$

The protons are relatively small and can pass through the electrolyte membrane (open block in center of Fig. 6.1). The electrons are relatively large and cannot pass through the electrolyte. They must pass by electrical conduction back through the anode, gas diffusion layer, and bipolar plate, through an electrical load (such as a lightbulb), and around to the other side of the fuel cell. The electrons then conduct through the bipolar plate, gas diffusion layer, and anode, where they recombine with the protons in the presence of oxygen, in a reduction reaction to form water:

$$O_2 + 4\,H^+ + 4\,e^- \rightarrow 2\,H_2O \quad (6.2)$$

The oxygen gets to the cathode via the same convection-diffusion mechanism that the hydrogen uses to reach the anode. The overall reaction is equal to the sum of the anode and cathode reactions:

$$2\,H_2 + O_2 \rightarrow 2\,H_2O \quad (6.3)$$

The direct-current electricity produced by the fuel cell will have a current I and a voltage E_{cell}. It is customary to "stack" n cells together to make a much larger system, called a "fuel cell stack," with stack voltage $E = nE_{cell}$ and current I. A cartoon of a fuel cell stack is seen in Fig. 6.2. Note that the current is the same regardless of the number of cells in the stack. Additional information on the fuel cell chemistry and construction is available in the textbook by Larminie and Dicks (2003).

To predict the heat generation, we need to know the efficiency of the fuel cell. Let us define *efficiency* as the rate of electrical work within

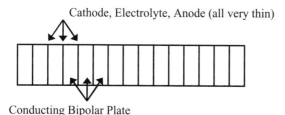

Figure 6.2 Side view of a fuel cell stack.

the fuel cell divided by the rate of thermal energy generation if the fuel were burned by combustion, such that

$$\eta = \frac{P}{C_{\text{combustion}}} \tag{6.4}$$

The rate of electrical work is also called *power* and is given as the product of the stack voltage E and current I within the fuel cell,

$$P = IE = InE_{\text{cell}} \tag{6.5}$$

and the rate of thermal energy generation due to combustion is given as

$$C_{\text{combustion}} = nM_{\text{fuel}} \Delta H \tag{6.6}$$

where M_{fuel} is the molar flow rate of hydrogen *per cell within the stack* and ΔH is the heat of formation, given as $\Delta H = -241.8$ kJ/mol if the water product is steam (lower heating value) or $\Delta H = -285.8$ kJ/mol if the water product is liquid (higher heating value). Fuel cells for commercial and/or automotive uses will often have a condenser to produce liquid product, so the higher heating value will be used here.

The electrical current I is also related to the hydrogen molar flow rate M_{fuel}. Equation (6.1) shows that for each mole of H_2 reacted at the anode, two moles of electrons are generated. The charge of a mole of electrons, in units of coulombs per mole, is given as Faraday's constant $F = 96{,}485$ C/mol. Thus, a molar flow rate M_{fuel} to each cell gives an electrical current of

$$I = \frac{M_{\text{fuel}} \text{ mol fuel}}{\text{s}} \frac{2 \text{ mol e}^-}{\text{mol fuel}} \frac{FC}{\text{mol e}^-} = 2FM_{\text{fuel}} \quad A \tag{6.7}$$

since one coulomb per second is equal to one ampere of current. Substituting Eq. (6.7) into Eq. (6.5) and dividing by Eq. (6.6) gives the efficiency η,

$$\eta = \frac{2FE_{\text{cell}}}{\Delta H} = \frac{E_{\text{cell}}}{1.48} \tag{6.8}$$

if the higher heating value of hydrogen is used.

6.4 Steady-State Calculations

Energy that is not converted into electrical power is released as heat. A simple energy balance on the process gives

$$C_{\text{combustion}} = P + Q \qquad (6.9)$$

Combining Eqs. (6.4) and (6.9) to eliminate $C_{\text{combustion}}$ gives the heat generation within the fuel cell:

$$Q = P\left(\frac{1}{\eta} - 1\right) \qquad (6.10)$$

We will now estimate the electrical energy generation and heat generation in a fuel cell stack.

Example calculation: efficiency, fuel usage, and heat generation

Consider a fuel cell stack with five cells of cross-sectional area $A_{\text{cell}} = 10\ \text{cm}^2$. Such a small fuel cell may be used in a cell phone or a laptop computer. Assume that the cell voltage E_{cell} is related to the current density $i = I/A_{\text{cell}}$ in units of mA/cm² according to the relationship

$$E_{\text{cell}} = E_{\text{oc}} - A\ln(i) - ir - m\exp(ni) \qquad (6.11)$$

where, for a certain type of proton-exchange membrane fuel cell, the parameters in this equation are given by Laurencelle et al. (2001) as $E_{\text{oc}} = 1.031$ V, $A = 0.03$ V, $r = 2.45 \times 10^{-4}$ kΩ cm², $m = 2.11 \times 10^{-5}$ V, and $n = 8 \times 10^{-3}$ cm²/mA. The first term estimates the voltage at zero current density, also called the *open-circuit voltage*. The second term represents *activation losses*, the third term represents *ohmic losses*, and the final term represents *mass-transfer losses*. For additional information, consult the text by Larminie and Dicks (2003). The components of Eq. (6.11) are plotted in Fig. 6.3. It is customary to operate a fuel cell within the "ohmic" region of the graph; thus for a current density of 500 mA/cm², the cell voltage $E_{\text{cell}} = 0.72$ V. The stack voltage is $E = nE_{\text{cell}} = 5(0.72) = 3.6$ V, and the current is given by $iA_{\text{cell}} = 500$ mA/cm² (10 cm²)(1 A/1000 mA) = 5 A. The total power is equal to the stack voltage multiplied by the current, $P = IE = (5\ \text{A})(3.6\ \text{V}) = 18$ W.

The molar hydrogen flow rate per cell required to produce this power can be calculated by rearranging Eq. (6.7) in terms of M_{fuel}:

$$M_{\text{fuel}} = \frac{I}{2F} \qquad (6.12)$$

Thus, $M_{\text{fuel}} = 5\ \text{A}/2/(96{,}485\ \text{C/mol}) = 2.6 \times 10^{-5}$ mol/s. The total hydrogen molar flow rate can be calculated by multiplying this value by the number of cells and is equal to 1.3×10^{-4} mol/s. Noting that as the

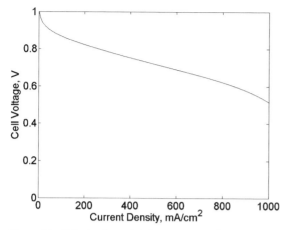

Figure 6.3 Polarization plot for proton-exchange membrane fuel cell.

molecular weight of hydrogen is 2 g/mol, it would take about one hour to use a gram of hydrogen in this fuel cell.

Using Eq. (6.8), the efficiency of the fuel cell is $\eta = 0.72/1.48 = 0.487$. Thus, the heat generation can be calculated from Eq. (6.10) as $Q = 18$ W $(1/0.487 - 1) = 19.0$ W. The heat generation per cell is thus $Q_{cell} = 19.0/5 = 3.8$ W. This heat will have to be removed by cooling air to maintain the fuel cell at a steady-state temperature, which we will now discuss.

Convective Cooling

Consider the fuel cell stack shown in Fig. 6.2. Larminie and Dicks (2003) state that fuel cells with $P > 100$ W require air cooling. To avoid drying out the membrane, separate reactant air and cooling air are required. This is achieved by placing cooling channels within the bipolar plate. A schematic of one cell with such a cooling channel is shown in Fig. 6.4. We assign the following variables, from the perspective taken from a side view of the fuel cell: L is the height of the plate and channel; d is the depth of the plate and channel (so the product Ld is the area A_{cell}), t is the thickness of the bipolar plate, cathode, electrolyte, and anode; and w is the width of the cooling channel.

Heat generation within the fuel cell occurs near the cathode, electrolyte, and anode (left and right sides of the schematic in Fig. 6.4). In essence, a total heat generation rate of $Q_{cell}/2$ W can be considered to be input at the left and right sides of the cell, as shown in the left portion of Fig. 6.5. Since the cathode, electrolyte, anode, and bipolar plate are made of materials with high thermal conductivity (about 20 W/m-K),

6.6 Steady-State Calculations

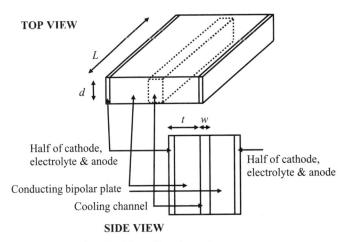

Figure 6.4 Single cell with cooling channel.

the heat source can be considered to be input directly into the cooling channel, as seen in the right portion of Fig. 6.5. Thus, it is appropriate to model this system as heat transfer to a channel with a constant wall flux at the boundaries (due to a uniform heat generation rate per unit area of the channel surface).

An energy balance over the fluid in the channel yields

$$\dot{m}_{\text{coolant}} C_p (T_{\text{coolant,out}} - T_{\text{coolant,in}}) = Q_{\text{cell}} \tag{6.13}$$

which can be used to estimate the exit temperature of the cooling air. The maximum temperature of the channel wall can be estimated using a heat-transfer coefficient. For channels with a constant heat flux at the boundary, fully developed laminar flow, and a very high channel

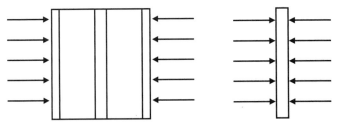

Figure 6.5 Heat generation within a single cell of a fuel cell stack. The picture on the left shows the heat source within the cathode, electrolyte, and anode of the fuel cell. The picture on the right shows an approximation where the heat is input directly into the cooling channel.

aspect ratio, Incropera and DeWitt (1996) report the following Nusselt number:

$$\text{Nu} = \frac{hD_h}{k_{gas}} = 8.23 \tag{6.14}$$

where D_h is the hydraulic diameter, given as

$$D_h = \frac{4A_{\text{cross section}}}{P_{\text{channel}}} \tag{6.15}$$

where $A_{\text{cross section}}$ and P_{channel} are the cross-sectional area and the perimeter of the cooling channel, respectively. For a channel aspect ratio of 1, 2, 4, and 8, the Nusselt number is equal to 3.61, 4.12, 5.33, and 6.49. The correlation for the Nusselt number is valid when the Reynolds number is less than 2300:

$$\text{Re} = \frac{4\dot{m}_{\text{coolant}}}{\mu_{\text{gas}} P_{\text{channel}}} \tag{6.16}$$

where μ_{gas} is the gas viscosity. As the coolant heats up as it travels along the channel, the location of maximum solid surface temperature will be at the exit. An energy balance yields that the temperature difference between the solid and gas is a constant, due to the fact that a uniform heat flux is present:

$$h(T_{\text{surface}} - T_{\text{gas}}) = \frac{Q_{\text{cell}}}{LP_{\text{channel}}} \tag{6.17}$$

Finally, an energy balance within the bipolar plate, cathode, and anode yields the relationship between the surface temperature and that at the cell edge where the heat is supplied

$$k_{\text{solid}} \frac{T_{\text{edge}} - T_{\text{surface}}}{t} = \frac{Q_{\text{cell}}}{LP_{\text{channel}}} \tag{6.18}$$

Again, this temperature difference is a constant along the entire channel length because of the constant-heat-flux approximation.

Example calculation: cooling the fuel cell

We will apply the principles in the preceding section to remove the 19 W of heat generated within the fuel cell. In general, fuel cells with power output $P < 100$ W can be cooled by natural convection. As this fuel cell may be placed inside a cell phone or a laptop computer, convective cooling may be justified. The procedure illustrated here may also be extended to a larger stack.

Proton-exchange membranes are typically operated about 50°C. Temperatures above 100°C can damage the fuel cell components. If we wish

6.8 Steady-State Calculations

TABLE 6.1 Dimensions and Thermal Properties

Item	Value
L	3.16 cm = 0.0316 m
d	3.16 cm = 0.0316 m
t	0.5 cm = 0.005 m
w	0.1 cm = 0.001 m
k_{solid}	20 W/m-K
k_{gas}	0.0263 W/m-K
μ_{gas}	1.84×10^{-4} g/cm-s
C_p	1.0 J/g-K

to maintain the maximum temperature within the fuel cell at 60°C, for example, with cooling air at 20°C, what air mass flow rate is required? Thermal properties and dimensions of the fuel cell are given along with air thermal properties in Table 6.1.

This problem is solved in a procedure reverse as that derived in the previous section. The channel perimeter $P_{channel}$ is equal to $2d + 2w = 0.0652$ m. Equation (6.18) is then used to estimate the solid surface temperature at the cooling channel exit:

$$T_{surface} = T_{edge} - \frac{Q_{cell}}{LP_{channel}} \frac{t}{k_{solid}}$$

$$= 60°C - \frac{3.8 \text{ W}}{0.0316 \text{ m}} \frac{1}{0.0652 \text{ m}} \frac{0.005 \text{ m}}{20 \text{ W/m} - \text{K}} \frac{1°C}{1 \text{ K}} = 59.5°C \quad (6.19)$$

The aspect ratio of the channel is $d/w = 31.6$, so it is safe to use the infinite aspect ratio value for the Nusselt number Nu = 8.23. The hydraulic diameter is given by Eq. (6.15):

$$D_h = \frac{4A_{cross\ section}}{P_{channel}} = \frac{4dw}{2(d+w)}$$

$$= \frac{4}{2}\left(\frac{0.0316 \text{ m}}{0.0316 \text{ m} + 0.001 \text{ m}}\right) 0.001 \text{ m} = 0.0019 \text{ m} \quad (6.20)$$

which is close to $2w$ as $d \gg w$. Equation (6.14) can then be used to calculate the heat-transfer coefficient:

$$h = 8.23 \frac{k_{gas}}{D_h} = 8.23 \left(\frac{0.0263 \text{ W/m-K}}{0.0019 \text{ m}}\right) = 111 \text{ W/m}^2\text{-K} \quad (6.21)$$

The heat-transfer coefficient can now be used to determine the gas exit temperature using Eq. (6.17):

$$T_{gas} = T_{surface} - \frac{Q_{cell}}{hLP_{channel}}$$
$$= 59.5°C - \frac{3.8 \text{ W}}{111 \text{ W/m}^2 - \text{K}} \left(\frac{1}{0.0316 \text{ m}}\right)\left(\frac{1}{0.0652 \text{ m}}\right)\frac{°C}{K} = 42.8°C \quad (6.22)$$

Finally, the air mass flow rate can be determined from Eq. (6.13), by setting $T_{coolant,out} = T_{gas}$. The resulting expression is

$$\dot{m}_{coolant} = \frac{Q_{cell}}{C_p(T_{coolant,out} - T_{coolant,in})}$$
$$= \frac{3.8 \text{ W}}{1.0 \text{ J/g-K}}\frac{1}{42.8 - 20°C} = 0.166 \text{ g/s} \quad (6.23)$$

To prove that this is laminar flow, the Reynolds number must be calculated from Eq. (6.16):

$$\text{Re} = \frac{4\dot{m}_{coolant}}{\mu_{gas} P_{channel}} = \frac{4}{1.84 \times 10^{-4} \text{ g/cm-s}} \frac{0.166 \text{ g/s}}{0.0652 \text{ m}} \frac{\text{m}}{100 \text{ cm}} = 550 \quad (6.24)$$

The total mass flow rate of coolant is equal to the value for one cell multiplied by the number of cells, so $\dot{m}_{coolant,total} = 0.83$ g/s.

Discussion and Conclusions

Analysis of the sample system presented above shows that the maximum cooling flow rate to maintain laminar flow within the cooling channel is 0.689 g/s per cell or 3.44 g/s total. In this circumstance, the maximum temperature within the fuel cell is $T_{edge} = 42.5°C$, the maximum surface temperature is $T_{surface} = 42.0°C$, and the exit coolant temperature is $T_{coolant,out} = 25.5°C$. The minimum cooling flow rate can also be determined by selecting a maximum value for T_{edge}, say, 90°C. The flow rate is 0.072 g/s per cell or 0.36 g/s total. The maximum surface temperature is $T_{surface} = 89.5°C$, and the exit coolant temperature is $T_{coolant,out} = 73.0°C$. The Reynolds number in this case is 240.

In practice, the power required to operate electronic equipment by the fuel cell will vary. During high loads, the fuel cell temperature will likely increase. Using a maximum possible temperature of 90°C and a Q_{cell} value of 5.7 W (representing a 50 percent increase over the base case considered here) yields a cooling mass flow rate of 0.128 g/s per cell or 0.64 g/s total.

6.10 Steady-State Calculations

This calculation applies convection and conduction heat transfer to the design of a fuel cell cooling system. The situation studied here can be considered a worst-case scenario as the cooling fluid must remove all the heat generated within the fuel cell. In practical applications, there are additional heat losses to the environment. The total heat generated per cell Q_{cell} can be multiplied by a percentage to account for these effects.

References

Incropera, F., and DeWitt, D. *Fundamentals of Heat and Mass Transfer*, 4th ed. Wiley, New York (1996).

Larminie, J., and Dicks, A. *Fuel Cell Systems Explained*, 2d ed. Wiley, Hoboken, N.J. (2003).

Laurencelle, F., Chahine, R., Hamelin, J., Fournier, M., Bose, T. K., and Lapperriere, A. "Characterization of a Ballard MK5-E Proton Exchange Membrane Stack," *Fuel Cells* **1**(1), 66–71 (2001).

Chapter 7

Turbogenerator Rotor Cooling Calculation

James T. McLeskey, Jr.
Department of Mechanical Engineering
Virginia Commonwealth University
Richmond, Virginia

Problem

Prior to 1960, large turbogenerator rotors were cooled indirectly and limited in output to approximately 200 MW. Heat generated in the windings was conducted to the surface of the rotor and then removed from the outer diameter by convection to the hydrogen gas. As the capability increased in the late 1950s, it became necessary to find more efficient methods of cooling the rotors. Manufacturers developed cooling systems where the cooling gas was directly in contact with the copper conductors. These direct-cooled rotors helped make possible the construction of turbogenerator units with capacities as high as 1300 MW.

In spite of the improved cooling techniques, rotor heating can still be a limiting factor under lagging power factor conditions. Determining the rotor temperature profile is necessary when designing a new machine. In addition, with improvements in steam turbine design, it is now possible to retrofit older steam turbines and provide power increases of as much as 10 percent over the original rating. It is often necessary to determine the ability of the generator rotor to tolerate this increase in power.

In order to better quantify these limits, it is necessary to calculate the cooling gas flow and temperature profile in the generator rotor. The

7.2 Steady-State Calculations

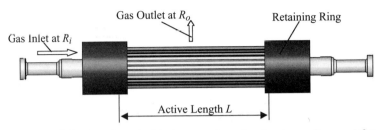

Figure 7.1 Side view of a typical turbogenerator rotor showing cooling gas inlet and outlet regions.

following calculation is for a direct-cooled turbogenerator rotor with axial flow passages in the windings. The dimensions and configuration are not for any particular machine, but are typical. The principles can be applied to other designs with appropriate adjustments.

Assumptions

1. All cooling is due to direct cooling of hydrogen in contact with the copper.
2. Radial and axial heat transfer within the copper is ignored.

Cooling principle

In a large turbogenerator, the rotor provides the rotating magnetic field. In order to generate this field, current passes through the windings of the rotor. Naturally, these windings can become quite hot. To prevent damage to the electrical insulation which surrounds the windings, it is necessary to cool the windings.

Cooling the winding is often done as shown in Figs. 7.1 to 7.4. Cold gas enters under the retaining ring, travels axially through the copper windings, and exits at the center of the rotor. The pressure difference driving the gas results from the different radii of the inlet and exit points. The inlet radius is less than the exit radius, so the inlet is traveling at a lower linear velocity and, in accordance with Bernoulli's principle, is at a higher pressure.

Calculation Method

The calculation consists of five primary steps:

1. Determine the cooling gas properties.
2. Determine the volume flow rate of the cooling gas through the rotor.

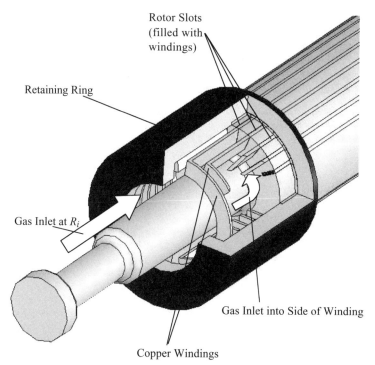

Figure 7.2 Cutaway view of generator winding head region (under the retaining ring) showing the gas inlet and filled rotor slots.

3. Determine the temperature rise in the gas passing through the rotor.
4. Determine the temperature difference between the cooling gas and the copper windings.
5. Determine the absolute temperature of the copper.

Determine the gas properties

Modern turbogenerators are capable of achieving nearly 300 MW of output using air as the cooling medium. However, all units larger than that and most older units greater than 80 MW use hydrogen as the cooling medium because of its low density and viscosity and relatively high heat capacity. Our calculation will assume that hydrogen is used.

The first step in calculating the cooling is to determine the gas properties based on the estimated operating conditions. In a computer model, this step might be repeated iteratively after the rest of the calculation is performed, but a generator designer can usually make a good first approximation based on the typical data.

7.4 Steady-State Calculations

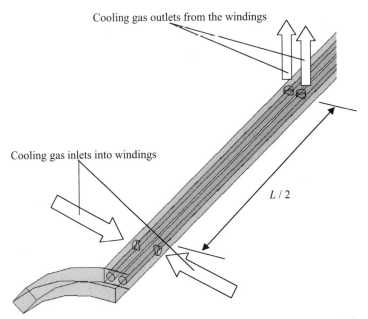

Figure 7.3 Schematic of a single axially direct-cooled generator rotor winding turn showing gas inlet and outlet holes.

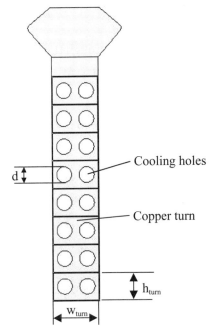

Figure 7.4 Cross-sectional view of a generator rotor winding slot showing the axial cooling holes in the direct-cooled rotor windings. This slot contains eight turns.

TABLE 7.1 Operating Conditions

Parameter	Variable	Value	Comment
Cooling medium	—	Hydrogen	—
H_2 pressure	P	75 psig = 517×10^3 Pa	—
Cold-gas temperature	T_{cold}	45°C = 318 K	This is the estimated temperature of the cold gas entering the rotor; it is based on the stator cooler design
H_2 purity	$H_{2,\text{pure}}$	98% = 0.98	The hydrogen in the machine will not be 100% pure
Estimated maximum temperature rise of rotor	ΔT_{max}	75°C	Limited by insulation
Maximum temperature of H_2	$T_{H_2,max}$	120	$T_{cold} + \Delta T_{max}$
Estimated average H_2 temperature	$T_{\text{avg gas}}$	82.5°C = 356 K	$(T_{cold} + T_{H_2,max})/2$

Gather the operating conditions. See Table 7.1.

Determine the gas properties at $T_{\text{avg gas}}$. It is necessary to find the properties for both the hydrogen and the air in the system and then use the hydrogen purity to find a weighted average.

Density. Using the ideal-gas law, we obtain

$$\rho_{H_2} = \frac{P}{R_{H_2} T_{\text{avg gas}}} \quad \text{where} \quad R_{H_2} = 4.124 \times 10^3 \text{ J/kg K} \quad (7.1)$$

$$\rho_{\text{air}} = \frac{P}{R_{\text{air}} T_{\text{avg gas}}} \quad \text{where} \quad R_{\text{air}} = 0.287 \times 10^3 \text{ J/kg K} \quad (7.2)$$

for the given conditions, $\rho_{H_2} = 0.352$ kg/m³ and $\rho_{\text{air}} = 5.06$ kg/m³. The weighted average is

$$\rho = \rho_{H_2} \cdot H_{2,\text{pure}} + \rho_{\text{air}} \cdot (1 - H_{2,\text{pure}})$$
$$= 0.446 \text{ kg/m}^3 \quad (7.3)$$

Absolute viscosity

$$\mu_{H_2} = 1.053 \times 10^{-5} \text{ Pa-s}$$
$$\mu_{\text{air}} = 2.155 \times 10^{-5} \text{ Pa-s}$$
$$\mu = \mu_{H_2} \cdot H_{2,\text{pure}} + \mu_{\text{air}} \cdot (1 - H_{2,\text{pure}})$$
$$= 1.075 \times 10^{-5} \text{ kg m}^{-1}\text{s}^{-1} \quad (7.4)$$

7.6 Steady-State Calculations

Kinematic viscosity

$$v_k = \frac{\mu}{\rho}$$

$$= 2.409 \times 10^{-5} \text{ m}^2\text{s}^{-1} \quad (7.5)$$

Specific heat

$$C_p = 10{,}000 \text{ W-s/kg-K}$$

Thermal conductivity

$$\lambda = 0.2 \text{ W/m-K}$$

Determine the volume flow rate of the cooling gas through the rotor

The primary cooling mode inside a direct-cooled rotor is forced convection of the heat from the copper to the cooling gas. Therefore, it is necessary to determine the volume flow rate and the velocity of the gas through the copper windings. The pressure driving the flow is generated as a result of the difference in the diameters of the gas inlet and outlet.

Gather the geometric data. See Table 7.2.

The self-generated pressure and the volume flow rate depend on the geometric configuration of the rotor. Although some approximations can be made when the exact dimensions are unknown, more fully defining the problem will lead to more accurate results.

Find the self-generated pressure. Because the hydrogen inlet and outlet are at different diameters, the linear velocities are different. The inlet is at a smaller diameter and is therefore traveling at a lower linear velocity and, in accordance with Bernoulli's principle, is at a higher pressure. This forces the cooling gas through the rotor:

$$\Delta P_{\text{self}} = \frac{\rho(2\pi \times \text{rpm}/60 \text{ s})^2(R_o^2 - R_i^2)}{2}$$

$$= 2.854 \times 10^3 \text{ Pa} \quad (7.6)$$

where rpm is the rotation speed in revolutions per minute (r/min). In some machines, additional pressure for moving the gas through the rotor is provided by external fans.

Turbogenerator Rotor Cooling Calculation

TABLE 7.2 Geometric Dimensions (See Figs. 7.1 to 7.5)

Parameter	Variable	Value	Comment
Rotor body radius	R_o	0.5 m	Outer radius of rotor body
H_2 inlet radius	R_i	0.4 m	Radius under retaining ring where H_2 enters the rotor
Active length	L	6.0 m	Length of iron through which the windings pass
Rotation speed	rpm	3600 min^{-1}	—
Number of slots	N_{slots}	20	—
Number of Cu turns per slot (layers)	Turns	8	—
Holes per turn	Hole$_{turn}$	2	—
Holes per slot	Hole$_{slot}$	16	Turns · Hole$_{turn}$
Turn thickness	h_{turn}	0.0148 m	—
Turn width	w_{turn}	0.030 m	—
Copper resistivity	R_{Cu}	0.01675×10^{-6} Ω-m	—
Cooling passage diameter	d	8 mm	—
Excitation current	I	2000 A	—

Determine gas velocity through the windings. The gas enters the rotor under the retaining rings at each end and moves to the center where it exits. This means that all the calculations are performed for half of the active length of the rotor as shown in Fig. 7.5.

The velocity depends on the pressure drop through the rotor. The pressure drop depends on the friction and the geometry:

$$\Delta P = \Delta P_{\text{friction}} + \Delta P_{\text{geometry}} \tag{7.7}$$

The pressure drop due to friction can be found from

$$\Delta P_{\text{friction}} = \lambda_f \frac{L/2}{d} \frac{\rho v^2}{2} \tag{7.8}$$

where $\lambda_f = 0.316/\text{Re}^{1/4} = 0.316/(vd/v_k)^{1/4}$ for smooth copper profiles and turbulent flow. The term $L/2$ is used because the gas enters at one

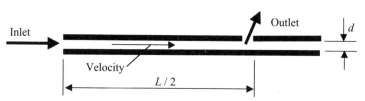

Figure 7.5 Schematic cross section of the side of a single cooling hole in the copper winding.

7.8 Steady-State Calculations

end and exits at the center of the rotor. The dynamic pressure is given by $\rho v^2/2$.

The pressure drop due to geometry is given by

$$\Delta P_{\text{geometry}} = c \frac{\rho v^2}{2} \tag{7.9}$$

where c is a geometric coefficient, typically between 5 and 15 depending on the rotor.

If we define

$$a = 0.316 \left(\frac{v_k}{d}\right)^{1/4} \frac{L/2}{d}$$

$$= 27.76 \, (\text{m/s})^{0.25} \tag{7.10}$$

and

$$b = 2\frac{\Delta P}{\rho} \quad \text{with} \quad \Delta P = \Delta P_{\text{self}}$$

$$= 12{,}789 \, \text{N-m/kg} \tag{7.11}$$

then we can write the equation for the pressure loss as

$$\frac{b}{a} = v^{1.75} + \frac{c}{a} \cdot v^2 \tag{7.12}$$

If we determine a and b using the values given above and assume a value for c ($c = 10$), the roots of the above equation can be found to give the velocity.

Using MathCad to find the roots, the value obtained was $v = 23.81$ m/s in each cooling hole. The volume flow rate in a single cooling hole can be found by multiplying this velocity by the area of each hole:

$$V_{\text{hole}} = \text{area} \times \text{velocity} = \frac{\pi d^2}{4} \cdot v$$

$$= 1.196 \times 10^{-3} \, \text{m}^3/\text{s} \tag{7.13}$$

Then the total volume flow rate in a single slot can be determined by multiplying this value by the number of holes in a single slot:

$$V_{\text{slot}} = V_{\text{hole}} \cdot \text{Hole}_{\text{slot}}$$

$$= 19.1 \times 10^{-3} \, \text{m}^3/\text{s} \tag{7.14}$$

Determine the temperature rise in the gas passing through the rotor

The rotor is assumed to be operating under steady-state conditions. Therefore, the hydrogen gas must remove any heat generated by the windings. The first step is to determine the amount of heat produced in the winding in each slot (actually in each half-slot since the cooling gas exits at the center). This is due to I^2R losses from the excitation current. The heat is found by first calculating the cross-sectional area of one turn of copper (Fig. 7.4) and multiplying by the resistivity to determine the total resistance. This value is then multiplied by the square of the excitation current. This, however, requires an estimate of the copper temperature rise (assumed here as a first guess to be $\Delta T = 75°C$).

The cross-sectional area of one turn is

$$A_{Cu} = (h_{turn} \times w_{turn}) - \frac{\pi d^2}{4} \times \text{Hole}_{turn}$$
$$= 343.5 \times 10^{-6} \text{ m}^3 \tag{7.15}$$

The total resistance of windings in one half-slot is

$$R_{slot} = R_{Cu}(1 + 0.004 \cdot \Delta T) \cdot \text{turns} \cdot \frac{L/2}{A_{Cu}}$$
$$= 1.52 \times 10^{-3} \, \Omega \tag{7.16}$$

Power (heat) produced in one half-slot is

$$P_{slot} = I^2 R_{slot} = 6085 \text{ W} \tag{7.17}$$

This power (heat) must be removed by the hydrogen. To find the temperature rise of the hydrogen, divide this power by the total heat capacity of the hydrogen flowing through the copper:

$$\Delta T_{H_2} = \frac{P_{slot}}{C_p \, \rho \, V_{slot}}$$
$$= 71.25°C \tag{7.18}$$

The temperature profile of the hydrogen gas is shown in Fig. 7.6. It enters from the end at the cold gas temperature of 45°C and reaches a maximum of 116.25°C at the center of the rotor.

7.10 Steady-State Calculations

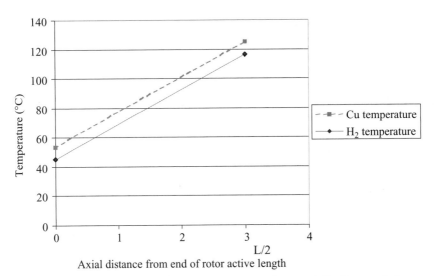

Figure 7.6 Temperature profiles of H_2 gas and copper for an axially direct-cooled generator rotor.

Determine the temperature difference between the cooling gas and the copper windings

Under steady-state conditions, all the heat generated in the copper is assumed to be transferred to the hydrogen. However, because there is some thermal resistance between the copper and the hydrogen, the copper is at a higher temperature. This temperature must be determined by finding the heat-transfer coefficient between the hydrogen and the copper.

Reynolds number of flow in holes

$$\text{Re} = \frac{vd}{v_k} = 7907 \tag{7.19}$$

Prandtl number of flow

$$\text{Pr} = \frac{v_k \rho C_p}{\lambda} = 0.538 \tag{7.20}$$

Heat-transfer coefficient for flow in smooth copper pipe

$$h = 0.023 \cdot \frac{\lambda \cdot \text{Re}^{0.8} \cdot \text{Pr}^{0.4}}{d}$$

$$= 589.2 \, \text{W/m}^2\text{-}°\text{C} \tag{7.21}$$

The temperature difference between the copper and the hydrogen is

$$\Delta T_{\text{CuH}_2} = \frac{P_{\text{slot}}}{h \cdot \text{Area}_{\text{surf}}} \quad (7.22)$$

where $\text{Area}_{\text{surf}}$ equals surface area of the copper in contact with the hydrogen (the perimeter of the hole times the length of the copper):

$$\text{Area}_{\text{surf}} = \pi d \cdot \frac{L}{2} \cdot \text{Hole}_{\text{slot}} = 1.206\,\text{m}^2 \quad (7.23)$$

$$\Delta T_{\text{CuH}_2} = 8.57°\text{C}$$

The temperature profile of the copper is shown in Fig. 7.6. The copper temperature remains 8.6°C above the hydrogen temperature across the length of the rotor and reaches its maximum at the center of the rotor.

Determine maximum temperature of copper

The maximum copper temperature occurs at the center of the rotor for a rotor with this type of cooling. It can be found from

$$T_{\text{Cu}} = T_{\text{cold}} + \Delta T_{\text{H}_2} + \Delta T_{\text{CuH}_2}$$

$$= 45 + 71.2 + 8.6 = 124.8°\text{C} \quad \text{maximum value} \quad (7.24)$$

In order to protect the insulation, the maximum desired temperature is typically 130°C, so this rotor design would be acceptable.

Summary and Conclusions

The calculation presented here provides a method for determining the temperature profile in a large turbogenerator rotor with direct axial cooling using hydrogen. For the particular parameters chosen, the hydrogen gas reaches a maximum of approximately 116°C. This results in a maximum copper winding temperature of approximately 125°C. If this were an actual rotor, the maximum temperature would be within the typically quoted class B temperature limit of 130°C. The maximum copper temperature occurs at the center of the rotor for this design.

A wide variety of direct-cooled rotor designs are produced by various manufacturers. The principles outlined in this calculation can be applied to other designs with the appropriate adjustments for such things as different cooling flow patterns (Costley and McLeskey, 2003) or substantial indirect cooling (McLeskey et al., 1995). This analysis of the thermal limits of the rotor will allow for maximum utilization under all generator operating conditions.

7.12 Steady-State Calculations

References

Costley, J. M., and J. T. McLeskey, Jr., "Diagonal-Flow, Gap Pickup Generator Rotor Heat Transfer Model," *Proceedings of IMECE'03: 2003 ASME International Mechanical Engineering Congress & Exposition*, Washington, D.C., vol. 374, no. 1, pp. 389–392, Nov. 16–21, 2003.

Incropera, F. P., and D. P. DeWitt, *Fundamentals of Heat and Mass Transfer*, 5th ed., Wiley, New York, 2002.

McLeskey, J. T., Jr., A. T. Laforet, D. R. Snow, and J. V. Matthews III, "Generator Rotor Replacement by a Third Party Manufacturer," *Conference Papers of Power-Gen Americas '95*, Anaheim, Calif., vol. 4, pp. 391–401, 1995.

Munson, B. R., D. F. Young, and T. H. Okiishi, *Fundamentals of Fluid Mechanics*, 4th ed., Wiley, New York, 2002.

Chapter 8

Heat Transfer through a Double-Glazed Window

P. H. Oosthuizen
Department of Mechanical Engineering
Queen's University
Kingston, Ontario, Canada

David Naylor
Department of Mechanical and Industrial Engineering
Ryerson University
Toronto, Ontario, Canada

Problem Statement

Consider heat transfer through a 1-m-high double-paned window under nighttime conditions, when the effects of solar radiation can be ignored. The heat transfer can be assumed steady and one-dimensional, and the effects of the window frame can be ignored. The window glass can be assumed to be 3 mm thick and to have a thermal conductivity of 1.3 W/m°C. The gap between the panes is 10 mm. The air temperature inside the building can be assumed to be 20°C and the outside air temperature, 5°C. On the basis of available results, the heat-transfer coefficients on the inner and outer surfaces of the window can be assumed to be 4 and 25 W/m²°C, respectively, and the emissivity of the glass can be assumed to be 0.84. Find the heat-transfer rate through the window per unit area for the case where the gas between the panes is air and for the case where it is argon. Also express these results in terms of an overall heat-transfer coefficient and in terms of the overall heat-transfer resistance.

8.2 Steady-State Calculations

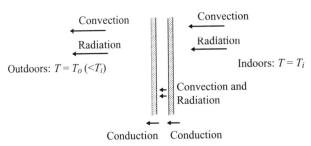

Figure 8.1 Heat-transfer situation under consideration.

Solution

The situation considered here is shown in Fig. 8.1. The following nomenclature is used here:

A	Frontal area of window
AR	Aspect ratio; H/w
H	Height of window
h_{cb}	Convective heat-transfer coefficient for flow between window panes
h_{ci}	Convective heat-transfer coefficient for inner surface of window
h_{co}	Convective heat-transfer coefficient for outer surface of window
h_r	Radiation heat-transfer coefficient
h_{rb}	Radiation heat-transfer coefficient for flow between window panes
h_{ri}	Radiation heat-transfer coefficient for inner surface of window
h_{ro}	Radiation heat-transfer coefficient for outer surface of window
k	Thermal conductivity
Nu	Nusselt number based on w
Nu_a	First Nusselt number value given by between-glass convection equations
Nu_b	Second Nusselt number value given by between-glass convection equations
Nu_c	Third Nusselt number value given by between-glass convection equations
Q	Heat-transfer rate through window
Q_b	Heat-transfer rate through gap between window panes
Q_i	Heat-transfer rate to inner surface of window
Q_{gi}	Heat-transfer rate through inner pane
Q_{go}	Heat-transfer rate through outer pane
Q_o	Heat-transfer rate from outer surface of window
Ra	Rayleigh number based on w and $T_2 - T_3$
R	Overall heat-transfer resistance

R_i Inside heat-transfer resistance
R_o Outside heat-transfer resistance
R_g Heat-transfer resistance of glass window pane
R_b Heat-transfer resistance of pane-to-pane gap
T_i Inside air temperature
T_o Outside air temperature
T_1 Temperature of inside surface of window
T_2 Temperature of outer surface of inner window pane
T_3 Temperature of inner surface of outer window pane
T_4 Temperature of outside surface of window
w Distance between the two panes of glass
β Bulk expansion coefficient
ν Kinematic viscosity of between-panes gas

Because the effects of solar radiation are ignored and because the glass is essentially opaque to the infrared radiation therefore involved in the situation considered, the glass panes will be assumed to behave as conventional opaque gray bodies. The radiant heat transfer will be treated by introducing radiant heat-transfer coefficients h_r.

Because it is assumed that the heat transfer is steady, it follows that the rate of heat transfer from the inside air to the inner surface of the window is equal to the rate of heat transfer through the inner glass pane, which in turn is equal to the rate of heat transfer across the gap between the panes, which in turn is equal to the rate of heat transfer across the outer glass pane, which in turn is equal to the rate of heat transfer from the outside surface of the outer glass pane to the outside air; thus considering the heat-transfer rate per unit window area, we obtain

$$\frac{Q_i}{A} = \frac{Q_{gi}}{A} = \frac{Q_b}{A} = \frac{Q_{go}}{A} = \frac{Q_o}{A}$$

The subscripts i, gi, b, go, and o refer to the heat-transfer rate from the inside air to the inside glass pane, across the inner glass pane, across the gap between the panes, across the outer glass pane, and from the outside glass pane to the outside air, respectively.

The heat transfer from the inside air to the inner glass surface is accomplished by a combination of radiation and convection. These will be expressed in terms of convective and radiative heat-transfer coefficients defined by

$$\frac{Q_{ci}}{A} = h_{ci}(T_i - T_1) \qquad \frac{Q_{ri}}{A} = h_{ri}(T_i - T_1)$$

8.4 Steady-State Calculations

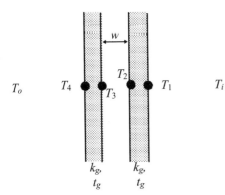

Figure 8.2 Temperatures used in obtaining solution.

where T_i is the inside air temperature and T_1 is the temperature of the inner surface of the inner glass pane as shown in Fig. 8.2. Now it is assumed that the window area is small compared to the size of the room to which it is exposed, so the following can be assumed:

$$\frac{Q_{ri}}{A} = \sigma\varepsilon(T_i^4 - T_1^4) = \sigma\varepsilon(T_i^2 + T_1^2)(T_i + T_1)(T_i - T_1)$$

Hence, since by definition

$$\frac{Q_{ri}}{A} = h_{ri}(T_i - T_1)$$

it follows that

$$h_{ri} = \sigma\varepsilon(T_i^2 + T_1^2)(T_i + T_1)$$

The total heat-transfer rate to the inner glass surface is therefore given by

$$\frac{Q}{A} = \frac{Q_{ci}}{A} + \frac{Q_{ri}}{A} = (h_{ci} + h_{ri})(T_i - T_1)$$

The heat is transferred across the inner pane by conduction so that since steady, one-dimensional conduction is being considered, it follows that

$$\frac{Q}{A} = k_g \frac{T_1 - T_2}{t_g}$$

where the temperature T_2 is as defined in Fig. 8.2.

The heat-transfer rate across the gap between the panes is given by

$$\frac{Q}{A} = (h_{cb} + h_{rb})(T_2 - T_3)$$

Because the gap between the panes, 1 cm, is small compared to the height of the window, 1 m, it will be assumed that the radiant heat transfer across the gap is given by the equation for radiant heat between two infinite gray surfaces

$$\frac{Q_{rb}}{A} = \frac{\sigma \left(T_2^4 - T_3^4\right)}{1/\varepsilon + 1/\varepsilon - 1} = \frac{\sigma \left(T_2^2 + T_3^2\right)(T_2 + T_3)(T_2 - T_3)}{1/\varepsilon + 1/\varepsilon - 1}$$

$$= \frac{\sigma \left(T_2^2 + T_3^2\right)(T_2 + T_3)(T_2 - T_3)}{2/\varepsilon - 1}$$

where the emissivities of the two surfaces are the same. From this equation it follows that

$$h_{rb} = \frac{\sigma \left(T_2^2 + T_3^2\right)(T_2 + T_3)}{1/\varepsilon + 1/\varepsilon - 1} = \frac{\sigma \left(T_2^2 + T_3^2\right)(T_2 + T_3)}{2/\varepsilon - 1}$$

For the reasons given when discussing the heat transfer across the inner glass pane, the heat-transfer rate across the outer pane is given:

$$\frac{Q}{A} = k_g \frac{T_3 - T_4}{t_g}$$

Finally, consider heat transfer from the outer surface of the outer glass pane to the outside air. For the same reasons as those discussed when considering the heat transfer from the inside air to the inside glass surface, it follows that

$$\frac{Q}{A} = (h_{co} + h_{ro})(T_4 - T_o)$$

where

$$h_{ro} = \sigma\varepsilon\left(T_4^2 + T_o^2\right)(T_4 + T_o)$$

The equations above give

$$\frac{Q}{A}\frac{1}{h_{ci} + h_{ri}} = T_i - T_1 \qquad \frac{Q}{A}\frac{1}{k_g/t_g} = T_1 - T_2$$

$$\frac{Q}{A}\frac{1}{h_{cb} + h_{rb}} = T_2 - T_3 \qquad \frac{Q}{A}\frac{1}{k_g/t_g} = T_3 - T_4$$

$$\frac{Q}{A}\frac{1}{h_{co} + h_{ro}} = T_4 - T_o$$

8.6 Steady-State Calculations

Adding these equations together and rearranging then gives

$$\frac{Q}{A} = \frac{T_i - T_o}{\dfrac{1}{h_{ci} + h_{ri}} + \dfrac{1}{k_g/t_g} + \dfrac{1}{h_{cb} + h_{rb}} + \dfrac{1}{k_g/t_g} + \dfrac{1}{h_{co} + h_{ro}}}$$

The heat-transfer rate is commonly expressed in terms of an overall heat-transfer coefficient U, which is defined by

$$\frac{Q}{A} = U(T_i - T_o)$$

From the preceding two equations it follows that

$$\frac{1}{U} = \frac{1}{(h_{ci} + h_{ri})} + \frac{1}{k_g/t_g} + \frac{1}{(h_{cb} + h_{rb})} + \frac{1}{k_g/t_g} + \frac{1}{(h_{co} + h_{ro})}$$

Alternatively, the overall heat transfer rate can be expressed in terms of the thermal resistance R of the window, which is defined by

$$\frac{Q}{A} = \frac{(T_i - T_o)}{R}$$

Comparison of these equations then gives

$$R = \frac{1}{h_{ci} + h_{ri}} + \frac{1}{k_g/t_g} + \frac{1}{h_{cb} + h_{rb}} + \frac{1}{k_g/t_g} + \frac{1}{h_{co} + h_{ro}}$$

$$= R_i + R_g + R_b + R_g + R_o$$

where

$$R_i = \frac{1}{h_{ci} + h_{ri}}, \ R_g = \frac{1}{k_g/t_g}, \ R_b = \frac{1}{h_{cb} + h_{rb}}, \ R_o = \frac{1}{h_{co} + h_{ro}}$$

The space between the two panes of glass forms a vertical high-aspect-ratio enclosure. In order to find the convective heat-transfer rate between the two glass panes, the heat-transfer rate across this enclosure is assumed to be expressed in terms of the Rayleigh number Ra, and the aspect ratio AR, which are defined by

$$\text{Ra} = \frac{\beta g w^3 (T_2 - T_1)}{\nu^2}$$

$$\text{AR} = \frac{H}{w}$$

respectively. The following three Nusselt numbers are then defined (see El Sherbiny, S. M., Raithby, G. D., and Hollands, K. G. T., 1982,

"Heat Transfer by Natural Convection across Vertical and Inclined Air Layers," *J. Heat Transfer*, vol. 104, pp. 94–102)

$$\text{Nu}_a = 0.0605\,\text{Ra}^{1/3}$$

$$\text{Nu}_b = \left[1 + \left\{\frac{0.104\,\text{Ra}^{0.293}}{1 + (6310/\text{Ra})^{1.36}}\right\}^3\right]^{1/3}$$

$$\text{Nu}_c = 0.242\left(\frac{\text{Ra}}{\text{AR}}\right)^{0.272}$$

and the Nusselt number for the heat transfer across the enclosure is taken as the largest of these values:

$$\text{Nu}_b = \text{maximum}\,(\text{Nu}_a, \text{Nu}_b, \text{Nu}_c)$$

The value of the convective heat-transfer coefficient for the gap between the glass panes is then given by

$$h_{ci} = \frac{\text{Nu}_b w}{k_b}$$

where k_b is the thermal conductivity of the gas in the enclosure between the glass panes.

The thermal conductivity k_b and the kinematic viscosity in the Rayleigh number are evaluated at the mean temperature in the gap, at

$$T_{bm} = \frac{T_2 + T_3}{2}$$

An iterative solution procedure is required; this procedure involves the following steps:

1. Guess the values of T_1, T_2, T_3, and T_4.
2. Using these values, calculate h_{ri}, h_{rb}, h_{ro}, and h_{cb}.
3. Calculate Q/A, U, and R.
4. Using this value of Q/A, calculate T_1, T_2, T_3, and T_4 using

$$T_i = T_1 - \frac{Q}{A}\frac{1}{h_{ci} + h_{ri}} \qquad T_2 = T_1 - \frac{Q}{A}\frac{1}{k_g/t_g}$$

$$T_3 = T_2 - \frac{Q}{A}\frac{1}{h_{cb} + h_{rb}} \qquad T_4 = T_3 - \frac{Q}{A}\frac{1}{k_g/t_g}$$

8.8 Steady-State Calculations

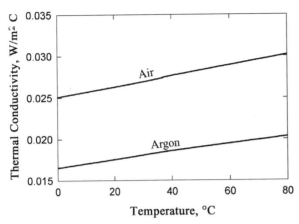

Figure 8.3 Variation of thermal conductivity k of air and argon with temperature.

5. Using these temperature values, repeat steps 1 to 3. If these new temperature values are essentially the same as the guessed values, the procedure can be stopped. If not, the process is repeated to again give new values for the temperatures.

This procedure can be implemented in a number of ways. The results given here were obtained by implementing this procedure in the spreadsheet software Excel.

In order to obtain the results, the values of ν and k for air and argon have to be known over the temperature range covered in the situation considered. The values assumed here are shown in Figs. 8.3 and 8.4.

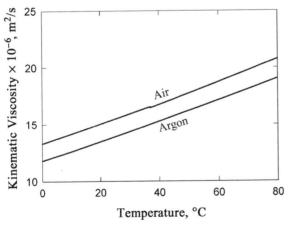

Figure 8.4 Variation of kinematic viscosity ν of air and argon with temperature.

TABLE 8.1 Thermal Resistance Values, m² C/W

Gas	R_i	R_g	R_b	R_g	R_o	R
Air	0.1154	0.00231	0.1605	0.00231	0.03432	0.3148
Argon	0.1152	0.00231	0.1951	0.00231	0.03433	0.3492

The calculated values of Q/A, U, and R for the two cases are

Air $Q/A = 47.65$ W/m² C, $U = 3.18$ W/m² C, $R = 0.315$ m² C/W
Argon $Q/A = 42.95$ W/m² C, $U = 2.86$ W/m² C, $R = 0.349$ m² C/W

It will be seen that the heat-transfer rate per unit area with argon between the panes is about 11 percent lower than it is with air between the panes. This is basically due to the fact that argon has a lower thermal conductivity than does air.

The R values for the two gases for each component of the window are shown in Table 8.1. From the values given in this table it will be seen that the biggest contributor to the overall thermal resistance is the gas space between the panes. The thermal resistance at the inside surface of the window is the next-highest contributor.

It was also found that the Nusselt number for the between-panes heat transfer was 1 in both cases considered, indicating that the heat transfer across the gas layer is effectively by conduction, since if this is true, it follows that

$$\frac{Q}{A} = k_b \frac{T_2 - T_3}{w} \quad \text{that is} \quad \frac{(Q/A)w}{k_b(T_2 - T_3)} = 1$$

(i.e., Nu = 1). This is because the gap between the panes is small.

Part 3

Transient and Cyclic Calculations

The contributors of the chapters in this part work in U.S. government defense installations, consulting firms, industry in the United States and Japan, and academia. The calculations they describe come out of real life:

- *Locating corrosion under paint layers on a metal surface*
- *The combustion tube of a prototype pulsed detonation engine*
- *Coke deposits in fired heaters processing hydrocarbon liquids, vapors, or gases*
- *Tape drives used as backup storage devices of computer data*
- *Infrared thermographic (IR) surveys of roofs to detect leaks*
- *A motor vehicle painting process that involves a series of paint application steps and paint drying/curing ovens*
- *An electronics cooling system*
- *Laminates of dissimilar materials, which are commonly found in industrial products, such as printed-circuit boards in electronic equipment*
- *Heat and mass transfer with phase change, which is encountered in drying operations, the chemical process industry, air conditioning and refrigeration, and manufacturing operations*
- *A method of optimizing vapor compression refrigeration cycles*
- *A Low Temperature, Low Energy Carrier (LoTEC©), a passive thermal carrier designed to permit payload transport*

9.2 Transient and Cyclic Calculations

to and from the International Space Station at approximately constant temperature without external power

Most of the calculations in this part are time-dependent, although there are several in which steady-state calculations do deal adequately with the problem at hand, even though, for example, intermittent heat generation may be involved. Heat-transfer-coefficient determination again plays a role in several calculations.

Chapter 9

Use of Green's Function to Solve for Temperatures in a Bimaterial Slab Exposed to a Periodic Heat Flux Applied to Corrosion Detection

Larry W. Byrd
AFRL/VASA
Wright-Patterson Air Force Base, Ohio

This problem demonstrates the use of separation of variables and Green's function to determine the transient temperature distribution in a layer of paint covering a corroded metal surface. It simulates a proposed noncontacting method of locating corrosion under paint. The following pages give the details of the analysis and a comparison with a finite-difference approximation.

The detection concept is based on the assumption that a corrosion layer between the metal substrate and the paint will have a larger thermal resistance than will areas with no corrosion. A sinusoidally varying heat flux is applied to the painted surface as shown in Fig. 9.1. It is desired to show that this will result in a periodic variation in surface temperature that is at the same frequency as the applied flux. The unique aspect of this approach is that the phase lag between the surface temperature and the incident flux, not a change in amplitude of the resulting sinusoidal variation in surface temperature, is used as the

9.4 Transient and Cyclic Calculations

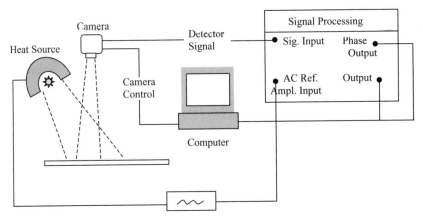

Figure 9.1 Corrosion detection system.

indicator of corrosion. The incident flux is chosen to be small enough to cause only a slight rise in the paint temperature. The only way to detect this temperature fluctuation is to filter the sinusoidal signal at the load frequency to eliminate the typical random fluctuations in temperature as shown in Fig. 9.2. This has been claimed to give a resolution as good as 0.001 K in commercial differential thermography systems [1].

A one-dimensional heat conduction model of an aluminum plate with corrosion at the interface between the aluminum and the paint is shown in Fig. 9.3. The effect of the corrosion layer is modeled by a change in the thermal conductance h_1 between the paint and the substrate. If there is no corrosion and the thermal contact was perfect, the conductance h_1 is infinite. This corresponds to zero resistance to heat conduction. In real

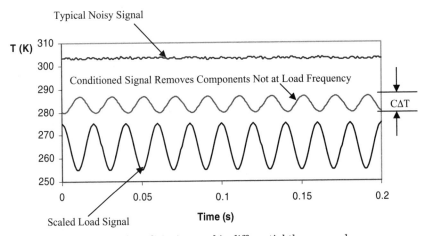

Figure 9.2 Typical signal conditioning used in differential thermography.

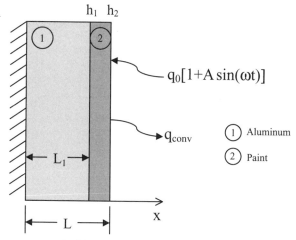

Figure 9.3 Problem geometry.

systems the conductance is a large, but not infinite, value. As the surface begins to corrode, the thermal conductance decreases and the layer acts as an insulator. This means that less energy will be conducted into the substrate and more will be lost to the environment. The conductance for no corrosion is assumed to be greater than 1000 W/m² C and is on the order of 100 W/m² C when corrosion is present. A mathematical model of the system solves the heat conduction equation in each layer with additional boundary conditions at the interface to account for the thermal conductance. The change in volume due to the corrosion is not modeled; thus the interface is shown as having a negligible thickness.

The heat conduction equation for each region and the associated boundary conditions are given as Eqs. (9.1a) to (9.1d) and (9.2a) and (9.2b). Equations (9.1b) and (9.1c) specify that the heat flux is continuous across the interface but the temperature is discontinuous, with the difference controlled by the value of h_1. It is assumed that the system and the surroundings are initially at a uniform temperature T_0. At $t = 0$, the painted surface at $x = L$ is heated by incident radiation such that the absorbed part is given by $q_0[1 + A \sin(\omega t)]$ and also loses heat by convection. The back surface at $x = 0$ is considered to be insulated as shown by Eq. (9.1a), but the results should not be significantly different if heat transfer by convection or radiation was included.

Region 1:

$$\alpha_1 \frac{\partial^2 T_1}{\partial x^2} = \frac{\partial T_1}{\partial t} \qquad 0 < x < L_1;\quad t > 0 \qquad (9.1)$$

9.6 Transient and Cyclic Calculations

$$\frac{\partial T_1}{\partial x} = 0 \qquad x = 0; \quad t > 0 \qquad (9.1a)$$

$$-k_1 \frac{\partial T_1}{\partial x} = h_1(T_1 - T_2) \qquad x = L_1; \quad t > 0 \qquad (9.1b)$$

$$k_1 \frac{\partial T_1}{\partial x} = k_2 \frac{\partial T_2}{\partial x} \qquad x = L_1; \quad t > 0 \qquad (9.1c)$$

$$T_1 = T_0 \qquad 0 < x < L_1; \quad t = 0 \qquad (9.1d)$$

Region 2:

$$\alpha_2 \frac{\partial T_2}{\partial x^2} = \frac{\partial T_2}{\partial t} \qquad L_1 < x < L; \quad t > 0 \qquad (9.2)$$

$$k_2 \frac{\partial T_2}{\partial x} + h_2 T_2 = h_2 \underbrace{\left[T_0 + \frac{q_0[1 + A\sin(\omega t)]}{h_2} \right]}_{f(t)}$$

$$\qquad x = L; \quad t > 0 \qquad (9.2a)$$

$$T_2 = T_0 \qquad x < L_1 < L; \quad t = 0 \qquad (9.2b)$$

Separation-of-Variables Technique

The separation-of-variables method is a well-known technique used to solve partial differential equations when the solution can be written as a product of functions of a single independent variable [2]. For one-dimensional heat conduction in a slab of width L, with no heat generation, the equation is

$$\frac{\partial^2 T}{\partial x^2} = \frac{1}{\alpha} \frac{\partial T}{\partial t} \qquad (9.3)$$

Substituting $T(x, t) = X(x)\Gamma(t)$ into Eq. (9.3) gives

$$\frac{X''(x)}{X(x)} = \frac{\Gamma'(t)}{\alpha \Gamma(t)} = -\beta^2 \qquad (9.4)$$

β is a constant called an *eigenvalue* which will be determined by the boundary conditions. $\Gamma(t)$ can be found in terms of β by integrating the second ordinary differential equation in Eq. (9.4):

$$\Gamma(t) = \Gamma_0 e^{-\alpha \beta^2 t} \qquad (9.5)$$

$X(x)$ is found by solving the first ordinary differential equation in Eq. (9.4) and will have the following form:

$$X(x) = a_0 \sin \frac{\beta x}{L} + a_1 \cos \frac{\beta x}{L} + b_0 \sinh \frac{\beta x}{L} + b_1 \cosh \frac{\beta x}{L} \tag{9.6}$$

$X(x)$ is referred to as an *eigenfunction*. To satisfy the initial condition, the solution is written as an infinite series of the eigenfunctions multiplied by $\Gamma(t)$ as

$$T(x, t) = \sum_{n=0}^{\infty} A_n X(\beta_n, x) \Gamma(\beta_n, t) \tag{9.7}$$

The constants A_n are found using the initial condition and the orthogonality of the eigenfunctions. *Orthogonality* means that the integral of the product of any two eigenfunctions $X(\beta_m, x)$ and $X(\beta_n, x)$ over the domain is zero unless $m = n$. The constants Γ_0 and those in $X(x)$ have been absorbed into A_n. The initial condition can be written as

$$T_0 = \sum_{n=0}^{\infty} A_n X(\beta_n, x) \tag{9.7a}$$

Multiplying both sides of Eq. (9.7) by $X(\beta_m, x)$ and integrating gives

$$\int_{x=0}^{L} T_0 X(\beta_m, x) dx = \int_{x=0}^{L} A_m X^2(\beta_m, x) dx = A_m N(\beta_m) \tag{9.8}$$

where $N(\beta_m)$ is the normalization integral. This is easily rearranged to give A_m for any value of m.

The present problem cannot be solved by separation of variables in its simplest form because the boundary condition at $x = L$ is not only nonhomogeneous but also time-dependent. It can be solved using a Green's function approach for composite media as given by Ozisik [3]. The problem is first transformed using

$$T_i(x, t) = \theta_i(x, t) + \xi_i(x) f(t) \tag{9.9}$$

to remove the nonhomogeneous boundary condition at $x = L$. This results in a time-dependent volumetric heat source term in the heat conduction equation. The subscript $i = 1, 2$ refers to the regions 1 and 2. $f(t)$ is defined in Eq. (9.2a), while θ and ξ are the solutions to the following auxiliary problems.

9.8 Transient and Cyclic Calculations

Region 1:

$$\alpha_1 \frac{\partial^2 \theta_1}{\partial x^2} - \xi_1 \frac{df(t)}{dt} = \frac{\partial \theta_1}{\partial t} \qquad 0 < x < L_1; \quad t > 0 \qquad (9.10)$$

$$\frac{\partial \theta_1}{\partial x} = 0 \qquad x = 0; \quad t > 0 \qquad (9.10a)$$

$$-k_1 \frac{\partial \theta_1}{\partial x} = h_1(\theta_1 - \theta_2) \qquad x = L_1; \quad t > 0 \qquad (9.10b)$$

$$k_1 \frac{\partial \theta_1}{\partial x} = k_2 \frac{\partial \theta_2}{\partial x} \qquad x = L_1; \quad t > 0 \qquad (9.10c)$$

$$\theta_1(0) = T_0 - f(0) \qquad 0 < x < L_1; \quad t = 0 \qquad (9.10d)$$

Region 2:

$$\alpha_2 \frac{\partial^2 \theta_2}{\partial x^2} - \xi_2 \frac{df(t)}{dt} = \frac{\partial \theta_2}{\partial t} \qquad L_1 < x < L; \quad t > 0 \qquad (9.11)$$

$$k_2 \frac{\partial \theta_2}{\partial x} + h_2 \theta_2 = 0 \qquad x = L; \quad t > 0 \qquad (9.11a)$$

$$\theta_2(0) = T_0 - f(0) \qquad L_1 < x < L; \quad t = 0 \qquad (9.11b)$$

Region 1:

$$\frac{\partial^2 \xi_1}{\partial x^2} = 0 \qquad 0 < x < L_1 \qquad (9.12)$$

$$\frac{\partial \xi_1}{\partial x} = 0 \qquad x = 0 \qquad (9.12a)$$

$$-k_1 \frac{\partial \xi_1}{\partial x} = h_1(\xi_1 - \xi_2) \qquad x = L_1 \qquad (9.12b)$$

$$k_1 \frac{\partial \xi_1}{\partial x} = k_2 \frac{\partial \xi_2}{\partial x} \qquad x = L_1 \qquad (9.12c)$$

Region 2:

$$\frac{\partial^2 \xi_2}{\partial x^2} = 0 \qquad L_1 < x < L \qquad (9.13)$$

$$k_2 \frac{\partial \xi_2}{\partial x} + h_2 \xi_2 = h_2 \qquad x = L_1 \qquad (9.13a)$$

For this problem, $\xi_1 = \xi_2 = 1$.

The solution for $\theta_i(x, t)$ can be written in terms of Green's function as

$$\theta_i(x, t) = \sum_{j=1}^{2} \left\{ \underbrace{\int_{x_j}^{x_{j+1}} G_{ij}(x, t|x', \tau)\big|_{\tau=0} F_j(x')dx'}_{\text{transient}} \right.$$

$$\left. + \underbrace{\int_{\tau=0}^{t} \int_{x_j}^{x_{j+1}} G_{ij}(x, t|x', \tau) \left[\frac{\alpha_j}{k_j} g_j(x', \tau)\right] dx' d\tau}_{\text{periodic}} \right\} \quad (9.14)$$

where Green's function is defined as

$$G_{ij}(x, t|x', \tau) = \sum_{n=1}^{\infty} \frac{e^{-\beta_n^2(t-\tau)} \left(\frac{k_j}{\alpha_j}\right) \Psi_{in}(x)\Psi_{jn}(x')}{N_n} \quad (9.15)$$

The normalization integral is

$$N_n = \frac{k_1}{\alpha_1} \int_{x=0}^{L_1} \Psi_{1n}^2(x)dx + \frac{k_2}{\alpha_2} \int_{x=L_1}^{L} \Psi_{2n}^2(x)dx$$

$$\quad (9.16)$$

$$F_j = \theta_j(x, 0) = \frac{-q_0}{h_2} \quad \text{and} \quad \frac{\alpha_j g_j(x', \tau)}{k_j} = -q_0 A \omega \cos(\omega \tau)$$

The second term in Eq. (9.14) which is marked periodic will be found to be associated with a sinusoidally varying component, while the first term will show a transient response that goes to a terminal value. The eigenfunctions Ψ_{in} are solutions to the following problems.

Region 1:

$$\frac{\partial^2 \Psi_{1n}}{\partial x^2} + \frac{\beta_n^2}{\alpha_1} = 0 \qquad 0 < x < L_1 \qquad (9.17)$$

$$\frac{\partial \Psi_{1n}}{\partial x} = 0 \qquad x = 0 \qquad (9.17\text{a})$$

$$-\frac{\partial \Psi_{1n}}{\partial x} = H_1(\Psi_{1n} - \Psi_{2n}) \qquad x = L_1 \qquad (9.17\text{b})$$

$$k_1 \frac{\partial \Psi_{1n}}{\partial x} = k_2 \frac{\partial \Psi_{2n}}{\partial x} \qquad x = L_1 \qquad (9.17\text{c})$$

9.10 Transient and Cyclic Calculations

Region 2:

$$\frac{\partial^2 \Psi_{2n}}{\partial x^2} + \frac{\beta_n^2}{\alpha_2} = 0 \quad L_1 < x < L \quad (9.18)$$

$$\frac{\partial \Psi_{2n}}{\partial x} + H_2 \Psi_{2n} = 0 \quad x = L \quad (9.18\text{a})$$

$$H_i = \frac{h_i}{k_i}$$

Solving these equations gives

$$\Psi_{1n}(x) = A_{in} \sin\left(\frac{\beta_n x}{\sqrt{\alpha_i}}\right) + B_{in} \cos\left(\frac{\beta_n x}{\sqrt{\alpha_i}}\right) \quad (9.19)$$

The eigenvalues β_n and constants A_{in} and B_{in} are determined from the boundary conditions to arrive at a solution for $\theta_i(x, t)$, which is then substituted into Eq. (9.9) for $T_i(x, t)$. Equation (9.17a) gives $A_{1n} = 0$. This results in a set of homogeneous algebraic equations:

$$\begin{bmatrix} \dfrac{\gamma_1}{H_1} \sin(\gamma_1 L_1) - \cos(\gamma_1 L_1) & \sin(\gamma_2 L_1) & \cos(\gamma_2 L_1) \\ \dfrac{k_1}{k_2}\sqrt{\dfrac{\alpha_2}{\alpha_1}} \sin(\gamma_1 L_1) & \cos(\gamma_2 L_1) & -\sin(\gamma_2 L_1) \\ 0 & X_2 & Y_2 \end{bmatrix} \begin{pmatrix} B_{1n} \\ A_{2n} \\ B_{2n} \end{pmatrix} = \begin{pmatrix} 0 \\ 0 \\ 0 \end{pmatrix}$$

(9.20)

$$X_2 = \gamma_2 \cos(\gamma_2 L) + H_2 \sin(\gamma_2 L) \quad (9.20\text{a})$$

$$Y_2 = -\gamma_2 \sin(\gamma_2 L) + H_2 \cos(\gamma_2 L) \quad (9.20\text{b})$$

$$\gamma_i = \frac{\beta_n}{\sqrt{\alpha_i}} \quad (9.20\text{c})$$

Since the set of equations is homogeneous, the constants can be determined only to a multiplicative constant. Thus, without a loss of generality, B_{1n} is set equal to 1 and A_{2n} and B_{2n} are

$$A_{2n} = \sin(\gamma_2 L_1)\left[\cos(\gamma_1 L_1) - \frac{\gamma_1}{H_1}\sin(\gamma_1 L_1)\right]$$

$$- \frac{k_1}{k_2}\frac{\sqrt{\alpha_2}}{\sqrt{\alpha_1}} \cos(\gamma_2 L_1)\sin(\gamma_1 L_1) \quad (9.21)$$

$$B_{2n} = \cos(\gamma_2 L_1)\left[\cos(\gamma_1 L_1) - \frac{\gamma_1}{H_1}\sin(\gamma_1 L_1)\right]$$

$$+ \frac{k_1}{k_2}\frac{\sqrt{\alpha_2}}{\sqrt{\alpha_1}} \sin(\gamma_2 L_1)\sin(\gamma_1 L_1) \quad (9.22)$$

The eigenvalues β_n are found by using the requirement for a nontrivial solution that the determinant of the coefficient matrix is equal to zero:

$$\begin{vmatrix} \dfrac{\gamma_1}{H_1}\sin(\gamma_1 L_1) - \cos(\gamma_1 L_1) & \sin(\gamma_2 L_1) & \cos(\gamma_2 L_1) \\ \dfrac{k_1}{k_2}\sqrt{\dfrac{\alpha_2}{\alpha_1}}\sin(\gamma_1 L_1) & \cos(\gamma_2 L_1) & -\sin(\gamma_2 L_1) \\ 0 & X_2 & Y_2 \end{vmatrix} = 0 \quad (9.23)$$

The normalization integral can now be written as

$$N_n = \frac{k_1}{\alpha_1}\int_0^{L_1}\cos^2(\gamma_{1n}x)\,dx + \frac{k_2}{\alpha_2}\int_{L_1}^{L}[A_{2n}\sin(\gamma_{2n}x) + B_{2n}\cos(\gamma_{2n}x)]^2\,dx$$

$$= \frac{k_1}{\alpha_1}\left[\frac{L_1}{2} + \frac{\sin(2\gamma_{1n}L_1)}{4\gamma_{1n}}\right]$$

$$+ \frac{k_2}{\alpha_2}\left\{A_{2n}^2\left[\frac{L-L_1}{2} - \frac{\sin(2\gamma_{2n}L)-\sin(2\gamma_{2n}L_1)}{4\gamma_{2n}}\right]\right.$$

$$+ B_{2n}^2\left[\frac{L-L_1}{2} + \frac{\sin(2\gamma_{2n}L)-\sin(2\gamma_{2n}L_1)}{4\gamma_{2n}}\right]$$

$$\left.+ \frac{2A_{2n}B_{2n}}{4\gamma_{2n}}\left[\frac{\cos(2\gamma_{2n}L_1)-\cos(2\gamma_{2n}L)}{4\gamma_{2n}}\right]\right\} \quad (9.24)$$

To demonstrate that the resulting surface temperature will provide a signal at the load frequency, only the periodic part of the solution for region 2 is needed because the rest will be filtered out. From Eq. (9.14) this is

$$\theta_{2p} = \int_{\tau=0}^{t}\frac{\sum_{n=1}^{\infty}e^{-\beta_n^2(t-\tau)}[-q_0 A\omega\cos(\omega\tau)]I_{2n}\Psi_{2n}(x)}{N_n h_2}\,d\tau \quad (9.25)$$

where

$$I_{2n} = \frac{k_1}{\alpha_1}\int_{x=0}^{L_1}\Psi_{1n}(x)\,dx + \frac{k_2}{\alpha_2}\int_{x=L_1}^{L}\Psi_{2n}(x)\,dx \quad (9.26)$$

9.12 Transient and Cyclic Calculations

Evaluating the integral over τ gives

$$\theta_{2p}(x,t) = \sum_{n=1}^{\infty} \frac{e^{-\beta_n^2 t}(-q_0 A\omega I_{2n})\left[\frac{e^{\beta_n^2 t}\sin(\omega t + \phi_n)}{\sqrt{\beta_n^4 + \omega^2}} - \frac{\beta_n^2}{\beta_n^4 + \omega^2}\right]\Psi_{2n}(X)}{N_n h_2} \quad (9.27)$$

The term with the negative exponential will die out to leave the periodic component:

$$\theta_{2p}(x,t) = \frac{-q_0 A\omega}{h_2} \sum_{n=1}^{\infty} \frac{I_{2n}\Psi_{2n}(x)\sin(\omega t + \phi_n)}{N_n\sqrt{\beta_n^4 + \omega^2}} \quad (9.28)$$

The periodic component of the temperature at $x = L$ is

$$T_{2p}(L,t) = T_{\infty} + \frac{q_0}{h_2}\left[1 - A\omega \sum_{n=0}^{\infty} C_n \sin(\omega t + \phi_n)\right] \quad (9.29)$$

where $C_0 = -1/\omega$, $\phi_0 = 0$, and $C_n = \left[I_{2n}\Psi_{2n}(L)/N_n\sqrt{\beta_n^4 + \omega^2}\right]$ for $n \geq 1$.

The sum can be rewritten using trigonometric identities to give

$$T_{2p}(L,t)$$
$$= T_{\infty} + \frac{q_0}{h_2}\left[1 - A\omega\left(\sqrt{\left(\sum_{n=0}^{\infty} C_n \cos(\phi_n)\right)^2 + \left(\sum_{n=0}^{\infty} C_n \sin(\phi_n)\right)^2}\right)\right.$$
$$\left. \times \sin(\omega t + \phi)\right] \quad (9.30)$$

$$\phi = \tan^{-1}\left(\frac{\sum_{n=0}^{\infty} C_n \sin(\phi_n)}{\sum_{n=0}^{\infty} C_n \cos(\phi_n)}\right) - \pi \quad (9.31)$$

The surface temperature can be seen to have the correct form to be used to detect a change in thermal conductance. Specifically, it is proportional to the single load frequency ω with a phase angle that is a function of the conductance between the paint and substrate. Figure 9.4 shows the surface temperature for a time range of 10 s. Also shown by the blue line is a finite-difference approximation of the surface temperature. This method used an explicit scheme to update the temperature. In Fig. 9.5, the time and temperature scales have been changed to show the temperature profile up to 5000 s. Note that at long times the

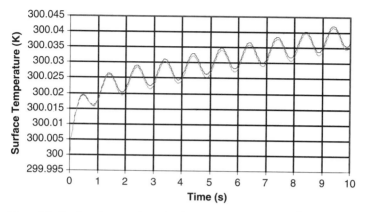

Figure 9.4 Surface temperature for short times: $q_0 = 10$ W/m², $h_1 = 1000$ W/m² C, $h_2 = 6$ W/m² C, $L_1 = 2$ mm, $L = 2.54$ mm.

exponential part of the solution disappears and the temperature approaches a steady-state temperature with a superimposed ripple. The ripple is not as apparent on this figure because of the scale used to illustrate the total change in temperature, which can be shown to approach $q_0/h_2 = (10 \text{ W/m}^2)/(6 \text{ W/m}^2 \text{ K}) = 1.7$ K. The ripple magnitude is small because of the low-level incident energy ($q_0 = 10$ W/m²) used for this particular simulation. Low levels are desirable because the surface temperature will change by only a few degrees, but this makes it harder to detect the ripple magnitude, so some control is required. Figure 9.6 shows the difference in phase lag between a segment with h_1 assumed to be either 1000 or 5000 W/m² C and the phase lag associated with corrosion ($h_1 = 100$ W/m² C) for a paint thickness of 0.254 mm. This is convenient for corrosion detection because if a plot of phase lag is

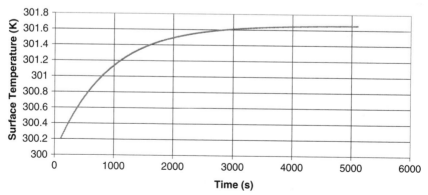

Figure 9.5 Surface temperature for long times: $q_0 = 10$ W/m², $h_1 = 1000$ W/m² C, $h_2 = 6$ W/m² C, $L_1 = 2$ mm, $L = 2.254$ mm.

9.14 Transient and Cyclic Calculations

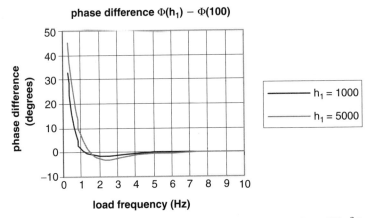

Figure 9.6 Phase lag referenced to a thermal conductance of 100 W/m²C: $q_0 = 10$ W/m², $L_1 = 2$ mm, $L = 2.254$ mm, $h_2 = 6$ W/m² C.

shown over the surface, the areas that are corroded will have a large contrast compared to the pristine regions. The phase lag is dependent on the thermal conductance h_1 and the heater frequency. If the heater frequency is too high, the phase lag is the same regardless of h_1, and it is not a good indicator of corrosion. Figure 9.7 shows the effect of the thickness of the paint layer. As can be seen, the thicker the paint is, the lower the frequency needed to detect a change in phase lag. This can be problematic because the filtering technique is faster and has less noise as the load frequency is increased.

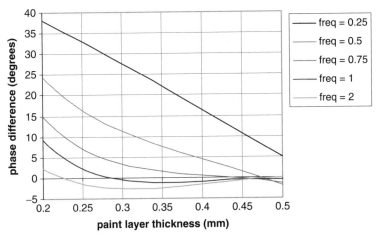

Figure 9.7 Effect of paint thickness: $q_0 = 10$ W/m², $h_1 = 1000$ W/m² C, $h_2 = 6$ W/m² C, $L_1 = 2$ mm.

In summary, this analysis gives the temperature distribution in a two-layer material exposed to a sinusoidally varying heat flux on one face and insulated on the other. The surface temperature shows the typical asymptotic rise of a slab exposed to a hot gas with a superimposed ripple. The phase lag between the incident heat flux and the surface temperature is a function of the material parameters and the thermal conductance between the material layers.

References

1. Lesniak, J. R., *Differential Thermography for Elevated Temperatures*, USAF SBIR Final Report, Contract F33165-95-C-2504, Aug. 25, 1997.
2. Berg, W. B., and McGregor, J. L., *Elementary Partial Differential Equations*, Holden-Day, San Francisco, 1966.
3. Ozisik, M. N., *Heat Conduction*, Wiley, New York, 1980.

Chapter 10

Lumped-Capacitance Model of a Tube Heated by a Periodic Source with Application to a PDE Tube

Larry W. Byrd
AFRL/VASA
Wright-Patterson AFB, Ohio

Fred Schauer
Pulsed Detonation Research
AFRL/PRTS
Wright-Patterson AFB, Ohio

John L. Hoke
Innovative Scientific Solutions, Inc.
Wright-Patterson AFB, Ohio

Royce Bradley
Innovative Scientific Solutions, Inc.
Wright-Patterson AFB, Ohio

Introduction

This problem considers estimation of temperature as a function of time of a circular tube heated internally by a hot gas with a periodically varying temperature. The lumped-capacitance method neglects the temperature variation in the tube wall, so it is described as simply $T(t)$. It is also assumed for this problem that the heat transfer is primarily in the radial direction, so axial conduction is neglected. The solution will be

10.2 Transient and Cyclic Calculations

Figure 10.1 Prototype pulsed detonation engine with four detonation tubes.

applied to the combustion tube of a prototype pulsed detonation engine (PDE) shown in Fig. 10.1. Data taken from a water-cooled aluminum tube and a free-convection–radiatively cooled steel tube is analyzed to determine the heat-transfer characteristics of the interior and exterior surfaces.

A PDE operates cyclically, completing a fill-detonation-exhaust cycle typically many times per second. Conventional engines chemically release heat through deflagrative combustion, resulting in slow burning and isobaric (constant-pressure) heat addition. Detonative combustion propagates at supersonic speeds, resulting in pressure gain combustion and resultant higher thrust and efficiency. The fill-burn-exhaust cycle of the tube consists of equal time intervals for each phase. The pressure in a PDE is initially low. The detonation processes increases the pressure substantially, and the resultant high-momentum flux from the detonation tube produces thrust and allows the PDE to self-aspirate as a result of the overexpansion observed near 4 ms in Fig. 10.3. In practice, the detonation-exhaust portion of the PDE cycle is followed by a purge cycle of cold air which serves as a buffer between hot products and the fresh reactants entering the detonation tube prior to the next detonation and to cool the hot tube walls. The PDE thrust can be controlled by regulating operating various operating parameters such as the frequency of the detonations.

Mathematical Formulation

Figure 10.2 shows a schematic a section of the tube wall with the convective heat transfer occurring at the inside and outside surfaces

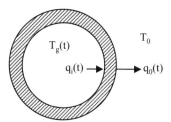

Figure 10.2 Schematic of the lumped-capacitance model.

characterized by $q_i(t) = h_i A_i [T_g(t) - T(t)]$ and $q_o(t) = h_o A_o [T(t) - T_0]$, $T_g(t)$ is the gas temperature in the tube, T_0 is the ambient temperature outside the tube, and h is the average heat-transfer coefficient. The subscripts i and o refer to inner and outer, respectively. An energy balance gives

$$mC_p \frac{dT}{dt} = h_i A_i (T_g - T) - h_o A_o (T - T_0) \qquad (10.1)$$

The gas temperature is modeled as

$$T_g = T_m + \Delta T \cos(\omega t) \qquad (10.2)$$

with $\Delta T = (T_f - T_0)/2$ and $T_m = (T_f + T_0)/2$, and $\omega = 2\pi f$ is the circular firing frequency.

As seen in Fig. 10.3, this is an approximation to the actual temperature, which is a function of position and shows a sharp spike as the detonation occurs. There are several ways to improve the results if needed. One is to include the gas temperature as a Fourier series which could be made to fit any expected time variation. Another method is to break the problem into discrete time increments with constant values of the gas temperature appropriate for the phase of the fill-burn-purge cycle and

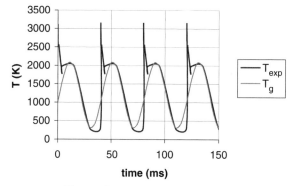

Figure 10.3 Expected gas temperature in PDE tube, $T_{g,\exp}$, and the approximation T_g.

10.4 Transient and Cyclic Calculations

sum up the changes in tube temperature as time progresses. The heat transfer would also be expected to be a function of the tube temperature, but any difference between the model and the actual apparatus will be absorbed into the values of h_i and h_o, which will be determined semiemperically. Thus they will be average coefficients over the timespan of interest.

Solution

Using $m = \rho \pi (r_o^2 - r_i^2) L$, $A_i = 2\pi r_i L$, and $A_o = 2\pi r_o L$, Eq. (10.1) becomes

$$\frac{dT}{dt} + aT = b + c \cos(\omega t) \tag{10.3}$$

$$a = \frac{2(h_i r_i + h_o r_o)}{\rho \left(r_o^2 - r_i^2\right) C_p} \tag{10.3a}$$

$$b = \frac{2(h_i r_i T_m + h_o r_o T_0)}{\rho \left(r_o^2 - r_i^2\right) C_p} \tag{10.3b}$$

$$c = \frac{2 h_i r_i \Delta T}{\rho \left(r_o^2 - r_i^2\right) C_p} \tag{10.3c}$$

The solution to this ordinary differential equation is given by the sum of a particular and homogeneous solution, $T = T_p + T_h$ [1].
  Temperature T_h is the solution to

$$\frac{dT_h}{dt} + a T_h = 0 \tag{10.4}$$

This can be solved by separating the variables and integrating to give

$$T_h = T_{h,0} e^{-at} \tag{10.5}$$

The particular solution T_p would be expected to have a form similar to the right-hand side of Eq. (10.3). It is assumed as

$$T_p = \alpha + \beta \cos(\omega t) + \gamma \sin(\omega t) \tag{10.6}$$

Substituting into Eq. (10.3) gives

$$[-\beta \omega + a \gamma] \sin(\omega t) + [\omega \gamma + a \beta] \cos(\omega t) + a \alpha = b + c \cos(\omega t) \tag{10.7}$$

Equating the coefficients of the sin(ωt) and $cos(\omega t)$ and constant terms gives

$$-\beta\omega + a\gamma = 0 \tag{10.7a}$$

$$\omega\gamma + a\beta = c \tag{10.7b}$$

$$a\alpha = b \tag{10.7c}$$

which gives

$$\alpha = \frac{b}{a} \tag{10.8a}$$

$$\beta = \frac{ca}{\omega^2 + a^2} \tag{10.8b}$$

$$\gamma = \frac{c\omega}{\omega^2 + a^2} \tag{10.8c}$$

Thus

$$T_p = \frac{b}{a} + c\left[\frac{a}{\omega^2 + a^2}\cos(\omega t) + \frac{\omega}{\omega^2 + a^2}\sin(\omega t)\right] \tag{10.9}$$

This form can be readily transformed into a sinusoidal function with a phase angle ϕ by substituting

$$\sin(\phi) = \frac{a}{\sqrt{\omega^2 + a^2}} \qquad \cos(\phi) = \frac{\omega}{\sqrt{\omega^2 + a^2}}$$

and using the trigonometric identity:

$$\cos(\omega t)\sin(\phi) + \sin(\omega t)\cos(\phi) = \sin(\omega t + \phi) \tag{10.10}$$

$$\phi = \tan^{-1}\left(\frac{a}{\omega}\right) \tag{10.10a}$$

The particular solution is then found to be

$$T_p = \frac{b}{a} + \frac{c\sin(\omega t + \phi)}{\sqrt{\omega^2 + a^2}} \tag{10.11}$$

The temperature can now be written as

$$T(t) = T_{h,0}e^{-at} + \frac{b}{a} + \frac{c\sin(\omega t + \phi)}{\sqrt{\omega^2 + a^2}} \tag{10.12}$$

10.6 Transient and Cyclic Calculations

The integration constant $T_{h,0}$ can be found by using the initial condition $T = T_0$ at $t = 0$, giving the temperature as

$$T(t) = \left[T_0 - \left(\frac{b}{a} + \frac{c\sin(\phi)}{\sqrt{\omega^2 + a^2}}\right)\right]e^{-at} + \frac{b}{a} + \frac{c\sin(\omega t + \phi)}{\sqrt{\omega^2 + a^2}} \quad (10.13)$$

This equation will be used to estimate values for a and b so that h_i and h_o can be determined from experimental data for the special case where c is small. When representative values of the thermal parameters are substituted into the definitions for the constants a, b, and c, it is seen that $c/\sqrt{\omega^2 + a^2} \ll T_0$ or b/a for $f = 24$ Hz, which is typical of the operating frequency for which data were available. It also means that as $t \to \infty$, then $T \to b/a \equiv T_\infty$. This allows a simplification of Eq. (10.13) to $T_*(t) = -at$, which is linear in time. This allows a straightforward determination of a from the slope of the line. T_* is given as

$$T_*(t) = \ln\left(\frac{T - b/a}{T_0 - b/a}\right) \approx \ln\left(\frac{T - T_\infty}{T_0 - T_\infty}\right) \quad (10.14)$$

Figure 10.4 shows the calculated and experimental temperatures using approximately 85 s of data. T_∞ is taken at 84.67 s and used to calculate b once a is found from Eq. (10.14). As can be seen, the model gives good agreement with experiment as the largest difference is less than 3.4 K. Equations (10.3a) and (10.3b) can now be used to determine $h_i = 46.1$ W/m^2 K and $h_o = 1100$ W/m^2 K. It should be noted that h_i is an averaged value over all phases of the firing cycle and reflects the sinusoidal approximation to the periodic but nonsinusoidal gas

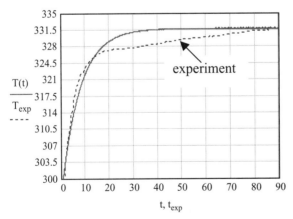

Figure 10.4 Comparison with experiment for cooled aluminum tube.

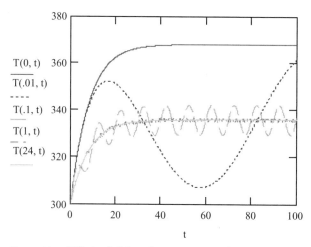

Figure 10.5 Effect of firing frequency on tube temperature $T(f, t)$ versus t.

temperature. These values of heat-transfer coefficients were then used to explore the temperature variation with time and firing frequency as shown in Fig. 10.5. Here a frequency value of zero models a continuous flow of hot-gas temperature at the flame temperature with no fill or purge phases. This special case shows the typical exponential behavior found in a first-order system. For nonzero frequency the temperature response consists of a transient going exponentially toward a steady-state value with a superimposed ripple due to the periodic nature of the gas temperature. As the frequency is increased, the ripple magnitude decreases and is almost negligible at the test frequency of 24 Hz. For these calculations it is assumed that the heat-transfer coefficients are not a function of the firing frequency. This is probably true for the water-cooled exterior of the tube but may be incorrect for h_i.

Steel Tube

Analysis of the steel tube data was carried out in a manner similar to that for the water-cooled aluminum tube. The firing frequency for these data was 25 Hz. There are some differences, however:

1. Data at several points along the tube demonstrated a marked temperature variation with length. This could be due to variations in heat-transfer coefficient or gas temperature inside the tube or axial heat conduction along the tube, especially near the head, which acted as a heat sink. Figure 10.6 shows the temperature at various points as a function of time.

10.8 Transient and Cyclic Calculations

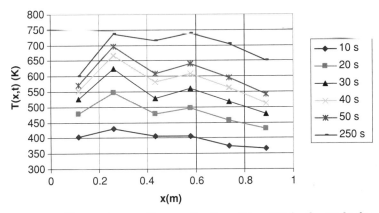

Figure 10.6 Temperature profiles as a function of time $T(t)$ for the steel tube.

2. The mechanism of heat loss to the surroundings differed from that observed for the outer-cooled aluminum tube. As shown in Fig. 10.7, radiative heat loss predominates after about 60 s. The free-convection heat-transfer coefficient and radiative loss to the ambient are calculated using Eqs. (10.15) [2] and (10.16):

$$h = \left(\frac{k}{D}\right)\left\{0.6 + \frac{0.387 \text{Ra}_D^{1/6}}{\left[1 + (0.559/\text{Pr})^{9/16}\right]^{8/27}}\right\}^2 \quad (10.15)$$

$$\text{Ra}_D = \frac{9.81 \cdot (T - T_0)D^3}{(T + T_0)\nu\alpha_a/2} \quad (10.15a)$$

$$q_r = \varepsilon \sigma A_o (T^4 - T_0^4) \quad (10.16)$$

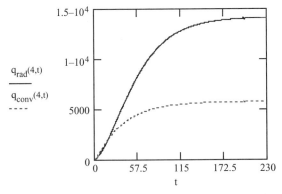

Figure 10.7 Heat flux due to convection and radiation for the steel tube $x = 0.432$ m.

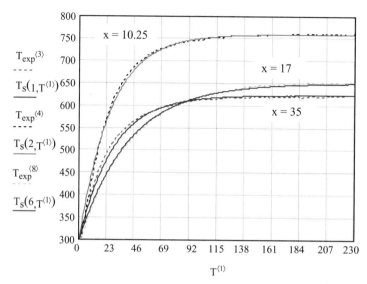

Figure 10.8 Comparison of theoretical and experimental temperatures.

where Pr is the Prandtl number, ν is the kinematic viscosity, and α_a is the thermal diffusivity of the air surrounding the tube and evaluated at $(T + T_0)/2$. The Stephan-Bolztmann constant is σ, and ε is the emissivity. The temperature must be in an absolute scale for both the free convection and radiation calculations.

The energy balance including radiation is given as

$$mC_p \frac{dT}{dt} = h_i A_i (T_g - T) - h_o A_o (T - T_0) - \varepsilon \sigma A_o (T^4 - T_0^4) \quad (10.17)$$

This is a nonlinear ordinary differential equation and can be readily solved by numerical means, but this is outside the scope of the present analysis. Figure 10.8 compares the theoretical and experimental results (dotted lines) at selected points along the tube. The other points show similar agreement. Table 10.1 gives the calculated interior and exterior

TABLE 10.1 Average Heat-Transfer Coefficients on Interior and Exterior of Steel Tube at Selected Locations

x, m	h_i, W/m² K	h_o, W/m² K
0.114	202	314
0.26	309	258
0.432	201	200
0.578	221	190
0.737	193	207
0.889	149	203

10.10 Transient and Cyclic Calculations

heat-transfer coefficients. Since the analysis did not include radiative effects, the heat-transfer coefficients must be considered as equivalent coefficients which lump both convective and radiative effects into a single coefficient. As can be seen from Fig. 10.8, this gives good agreement between theory and experiment.

Summary

A lumped-capacitance model of a pulse detonation engine exhaust tube was modeled and good agreement was found between experiment and theory when empirically determined constants were used. The temperature response consisted of a transient going exponentially to a steady-state value with a superimposed ripple due to the periodic nature of the gas temperature. At firing frequencies above 1 Hz, the ripple magnitude becomes negligible compared to the mean temperature. The method allowed the calculation of a semiempirical equivalent heat-transfer coefficient which lumped the effect of radiative and convective losses.

Nomenclature

a	Constant given by Eq. (10.3a), s^{-1}
A	$2\pi r L$ surface area of tube, m^2
b	Constant given by Eq. (10.3b), K/s
c	Constant given by Eq. (10.3c), K/s
C_p	Specific heat, J/kg K
D	Diameter, m
f	Frequency, Hz
h	Convective heat-transfer coefficient, W/m^2 K
L	Length of tube, m
m	Mass of tube, kg
Pr	Prandtl number for air $= \nu/\alpha_a$
r	Radius, m
Ra$_D$	Rayleigh number given by Eq. (10.15a)
t	Time, s
T	Temperature of tube, K
T_f	Flame temperature, K
T_g	Gas temperature, K
T_h	Homogeneous solution to the differential equation for temperature, K
$T_{h,0}$	Initial value of the homogeneous solution, K
T_m	Mean gas temperature, K

T_0	Ambient temperature, K
T_p	Particular solution to the differential equation for temperature, K
T_*	Natural logarithm of dimensionless temperature difference given by Eq. (10.14)
T_∞	Temperature at $t = \infty$, K
q	Heat-transfer rate, W
q_r	Heat flux due to radiation, W/m^2
x	Distance along tube measured from engine head, m

Greek

α, β, γ	Constants in the particular solution given by Eq. (10.6), K
α_a	Thermal diffusivity of air, m^2/s
ΔT	Magnitude of the gas temperature fluctuation, K
ε	Emissivity
σ	Stephan-Boltzmann constant 5.67×10^{-8}, W/m^2 K^4
ν	Kinematic viscosity of air, m^2/s
ϕ	Phase lag given by Eq. (10.10a), radians
ρ	Density, kg/m^3
ω	Circular frequency $\omega = 2\pi f$, rad/s

Subscripts

i	Inner
o	Outer

References

1. Simmons, G. F., *Differential Equations with Applications and Historical Notes*, McGraw-Hill, New York, 1972.
2. Incropera, F. P., and DeWitt, D. P., *Fundamentals of Heat and Mass Transfer*, 4th ed., Wiley, New York, 1996, p. 502.

Chapter

11

Calculation of Decoking Intervals for Direct-Fired Gas and Liquid Cracking Heaters

Alan Cross
Little Neck, New York

Introduction

Fired heaters processing hydrocarbon liquids, vapors, or gases are subject to the formation of coke deposits at the inner walls of the tubular heating coils contained within the radiant and convection sections of the heater, if the film temperatures at these locations are high enough. The thickness of the coke deposits increase with increasing temperature and time, and because such deposits have relatively low conductivity, the insulating effect limits cooling of the tube wall by the process fluid and results in an increase in tube metal temperature with time. When the tube metal temperature reaches the design temperature of the tubes, the heater must be shut down for decoking or otherwise decoked, using on-stream procedures, if coil damage is to be avoided. Since decoking represents lost production time and requires operator intervention, it is desirable to keep the interval between decokings as long as possible so as to minimize overall operating costs. Since the frequency of decoking is an important operating characteristic, potential users of fired cracking heaters will often request that on-stream time be specified and in some cases guaranteed. Despite these requirements, very little has been published regarding coke deposition rates and calculation of

11.2 Transient and Cyclic Calculations

same, nor has any information on this subject been provided in such heater design documents as the American Petroleum Institute (API) fired heater standards. Likewise, provision has not been made for incorporating specification of run length in API data sheets. The following analysis is designed to indicate how run length may be calculated for typical fired cracking heaters, using available thermal decomposition data for hydrocarbon and petroleum fractions.

Calculation of Maximum Heat Flux

Reference 1 provided a correlation among fired heater flue gas temperatures at the outlet of the radiant section, burner mass flow rate, burner flame temperature, burner flame diameter, tube and flue gas emissivity, and flame length. The correlation assumed that the flue gas temperature at the burner outlet was equal to the adiabatic flame temperature and in the case of burners firing vertically upward, that the flue gas progressively cools as a result of radiation to the tubular process coil and eventually reaches the bridgewall temperature at the outlet of the radiant section. Thus

$$W_B C_p (0.5) \left[\frac{1}{(T_{\text{low}})^2} - \frac{1}{(T_{\text{high}})^2} \right] = 0.173 \, \gamma \epsilon_2 \lambda \pi D_F L_F (10^{-8}) \qquad (11.1)$$

$$\gamma = 544 \, [D_F P (\text{mf}_{CO_2} + \text{mf}_{H_2O})]^{0.417} \qquad (11.1a)$$

Once the bridgewall temperature, the burner diameter, and the height of the radiant section have been fixed, the following equations may be used in conjunction with Eq. (11.1) to evaluate such pertinent variables as burner mass flow rate, heat absorbed by the process coil, burner liberation, tubular heat-transfer surface, burner and tube spacing, and the number of tubes irradiated by a single burner. It should be noted that fixation of the bridgewall temperature, burner diameter, and radiant section height should be sufficient to yield a burner spacing of 3 to 5 ft and a burner velocity of about 20 ft/s—which will be somewhat higher if pressurized air is used. In other words, pressure drop across the burner should not exceed the available draft at the burner location.

$$\frac{Q_{\text{abs}}}{\text{Burner}} = W_B C_p (T_F - T_{\text{BWT}}) \qquad (11.2)$$

$$\frac{Q_{\text{lib}}}{\text{Burner}} = W_B C_p (T_F - 60) \qquad (11.3)$$

$$\frac{Q_{\text{abs}}/\text{burner}}{A_o} = (RR)_{\text{avg}} \qquad (11.4)$$

$$CC_{burner} = \frac{N_T}{row} \text{ (CC tubes)} \tag{11.5}$$

$$\frac{N_T}{Row} \pi D_o (L_F) = \frac{A_o}{N_R} \tag{11.6}$$

For purposes of illustration, it will be assumed that the heater will be of the horizontal tube cabin type with burners firing vertically upward between two rows of tubes backed by refractory walls and fired from one side only. If this were the case, the effective overall tube length would be given by

$$N_{BT} = \frac{Q_R}{Q_{abs}/\text{burner}} \tag{11.7}$$

$$L_{FT} = N_{BT}(CC_{burner}) \tag{11.8}$$

To calculate maximum heat flux, Eq. (11.1) is solved for a 50°F flue gas temperature differential, thereby allowing for calculation of a corresponding differential flame length. The equations which follow are then again used in conjunction with Eq. (11.1) to solve for the maximum heat flux, noting that at the bottom of the heater T_{high} = flame temperature and T_{low} = (flame temperature – 50°F) and that at the top of the heater T_{low} = bridgewall temperature and T_{high} = (bridgewall temperature + 50°F). Further note that the maximum flux occurs at the outlet end of the coil, which in the case of concurrent flow locates the coil outlet adjacent to the bridgewall and which in the case of countercurrent flow locates the coil outlet adjacent to the burner outlet. Calculations indicate that concurrent flow is to be preferred because of the much longer run length that results with this arrangement.

$$A_{o,I} = \frac{N_T}{row} N_R (\pi) D_o L_I \tag{11.9}$$

$$A_{C_p} = CC_{burner}(L_I)(N_R) \tag{11.10}$$

$$Q_{I,abs} = W_B C_p (50) \tag{11.11}$$

$$\left(\frac{Q}{A}\right)_{max} = \frac{Q_{I,abs}}{A_{C_p,I}} \tag{11.12}$$

$$\left(\frac{Q}{A}\right)_{avg} = \frac{Q_{I,abs}}{A_{o,I}} \tag{11.13}$$

Hydrocarbon and Petroleum Fraction Decomposition Rates

Examination of the reaction velocity data of Fig. 11.1, obtained from Ref. 2, indicates that hydrocarbon and petroleum fraction thermal

11.4 Transient and Cyclic Calculations

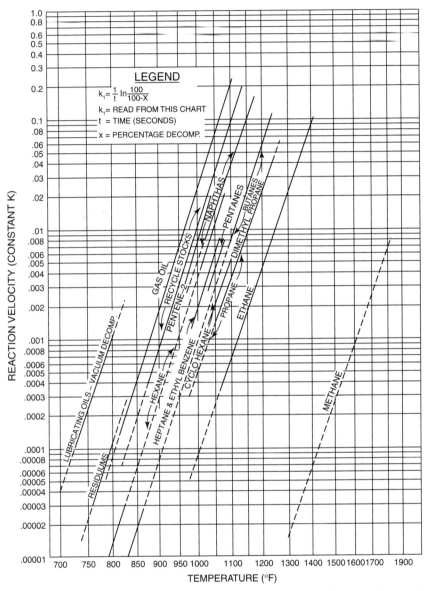

Figure 11.1 Constants for initial rates of thermal decomposition of hydrocarbons and petroleum fractions.

decomposition may be considered to vary as the reciprocal of the absolute temperature in accordance with an Arrhenius equation for a first-order reaction:

$$K = Ae^{-B/T_{\text{film}}} \qquad (11.14)$$

The resulting reaction velocity constant equations determined for a typical liquid feedstock, gas oil, and a typical gaseous feedstock, ethane, are as follows, as is the equation for calculating film temperature:

$$\left(\frac{Q}{A}\right)_{max} \frac{1}{h_{10}} = T_{film} - T_{fluid\,out} \qquad (11.15)$$

$$\ln(K_{GO}) = 30.7 - \frac{50,248}{T_{film}} \qquad (11.16)$$

$$\ln(K_{eth}) = 22.88 - \frac{47,025}{T_{film}} \qquad (11.17)$$

The variation of reaction time with the reaction velocity constant is, in accordance with the data source, equal to

$$\Theta = \frac{1}{K}\ln\left(\frac{100}{100-X}\right) \qquad (11.18)$$

For liquid feedstocks, such as gas oil, X is considered equal to the percentage yield of 400°F endpoint gasoline and in the case of gaseous feedstocks, such as ethane, to the percentage conversion. In the case of gas oil cracking, the percentage yield of gasoline was assumed equal to 16 percent and the yield of coke was assumed equal to 25 percent, as these are typical yields of products leaving a delayed coking drum. In the case of ethane cracking, 70 percent conversion and 80 percent coke yield was assumed. With regard to the liquid reaction velocity constants used, it must be assumed that the data are applicable to typical hydrocarbon fractions and no doubt vary in accordance with the source and chemical composition of the crude oil from which the fractions are derived. It would appear from the variations in the data between light and heavy fractions, such as naphtha and residua, that the velocity constants vary directly with molecular weight of the feedstock. Thus, in the absence of experimental reaction rate data, such a relationship might prove useful.

Coke Deposition Rates

The rate of coil coke deposition is considered equal to the thickness of a stationary layer of liquid or gaseous material at the inside wall of the process coil after having been converted to coke, divided by the time required for conversion, as defined by Eq. (11.18). As a first approximation, it is assumed that the thickness of the stationary layer referred to is equal to the molecular diameter, as determined by the Avogadro

11.6 Transient and Cyclic Calculations

number, which specifies the number of molecules in 1 gram mole of material:

$$N_{\text{Avog}} = 6.03(10^{23}) \tag{11.19}$$

$$\frac{(\text{MW})\bar{V}_F}{454(N_{\text{Avog}})} = V_F \tag{11.20}$$

$$M_F = \frac{\text{MW}}{454(N_{\text{Avog}})} \tag{11.21}$$

$$M_F(C_Y)\bar{V}_C = V_C \tag{11.22}$$

$$t_F = (V_F)^{1/3} \tag{11.23}$$

The thickness of the coke deposit based on a stationary monomolecular layer of feedstock is then

$$t_C = \frac{V_C}{V_F} t_F \tag{11.24}$$

and the coke deposition rate is

$$R_D = \frac{t_C}{\Theta} \tag{11.25}$$

The *run length* or interval between decokings is equal to the maximum allowable coke thickness, or that thickness at which the design tube metal is reached, divided by the coke deposition rate. The maximum allowable coke thickness is given by

$$\left(\frac{Q}{A}\right)_{\text{max}} \left(\frac{t_W}{K_W} + \frac{1}{h_{10}} + \frac{L_C}{K_C}\right) = [T_{\text{design}} - T_{\text{out}}] \tag{11.26}$$

and the run length, by

$$R_L = \frac{L_C/12}{R_D} \tag{11.27}$$

Pertinent calculated run length data for gas oil and ethane cracking are summarized in Table 11.1.

Concluding Remarks

A method for calculating run length on the basis of fired heater operating conditions and design variables has been proposed. Run length so calculated is based on a stationary monomolecular layer of fluid at the tube wall, and the thickness of the initial layer is calculated on the basis of feedstock molecular weight, density, and the Avogadro number.

TABLE 11.1 Run-Length Data for Direct-Fired Liquid and Gas Cracking Heaters

Cracking heater feed	Gas oil	Ethane
Flow arrangement	Countercurrent	Countercurrent
Decoking interval, years	0.25	0.36
Process fluid temperature inlet/outlet, °F	750/920	1250/1550
Burner diameter/flame diameter, ft	1.0/2.7	1.0/2.7
Flame temperature/bridgewall temperature, °F	3500/1600	3500/2000
Burner flame and radiant section height, ft	25	30
Burner flow rate, lb/h	4673	7350
Process tube heat absorption, Btu/h-burner	2.66	3.31
Burner heat liberation, Btu/h-burner	4.82	7.59
Burner center-to-center spacing, ft	3.0	7.66
Tube outer diameter, in./tube center-to-center spacing, in.	4.5/8.0	5.5/12
Number of tubes per burner	4.52	7.66
Incremental flame length for 50°F ΔT, ft	0.234	0.72
Incremental tube area, ft^2	2.49	7.94
Incremental cold-plane area, ft^2	1.404	5.52
Overall average flux, Btu/h-ft^2	10,000	20,000
Average flux/maximum flux at outlet, Btu/h-ft^2	28,100/50,000	27,800/39,900
Film temperature, °F	1032	2019
Reaction velocity constant	0.053	50.4
Coke formation rate, ft/s	$0.0214(10^{-8})$	$0.0163(10^{-8})$
End-of-run coke thickness, in.	0.0207	0.022

The run length so calculated must be considered as a maximum since the stationary layer at the wall might be more than one molecule thick, in which case the run length would be shorter than predicted. Thus, a generous negative allowance should be provided if a guaranteed run length is to be provided. Furthermore, the heater should be designed for concurrent flow when hydrocarbon feedstock tube metal temperatures exceed approximately 700°F, as the concurrent arrangement in such cases would very significantly increase run length. At temperatures in this range, however, it is likely that run length would be limited by pressure drop rather than high tube metal temperature, particularly if relatively small-diameter tubes are used.

Regarding the lack of reaction velocity data for feedstocks with a wide boiling-point range, for example, crude oil, run-length calculation would require that flash curve data be obtained so that the boiling range and composition of the tube wall fluid at the coil outlet might be categorized as being similar to gas oil, residuum, or some other fraction for which data are available. Furthermore, if a cracked product slate is unavailable, it would be prudent to assume that feedstock decomposition of the fluid at the wall results in deposition of 100 percent of the carbon in the feed, in which case the calculated run length would be shorter than would otherwise be predicted.

11.8 Transient and Cyclic Calculations

Nomenclature

A	Constant
$A_{C_p,I}$	Incremental cold-plane area, ft^2
A_o	Outside tube surface per burner, ft^2
$A_{o,I}$	Incremental tube surface, ft^2
B	Constant
C_P	Flue gas specific heat, Btu/lb-°F
C_Y	Pounds of coke deposited per lb fluid
CC_{burner}	Burner center-to-center spacing, ft
CC_{tube}	Tube center-to-center spacing, ft
D_F	Flame diameter, ft = $D_{I,D} (T_F/T_{fuel\ +\ air})^{1/2}$
$D_{I,D}$	Burner inner diameter, ft
D_o	Tube outer diameter, ft
e	2.718
h_{10}	Corrected inside heat-transfer coefficient, Btu/h-ft^2-°F
K	Reaction velocity constant
K_C	Thermal conductivity of coke, 13 Btu/h-ft^2-°F/in.
K_{eth}	Ethane reaction velocity constant
K_{GO}	Gas oil reaction velocity constant
K_W	Thermal conductivity of tube, Btu/h-ft^2-°F/in.
L_C	Coke thickness, in.
L_F	Burner flame or radiant section height, ft
L_{FT}	Total heater length, ft
L_I	Incremental flame length, ft
ln	Natural logarithm
M_F	Pounds of fluid per molecule fluid
mf_{CO_2}	mol fraction CO_2
mf_{H_2O}	mol fraction H_2O
MW	Molecular weight
N_{Avog}	Avogadro number, 6.03 (10^{23})
N_{BT}	Total number of burners per heater
N_R	Number of rows of tubes: 2 as for 2 rows of tubes straddling single row of burners in liquid cracking heater; 1 as for 2 rows of burners straddling single row of tubes in gas cracking heater
(N_T/row)	Number of tubes per row per burner
P	Gas pressure, atm
Q_{abs}/burner	Coil heat absorption per burner, Btu/h
$Q_{I,abs}$	Incremental heat absorption, Btu/h
Q_{lib}/burner	Heat liberation per burner, Btu/h

Q_R	Total radiant section absorption, Btu/h
$(Q/A)_{avg}$	Average heat flux, Btu/h-ft^2
$(Q/A)_{max}$	Maximum heat flux, Btu/h-ft^2
°R	°F + 460
R_D	Coke deposition rate, ft/s
R_L	Run length, s
$(RR)_{avg}$	Average radiant flux per burner, Btu/h-ft^2
t_C	Coke thickness, ft
t_F	Fluid thickness, ft
t_W	Tube wall thickness, in.
T_{design}	Tube wall design temperature, °R
T_F	Flame temperature, °R
T_{film}	Film temperature, °R
$T_{fluid\ out}$	Outlet fluid temperature, °R
$T_{fuel+air}$	Temperature of air and fuel entering burner, °R
T_{high}	Higher flue gas temperature, °R
T_{low}	Lower flue gas temperature, °R
T_{out}	Fluid outlet temperature, °R
V_C	Cubic feet of coke per molecule fluid
$\bar{V}_C$	Coke specific volume, ft^3/lb
V_F	Cubic feet of fluid per molecule fluid
$\bar{V}_F$	Fluid specific volume, ft^3/lb
W_B	Burner flue gas mass flow rate, lb/h
X	Percent reacted for gaseous feed or −400°F E.P. gasoline yield for liquid feed

Greek

γ	$544[D_F P\ (mf_{CO_2} + mf_{H_2O})]^{0.417}$
ϵ_2	Tube wall emissivity, 0.8
Θ	Reaction time, s
λ	Flue gas, tube temperature correction factor ($T^4_{low\ avg} - T^4_{high\ avg}/T^4_{w\ avg}$): λ for gas oil cracking = 0.9, λ for ethane cracking = 0.7

References

1. Cross, A., "Evaluate Temperature Gradients in Fired Heaters," *Chemical Engineering Progress*, June 2002, p. 42.
2. Nelson, W. L., "Constants for Rates of Thermal Decomposition of Hydrocarbons and Petroleum Fractions," in *Fuels, Combustion, and Furnaces*, edited by J. Griswold, McGraw-Hill, New York, 1946, p. 164.

Chapter 12

Transient Heat-Transfer Problem: Tape Pack Cooling

David R. Dearth
Applied Analysis & Technology
Huntington Beach, California

Introduction

Early in my engineering consulting career, development work for tape drives used as backup storage devices of computer data was just beginning. I was assigned the responsibility for developing performance and operational standards. One problem associated with tape drives is that data reliability is sensitive to temperature extremes. As part of the work performed on improving data reliability it was of interest to estimate the time required for tape packs to return to room temperature (RT) after extended exposure to high temperatures, such as when a tape pack is left in an automobile all day or exposed to direct sunlight. In this investigation both conventional hand solutions and finite-element analysis (FEA) techniques were utilized to estimate midpack temperature response (function of time) as the tape material (polyester) cools to room temperature. Figure 12.1 shows a typical tape pack design concept. It was desired to estimate the time for the tape pack center to return to room temperature (72°F) after a temperature "soak" at 130°F.

The presence of the aluminum hub on which the tape is wound made hand solutions for the transient heat-transfer solution quite difficult to solve. Therefore, FEA techniques were decided to be the most viable

12.2 Transient and Cyclic Calculations

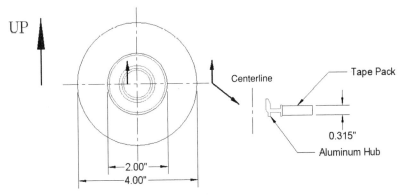

Figure 12.1 Tape pack with aluminum hub.

technique. Before beginning development of the FEA idealization, I recommend doing sample "warmup" problems with known textbook solutions to a simplified version of the actual geometry shown in Fig. 12.1. This test problem was extensively investigated prior to tackling the real problem. This sample problem was also a "sanity" check, a hand solution was created to ensure that I was using the correct units on the heat-transfer parameters *before* I invested time on the actual FEA mathematical idealizations.

The representative warmup problem selected for investigating this heat-transfer project was a simplified tape pack idealized as a flat vertical plate. A search through the engineering literature addressing heat-transfer problems located various portions of each individual aspect of these types of heat-transfer problems. However, none of the engineering references contained a complete approach to the problem from beginning to end. This sample transient heat-transfer problem contains all the features of any real-life problem as a comprehensive, step-by-step solution to the simplified version of the tape pack with an FEA model to cross-check the results. Once we were confident of our methodology, techniques, and procedures from investigating this sample problem, we then proceeded to address the real-life problem.

Sample Problem: Heat-Transfer Solution for Transient Tape Pack Cooling

Figure 12.2 shows a sketch of the simplified tape pack transient cooling problem with the tape pack initially at high-temperature soak $T_i = 130°F$ and suddenly exposed to surrounding still air at $T_\infty = 72°F$. The magnetic tape pack is essentially polyester plastic (Mylar) with

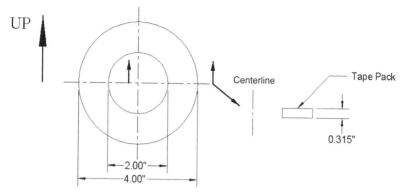

Figure 12.2 Simplified tape pack transient heat-transfer cooling with aluminum reel hub neglected.

thermal conductivity $K = 0.02168$ Btu/h-ft-°F. The presence of the tape oxide coating was neglected.

For this problem, it is desired to estimate the time required for the tape pack to cool down to 99 percent of room temperature (72.72°F after temperature soak at 130°F.

The first step in the analysis is to determine an average heat-transfer film coefficient h_{avg} for free (natural) convection to the surrounding ambient air (fluid). Assuming the vertical tape pack to be similar to a vertical plane an estimate for average heat-transfer film coefficient is $h_{avg} = 1.6985$ Btu/h-ft²-°F. The arithmetic for estimating h_{avg} can be found in the detailed hand calculations that follow.

To gain confidence in our solution techniques, we will present two approaches: (1) a hand solution using conventional equations found in most engineering textbooks on heat transfer and (2) correlate results to a finite-element idealization.

Answers

A reel of magnetic tape is assumed to have the following thermal properties for polyester material (trade name Mylar):

Tape density $\rho = 87.09$ lb/ft³

Tape specific heat $C_p = 0.315$ Btu/lb-°F

Tape thermal conductivity $K = 0.02168$ Btu/h-ft-°F

The tape reel physical dimensions are inside diameter of 2.00 in., outside diameter 4.00 in., and thickness 0.315 in., and the tape reel is initially at uniform elevated temperature 130°F. The tape pack is suddenly

12.4 Transient and Cyclic Calculations

exposed to surrounding free ambient still air at RT (72°F) and the average free convective heat-transfer film coefficient $h_{avg} = 1.6985$ Btu/h-ft²-°F.

Determine the time required for the center region of the tape pack to return to 99 percent of RT (72.72°F).

Step 1: Compute average film coefficient from mean film surface temperature T_{film}

The mean film temperature is

$$T_{film} = \frac{T_h + T_\infty}{2} \tag{12.1}$$

Substituting into Eq. (12.1), we obtain

$$T_{film} = \frac{130 + 72}{2} = 101°F$$

Table 12.1 lists properties of air at $T_{film} = 101°F$ using linear interpolation from data in table A-4 of Ref. 1.

Step 2: Compute Grashof Gr number and check for laminar or turbulent free convection

Let $T_{film} = T_{surface} = T_s$:

$$\text{Gr} = \frac{g\beta(T_s - T_\infty)L^3}{v^2} \tag{12.2}$$

For gas (air), we obtain

$$\beta = \frac{1}{T, °R} = \frac{1}{101 + 460} = 1.783 \times 10^{-2} \, °R^{-1} \tag{12.3}$$

Let the characteristic length equal surface area divided by the perimeter: $L = L_c = (\pi/4)(4^2 - 2^2)/(\pi 4) = 0.75$ in. Substituting into

TABLE 12.1 Properties of Air at 101°F

Density $\rho = 0.0709$ lb/ft³
Specific heat $C_p = 0.2404$ Btu/lb-°F
Dynamic viscosity $\mu = 1.2767 \times 10^{-5}$ lb/s-ft
Kinematic viscosity $v = 18.16 \times 10^{-5}$ ft²/s
Thermal conductivity $k = 0.01567$ Btu/h-ft-°F
Thermal diffusivity $\alpha = 0.9281$ ft²/h
Prandtl number $\text{Pr} = 0.705$

Eq. (12.2), we obtain

$$\text{Gr} = \frac{(32.2 \text{ ft/s}^2)(1.783 \times 10^{-3} \, ^\circ\text{R}^{-1})(130 - 72)\,^\circ\text{R} \, (0.75/12 \text{ ft})^3}{(18.16 \times 10^{-5} \text{ ft}^2/\text{s})^2}$$

$$= 2.464 \times 10^4$$

Compute the Rayleigh number:

$$\text{Ra} = \text{Gr Pr} = (2.464 \times 10^4)(0.705) = 1.738 \times 10^4$$

Since Ra = Gr Pr < 10^9, laminar flow conditions may be assumed.

Step 3: Compute average film heat-transfer coefficient $h_{avg} = \bar{h}$

First compute the Nusselt number:

$$\overline{\text{Nu}} = \frac{\bar{h}L}{k} \tag{12.4}$$

The Nusselt number can also be represented using empirical relations depending on the geometry of the free convective system:

$$\overline{\text{Nu}} = C(\text{Gr Pr})^a \tag{12.5}$$

Then

$$\frac{\bar{h}L}{k} = \overline{\text{Nu}} = C(\text{Gr Pr})^a \tag{12.6}$$

To estimate the heat-transfer film coefficient, neglect radial effects and treat the tape pack similar to a vertical plate with thickness 0.315 in. For laminar flow, select the following coefficients: $C = 0.59$ and $a = 1/4$, from table 7-1 of Ref. 2. Substituting and solving for $\bar{h}$, we obtain

$$\bar{h} = \frac{0.01567 \text{ Btu/h-ft-}^\circ\text{F}}{0.75/12 \text{ ft}} (0.59)(1.738 \times 10^4)^{1/4}$$

$$= 1.6985 \text{ Btu/h-ft}^2\text{-}^\circ\text{F}$$

Step 4: Determine solutions for time-dependent temperatures at the tape pack center

The partial differential equation for transient cooling in one spatial direction, x in this case, can be found in Ref. 2:

$$\frac{\partial^2}{\partial x^2} T(x, t) = \frac{1}{\alpha} \frac{\partial}{\partial t} T(x, t) \tag{12.7}$$

12.6 Transient and Cyclic Calculations

where α represents thermal diffusivity for the tape material:

$$\alpha = \frac{K}{\rho \cdot C_p} \qquad (12.8)$$

Substituting into Eq. (12.8), we obtain

$$\alpha = \frac{0.02168 \text{ Btu/h-ft-}°\text{F}}{(87.09 \text{ lb/ft}^3)(0.315 \text{ Btu/lb-}°\text{F})}$$

$$= 7.903 \times 10^{-4} \text{ ft}^2/\text{h}$$

The solution to the partial differential equation can be determined using the method of separation of variables. In both undergraduate and graduate classes I've taught at California State University at Long Beach (CSULB), considerable class time has been spent investigating the exact solution to Eq. (12.7). One can find numerous formulations to the solution of Eq. (12.7).

Without consuming a great deal of time on the details of how to solve Eq. (12.7), one formulation for the exact solution can be represented by a series solution with the following nondimensional definitions used to simplify writing of the solution. Define the nondimensional (normalized) temperature function:

$$u(x, t) = \frac{T(x, t) - T_\infty}{T_i - T_\infty} \qquad (12.9)$$

Define the nondimensional distance parameter x', where x is a local coordinate direction from the pack centerline toward the outside surface:

$$x' = \frac{x}{L} \qquad (12.10)$$

Define the nondimensional time parameter, called the Fourier modulus Fo:

$$\text{Fo} = \left(\frac{\alpha}{L_c^2}\right) \cdot t \qquad (12.11)$$

Define the nondimensional temperature constant, called the Biot number Bi:

$$\text{Bi} = \frac{\overline{h} \cdot L_c}{K} \qquad (12.12)$$

Using the nondimensional variables above, the exact solution to Eq. (12.7) can be represented by the following series solution:

$$u(x, t) = \sum_{n=1}^{\infty} a_n \cos(\lambda_n x) e^{-\lambda_n^2 \text{Fo}} \tag{12.13}$$

where λ_n are roots to

$$\text{Bi} \cdot \cos \lambda_n - \lambda_n \cdot \sin \lambda_n = 0 \tag{12.14}$$

with a_n and N_n given by Eqs. (12.15a) and (12.15b), respectively:

$$a_n = \frac{1}{N_n} \int_0^1 \cos \lambda_n x \, dx \tag{12.15a}$$

$$N_n = \int_0^1 \cos^2 \lambda_n x \, dx \tag{12.15b}$$

Even with all the various approaches available for solving Eq. (12.7), one still ends up with quite a task remaining before completing the solution arithmetic. Further simplifications can be made depending on the magnitude of Bi. For Bi < 0.1, one can use a lumped-parameter solution that assumes the tape pack approximated by a single concentrated mass. For Bi > 0.1, one can use chart solutions, called Heisler charts [5]. Figure 12.3 shows a chart solution for an infinite slab at the midplane location ($x' = x/L = 0$ at $x = 0$).

The chart solution method can be somewhat cumbersome to read if one has a lot of points to plot, but it represents an easy way to address Eq. (12.13). An alternate method to chart solutions is to compute values for Eq. (12.13) directly. After some investigation, one can reduce Eq. (12.13) and formulate a one-term, time-dependent approximate solution at the midplane location ($x' = x/L = 0$ at $x = 0$) with < 2 percent error. The one-term approximation to Eq. (12.13) is shown as Eq. (12.16) for $u(x, t)$ as

$$u(x, t) = A_1 e^{-\lambda_1^2 \text{Fo}} \tag{12.16}$$

For the problem at hand, let the characteristic length L_c for Eqs. (12.11) and (12.12) equal one-half the tape reel width $L_c = (0.315/2)$ in. = 0.1575 in.

Substituting into Eq. (12.11) for the Fourier modulus Fo, we obtain

$$\text{Fo} = \frac{7.903 \times 10^{-4} \text{ ft}^2/\text{h}}{\left(\dfrac{0.1575 \text{ in.}}{12 \text{ in./ft}} \right)^2} \cdot (t, \text{h})$$

$$= 4.5876 \cdot (t, \text{h})$$

$$= 0.07646 \cdot (t, \text{min})$$

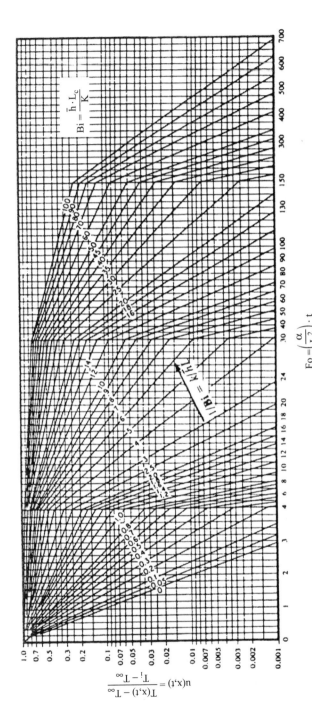

Figure 12.3 Heisler chart: centerline temperatures for infinite plate of thickness 2L [5].

Substituting into Eq. (12.12) for the Biot number Bi, we have

$$\text{Bi} = \left(\frac{1.6985 \text{ Btu/h-ft}^2\text{-}°\text{F}}{0.02168 \text{ Btu/h-ft-}°\text{F}}\right)\left(\frac{0.1575 \text{ in.}}{12 \text{ in./ft}}\right)$$

$$= 1.0283$$

For Bi = 1.028, solve for the coefficients A_1 and λ_1 using Eqs. (12.14), (12.15a), and (12.15b):

$$A_1 = 1.12078 \quad \text{and} \quad \lambda_1 = 0.86730$$

Step 5: Estimate time required for midpack to cool down to 99 percent of RT

Substituting into Eq. (12.16) and solving for the time required for the center region of the tape pack ($x' = x/L = 0$ at $x = 0$) to return to 99 percent of RT (72.72°F), t in minutes, we obtain

$$u(x, t) = \frac{T(0, t) - T_\infty}{T_i - T_\infty} = \frac{72.72 - 72}{130 - 72} = 1.12078 \cdot e^{-(0.86730)^2 \cdot (0.07646) \cdot (t, \text{min})}$$

$$0.012414 = 1.12078 \cdot e^{-0.05751 \cdot (t, \text{min})}$$

$$\ln\left(\frac{0.012414}{1.12078}\right) = -0.05751 \cdot (t, \text{min})$$

$$\frac{-4.50297}{-0.05751} = t, \text{min}$$

$$t = 78.29 \text{ min}$$

A spreadsheet is used to minimize round off in the arithmetic.

Step 6: Evaluate T(0, t) at 5-min intervals for comparison to FEA results

Table 12.2 lists estimates for internal temperature of the tape pack at the midplane location ($x' = x/L = 0$ at $x = 0$) as a function of time t. A spreadsheet is used in this table to minimize roundoff in the arithmetic of Eq. (12.16).

Step 7: Verify hand calculations using FEA idealization

Table 12.3 lists a comparison of estimates for temperatures at the tape pack center versus time when the geometry shown in Fig. 12.2 is analyzed using both hand solutions and an FEA mathematical idealization

12.10 Transient and Cyclic Calculations

TABLE 12.2 Hand Solution: Tape Pack Cooling Center Temperatures versus Time

	Time, min	Fo{t}	$u(0,t)$	$T(0,t)$	
Start	0.00	0.000	1.000	130.00	
	5.00	0.382	0.841	120.76	
	10.00	0.765	0.631	108.57	
	15.00	1.147	0.473	99.43	
	20.00	1.529	0.355	92.58	
	25.00	1.911	0.266	87.44	
	30.00	2.294	0.200	83.58	
	35.00	2.676	0.150	80.68	
	40.00	3.058	0.112	78.51	
	45.00	3.441	0.084	76.89	
	50.00	3.823	0.063	75.66	
	55.00	4.205	0.047	74.75	
	60.00	4.588	0.036	74.06	
	65.00	4.970	0.027	73.55	
	70.00	5.352	0.020	73.16	
	75.00	5.734	0.015	72.87	Range
	80.00	6.117	0.011	72.65	99% RT
	85.00	6.499	0.008	72.49	
	90.00	6.881	0.006	72.37	
	95.00	7.264	0.005	72.28	
End	100.00	7.646	0.004	72.21	

TABLE 12.3 Tape Pack Cooling Center Temperatures versus Time Comparison of FEA versus Hand Solutions

	Time, min	FEA	Hand	Percent difference	
Start	0	130.000	130.000	0.000	
	5	120.876	120.760	−0.001	
	10	108.692	108.574	−0.001	
	15	99.484	99.434	−0.001	
	20	92.586	92.578	0.000	
	25	87.419	87.435	0.000	
	30	83.549	83.578	0.000	
	35	80.650	80.684	0.000	
	40	78.479	78.514	0.000	
	45	76.853	76.886	0.000	
	50	75.635	75.665	0.000	
	55	74.723	74.749	0.000	
	60	74.039	74.062	0.000	
	65	73.528	73.547	0.000	
	70	73.144	73.160	0.000	
	75	72.857	72.870	0.000	Range
	80	72.642	72.653	0.000	99% RT
	85	72.481	72.490	0.000	
	90	72.360	72.367	0.000	
	95	72.270	72.275	0.000	
End	100	72.202	72.207	0.000	

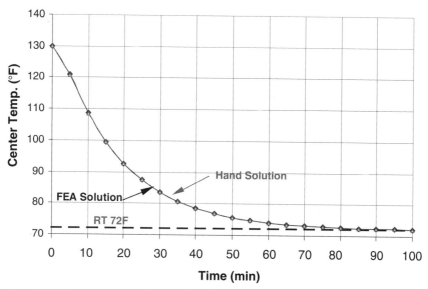

Figure 12.4 Summary comparison results of hand versus FEA solution, transient thermal response.

processed using MSC/Nastran. A comparison between the solutions estimates a percentage difference no larger then ±0.001 percent during the initial transient response with virtually no difference between the solutions after approximately 15 min of cooling. Figure 12.4 shows a summary comparison between the hand solution and the FEA model solutions as a graph.

Applying the Principles

With a high degree of confidence in defining and solving a simplified version of the transient heat-transfer problem, the final step is to apply the principles. Because of the complexity of the geometry that includes an aluminum hub in addition to the wound tape, FEA methods are better suited for addressing the solutions. Before we discuss the finite-element solution, a short introduction to FEA follows.

What Is Finite-Element Analysis, and How Does It Work?

Finite-element analysis methods were first introduced in 1943. Finite-element analysis uses the Ritz method of numerical analysis and minimization of variational calculus to obtain approximate solution to systems. By the early 1970s, FEA was limited to high-end companies involved in the aerospace, automotive, and power plant industries that

12.12 Transient and Cyclic Calculations

could afford the cost of having mainframe computers. After the introduction of desktop personal computers, FEA was soon made available to the average user. FEA can be used to investigate new product designs and improve or refine existing designs.

The FEA approach is a mathematical idealization method that approximates physical systems by idealizing the geometry of a design using a system of grid points called *nodes*. These node points are connected together to create a "finite" number of regions or mesh. The regions that define the mesh are assigned properties such as the type of material and the type of mesh used. With the mesh properties assigned, the mesh regions are now called "elements." The types of mesh include rod elements, beam elements, plate elements, and solid elements. Constraints or reactions are entered to simulate how the idealized geometry will react to the defined loading.

Analysis models can represent both linear and nonlinear systems. *Linear* models use constant parameters and assume that the material will return to its orignal shape after the loading is removed. *Nonlinear* models consist of stressing the material past its elastic capabilities, and permanent deformations will remain after the loading is removed. The definitions for the materials in these nonlinear models can be complex.

For heat transfer, FEA models can be idealized for both steady-state and transient solutions. The mesh elements are defined using the material conductivity or thermal dynamics to represent the system being analyzed. *Steady-state transfer* refers to materials with constant thermoproperties and where heat diffusion can be represented by linear equations. Thermal loads for solutions to heat-transfer analysis include parameters such as temperature, internal heat generation, and convection.

FEA Results: Tape Pack with Aluminum Hub

Figure 12.5 shows an FEA model idealization of a tape pack with the aluminum hub included as shown in Fig. 12.1. The FEA model shown in Fig. 12.5 is analyzed for the case of a tape pack that is initially at temperature soak of 130°F and shows the transient temperature response after 10 min of cooling.

Table 12.4 lists numerical values for estimates of tape pack cooling as a function of time when the FEA model shown in Fig. 12.5 is analyzed using MSC/Nastran. Figure 12.6 shows the results graphically.

Using linear interpolation of the data shown in Fig. 12.6 from $t = 65$ min to $t = 70$ min, one can estimate the time required for the tape pack center to return to 99 percent of RT $t_{99\%} = 69.1$ min.

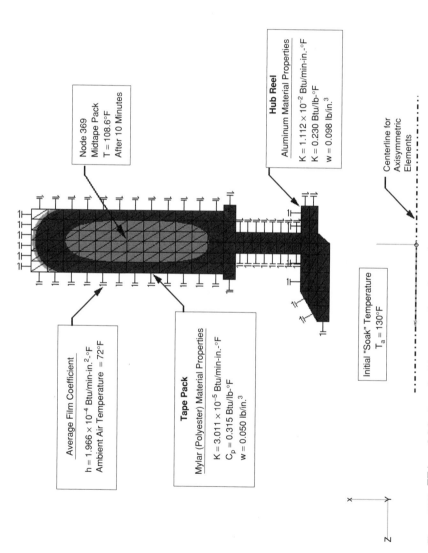

Figure 12.5 FEA model for transient heat-transfer solution, tape pack with aluminum hub.

12.13

12.14 Transient and Cyclic Calculations

TABLE 12.4 Summary of Tabular Results of FEA Solution: Transient Time History Thermal Response for Tape Pack with Aluminum Hub at Midpack Location (Node 369)

	Time, min	FEA	
Start	0	130.000	
	5	120.872	
	10	108.545	
	15	98.960	
	20	91.666	
	25	86.224	
	30	82.227	
	35	79.323	
	40	77.228	
	45	75.725	
	50	74.649	
	55	73.883	
	60	73.337	
	65	72.948	Range
	70	72.673	99% RT
	75	72.477	
	80	72.338	
	85	72.240	
	90	72.170	
	95	72.120	
End	100	72.085	

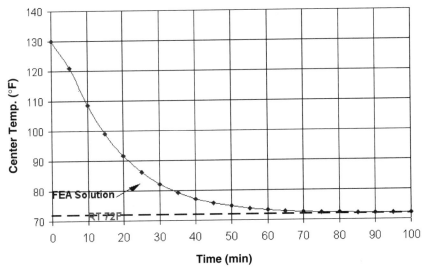

Figure 12.6 Summary graphical results of FEA solution, transient time history thermal response for tape pack with aluminum hub at midpack location (node 369).

As a result of this transient heat-transfer analysis, it was conservatively recommended that tape packs suspected of previous exposure to temperature extremes be allowed to stabilize to room temperature for a period of least 90 min before making new backup copies or restoring data from a previous recording.

References

1. E. R. G. Eckert and Robert M. Drake, Jr., *Heat and Mass Transfer*, 2d ed., McGraw-Hill, 1959.
2. J. P. Holman, "Conduction-Convection Systems," in *Heat Transfer*, 5th ed., McGraw-Hill, 1981, chap. 2-9.
3. W. M. Rohsenow and J. P. Hartnett, *Handbook of Heat Transfer*, McGraw-Hill, 1973.
4. Donald R. Pitts and Leighton E. Sissom, *Heat Transfer*, 2d ed., Schaum's Outline Series, McGraw-Hill, 1977.
5. M. P. Heisler, *Transactions of the American Society of Mechanical Engineers (Trans. ASME)*, 69: 227 (1947).

Chapter 13

Calculation of the Transient Response of a Roof to Diurnal Heat Load Variations

Jack M. Kleinfeld
Kleinfeld Technical Services, Inc.
Bronx, New York

Background

The calculations presented here are for the transient heat transfer of a roof under diurnal variation of environmental temperature changes and solar loading. They include an estimate of the impact of water in the roof, due to a leak, on the surface temperature of the roof.

This problem is of particular interest for infrared (IR) thermographic surveys of roofs to detect leaks evidenced by the presence of moisture in the roof insulation under the membrane or surface of the roof. Infrared thermography is an established method[1] for detecting the presence of wet insulation in roofs that depends on differences in the response of wet and dry roof areas to transient changes of the roof temperature in response to the daily loading cycle. In essence, the wet areas have a higher thermal mass than do the dry areas and lag behind the dry areas in exhibiting temperature changes as the environment changes. In practice, infrared thermographic roof moisture surveys are generally carried out at night while the roof is cooling after the day's heating. Under those conditions, the wet areas, which are cooling more slowly, will show as relatively warmer than the dry areas.

13.2 Transient and Cyclic Calculations

The calculations as shown are typical of the methodology for generating a transient response to a variable input. The methodology shown is based on using finite-element analysis (FEA) software to perform the calculations. The presentation is kept as generic as possible, but the work shown used Cosmos/M software and may have some details specific to it. This chapter is not, however, a software tutorial.

Problem Statement

Determine the temperature history and characteristics of a roof surface for dry and wet conditions of the roof's insulation, corresponding to a localized leak into the roof from the surface covering. Conditions selected correspond to August in the New York City area at 40°N latitude.

Approach

A small through-section of the roof is modeled as representative of the entire roof. Those components which will not have an appreciable effect on the temperature history of the roof are omitted from the model. Where possible, the boundary conditions and properties of the materials are determined from literature sources. Where such are not readily available, estimated values are used for illustration. Separate calculations for the performance of dry and wet roof conditions are carried out, giving the behavioral characteristics of each, but not providing the interaction between wet and dry sections of the roof. The transient boundary conditions are developed as locally linearized approximations of the continuously varying actual conditions.

Calculation Details

Model

Figure 13.1 shows the model as developed for the analysis. It is a representation of the roof structure. Using an FEA approach means that it is set up on a computer. If only one condition is evaluated per calculation, the problem is, in fact, only a one-dimensional (1D) analysis. It is easier, however, to use a two-dimensional (2D) model, in terms of the setup in the modeling software. The varying conditions that are run are different moisture contents of the roof, in this case dry and, as an example, 20 percent water by weight in the insulation layer. Additional cases can be run for differing boundary conditions (BCs), such as time of year, or weather conditions. If a different type of roof is of interest, an additional model is needed.

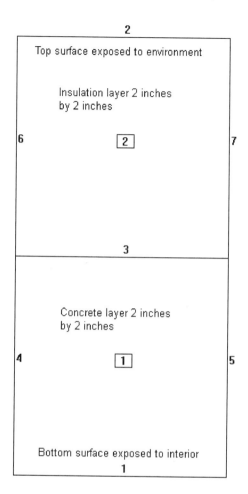

Figure 13.1 Outline of model used in the FEA software, showing curves 1 to 7 and surfaces or regions 1 and 2.

In some cases, it is desirable to understand the interaction on a roof of wet and dry areas. This would call for a three-dimensional (3D) model, since the lateral or horizontal heat transfer between the wet and dry areas, potentially of different sizes, is now of interest. Such a 3D model would be set up with both wet and dry areas designated and run under otherwise similar boundary conditions. The diffusion of the thermal signature or surface temperature profile between the wet and dry areas could then be determined. Since the approach is essentially identical for the 2D and 3D models, only the 2D model, evaluated for two roof conditions, is presented here.

The model as configured represents a 2 in. thickness of insulation on a 2-in.-thick concrete roof deck. The roof membrane is not modeled, since

13.4 Transient and Cyclic Calculations

it is assumed to have negligible thermal properties. However, its emissivity is used for the radiational behavior of the top surface of the roof.

Assumptions in the model

The interface between the insulation and the roof deck is assumed to have full thermal contact between the materials. This is represented by curve 3 in Fig. 13.1. No air gap between them is considered. If the roof being modeled were known to have air gaps due to unevenness of the deck or of the insulation installation, then either an additional region for the air between the insulation and the deck or a contact gap in the FEA software would be used to represent them. The water in the wet insulation is assumed to be uniformly distributed in the insulation and to not puddle either on top or under the insulation. If puddling were suspected, additional regions in the model would be needed.

Boundary conditions

The boundary conditions for the calculation occur at the top and bottom surfaces of the roof. These are represented by curves 2 and 1, respectively, in Fig. 13.1. The sides of the modeled domain, which would be connected to more roof, are treated as adiabatic. These are represented by curves 4 to 7 in Fig. 13.1. This is equivalent to assuming that the adjacent portions of the roof are the same as the modeled portion and that the roof is much larger than 2 in. wide. It also assumes that the portion modeled is not at the edge of the roof or near a structure on the roof.

The boundary conditions on the top surface of the roof account for direct solar heating of the roof, radiational transfer between the roof and the sky, and convective transfer between the roof and the surrounding air. While the solar heating is, in fact, a radiation process, it can be described and modeled as a heat flux applied to the roof. This has a significant advantage in FEA modeling, since it removes one level of nonlinearity from the model, making it solve more quickly, and also allows the radiation BC to be used to describe the interchange with the sky. The software used would not allow two radiation BCs at the same location. In addition, the solar loading information needed is available in the literature in the form of heat flux data.

The boundary conditions on the bottom surface of the roof are primarily convection to the inside air temperature and radiation to the building interior. Radiation is often approximated as a linear phenomenon over a short temperature span and represented as an additional convective-style heat-transfer coefficient in the convective equation. This simplifies solutions for hand calculations, but is not necessary or desirable for the system as modeled in an FEA package.

TABLE 13.1 Solar Flux

Time, h	Flux from ASHRAE, Btu/h ft^2	Flux as extended and used with 0.915 absorptivity, Btu/h ft^2
0		0
5.5		0
6	12	10.98
7	62	56.73
8	122	111.63
9	174	159.21
10	214	195.81
11	239	218.685
12	247	226.005
13	239	218.685
14	214	195.81
15	174	159.21
16	122	111.63
17	62	56.73
18	12	10.98
18.5		0
24		0

The solar flux or solar heat gain can be found in the literature. ASHRAE[2] presents it for average cloudless days in tabular form for both vertical and horizontal surfaces, as well as for normal incident radiation. It is tabulated for each month of the year at different latitudes. A single set of values was used here, shown in Table 13.1, for August at 40°N latitude. The values were adjusted by a factor of 0.915 representing the average emissivity or absorptivity of roofing-type materials.[3] The appropriate value for this and for the degree of cloudiness should be used for a particular situation. In addition, since the software interpolates the flux, and other values, from a linearized tabular curve, it is necessary to extend the table both down to zero flux and out to zero time and 24 h, in order to cover the entire calculational range. This was done by adding points with zero value at 0 and 24 h, representing the end of the day, and at 0530 and 1830 h, representing 1/2 h beyond the end of the data. The extended curve is shown in Fig. 13.2 and its constituent data, in Table 13.1.

The roof surface exchanges energy with the surroundings by radiation to the sky, provided there are no significant physical objects in the way, as was assumed. This radiation BC requires a value for the temperature of the sky. Clear skies are actually quite cold, especially at night. In the absence of any hard data, values were assumed for the purposes of this demonstration calculation. They are shown in Table 13.2 and Fig. 13.3. An emissivity value for the roof was used for the radiation BC as it was used for the solar flux.

13.6 Transient and Cyclic Calculations

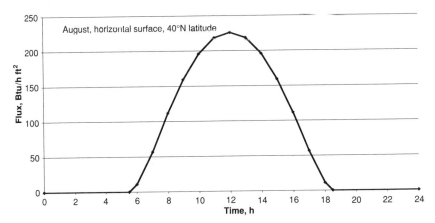

Figure 13.2 Solar flux boundary condition.

The interior or underside surface of the roof also needs to have a radiation BC to represent its interchange with the building's interior. The building was assumed to be at a constant 70°F for both this BC and the convective transfer BC. An emissivity of 0.92, corresponding to flat-finish paint, was used for the radiation BC.

In many ways the convective BCs, on both the top and bottom surfaces, are the hardest to specify. The heat-transfer coefficients are not readily determined. The situation is complicated by the fact, which can be seen from the results, that the convective transport reverses direction over the course of the day. Because of the radiational cooling to the night sky, the roof temperatures drop below the surrounding air temperature. This is true on both the top and bottom surfaces of the roof. The convection can potentially be either free or forced. If free convection pertains to the situation being modeled, determination of the heat-transfer coefficient is complicated by the reversal of the heat flow. For a downward-facing hot surface, ASHRAE[4] points out that free convection should not actually take place, since the hotter, lighter layer is above the cooler bulk. They report that some, however, does occur, and

TABLE 13.2 Sky Temperature

Time, h	Sky temperature, °F
0	25
6	25
7	55
18	55
19	25
24	25

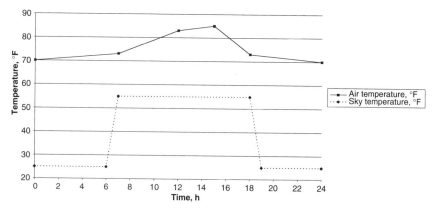

Figure 13.3 Environmental temperatures used for BCs.

suggest that the coefficient is less than half that of the coefficient for the same case but with the temperatures reversed. Perry et al.[5] and ASHRAE[4] present similar correlations for these cases.

A temperature- and therefore time-dependent formulation of the convective heat-transfer coefficient should be used for the most accuracy. This has not been done for this calculation. It becomes difficult to evaluate the various coefficients required. For the top of the roof, where there is the complication of a varying air temperature, it also would be difficult to code the heat-transfer coefficient into the FEA software, since the formulations generally use the temperature difference between the surface and the bulk air. Approximate values for the heat-transfer coefficients were estimated for free convection. Values of 1 Btu/h ft^2°F for the exterior and 0.8 Btu/h ft^2°F for the interior were selected. Were the system considered to be in forced convection, a less surface-temperature-dependent solution should be applicable. The air temperatures assumed for the outside air are shown in Table 13.3 and Fig. 13.3. If data of hourly air temperatures were available, they would be used here.

TABLE 13.3 Air Temperature

Time, h	Air temperature, °F
0	70
7	73
12	83
15	85
18	73
24	70

13.8 Transient and Cyclic Calculations

The 24-h cycle temperatures and fluxes were repeated without change for the second 24-h cycle in the calculations. This simulates two identical days. The first, which gives results quite similar to those for the second except for some variation in the first few hours, was treated as a break-in period and would not be used for the final results.

Properties

For a transient thermal solution, the heat capacity and the density of the materials are needed in addition to the thermal conductivity that is necessary for a steady-state solution. Properties for dry materials are relatively available in the literature. Bench testing can also be used to obtain them if necessary. The properties of wet materials, such as wet insulation, are not readily available. They can be approximated for low to moderate water contents by

Wet material property = dry material property
$$+ \text{(fraction water)} \cdot \text{(water material property)}$$

The approach is predicated on the dry material having open pores that are filled with the water as the water content increases, essentially adding the properties of the water in parallel to those of the dry material. For insulation materials, such as mineral wool or fiberglass batts, this should be a good approximation. This approximation fails completely at 100 percent water, where it predicts that water's properties are equal to the dry material's plus the water's property. It also should not be used at high water concentrations. For the current example, with 20 percent water content, the approximation should be quite good.

Initial conditions

An approximate, uniform temperature for the entire roof was set before beginning the calculations. In order to reduce the effect of errors in this value, the simulation was run for a 48-h period and only the second 24-h period would be reported as the result. The full 48-h calculation results are generally shown here.

Time steps

Selection of the time increments to use in calculating the behavior of the roof will affect the run time of the model and its accuracy. Short time steps will cause long run times, with excessively short steps not necessarily improving the accuracy of the results. Long time steps will cause inaccuracies in the calculations. Another consideration in selecting time steps is the size of the finite elements making up the model. The times must be such that changes in the far boundary of the elements occur over the time interval selected. If this is not the case, that is, if the elements are too large or the time steps too small, the calculations will fail and the software will not converge. One indication of whether the time steps are sufficiently short is the shape and smoothness of the results. An explicit check is to run the simulation with a shorter or longer time step, varying by about 20 percent, to see if the results change significantly. If not, then the time step is sufficiently small. The step used for the calculations here was 900 s (15 min).

Calculation Results

General

The primary results of the calculations are the temperature history of the surface of the roof. Temperature histories of other places on the roof, such as the ceiling surface or the top of the concrete deck, are also of interest. The results from a given run of the FEA model provides values for a particular set of roof conditions and environmental conditions. As discussed above, a larger model could incorporate more than one roof condition, for example, by including both wet and dry areas in the same model. The comparison between the wet and dry results in the approach used is made by taking results from two runs.

Additional information can also be obtained using the FEA results. For example, once the conditions for a particular time are obtained, the relative importance of the various heat-transfer mechanisms, solar flux, convection, and radiation can be compared. This can be done as a hand calculation using the temperature condition at the time selected and the boundary conditions specific for that time.

Figures 13.4 and 13.5 show the temperature histories of the top and bottom, or exterior and interior surfaces, of the roof for dry and 20 percent wet conditions. They are plotted on the same scale. The wet roof exterior surface exhibits a peak temperature noticeably lower than that of the dry roof. The exterior surface becomes colder than the interior surface at night and hotter during the day. Both the exterior and interior surfaces drop below their local ambient air temperatures at night, driven by the radiational exchange between the exterior surface and the cold night sky. The temperature history of the top of the concrete deck, under the insulation, is not plotted on these two figures because it is almost identical to the interior surface temperature.

The peak temperatures of the exterior and interior surfaces occur at different times. The figures show the full 48-h calculation. The first few hours differ from the comparable time period in the second day's results.

13.10 **Transient and Cyclic Calculations**

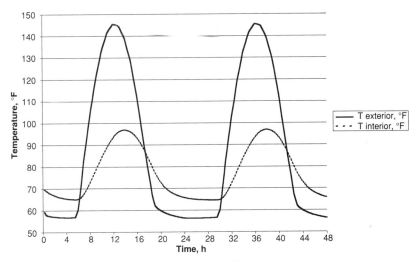

Figure 13.4 Temperature history of dry roof surfaces.

In this case, the differences are not large and the effects of the initial conditions are overcome by a single 24-h cycle of calculation; this is evident on comparison of the endpoints of the 48-h calculation with the beginning of the second 24-h cycle. Since they match, additional cycles are not necessary to evaluate the performance. For deeper, or more massive, structures, additional cycles would be needed to overcome the

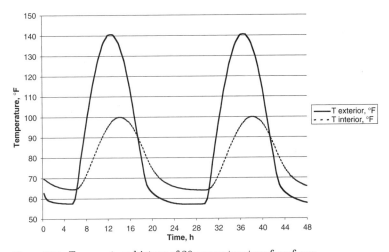

Figure 13.5 Temperature history of 20 percent wet roof surfaces.

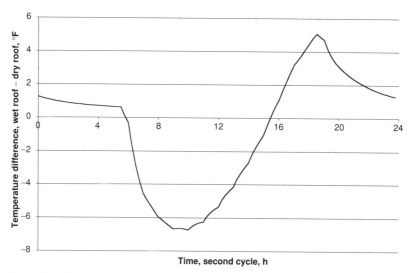

Figure 13.6 Surface temperature contrast between 20 percent wet and dry roof areas.

impact of the initial conditions selected. If day-to-day conditions vary widely, some consideration would have to be given to running a set of calculations with varying conditions over multiple day cycles.

For application to IR surveys, a key value is the contrast between wet and dry areas on the exterior surface, which is the one usually surveyed. Figure 13.6 shows the contrast between the wet and dry exterior surfaces. At night the wet areas are warmer than the dry areas. This reverses during the day, when the wet areas are cooler. The peak differences do not occur at the same time as the peaks in the temperature histories, but are offset by several hours. This indicates that the difference is maximized during the steepest transient of the roof.

Impact of variations in BCs and conditions

An issue raised above in the discussion of BC selection is the value for the convective heat-transfer coefficient. Values were estimated on the basis of natural convection, but the confidence level in the values is not high. The model can be used to evaluate the sensitivity of the results to the heat-transfer coefficient values, as well as to other values. Results were generated for an additional case where the heat-transfer coefficient was increased by a factor of 1.25 for both the exterior and interior surfaces of a dry roof. The resulting temperature history is shown on Fig. 13.7, which includes the base case as well as the air temperature above the roof. There is a noticeable impact on the results. Where the

13.12 Transient and Cyclic Calculations

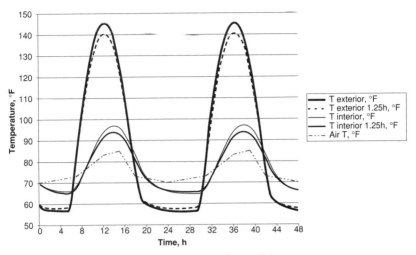

Figure 13.7 Sensitivity to assumed heat-transfer coefficient.

primary interest is the ability to find wet versus dry roof areas, the impact will be less important, since the same revised coefficients would be used for both the wet and dry cases. If the primary interest in performing the work is the actual temperatures of the roof, for example, if the material life were being evaluated, these levels of differences could be important.

Evaluations of differing conditions, such as time of year, cloud cover, location, roof surface emissivity, interior temperatures, roof construction, and other levels of water content, could be done in exactly the same manner as the work shown. For most of these, the changes would be in the BCs. For changes in the interior temperature, if, for example the building used a programmable thermostat, a time curve could be introduced. For changes in the roof surface, for example, from a black to a white surface, the solar flux would be adjusted by the solar emissivity of the surface. The white surface would probably have the same emissivity as the black in the infrared wavelength region, where the transfer by radiation occurs, so the emissivity for the radiation BC would not be changed. For a silver surface, with lower emissivity at both solar wavelengths and infrared wavelengths, the values of emissivity for the radiation BC would also need to be changed. Changes in the water content of the roof would require changes in the physical properties, as outlined above. Changes in the roof construction would require a new model representing the new roof design, but the BCs would probably not change with the possible exception of emissivity.

References

1. ASTM C1153-90, *Standard Practice for the Location of Wet Insulation in Roofing Systems Using Infrared Imaging*, American Society for Testing and Materials, Philadelphia, 1990.
2. *1993 Fundamentals Handbook* (I-P ed.), American Society of Heating, Refrigerating, and Air-Conditioning Engineers, Atlanta, 1993, chap. 27, tables 12–18.
3. Ibid., chap. 3, table 3.
4. Ibid., chap. 3.
5. Perry, Robert H., et al., *Chemical Engineer's Handbook*, 4th ed., McGraw-Hill, New York, 1963, chap. 10.

Chapter 14

Transient Heating of a Painted Vehicle Body Panel in an Automobile Assembly Plant Paint Shop Oven

Thomas M. Lawrence
Department of Biological and Agricultural Engineering
Driftmier Engineering Center
University of Georgia
Athens, Georgia

Scott Adams
Paint Engineering
Ford Motor Company
Vehicle Operations General Office
Allen Park, Michigan

Introduction

Production of automobiles, light trucks, and sport utility vehicles currently (as of 2005) averages around 12 million vehicles per year in the United States. Each of these vehicles undergoes a painting process that involves a series of paint application steps and paint drying-curing ovens. Since automotive assembly plants are automated systems with continuous-motion carriers, the ovens are designed as long enclosed spaces that control the heating process in both duration and temperature. Quality control concerns require strict control of the heating rate,

14.2 Transient and Cyclic Calculations

maximum temperature exposure, and time duration for exposure above the minimum cure temperature. Since a typical vehicle body is a complex shaped structure with varying part thicknesses and orientations, it is a difficult job to ensure that each portion of the body is exposed to heating conditions that will meet the strict quality control standards. Each assembly plant paint shop will regularly send instrumented vehicle bodies or similar devices to determine the actual air and body temperatures at key points over time while the body travels through an oven.

A typical automotive plant paint oven will be 100 to 130 m (300 to 400 ft) in length. The goals of this process heating are first to bring the painted vehicle body up from room temperature to a paint cure point of around 150 to 160°C (300 to 325°F), and then hold the body at the cure temperature for a period of time ranging from approximately 15 to 25 min. A quality curing job is achieved if the body is exposed to a combination of time at temperature that meets the paint manufacturer's criteria. Because of the wide range in thickness of a vehicle body part, orientation (horizontal, vertical), distance from the oven walls and ceiling, and the potential for blockage of exposure from other parts of the body, it is a difficult task to ensure that all portions of the painted surfaces reach the proper cure temperature and time.

There is potential for dirt being trapped in the wet surface when the freshly painted body first enters the oven. Therefore, a typical oven will first heat the painted vehicle body in zones with minimal air movement to avoid stirring up any dirt in the area. These heatup zones use radiant heating from either infrared heating lamps or radiant panels that are heated from the backside with hot air. Once the body has been in the oven for a while and the outer surface of the paint has dried such that the risk of dirt entrapment is minimal, the oven will typically employ convection—or a combination of both radiant and convection—heating.

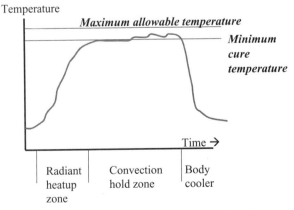

Figure 14.1 Plot of typical temperature versus time for representative point on a painted vehicle body.

TABLE 14.1 Typical Paint Shop Oven Specifications

Zone*	1	2	3	4	5	6
Length, m (ft)	12.2 (40)	12.2 (40)	12.2 (40)	15.2 (50)	18.3 (60)	21.3 (70)
Temperature, °C (°F)	107 (225)	171 (340)	160 (320)	154 (310)	152 (305)	152 (305)
Time, min	4	4	4	5	6	7
Burner capacity, kW (10^6 Btu/h)	879 (3)	879 (3)	879 (3)	732 (2.5)	732 (2.5)	732 (2.5)
Airflow, m³/s (cfm)	14.16 (30,000)	14.16 (30,000)	14.16 (30,000)	16.52 (35,000)	14.16 (30,000)	14.16 (30,000)

*Zones 1 to 3 are radiant heatup; zones 4 to 6 are convection hold zones.

A representative plot of the temperature versus time (and hence distance traveled in the oven) is given in Fig. 14.1. Key technical specifications for a typical oven are listed as an example in Table 14.1.

A modern car or truck body is a complex structure with a combination of thin and thicker metal elements. The orientation of a particular location on the vehicle body can greatly influence the heating exposure and hence the resulting paint temperature. Thus, there are a myriad of potential variables and combinations that must be accounted for when attempting to analyze the transient temperatures for a typical body as it passes through an oven. The paint shop will also occasionally send a carrier loaded with replacement parts through the painting booth and curing oven; for example, see the parts carrier shown in Fig. 14.2. These

Figure 14.2 Photo of typical spare-parts carrier.

14.4 Transient and Cyclic Calculations

parts carriers are exposed to the same temperatures as the regular vehicle bodies since they pass through the oven on the same conveyors.

Example Description

Metal thicknesses and heat-transfer effects vary by location on a vehicle body. As a relatively simple example, this problem will analyze the heatup of a painted door part after it first enters the oven and passes through the initial radiant heatup zones. The problem will consider how a simplified transient heat-transfer analysis can be applied to the lower section of a painted part below the window cutout area shown in Fig. 14.2. This section is essentially vertically oriented on the parts carrier. A simplified cross section of the oven in this example is given in Fig. 14.3. Tables 14.2 and 14.3, along with Fig. 14.4, contain a summary of the key parameters and assumptions made for this analysis.

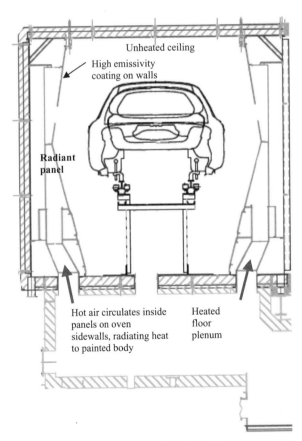

Figure 14.3 Simplified cross section of typical oven zone using radiant heat.

Transient Heating of a Painted Vehicle Body Panel

TABLE 14.2 Assumptions and Design Conditions Used in Example Analysis

Parameter	Metric	English
Oven air temperature, zone 1	107°C	225°F
Oven air temperature, zone 2	171°C	340°F
Oven air temperature, zone 3	160°C	320°F
Air temperature behind radiant panel, zone 1	152°C	305°F
Air temperature behind radiant panel, zone 2	216°C	420°F
Air temperature behind radiant panel, zone 3	204°C	400°F
Estimated radiant panel wall temperature, zone 1	131°C	268°F
Estimated radiant panel wall temperature, zone 2	194°C	381°F
Estimated radiant panel wall temperature, zone 3	183°C	361°F
Residency time in each zone, min	4 min	4 min
Radiant panel to zone air convection coefficient	5.1 W/m²·K	0.9 Btu/h·ft²·F
Radiant panel wall thickness	3 mm	1/8 in.
Radiant panel wall conductivity	53.5 W/m·K	30.9 Btu/h·ft·F
Length of each zone	12.2 m	40 ft
Airflow behind radiant panel	14.16 m³/s	30,000 cfm
Initial body temperature	27°C	80°F
Vehicle body material properties: thermal conductivity k	60 W/m·K	34.7 Btu/h·ft·F
Vehicle body material properties: thermal diffusivity α	1.7×10^{-5} m²/s	0.66 ft²/h
Painted part panel thickness	2.4 mm	3/32 in.
Painted part thermal emissivity	0.9	0.9
Painted part thermal conductivity	53.5 W/m·K	30.9 Btu/h·ft·F
Painted part specific heat	0.485 kJ/kg·K	0.12 Btu/lb$_m$·F
Painted part material density	7850 kg/m³	490 lb$_m$/ft³

This example will estimate the transient heating of the painted panel as it proceeds through the first three radiant heating zones of a typical oven. The residency time of the panel in each zone is 4 min, so the total time for this transient simulation will be 12 min.

The heat-transfer paths for this problem are summarized graphically in Fig. 14.5. The problem analysis consists of two basic steps:

1. Determination of the heat-transfer characteristics, rate equations, and reasonable assumptions for the paths shown in Fig. 14.5.

TABLE 14.3 Thermal Properties of Heated Air Inside Radiant Panel, Zone 3

Parameter	Metric	English
Kinematic viscosity ν	3.55×10^{-5} m²/s	3.77×10^{-4} ft²/s
Prandtl number Pr	0.697	0.697
Thermal conductivity k	0.03805 W/m-K	0.022 Btu/h-ft-°R

14.6 Transient and Cyclic Calculations

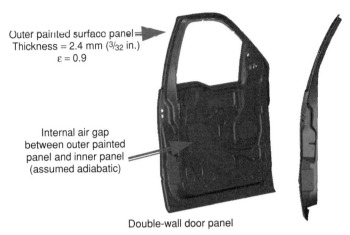

Outer painted surface panel
Thickness = 2.4 mm (3/32 in.)
$\varepsilon = 0.9$

Internal air gap between outer painted panel and inner panel (assumed adiabatic)

Double-wall door panel

Figure 14.4 Painted door panel dimensions and key parameters.

2. Combining these into a transient analysis model to simulate the painted part temperature during the initial heating.

The heat-transfer characteristics, coefficients, overall rate equations, and assumptions will be analyzed individually in the following sections.

Heat Transfer on Panel Inside Wall Surface

Heat transfer on the inside surface of the radiant panel is between the heated air supplied by the oven burner and the inner panel wall. The first step is to estimate the convection heat-transfer coefficient. To do so, the airflow velocity must be determined.

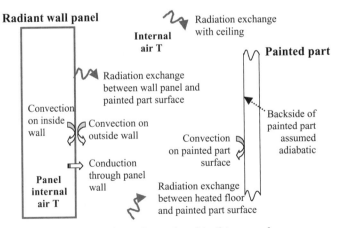

Figure 14.5 Heat-transfer paths analyzed in this example.

From the information in Table 14.2, a total airflow of 14.16 m³/s [30,000 ft³/min (cfm)] is provided to each radiant zone. The zones are 12.2 m (40 ft) in length, with radiant panels (and thus airflow) on both sides of the oven. Therefore, the total airflow per unit panel length along the oven is

$$\dot{\text{Vol}} = \frac{\text{total zone flow}/2}{\text{zone length}} = \frac{14.16 \text{ m}^3/\text{s}}{2 \cdot (12.2 \text{ m})} = 0.58 \frac{\text{m}^3/\text{s}}{\text{m}} \left(375 \frac{\text{cfm}}{\text{ft}}\right) \quad (14.1)$$

Various airflow ducting designs may be employed at different assembly plants. For this example, we will consider a case where the radiant panels are subdivided into sections every 1.5 m (5 ft) of length along the oven path. A diagram of this duct section and dimensions is shown in Fig. 14.6. Therefore, the total airflow in each section is 1.5 × 0.58 = 0.88 m³/s (1875 cfm).

On the basis of the heated air temperature inside the radiant panel in zone 3 of 204°C (400°F), the air properties listed in Table 14.3 were determined. A similar analysis process can be followed to compute the corresponding heat-transfer characteristics for zones 1 and 2. The next step is to determine the Reynolds number for airflow inside the radiant panel. Since the flow path is a rectangular duct, the Reynolds number will be based on the hydraulic diameter, or

$$D_h = \frac{4 \cdot A}{P} = \frac{4 \cdot 1.5 \text{ m} \cdot 0.3 \text{ m}}{3.6 \text{ m}} = 0.5 \text{ m} (1.65 \text{ ft}) \quad (14.2)$$

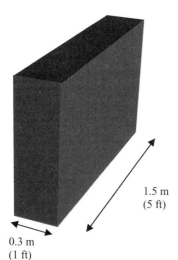

Figure 14.6 Radiant panel dimensions.

14.8 Transient and Cyclic Calculations

The airflow velocity inside the radiant panel ducting is

$$V = \frac{\text{total airflow}}{A} = \frac{0.88 \text{ m}^3/\text{s}}{(1.5 \text{ m})(0.3 \text{ m})} = 1.96 \text{ m/s } (6.25 \text{ ft/s}) \quad (14.3)$$

and the resulting Reynolds number is

$$\text{Re}_{D_h} = \frac{VD_h}{\nu} = \frac{(1.96 \text{ m/s}) \, 0.5 \text{ m}}{3.55 \times 10^{-5} \text{ m}^2/\text{s}} = 27{,}606 \quad (14.4)$$

Thus, the heated airflow inside the radiant panel is turbulent since the Reynolds number exceeds the generally accepted transition level of around 2300 for internal flow.

The convective heat-transfer coefficient on the internal panel walls can be estimated using the Dittus-Boelter equation [Eq. (8.60) in Incropera and DeWitt (2002)] with the airflow being "cooled." This equation states that

$$\text{Nu}_{D_h} = 0.023 \, \text{Re}_{D_h}^{4/5} \, \text{Pr}^n \quad \text{where} \quad n = 0.3 \quad (14.5)$$

Applying Eq. (14.5) to this problem gives an estimate of the Nusselt number of

$$\text{Nu}_{D_h} = \frac{hD_h}{k} = 0.023(27{,}606)^{4/5}(0.697)^{0.3} = 73.7 \quad (14.6)$$

with a resulting convection coefficient

$$h_i = \frac{\text{Nu}_{D_h} k}{D_h} = \frac{73.7 \cdot (0.03805 \text{ W/m} \cdot \text{K})}{0.5 \text{ m}}$$

$$= 5.6 \text{ W/m}^2 \cdot \text{K } [1.0 \text{ Btu/h} \cdot \text{ft}^2 \cdot {}^\circ\text{R}] \quad (14.7)$$

Convective Heat Transfer on Outer Wall Surface

Estimating the convection heat-transfer coefficient on the outer surface of the radiant panel wall (facing the oven interior) is a difficult process. The air inside the oven, although ideally "still," will have motion due to the movement of vehicle bodies inside the oven, slight pressure differences between zones, temperature gradients, and other factors. However, one can safely assume that the convection heat-transfer coefficient on the outer wall is less than on the inside surface. The convective heat-transfer rate on the outer radiant panel wall would also be expected to be slightly higher than for purely free (natural) convection; thus bounds

Transient Heating of a Painted Vehicle Body Panel

for the convection coefficient range are known. This example will assume a convection coefficient value of $h_o = 5.1$ W/m²-K (0.9 Btu/h-ft²-°R) for the outer panel wall.

Heat Transfer through Panel Wall

The conduction resistance through the panel wall is computed by

$$R = \frac{L}{kA} = \frac{0.0015 \text{ m}}{(53.5 \text{ W/m} \cdot \text{K}) 1 \text{ m}^2} = 2.8 \times 10^{-5} \text{ K/W } (3.4 \times 10^{-4} \text{ h-°R/Btu})$$

(14.8)

Therefore, the resistance to heat flow through the radiant panel wall is negligible compared to the convective heat transfer on both the inner and outer wall surfaces.

Estimation of Radiant Panel Outer Wall Surface Temperature

The two convection coefficients can be combined to obtain the overall thermal resistance for heat transfer between the heated air inside the radiant panel and the oven air, as shown below based on a unit panel area:

$$R_t = \frac{1}{\frac{1}{h_i A} + \frac{1}{h_o A}} = \frac{1}{\frac{1 \text{ m}^2 \cdot \text{K}}{(5.6 \text{ W}) 1 \text{ m}^2} + \frac{1 \text{ m}^2 \cdot \text{K}}{(5.1 \text{ W}) 1 \text{ m}^2}}$$

$$= 2.7 \text{ K/W } (0.47 \text{ h-°R/Btu})$$

(14.9)

An estimate of the outer panel wall temperature is necessary for use in calculation of the radiation heat transfer to the painted part panel for the transient modeling. Heat transfer on the internal wall from the heated air supply will be lost on the outer wall by both convection and radiation. Computing the energy lost through radiation is a difficult task to simplify, since it depends on the effective surrounding temperature as viewed from the outer panel wall. The effective temperature of the surroundings oscillates as the painted panels or vehicle bodies roll past through the oven and down the length of the oven as the panel heats up. For a good portion of time, the radiant panel views the panel on the opposite side of the oven at essentially the same temperature. Therefore, for this simplified analysis to estimate the panel outer wall temperature, we will first ignore the effect of radiation.

The wall temperature is first estimated using the conditions in zone 3, where the oven air temperature is 160°C, as noted in Table 14.3. The

14.10 Transient and Cyclic Calculations

outer panel wall temperature can be estimated from the following relation

$$Q_{\text{to wall}} = Q_{\text{from wall}}$$

$$h_i A(T_{\text{heated air}} - T_{\text{wall}}) = h_o A(T_{\text{wall}} - T_{\text{oven}}) \quad (14.10)$$

$$5.6 \frac{W}{m^2 K}(1\ m^2)(204°C - T_{\text{wall}}) = 5.1 \frac{W}{m^2 K}(1\ m^2)(T_{\text{wall}} - 160°C)$$

or

$$T_{\text{wall}} = 183°C\ (361°F) \quad (14.11)$$

Using a similar analysis, the radiant panel wall temperatures in zones 1 and 2 are estimated to be 131°C (268°F) and 194°C (381°F), respectively. Table 14.3 includes a listing of the estimated radiant wall panel temperatures for the three zones.

View Factors for Radiation Heat Transfer with Vertical Painted Panel

From an examination of the diagram in Figs. 14.3 and 14.5, three potential radiation heat-transfer paths exist with a vertically mounted painted part. These are with the oven radiant wall panels, plus the floor and ceiling. For the view factor to the floor, the panel "sees" mostly the section of floor that is heated, and it can be assumed that the entire floor viewed is heated without introducing any significant error. The oven is also much longer than one individual painted part; therefore, the view factor relationship between the painted part and the oven wall, ceiling, and floor can be considered to be two-dimensional. The typical oven cross section given in Fig. 14.3 shows radiant panel walls that are angled slightly to provide better radiation heat-transfer exposure for the painted body surfaces. In computing the radiation view factors, the radiant panel walls can be considered flat vertical plates facing the painted part as it moves through the oven, as shown in Fig. 14.7.

The view factor between the vertical section of interest on the painted part and the unheated oven ceiling can be determined from the view factor relations for perpendicular plates with a common edge, as shown in Fig. 14.7b. The view factor relation is as follows:

$$F_{1 \rightarrow 3} = F_{(1+2) \rightarrow 3} - F_{2 \rightarrow 3} \quad (14.12)$$

The two-dimensional view factor for perpendicular plates with a common edge is found using the method outlined in sec. 13.1 of Incropera and DeWitt (2002):

$$F_{i \rightarrow j} = \frac{1 + (w_i/w_j) - \sqrt{1 + (w_i/w_j)^2}}{2} \quad (14.13)$$

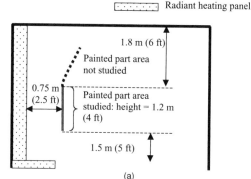

(a)

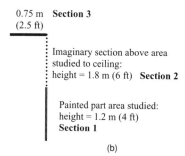

(b)

Figure 14.7 Radiation heat-transfer view factor estimation (not drawn to scale): (a) dimensions; (b) panel ceiling view factor determination.

In this equation, w_j and w_i are the lengths of sides i and j, respectively. Using the information in Fig. 14.7b, the view factors are computed as follows:

$$F_{(1+2) \to 3} = \frac{1 + (3.0/0.75) - \sqrt{1 + (3.0/0.75)^2}}{2} = 0.44 \quad (14.14)$$

$$F_{2 \to 3} = \frac{1 + (1.8/0.75) - \sqrt{1 + (1.8/0.75)^2}}{2} = 0.40 \quad (14.15)$$

$$F_{\text{panel} \to \text{ceiling}} = F_{1 \to 3} = F_{(1+2) \to 3} - F_{2 \to 3} = 0.44 - 0.40 = 0.04 \quad (14.16)$$

Radiation Heat-Transfer Calculations

If the heated floor and sidewall panel surfaces are assumed to be at the same temperature (the wall temperatures computed above), they can be lumped together in the radiation heat-transfer analysis, and this is the method chosen for this simple example. Thus, the radiation heat transfer to the painted panel in the example can be summarized by

$$Q_{\text{radiation}} = Q_{\text{rad, wall}} + Q_{\text{rad, floor}} + Q_{\text{rad, ceiling}} \quad (14.17)$$

14.12 Transient and Cyclic Calculations

One must next decide how to treat the radiation heat-transfer exchange between the painted part and the oven wall surfaces. Optimal heat transfer will be obtained when the oven surfaces are as near a blackbody as possible and are designed as such. The walls can be assumed to be essentially a blackbody for the purposes of this example, thus simplifying the analysis so that the net radiation heat transfer to the panel is calculated by

$$Q_{\text{rad, ceiling}} = \sigma F_{\text{panel-ceiling}} \, \varepsilon_{\text{panel}} \, A_{\text{panel}} \left(T_{\text{ceiling}}^4 - T_{\text{panel}}^4 \right) \quad (14.18)$$

$$Q_{\text{rad, wall}} + Q_{\text{rad, floor}} = \sigma (1 - F_{\text{panel-ceiling}}) \varepsilon_{\text{panel}} \, A_{\text{panel}} \left(T_{\text{wall}}^4 - T_{\text{panel}}^4 \right) \quad (14.19)$$

A more detailed heat-transfer analysis would consider this situation as an enclosure with multiple diffuse, gray surfaces, using the known (or estimated) emissivity values for the painted part, walls, ceiling, and floor surfaces to compute the net radiation exchange with the painted part.

Convection Heat Transfer to Painted Part

The other mode of heat transfer involved is convection between the painted part and the oven air. In the radiant heating zone, air movement is not desired so as to avoid potential dirt contamination in the wet paint. There will be a small amount of air motion across the painted surface due to the movement of the painted part carrier through the oven and various pressure imbalances that may exist. The convective heat-transfer coefficient will be relatively small, and the assumed value used for the outer radiant panel wall ($h = 5.1$ W/m^2-K) should also be a reasonable assumption for the painted part surface. The convective heat transfer to be used in the example calculation is given in the following equation:

$$Q_{\text{convection}} = h A_{\text{panel}} (T_{\text{oven}} - T_{\text{panel}}) \quad (14.20)$$

Transient Panel Temperature Analysis

The transient modeling will track the panel temperature starting from ambient during the 12 min it needs to pass through the heatup section. The painted panel is actually a double-wall structure, with the backside of the panel assumed to be an adiabatic surface (see Fig. 14.4). Recognize that there will be some heat transfer through the panel inner wall and into the air gap, but this assumption simplifies greatly the rough level calculations in this example.

Transient Heating of a Painted Vehicle Body Panel

A check should be made to see if the panel can be treated at a uniform temperature through the thickness of the part. Since this panel is composed of relatively thin [2.4 mm (3/32 in.)] material over much of the part, this would be a reasonable expectation. To check this, calculate the Biot number, which is used to determine whether a lumped-capacitance transient heat transfer is applicable. The Biot number is calculated using

$$\text{Bi} = \frac{hL_c}{k} \qquad (14.21)$$

where L_c is the characteristic length. The characteristic length used should be the entire thickness of the painted metal (2.4 mm), since the backside of the painted surface is assumed to be adiabatic in this analysis (see Fig. 14.4). The Biot number is therefore calculated as

$$\text{Bi} = \frac{(5.1 \text{ W/m}^2 \cdot \text{K})0.0024 \text{ m}}{53.5 \text{ W/m} \cdot \text{K}} = 2.3 \times 10^{-4} \qquad (14.22)$$

Since the Biot number is much less than 0.1, it is proper to assume that the temperature distribution through the thickness of the metal can be considered approximately uniform.

Calculation of the transient temperature will be conducted for this example using an explicit (time-marching) process with time steps of 30 s. The choice of time step made was for convenience, but the engineer should always check for potential stability problems in an explicit numerical solution. For example, sec. 5.9 in Ref. 1 describes the stability criteria for a finite-difference method solution.

The basic equation used for the transient simulation analysis is net energy into panel = rate of change of panel temperature, or

$$Q_{\text{net}} = mC_p \frac{\Delta T}{\Delta t} \qquad (14.23)$$

$$Q_{\text{rad, wall}} + Q_{\text{rad, floor}} + Q_{\text{rad, ceiling}} + Q_{\text{convection}} = mC_p \frac{T^{n+1} - T^n}{\text{time step, s}}$$

$$T^{n+1} = T^n + \frac{\text{time step}}{mC_p}(Q_{\text{rad, wall}} + Q_{\text{rad, floor}} + Q_{\text{rad, ceiling}} + Q_{\text{convection}})$$

$$(14.24)$$

where T^{n+1} = panel temperature at end of time step
T^n = panel temperature at beginning of time step
m = panel mass = $\rho V = \rho A \cdot$ thickness

14.14 Transient and Cyclic Calculations

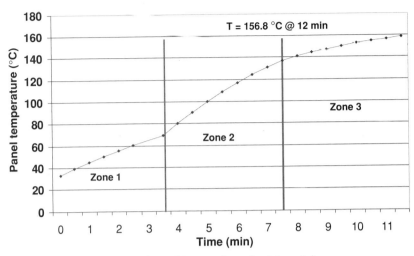

Figure 14.8 Predicted painted panel temperatures (metric units).

Equations (14.18) to (14.20) and (14.24) were incorporated into a spreadsheet solution of the estimated painted panel temperature over time as it passes through the first three radiant heating zones. Calculation results for the temperatures are given in Figs. 14.8 and 14.9 for the simulation in metric [International System (SI)] and English units, respectively. The temperature values differ slightly because of the rounding differences in the unit conversions for heat-transfer coefficients, panel thickness, and other variables. A shift in the temperature trends occurs as the painted part enters the next zone at the 4- and 8-min marks. Figures 14.10 and 14.11 show a breakdown between

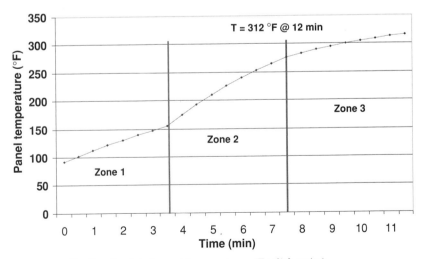

Figure 14.9 Predicted painted panel temperatures (English units).

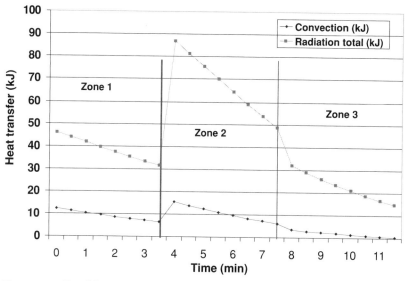

Figure 14.10 Panel heat-transfer breakdown (metric units).

convection and total radiation heat transfer to the panel over time for the metric and English units calculations, respectively.

Summary

This example problem used a typical industrial scenario to demonstrate the application of convection and radiation heat transfer. Several

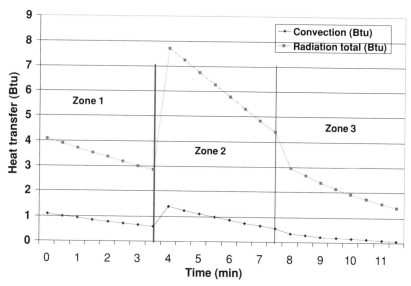

Figure 14.11 Panel heat-transfer breakdown (English units).

14.16 Transient and Cyclic Calculations

simplifying assumptions, and the logic behind these choices, were made to keep the analysis to a reasonable level of effort. The estimated panel temperatures are considered only initial guesses, and a more detailed modeling effort would be required to obtain more accurate results.

Nomenclature

A	Area
Bi	Biot number
C_p	Specific heat
D_h	Hydraulic diameter
h	Convection heat-transfer coefficient
F	View factor
k	Thermal conductivity
L_c	Characteristic length
m	Mass
Nu	Nusselt number
P	Perimeter
Q	Heat-transfer rate
R	Thermal resistance
Re	Reynolds number
t	Time
T	Temperature
V	Air velocity
$\dot{V}ol$	Volumetric airflow per unit length
w	Length of side in view factor calculation

Greek

Δ	Delta (change in value)
ε	Thermal emissivity
σ	Stephan-Boltzmann constant

Reference

Incropera, F. P., and DeWitt, D. P. *Fundamentals of Heat and Mass Transfer*, 5th ed. Wiley, New York, 2002.

Chapter 15

Thermal System Transient Response

James E. Marthinuss, Jr., and George Hall
Northrop Grumman Electronics Systems
Linthicum, Maryland

Introduction

The focus of this chapter will be transient heat-transfer calculations of an entire electronics-cooling system. It will essentially work from the small to the large aspects in transient calculations, first starting with the heat-dissipating parts, then with how the parts are packaged and tied to a heat sink, continuing with how to handle a coldplate transient, and then finishing up with a total system transient response. Topics to be discussed will include when to apply lumped capacitance to a transient calculation; how to use a detailed steady-state analysis to perform a simple transient analysis; and finally when and how to approach transient heat-transfer calculations, how to calculate a time constant, and how to properly apply a time constant to obtain a temperature response of a thermal system. Simple approximations will be made to solve complex transients without the aid of a finite-difference solver. Practical solutions to transient thermal problems will be discussed in detail.

A typical electronics-cooling system consists of electrical components generating heat, cooled by conduction across a board to a rack. The rack in turn is cooled by fluid flowing through it. It is desirable to know not only whether the components will remain cool enough to operate continuously under normal conditions but also if they can operate during transient conditions. Transient thermal events include power surges, coolant flow interruptions, and spikes in coolant temperature.

15.2 Transient and Cyclic Calculations

Problem Definition

Given a thermal system, how to determine the response to a transient condition

The modern, brute-force approach to analyzing any thermal system is to create a detailed all-encompassing automeshed finite-difference analysis/finite-element analysis (FDA/FEA) thermal model, input the transient event, and let the computer grind away. Depending on the computational resources available and with a little bit of luck in obtaining a proper mesh, a response can be predicted. But as with most computer modeling, a little bit of insight into the problem can easily offset deficiencies in computational power and the need for luck.

A more thoughtful approach is to quantify the approximate transient characteristics of a thermal system and then decide how and what to model, considering the transient event. This can be accomplished by determining the approximate transient characteristics of each part of the system and then determining how each part of the system will respond to the transient event. Once the general nature of the system's response is understood, detailed computer modeling can be devoted to the transient-critical sections of the system as necessary.

Typical electronics cooling

An electronics system consisting of numerous parts will be analyzed over several transient temperature profiles. Using this as an example, it will be easy to understand the important factors when performing a transient heat-transfer calculation that can be applied to any system. The basic heat path will be described in detail. We will start with the electronics, where the heat is dissipated which consists of commercial

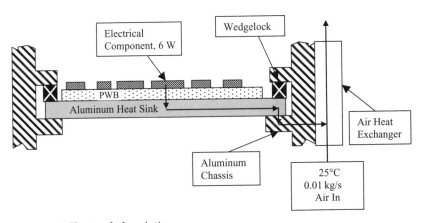

Figure 15.1 Heat path description.

TABLE 15.1 Description of Thermal System

Description	Width, cm	Depth, cm	Thermal path length, cm	Density, kg/cm³	Specific heat, W-s/kg-°C	Conductivity, W/cm-°C
Electrical component	2.5	2.5	0.125	0.0014	1200	$\Theta_{jc} = 2°C/W$
Airgap under component	2.5	2.5	0.05	1.15×10^{-6}	1090	0.0003
Electrical leads from component to PWB (40 leads)	0.025	0.0125	0.1	0.00785	455	2.87
PWB	2.5	2.5	0.15	0.0015	1100	0.0078
Heat sink	2.5	0.13	7.5	0.003	1050	1.7
Wedgelock	5	0.5	N/A	N/A	N/A	$R = 3.2$ cm²-°C/W
Chassis	1.0	2.5	1.2	0.003	1050	1.7
Air heat exchanger	1.0	3.0	5	N/A	N/A	N/A

electronics mounted to a printed wiring board (PWB). This PWB is then attached to a heat sink; the heat sink is then attached to a chassis sidewall using a wedgelock that provides a pressure contact to the chassis coldplate heat exchanger. This chassis coldplate heat exchanger is then cooled with forced air. A schematic showing the heat path as described above is shown in Fig. 15.1. The details of the system are listed in Table 15.1 and shown in Fig. 15.2.

Transient Characteristics

Governing equations

The governing equation in thermal systems is the first law of thermodynamics, which states that at any time the total energy in a system is conserved. For typical electronic cooling systems this can be written as[1]

$$\sum \dot{Q}_i = \sum \dot{Q}_{i \text{ conducted out}} + \dot{Q}_{i \text{ stored}} = \dot{Q}_{i \text{ impressed}}$$

For materials with constant thermal conductivies defined by the constitutive equation[1]

$$R_{12} = \frac{T_1 - T_2}{\dot{Q}_{12 \text{ conducted}}}$$

15.4 Transient and Cyclic Calculations

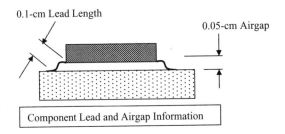

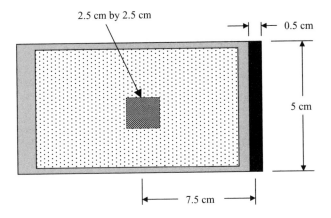

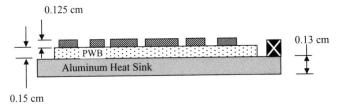

Figure 15.2 Dimensions.

and constant specific heats defined by[1]

$$C_p = \frac{\dot{Q}_{\text{stored}} \Delta t}{\rho V \Delta T}$$

Then the energy balance can be written as

$$\sum \dot{Q}_i = \sum_j \frac{T_i - T_j}{R_{ij}} + \rho_i V_i C_{pi} \frac{dT_i}{dt} = \dot{Q}_{i \text{ impressed}}$$

In order to solve for the temperature of each part of a system over time $T_i(t)$, it is first necessary to determine the resistances between parts and the capacitance of each part.

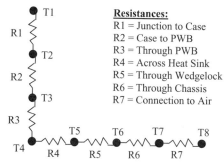

Figure 15.3 Resistor network.

Characterizing a Thermal System

Determine resistances

In order to calculate the transient response of the system, the thermal resistances must be calculated from the heat source to the boundary sink. To fully understand exactly what needs to be calculated, a simple resistor network is shown in Fig. 15.3. All of these resistances will be expressed in terms of temperature rise per power dissipated or °C/W.

The part manufacturer provided the resistance through the electronic part from junction (hottest spot on the part) to the case of the part as 2°C/W. This thermal resistance will be used for R_1. Next we must calculate the thermal resistance from the case of the part to the PWB. This part is not bonded to the PWB, but it has 40 electrical leads that connect the case of the part to the board. Knowing that the leads are made of copper and have a cross section of 0.025 × 0.0125 cm and are 0.1 cm long, we can calculate this thermal resistance as[1]

$$R = \frac{L}{kA}$$

where R = resistance from case to PWB, °C/W
L = length of the leads 0.1, cm
k = conductivity of copper, W/cm-°C
A = area, cm^2

Since there are 40 leads, which are 0.025 × 0.0125 cm, the total area of the leads can be calculated as

$$A = 0.025 \text{ cm} \times 0.0125 \text{ cm} \times (40 \text{ leads})$$
$$= 0.0125 \text{ cm}^2$$

15.6 Transient and Cyclic Calculations

Now that the area is known, it can be used as input to the resistance equation:

$$R_a = \frac{0.1 \text{ cm}}{(2.87 \text{ W/cm-}°\text{C}) \times (0.0125 \text{ cm}^2)}$$

$$= 2.79°\text{C/W}$$

This resistance is only for the leads. The thermal resistance for the airgap must also be included. Using the resistance equation and inputting the proper area of the part of 2.5×2.5 cm and conductivity of 0.0003 W/cm-°C for the airgap, the thermal resistance of the 0.05-cm-long airgap can be calculated:

$$R_b = \frac{0.05 \text{ cm}}{(0.0003 \text{ W/cm-}°\text{C}) \times (6.25 \text{ cm}^2)}$$

$$= 26.7°\text{C/W}$$

The resistance through the leads and the resistance through the air are parallel resistances, so these resistances need to be added up in parallel using the following equation:

$$R_2 = \frac{1}{\left(\frac{1}{R_a} + \frac{1}{R_b}\right)} = \frac{1}{\left(\frac{1}{2.79°\text{C/W}}\right) + \left(\frac{1}{26.7°\text{C/W}}\right)}$$

$$= 2.5°\text{C/W}$$

Even though air has an extremely low conductivity, it still lowers the overall resistance from case to PWB by roughly 10 percent. This is due mainly to the short distance between the case and the PWB. The thermal resistance calculation that was just done is for R_2. This same process needs to be done for the thermal resistance through the printed wiring board (PWB), which is R_3. This resistance is calculated by using the length of the heat path, which is the thickness of the PWB at 0.15 cm. The area used is the footprint of the electronic part, which is 6.25 cm², and the conductivity is 0.0078 W/cm-°C. Using the resistance equation once again, the thermal resistances can be calculated:

$$R_3 = \frac{0.15 \text{ cm}}{(0.0078 \text{ W/cm-}°\text{C}) \times (6.25 \text{ cm}^2)}$$

$$= 3.08°\text{C/W}$$

The next thermal resistance that must be calculated is the resistance across the aluminum heat sink, which is R_4. This resistance is calculated by using the length of the heat path, from the center of the heat sink to the edge of the heat sink, which is 7.5 cm. The area used is just the thickness of the aluminum, which is 0.13 cm, and this is multiplied by the electronic part width of 2.5 cm. The conductivity of aluminum at 1.7 W/cm-°C is also needed in order to solve the thermal resistance equation:

$$R_4 = \frac{7.5 \text{ cm}}{(1.7 \text{ W/cm-°C}) \times (0.13 \text{ cm} \times 2.5 \text{ cm})}$$

$$= 13.57°\text{C/W}$$

Next the thermal resistance through the wedgelock interface will be calculated. This thermal resistance is calculated by using an interface thermal resistance, which is 3.2°C-cm^2/W. The interface resistance is based on measured results and can be obtained from many different sources. This area resistance is divided by the interface area, which is 5 cm long by 0.5 cm wide:

$$R_5 = (3.2°\text{C-cm}^2/\text{W})/(5.0 \text{ cm} \times 0.5 \text{ cm})$$

$$= 1.28°\text{C/W}$$

The conduction path to the air must be calculated, which is R_6. This resistance is calculated by using the length of the heat path, from the wedgelock interface to the air heat exchanger face, which is 1.2 cm. The area used is just the cross-sectional area of the aluminum, which is 1.0 × 2.5 cm. Using the resistance equation and substituting the proper conductivity, length, and area, this resistance can be calculated:

$$R_6 = \frac{1.2 \text{ cm}}{(1.7 \text{ W/cm-°C}) \times (1.0 \text{ cm} \times 2.5 \text{ cm})}$$

$$= 0.28°\text{C/W}$$

The last thermal resistance to be calculated is for the convective heat transfer to the air in the heat exchanger, which is R_7 (Ref. 1):

$$R_7 = \frac{1}{hA}$$

where h = heat-transfer coefficient, W/cm^2-°C
A = heat exchanger surface area, cm^2

In order to calculate the heat-transfer coefficient in the heat exchanger, the Reynolds number must be calculated to determine whether the flow is turbulent or laminar. To calculate the Reynolds number, the hydraulic diameter must first be calculated on the 1.0 × 3.0 cm heat exchanger

15.8 Transient and Cyclic Calculations

duct as follows[1]:

$$D_h = \frac{4A}{P}$$

where D_h = hydraulic diameter, cm
A = area of passage, cm^2
P = wetted perimeter of passage, cm

Then

$$D_h = \frac{4 \times (1\,\text{cm} \times 3\,\text{cm})}{2 \times (1\,\text{cm} + 3\,\text{cm})}$$

$$= 1.5\,\text{cm}$$

The airflow velocity must be calculated on the basis of a mass flow rate of 0.01 kg/s and using air fluid properties at a temperature of 25°C. The density of air at 25°C is 1.15×10^{-6} kg/cm^3. The velocity can be calculated from the duct size of 1×3 cm by using the following equation[1]:

$$V_a = \frac{\dot{m}}{\rho A}$$

where V_a = air velocity, cm/s
$\dot{m}$ = mass flow rate, kg/s
ρ = density of air, kg/cm^3
A = area of air passage, cm^2

Then

$$V_a = \frac{0.01\,\text{kg/s}}{(1.15 \times 10^{-6}\,\text{kg/cm}^3) \times (1\,\text{cm} \times 3\,\text{cm})}$$

$$= 2898.6\,\text{cm/s}$$

Now the Reynolds number can be calculated from the following equation[1] by inputting the velocity of 2898.6 cm/s, the hydraulic diameter of 1.5 cm, and the kinematic viscosity of air of 0.156 cm^2/s:

$$\text{Re} = \frac{V_a D_h}{\nu}$$

where Re = Reynolds number
V_a = air velocity, cm/s
D_h = hydraulic diameter, cm
ν = kinematic viscosity of air, cm^2/s

Then

$$\text{Re} = \frac{(2898.6 \text{ cm/s}) \times (1.5 \text{ cm})}{0.156 \text{ cm}^2/\text{s}} = 27{,}871$$

This Reynolds number is turbulent because it is well above the transition Reynolds number of 2300, so we can now calculate the Nusselt number using the turbulent flow equation below. The Reynolds number of 27,871 and the Prandtl number for air at 25°C of 0.7 are input into the following equation[1]:

$$\text{Nu} = 0.023 \, \text{Re}^{4/5} \, \text{Pr}^{1/3}$$

where Nu = Nusselt number
Re = Reynolds number
Pr = Prandtl number for air

Then

$$\text{Nu} = 0.023 \times (27{,}871)^{4/5}(0.7)^{1/3} = 73.5$$

Once the Nusselt number is calculated, the heat-transfer coefficient can be calculated from the following equation,[1] using the Nusselt number of 73.5, the hydraulic diameter of 1.5 cm, and the conductivity of 0.0003 W/cm-°C:

$$h = \frac{\text{Nu} \, k}{D_h}$$

where h = heat-transfer coefficient, W/cm²-°C
Nu = Nusselt number
k = conductivity for air 0.0003, W/cm-°C
D_h = hydraulic diameter, cm

Then,

$$h = \frac{73.5 \times 0.0003 \text{ W/cm-°C}}{1.5 \text{ cm}} = 0.0147 \text{ W/cm}^2\text{-°C}$$

This is the heat-transfer coefficient, which can now be used as input to the equation below to calculate the resistance of the heat exchanger R_7. The active heat-exchanger area is 3 × 5 cm. This resistance is the connection between the air and chassis:

$$R_7 = \frac{1}{hA} = \frac{1}{(0.0147 \text{ W/cm}^2\text{-°C}) \times (3 \text{ cm} \times 5 \text{ cm})} = 4.54 \text{°C/W}$$

15.10 Transient and Cyclic Calculations

We have now calculated the thermal resistances in the entire heat path. These resistances are shown below:

$$R_1 = 2°C/W$$
$$R_2 = 2.5°C/W$$
$$R_3 = 3.08°C/W$$
$$R_4 = 13.57°C/W$$
$$R_5 = 1.28°C/W$$
$$R_6 = 0.28°C/W$$
$$R_7 = 4.54°C/W$$

Calculate steady-state temperatures

Using the thermal resistances calculated above, the steady-state temperatures along the heat path can be calculated. Knowing that the power dissipated is 6.0 W, the temperatures are shown below under steady-state conditions:

$$\Delta T = R \times Q$$

$$T_8 - T_7 = R_7 Q = (4.54°C/W) \times (6.0 \text{ W}) = 27.2°C$$
$$T_7 - T_6 = R_6 Q = (0.28°C/W) \times (6.0 \text{ W}) = 1.7°C$$
$$T_6 - T_5 = R_5 Q = (12.8°C/W) \times (6.0 \text{ W}) = 7.7°C$$
$$T_5 - T_4 = R_4 Q = (13.57°C/W) \times (6.0 \text{ W}) = 81.4°C$$
$$T_4 - T_3 = R_3 Q = (3.08°C/W) \times (6.0 \text{ W}) = 18.5°C$$
$$T_3 - T_2 = R_2 Q = (2.5°C/W) \times (6.0 \text{ W}) = 15.0°C$$
$$T_2 - T_1 = R_1 Q = (2°C/W) \times (6.0 \text{ W}) = 12.0°C$$

The temperature differences calculated above are used to fill in the list below (temperature values for T_7 down to T_1) to obtain temperatures T_1 through T_7. Temperature T_8 is the boundary air node, which is fixed at 25°C. Starting from this boundary node we can calculate the temperature for T_7, then T_6, and so on until we reach the temperature of T_1. This calculation is shown below:

$$T_7 = T_8 + (T_8 - T_7) = 25°C + 13.6°C = 52.2°C$$
$$T_6 = T_7 + (T_7 - T_6) = 52.2°C + 1.7°C = 53.9°C$$

$$T_5 = T_6 + (T_6 - T_5) = 53.9°C + 7.7°C = 61.6°C$$

$$T_4 = T_5 + (T_5 - T_4) = 61.6°C + 81.4°C = 143.0°C$$

$$T_3 = T_4 + (T_4 - T_3) = 143.0°C + 18.5°C = 161.5°C$$

$$T_2 = T_3 + (T_3 - T_2) = 161.5°C + 15.0°C = 176.5°C$$

$$T_1 = T_2 + (T_2 - T_1) = 176.5°C + 12.0°C = 188.5°C$$

If this unit were allowed to run until steady state is reached, then the resulting temperature would be 189°C, which is too hot for most electronics. A common junction temperature limit is 150°C.

Distributed thermal mass

The *thermal capacitance* or *thermal mass* of an object is the product of the specific heat and the mass of the object. In general, the larger the thermal mass, the slower the response of an object to thermal events such as heat inputs and changes to the boundary conditions. The simple lumped-capacitance approach assumes that the entire object is at an approximately uniform temperature and that all temperature gradients are outside the object.[1]

Heat conducting through a material causes temperature gradients in the material. If these gradients are significant, then portions of the thermal mass of the material are at significantly different temperatures. Where internal gradients are significant, the lumped-capacitance assumption is not valid.[1] The transient response of an object with significant internal temperature gradients can be approximated by treating the thermal mass as distributed throughout the object. By dividing up the object into multiple parts, each with its own thermal mass and thermal resistance to adjacent parts, the thermal mass of the object can be distributed and the correct thermal response can be approximated.

Lump masses where possible

Once the thermal resistances are known and the steady-state temperature gradients are determined, the thermal mass distribution can be considered. It is easy to lump thermal mass when the predominant temperature gradients are not caused by conduction but by convection, radiation, contact resistance, or thin interface layers (bonds). Thermal mass should be distributed where a relatively large conduction temperature gradient exists across a part that has significant thermal mass, such as a heat sink on a PWB.

For the thermal system shown in Fig. 15.4, there are several temperature gradients that occur across relatively short distances, indicating

15.12 Transient and Cyclic Calculations

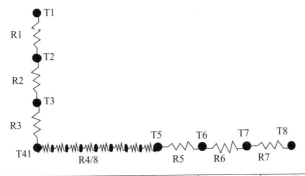

		Resistance	Temperature Q1= 6 W	Temp Gradient
T1 = Junction	R1 = Junction to Case	2.00°C/W	188.50°C	12.000°C
T2 = Case	R2 = Case to PWB	2.50°C/W	176.50°C	15.000°C
T3 = PWB Top	R3 = Through PWB	3.08°C/W	161.50°C	18.480°C
T41 = Heat Sink Top	R41 = Across Heat Sink	1.70°C/W	143.02°C	10.178°C
T42	R42	1.70°C/W	132.84°C	10.178°C
T43	R43	1.70°C/W	122.67°C	10.178°C
T44	R44	1.70°C/W	112.49°C	10.178°C
T45	R45	1.70°C/W	102.31°C	10.178°C
T46	R46	1.70°C/W	92.13°C	10.178°C
T47	R47	1.70°C/W	81.96°C	10.178°C
T48	R48	1.70°C/W	71.78°C	10.178°C
T5 = At Wedgelock	R5 = Through Wedgelock	1.28°C/W	61.60°C	7.680°C
T6 = Chassis	R6 = Through Chassis	0.28°C/W	53.92°C	1.680°C
T7 = Bottom of Chassis	R7 = Connection to Air	4.54°C/W	52.24°C	27.240°C
T8 = Air			25.00°C	

Figure 15.4 Distributed resistor network.

an appropriate location to divide thermal mass. These include R_2, the component leads; R_5, the wedgelock; R_6, the resistance through the chassis; and R_7, the fluid convection. These gradients allow the thermal mass of several parts to be lumped.

Several other conduction gradients occur across parts with significant thermal mass and require the mass to be distributed along the resistance. These include R_1, the junction to case resistance; R_3, the resistance through the PWB; and R_4, the resistance across the heat sink. The thermal mass and resistance of each of these sections should be divided up into subsections such that the temperature gradient across each subsection is relatively small. As with general meshing practices, it is good practice to divide up large temperature gradients into relatively insignificant ones in order to reduce the error associated with the discrete thermal model. In the case of this example, the gradient across the heat sink $(T_4 - T_5 = 82°C)$ should be divided into less significant gradients, say, 10°C. The heat sink should be divided into eight roughly

10°C parts, each with one-eighth the total heat-sink thermal mass. So the thermal model would now be as shown in Fig. 15.4.

Calculate capacitances

The capacitance must now be calculated. By using the areas already calculated in the thermal resistance calculation section, a volume may be easily generated, and therefore the capacitance can be paired up to the thermal resistance. By pairing up the thermal resistance to the proper capacitance, a thermal time constant can be calculated for each segment of the system. The first capacitance that we can calculate is for the electronic part. This part is a plastic component with a specific heat of 1200 W-s/kg-°C and a density of 0.0014 kg/cm³, which is 2.5 cm wide by 2.5 cm long by 0.125 cm high. By inputting this information into the equation below, the capacitance can be calculated for the electronic part C_1 (Ref. 1):

$$C_1 = \rho V C_p$$

where C_1 = capacitance, W-s/°C
C_p = specific heat, W-s/kg-°C
ρ = density, kg/cm³
V = volume, cm³

Then,

$$C_1 = (0.0014 \text{ kg/cm}^3) \times (2.5 \times 2.5 \times 0.125 \text{ cm}) \times (1200 \text{ W-s/kg-°C})$$

$$= 1.3125 \text{ W-s/°C}$$

Next we must calculate the capacitance associated with the connection between the electronic part and the PWB. This consists of the electronic part leads and the airgap. The density of the copper leads is 0.00785 kg/cm³, and the specific heat is 455 W-s/kg-°C. Using the area calculated under the resistance section for R_2 of 0.03125 cm² and the lead length of 0.1 cm, the volume can be calculated. The thermal capacitance of the leads can then be calculated as shown below:

$$C_a = \rho V C_p$$

$$= (0.00785 \text{ kg/cm}^3) \times (0.03125 \text{ cm}^2 \times 0.1 \text{ cm}) \times (455 \text{ W-s/kg-°C})$$

$$= 0.0112 \text{ W-s/°C}$$

The capacitance of the airgap must also be calculated. Using the dimensions of the airgap—2.5 cm long, 2.5 cm wide, and 0.05 cm thick—the volume can be calculated. Using the air density of 1.15×10^{-6} kg/cm³

15.14 Transient and Cyclic Calculations

and the specific heat of 1090 W-s/kg-°C, the capacitance of the airgap can be calculated as shown below:

$$C_b = \rho V C_p$$
$$= (1.15 \times 10^{-6}\,\text{kg/cm}^3) \times (2.5 \times 2.5 \times 0.05\,\text{cm}) \times (1090\,\text{W-s/kg-°C})$$
$$= 0.000392\,\text{W-s/°C}$$

We need to then add together the capacitance of the leads and the airgap, because the thermal path is treated as a single resistance. Adding the two capacitances C_a and C_b together, we can obtain the capacitance C_2 as follows:

$$C_2 = C_a + C_b = 0.0112\,\text{W-s/°C} + 0.000392\,\text{W-s/°C}$$
$$= 0.0116\,\text{W-s/°C}$$

The next capacitance that needs to be calculated is for the PWB or C_3. The density of the PWB is 0.0015 kg/cm³, and the specific heat is 1100 W-s/kg-°C. Using the area of the component of 6.25 cm² with a thickness of 0.15 cm, the capacitance can be calculated as shown below:

$$C_3 = \rho V C_p$$
$$= (0.0015\,\text{kg/cm}^3) \times (6.25\,\text{cm}^2 \times 0.15\,\text{cm}) \times (1100\,\text{W-s/kg-°C})$$
$$= 1.547\,\text{W-s/kg-°C}$$

The capacitance of the aluminum heat sink must also be calculated, which is C_4. Using the previously calculated area of the heat sink of 0.325 cm² in the thermal resistance section along with the length of 7.5 cm used in the same section for R_4, the volume of the heat sink is known. Using the aluminum density of 0.003 kg/cm³ and the specific heat of 1050 W-s/kg-°C, the capacitance of the aluminum heat sink can be calculated as shown below:

$$C_4 = \rho V C_p$$
$$= (0.003\,\text{kg/cm}^3) \times (0.325\,\text{cm}^2 \times 7.5\,\text{cm}) \times (1050\,\text{W-s/kg-°C})$$
$$= 7.68\,\text{W-s/°C}$$

The high-temperature gradients in the heat sink require that its capacitance be distributed. Dividing C_4 among the eight heat-sink nodes gives

$$C_{41} = C_4/8 = \frac{7.68\,\text{W-s/°C}}{8} = 0.96\,\text{W-s/°C}$$

The capacitance of the wedgelock C_5 must also be calculated. The area of the wedgelock interface calculated in the thermal resistance section for R_5 is 6.5 cm². Some assumptions must be made in order to calculate the volume, because at this interface there is no specific thickness. An interface resistance of 3.2 cm²-°C/W was used, so in order to back out what thickness to use to calculate the volume, an airgap thickness will be calculated as shown below on the basis of the interface resistance per unit area:

$$L = (3.2\,\text{cm}^2\text{-}°\text{C/W}) \times k = (3.2\,\text{cm}^2\text{-}°\text{C/W}) \times (0.0003\,\text{W/cm-}°\text{C})$$

$$= 0.00096\,\text{cm}$$

This calculated length of 0.00096 cm and the area of 6.5 cm² are used to calculate the volume of the wedgelock, which will be used to calculate the capacitance. Using the air density of 1.15×10^{-6} kg/cm³ and the specific heat of 1090 W-s/kg-°C, the capacitance of the wedgelock can be calculated as follows:

$$C_5 = \rho V C_p$$

$$= (1.15 \times 10^{-6}\,\text{kg/cm}^3) \times (6.5\,\text{cm}^2 \times 0.00096\,\text{cm}) \times (1090\,\text{W-s/}°\text{C})$$

$$= 7.82 \times 10^{-6}\,\text{W-s/}°\text{C}$$

The second-to-last (penultimate) capacitance that needs to be calculated is the conductive path between the wedgelock and the heat-exchanger face, which is C_6. Using the chassis area of 2.5 cm², along with the length of 1.2 cm for R_6, the volume of the chassis conductive path is known. Using the aluminum density of 0.003 kg/cm³ and the specific heat of 1050 W-s/kg-°C, the capacitance of the aluminum chassis can be calculated as shown below:

$$C_6 = \rho V C_p$$

$$= (0.003\,\text{kg/cm}^3) \times (2.5\,\text{cm}^2 \times 1.2\,\text{cm}) \times (1050\,\text{W-s/kg-}°\text{C})$$

$$= 9.45\,\text{W-s/kg-}°\text{C}$$

The final capacitance to be calculated is for the convective path to the air in the heat exchanger:

$$C_7 = \rho V C_p$$

$$= (1.15 \times 10^{-6}\,\text{kg/cm}^3) \times (1 \times 3 \times 5\,\text{cm}) \times (1090\,\text{W-s/kg-}°\text{C})$$

$$= 0.019\,\text{W-s/}°\text{C}$$

15.16 Transient and Cyclic Calculations

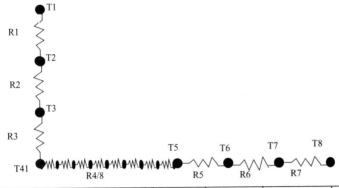

		Resistance	Temperature Q1= 6 W	Temp Gradient	Capacitance
T1 = Junction	R1 = Junction to Case	2.00°C/W	188.50°C	12°C	1.313 W-s/°C
T2 = Case	R2 = Case to PWB	2.50°C/W	177°C	15°C	0.012 W-s/°C
T3 = PWB Top	R3 = Through PWB	3.08°C/W	162°C	18°C	1.547 W-s/°C
T41 = Heat Sink Top	R41 = Across Heat Sink	1.70°C/W	143°C	10°C	0.96 W-s/°C
T42	R42	1.70°C/W	133°C	10°C	0.96 W-s/°C
T43	R43	1.70°C/W	123°C	10°C	0.96 W-s/°C
T44	R44	1.70°C/W	112°C	10°C	0.96 W-s/°C
T45	R45	1.70°C/W	102°C	10°C	0.96 W-s/°C
T46	R46	1.70°C/W	92°C	10°C	0.96 W-s/°C
T47	R47	1.70°C/W	82°C	10°C	0.96 W-s/°C
T48	R48	1.70°C/W	72°C	10°C	0.96 W-s/°C
T5 = At Wedgelock	R5 = Through Wedgelock	1.28°C/W	62°C	8°C	8E-06 W-s/°C
T6 = Chassis	R6 = Through Chassis	0.28°C/W	54°C	2°C	9.45 W-s/°C
T7 = Bottom of Chassis	R7 = Connection to Air	4.54°C/W	52°C	27°C	0.019 W-s/°C
T8 = Air			25°C		

Figure 15.5 Distributed capacitances.

All the capacitances have been calculated for the system. These values are listed below and are included in Fig. 15.5:

$$C_1 = 1.3125 \text{ W-s/}°\text{C}$$

$$C_2 = 0.0116 \text{ W-s/}°\text{C}$$

$$C_3 = 1.547 \text{ W-s/}°\text{C}$$

$$C_{41-48} = 0.96 \text{ W-s/}°\text{C}$$

$$C_5 = 7.82 \times 10^{-6} \text{ W-s/}°\text{C}$$

$$C_6 = 9.45 \text{ W-s/}°\text{C}$$

$$C_7 = 0.019 \text{ W-s/}°\text{C}$$

How to Evaluate Response

Once the resistances and capacitances are determined, there are several methods for determining the transient response based on the governing equation[1]:

$$\sum \dot{Q}_i = \sum_j \frac{T_i - T_j}{R_{ij}} + \rho_i V_i C_{pi} \frac{dT_i}{dt} = \dot{Q}_{i \text{ impressed}}$$

It would appear that since the governing equation for each part is a first-order ordinary differential equation, the solution would be an exponential function of the form[1]

$$T_i(t) \approx e^{-t/\tau}$$

Although a single mass resistor system does have an exponential response it is a common misconception that a thermal system with distributed mass can be characterized with a single time constant.

Single-time-constant fallacy

An electrical circuit analogy is often used to aid in understanding transient thermal characteristics. Treating a thermal system as a discrete electrical component, with a lumped capacitance, connected with a simple resistor to a known temperature, the transient thermal characteristics of the system can be reduced to a single time constant given by $\tau = RC$.[1]

The transient thermal response of a lumped-capacitance system is a simple exponential. Given a simple system without internal temperature gradients, this may be a valid characterization. Most actual thermal systems have significant internal temperature gradients, that effectively distribute the thermal capacitance. There is not a single time constant that can be associated with the response because of the large temperature gradients through the system. Many electrical component manufacturers will try to use a single time constant. This is not valid and will give erroneous results.

Multiple-transient response

Treating the sample as a single lumped-mass/single-resistor thermal system is characterized by a classic RC response to a thermal event as shown in Fig. 15.6. The response of this system is the same exponential response regardless of the nature of the thermal event. Figure 15.6 shows the response of the single lumped-mass system to the 6 W input shown as percent of response. The classic exponential response is

15.18 Transient and Cyclic Calculations

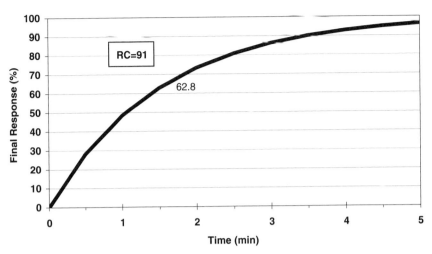

Figure 15.6 Lumped mass response.

$1/e = 63.2$ percent in one time constant. This response can be readily curve-fit with an exponential form and extrapolated.

A distributed mass system is made up of multiple parts with different time constants and thus does not respond with the same exponential decay response to every thermal event. Figure 15.7 shows the response

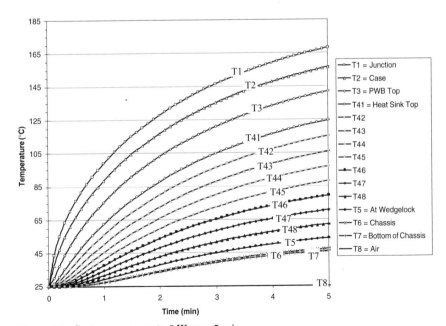

Figure 15.7 System response to 6 W over 5 min.

of the system, but in this case the thermal mass is distributed. The response of the distributed mass system is significantly different. The response of the outer part is only 40.8 percent, while the response of the more inner parts is significantly slower. *When the thermal mass of a system is distributed, the portions close to the transient event respond quickly while those farther away have to wait until the closer portions respond.* This characteristic results in an overall response that appears to have a rapid partial response to fast inputs and yet has a slower full response to longer events. In a short transient the adjacent parts will respond, but the more remote ones will have an insignificant response. In a long transient the adjacent parts will form stable temperature gradients with their adjacent parts while the more remote parts continue to slowly respond until they also come to equilibrium. Figure 15.8 shows how close the internal temperature gradients, the temperature difference from one part to the next, reach their final values during the transient. The adjacent parts quickly form constant temperature gradients, while the temperature gradients between remote parts take longer to form. The resulting long-term response, even for the parts near the thermal event, is the sum of their transient response plus the response of all the parts in the system. In general the partial transient response of a distributed mass system cannot be curve-fit and cannot be extrapolated. To characterize the transient response of a distributed mass system, thermal modeling must be used with particular detail to the portions partially responding to the transient.

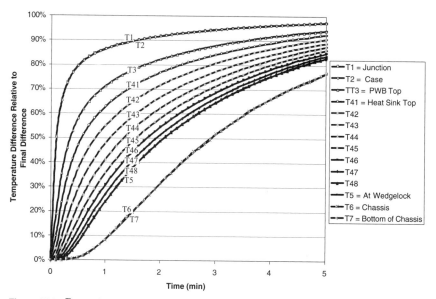

Figure 15.8 Percentage response to 6 W over 5 min.

15.20 Transient and Cyclic Calculations

Iterative solution

One method for solving for temperatures is to input the resistances, capacitances, powers, and boundary temperatures into a finite-difference solver and let the solver iterate through the time steps to solve for the temperatures. At each time step the solver will calculate the temperature rise of each node on the basis of the net heat on the node.

This method can also be solved using a spreadsheet with a solver function. The energy imbalances at each node and at each time step are summed up. The solver is used to iterate the temperatures in order to minimize the energy imbalance:

$$\sum \dot{Q}_i = \sum_j \frac{T_i - T_j}{R_{ij}} + \rho_i V_i C_{pi} \frac{dT_i}{dt} = \dot{Q}_{i\text{ impressed}}$$

Lumped-mass approximation

Another solution is to approximate the energy balance within each part by scaling the net energy flow down to the next part. The advantage to this method is that no iterations are required and the results are generally faster than with the actual response. In this approximation the response of each part over a short time is approximated by an exponential function. To determine the approximate exponential response of each part, a local time constant is calculated for each part relative to its neighboring parts.

Determine time constants. Although the long-term response of each part of the system cannot be described with a single time constant, the short-term response can be approximated as exponential with a time constant.

We have calculated the thermal resistances as well as the thermal capacitances of the entire system, which can now be used to generate the thermal time constants of the distributed thermal system. The time constant for each part is different depending on how the temperature of the part is being changed. By pairing up the thermal resistance to the proper capacitance, a thermal time constant can be calculated for each segment of the system. The equation shown below[1] will be used to calculate the time constant for each section:

$$\tau_i = \frac{C_i}{\sum_j (1/R_{ij})}$$

$$R_1 = 2°\text{C/W}$$

$$R_2 = 1.1°\text{C/W}$$

$$R_3 = 3.08°\text{C/W}$$

$$R_{41-48} = 1.7°C/W$$
$$R_5 = 0.49°C/W$$
$$R_6 = 0.28°C/W$$
$$R_7 = 4.54°C/W$$
$$C_1 = 1.3125 \text{ W-s/}°C$$
$$C_2 = 0.0116 \text{ W-s/}°C$$
$$C_3 = 1.547 \text{ W-s/}°C$$
$$C_{41-48} = 0.96 \text{ W-s/}°C$$
$$C_5 = 7.82 \times 10^{-6} \text{ W-s/}°C$$
$$C_6 = 9.45 \text{ W-s/}°C$$
$$C_7 = 0.019 \text{ W-s/}°C$$

The thermal time constant calculations and results are shown below:

$$\tau_1 = C_1 \times R_1 = (1.3125 \text{ W-s/}°C) \times (2°C/W) = 2.625 \text{ s}$$

$$\tau_2 = C_2 \times \frac{1}{1/R_1 + 1/R_2} = (0.0116 \text{ W-s/}°C)$$
$$\times \frac{1}{1/2°C/W + 1/2.5°C/W} = 0.013 \text{ s}$$

$$\tau_3 = C_3 \times \frac{1}{1/R_2 + 1/R_3} = (1.547 \text{ W-s/}°C)$$
$$\times \frac{1}{(1/2.5°C/W + 1/3.08°C/W)} = 2.135 \text{ s}$$

$$\tau_{41} = C_{41} \times \frac{1}{1/R_3 + 1/R_{41}} = (0.96 \text{ W-s/}°C)$$
$$\times \frac{1}{1/3.08°C/W + 1/1.7°C/W} = 1.050 \text{ s}$$

$$\tau_{42-48} = C_{42} \times \frac{1}{1/R_{41} + 1/R_{42}} = (0.96 \text{ W-s/}°C)$$
$$\times \frac{1}{1/1.7°C/W + 1/1.7°C/W} = 0.814 \text{ s}$$

15.22   Transient and Cyclic Calculations

$$\tau_5 = C_5 \times \frac{1}{1/R_{48} + 1/R_5} = (7.82 \times 10^{-6} \text{ W-s/}^\circ\text{C})$$

$$\times \frac{1}{1/1.7^\circ\text{C/W} + 1/1.28^\circ\text{C/W}} = 5.7 \times 10^{-6} \text{ s}$$

$$\tau_6 = C_6 \times \frac{1}{1/R_5 + 1/R_6} = (9.45 \text{ W-s/}^\circ\text{C})$$

$$\times \frac{1}{1/1.28^\circ\text{C/W} + 1/0.28^\circ\text{C/W}} = 2.17 \text{ s}$$

$$\tau_7 = C_7 \times \frac{1}{1/R_6 + 1/R_7} = (0.019 \text{ W-s/}^\circ\text{C})$$

$$\times \frac{1}{1/0.28^\circ\text{C/W} + 1/4.54^\circ\text{C/W}} = 0.005 \text{ s}$$

These time constants indicate the order of magnitude of the response of each section. A given part will generally reach equilibrium with its adjacent parts within three time constants. The overall system response, however, depends on much more than the individual time constants. The system's thermal characteristics are summarized in Fig. 15.9.

Approximating transient response.   Using the thermal capacitances and resistances that were calculated above, an approximation of the thermal system response can be performed. For this example the temperature at the end of 30 s of operation is to be calculated, but this calculation may be applied at any time in the transient. By knowing that the temperature gradients in the layers above affect the temperature response of the layers below, we can approximate this interaction by first calculating the response of the first layer in the thermal mass chain[1]:

$$\Delta T = \left[1 - \exp\left(\frac{-t}{RC}\right)\right](\Delta T_f)$$

where ΔT = time-related temperature change, $^\circ\text{C} = T(t) - T(0)$
t = time, s
ΔT_f = final or steady-state temperature difference, $^\circ\text{C} = T(t \to \infty) - T(0)$

Thermal System Transient Response 15.23

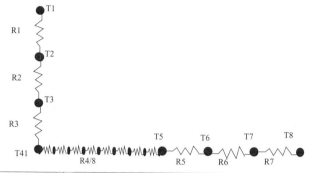

		Resistance	Temperature	Temp Gradient	Capacitance	Tau
			Q1= 6 W			
T1 = Junction	R1 = Junction to Case	2.00°C/W	188.50°C	12°C	1.313 W-s/°C	2.625 s
T2 = Case	R2 = Case to PWB	2.50°C/W	177°C	15°C	0.012 W-s/°C	0.013 s
T3 = PWB Top	R3 = Through PWB	3.08°C/W	162°C	18°C	1.547 W-s/°C	2.135 s
T41 = Heat Sink Top	R41 = Across Heat Sink	1.70°C/W	143°C	10°C	0.96 W-s/°C	1.050 s
T42	R42	1.70°C/W	133°C	10°C	0.96 W-s/°C	0.814 s
T43	R43	1.70°C/W	123°C	10°C	0.96 W-s/°C	0.814 s
T44	R44	1.70°C/W	112°C	10°C	0.96 W-s/°C	0.814 s
T45	R45	1.70°C/W	102°C	10°C	0.96 W-s/°C	0.814 s
T46	R46	1.70°C/W	92°C	10°C	0.96 W-s/°C	0.814 s
T47	R47	1.70°C/W	82°C	10°C	0.96 W-s/°C	0.814 s
T48	R48	1.70°C/W	72°C	10°C	0.96 W-s/°C	0.814 s
T5 = At Wedgelock	R5 = Through Wedgelock	1.28°C/W	62°C	8°C	8E-06 W-s/°C	5.70E-6 s
T6 = Chassis	R6 = Through Chassis	0.28°C/W	54°C	2°C	9.45 W-s/°C	2.171 s
T7 = Bottom of Chassis	R7 = Connection to Air	4.54°C/W	52°C	27°C	0.019 W-s/°C	0.005 s
T8 = Air			25°C			

Figure 15.9 Resistance network summary.

Inputting the resistance of 2°C/W (R_1) and the capacitance of 1.313 W/s-°C (C_1) along with the steady-state temperature difference of 12°C at the 30 s, the equation reduces as shown below:

$$\Delta T_1 = \left[1 - \exp\left(\frac{-30\,\text{s}}{(2°\text{C/W}) \times (1.313\,\text{W/s-°C})}\right)\right](12°\text{C})$$
$$= 12°\text{C}$$

This first layer has obviously already reached its steady-state temperature gradient in 30 s. The next layer down or the case of the component will see all the power propagate to itself. Knowing that, in order to change the temperature of the component junction as well as the case temperature, the entire thermal capacitance of the layers above the case must be considered. The capacitance of the junction and the case are added together as shown below, along with the time constant or RC associated with the temperature response of the junction and the case of the component combined. This calculation will be continued until all the temperature gradients through each section are calculated at the

15.24   Transient and Cyclic Calculations

30-s time period:

$$RC_2 = R_2(C_1 + C_2) + R_1C_1 = 2.5°\text{C/W} \ (1.313 \text{ W-s/}°\text{C}$$
$$+ \ 0.012 \text{ W-s/}°\text{C}) + (2.0°\text{C/W} \times 1.313 \text{ W-s/}°\text{C})$$
$$= 5.9 \text{ s}$$

$$\Delta T_2 = \left[1 - \exp\left(\frac{-t}{RC}\right)\right](\Delta T) = \left[1 - \exp\left(\frac{-30 \text{ s}}{5.9 \text{ s}}\right)\right](15°\text{C})$$
$$= 15°\text{C}$$

$$RC_3 = R_3 \ (C_1 + C_2 + C_3) + RC_2 = 3.08°\text{C/W} \ (1.313 \text{ W-s/}°\text{C}$$
$$+ \ 0.012 \text{ W-s/}°\text{C} + 1.55 \text{ W-s/}°\text{C}) + (5.9 \text{ s})$$
$$= 14.8 \text{ s}$$

$$\Delta T_3 = \left[1 - \exp\left(\frac{-t}{RC}\right)\right](\Delta T) = \left[1 - \exp\left(\frac{-30 \text{ s}}{14.8 \text{ s}}\right)\right](18.5°\text{C})$$
$$= 16.1°\text{C}$$

Now in layer 3, which is the PWB, the heat has not propagated completely through but just a certain percentage of heat has traveled through the PWB to the heat sink. Knowing this, we can adjust the temperature gradient of the next layer according to the lower percentage of power dissipation, which propagates to it.

With the calculated temperature gradient of only 16.1°C as compared to the steady-state temperature gradient of 18.5°C, only 87 percent of the dissipated power has moved to the heat sink from the PWB. In other words, the PWB has stored 13 percent of the power dissipated by the component.

Accordingly, the gradient through the heat sink layer will be adjusted directly by this percentage as shown below:

$$RC_{41} = R_{41}(C_1 + C_2 + C_3 + C_{41}) + RC_3$$
$$= 1.7°\text{C/W} \ (1.313 \text{ W-s/}°\text{C} + 0.012 \text{ W-s/}°\text{C} + 1.55 \text{ W-s/}°\text{C}$$
$$+ \ 0.96 \text{ W-s/}°\text{C}) + (14.8 \text{ s})$$
$$= 21.3 \text{ s}$$

$$\Delta T_{41} = \left[1 - \exp\left(\frac{-t}{RC}\right)\right](\Delta T) \times (\% \text{ of power}) = \left[1 - \exp\left(\frac{-30 \text{ s}}{21.3 \text{ s}}\right)\right]$$
$$\times \ (10.2°\text{C}) \times (0.87)$$
$$= 6.7°\text{C}$$

The steady-state temperature gradient through the first section of the heat sink is 10.2°C, while after 30 s of operation the temperature gradient is only 6.7°C, therefore a 65 percent power dissipation (energy storage) factor will be applied to the calculated temperature gradient for the next part of the heat sink (T_{42}):

$$RC_{42} = R_{42}(C_1 + C_2 + C_3 + C_{41} + C_{42}) + RC_{41}$$

$$= 1.7°\text{C/W} \, (1.313 \text{ W-s/°C} + 0.012 \text{ W-s/°C} + 1.55 \text{ W-s/°C}$$

$$+ 0.96 \text{ W-s/°C} + 0.96 \text{ W-s/°C}) + (21.3 \text{ s})$$

$$= 29.4 \text{ s}$$

$$\Delta T_{42} = \left[1 - \exp\left(\frac{-t}{RC}\right)\right] (\Delta T) \times (\% \text{ of power})$$

$$= \left[1 - \exp\left(\frac{-30 \text{ s}}{29.4 \text{ s}}\right)\right] (10.2°\text{C}) \times (0.65)$$

$$= 4.2°\text{C}$$

The percentage of heat which travels through this section of the heat sink is 4.2°C divided by 10.2°C, so for the next section of the heat sink (T_{43}) a power factor of 41 percent will be used:

$$RC_{43} = R_{43}(C_1 + C_2 + C_3 + C_{41} + C_{42} + C_{43}) + RC_{42}$$

$$= 1.7°\text{C/W} \, (1.313 \text{ W-s/°C} + 0.012 \text{ W-s/°C} + 1.55 \text{ W-s/°C}$$

$$+ 0.96 \text{ W-s/°C} + 0.96 \text{ W-s/°C} + 0.96 \text{ W-s/°C}) + (29.4 \text{ s})$$

$$= 39.2 \text{ s}$$

$$\Delta T_{43} = \left[1 - \exp\left(\frac{-t}{RC}\right)\right] (\Delta T) \times (\% \text{ of power})$$

$$= \left[1 - \exp\left(\frac{-30 \text{ s}}{39.2 \text{ s}}\right)\right] (10.2°\text{C}) \times (0.41)$$

$$= 2.2°\text{C}$$

This calculation will continue until all the temperature gradients through each layer are calculated. These calculations are shown below:

$$\text{Percent of power} = \frac{\Delta T_{43}}{\Delta T_{43f}} = \frac{2.2°\text{C}}{10.2°\text{C}} = 0.22$$

$$RC_{44} = R_{44}(C_1 + C_2 + C_3 + C_{41} + C_{42} + C_{43} + C_{44}) + RC_{43}$$

$$= 1.7°\text{C/W} \,(1.313 \text{ W-s/°C} + 0.012 \text{ W-s/°C}$$
$$+ 1.55 \text{ W-s/°C} + 0.96 \text{ W-s/°C} + 0.96 \text{ W-s/°C}$$
$$+ 0.96 \text{ W-s/°C} + 0.96 \text{ W-s/°C}) + (39.2 \text{ s})$$
$$= 50.6 \text{ s}$$

$$\Delta T_{44} = \left[1 - \exp\left(\frac{-t}{RC}\right)\right] (\Delta T) \times (\% \text{ of power})$$

$$= \left[1 - \exp\left(\frac{-30 \text{ s}}{50.6 \text{ s}}\right)\right] (10.2°\text{C}) \times (0.22)$$

$$= 1.0°\text{C}$$

$$\text{Percent of power} = \frac{\Delta T_{44}}{\Delta T_{44f}} = \frac{1.0°\text{C}}{10.2°\text{C}} = 0.10$$

$$RC_{45} = R_{45}(C_1 + C_2 + C_3 + C_{41} + C_{42} + C_{43} + C_{44} + C_{45})$$
$$+ RC_{44}$$

$$= 1.7°\text{C/W} \,(1.313 \text{ W-s/°C} + 0.012 \text{ W-s/°C}$$
$$+ 1.55 \text{ W-s/°C} + 0.96 \text{ W-s/°C} + 0.96 \text{ W-s/°C}$$
$$+ 0.96 \text{ W-s/°C} + 0.96 \text{ W-s/°C} + 0.96 \text{ W-s/°C})$$
$$+ (50.6 \text{ s})$$
$$= 63.6 \text{ s}$$

$$\Delta T_{45} = \left[1 - \exp\left(\frac{-t}{RC}\right)\right] (\Delta T) \times (\% \text{ of power})$$

$$= \left[1 - \exp\left(\frac{-30 \text{ s}}{63.6 \text{ s}}\right)\right] (10.2°\text{C}) \times (0.10)$$

$$= 0.4°\text{C}$$

$$\text{Percent of power} = \frac{\Delta T_{45}}{\Delta T_{45f}} = \frac{0.4°\text{C}}{10.2°\text{C}} = 0.04$$

$$RC_{46} = R_{46}(C_1 + C_2 + C_3 + C_{41} + C_{42} + C_{43} + C_{44}$$
$$+ C_{45} + C_{46}) + RC_{45}$$

$$= 1.7°\text{C/W} \,(1.313 \text{ W-s/°C} + 0.012 \text{ W-s/°C}$$
$$+ 1.55 \text{ W-s/°C} + 0.96 \text{ W-s/°C} + 0.96 \text{ W-s/°C}$$
$$+ 0.96 \text{ W-s/°C} + 0.96 \text{ W-s/°C} + 0.96 \text{ W-s/°C}$$
$$+ 0.96 \text{ W-s/°C}) + (63.6 \text{ s})$$
$$= 78.2 \text{ s}$$

Thermal System Transient Response

$$\Delta T_{46} = \left[1 - \exp\left(\frac{-t}{RC}\right)\right](\Delta T) \times (\% \text{ of power})$$

$$= \left[1 - \exp\left(\frac{-30\text{ s}}{78.2\text{ s}}\right)\right](10.2°\text{C}) \times (0.04)$$

$$= 0.13°\text{C}$$

It can be seen that farther away from the power dissipation source the thermal gradients are less established. Close to the end on the heat sink the temperature rise is only a fraction of the total steady-state temperature rise, or only about 1.5 percent. We will continue to calculate the temperature rise through the rest of the thermal transient model:

$$\text{Percent of power} = \frac{\Delta T_{46}}{\Delta T_{46f}} = \frac{0.13°\text{C}}{10.2°\text{C}} = 0.013$$

$$RC_{47} = R_{47}(C_1 + C_2 + C_3 + C_{41} + C_{42} + C_{43} + C_{44}$$
$$+ C_{45} + C_{46} + C_{47}) + RC_{46}$$

$$= 1.7°\text{C/W } (1.313 \text{ W-s/°C} + 0.012 \text{ W-s/°C}$$
$$+ 1.55 \text{ W-s/°C} + 0.96 \text{ W-s/°C} + 0.96 \text{ W-s/°C}$$
$$+ 0.96 \text{ W-s/°C} + 0.96 \text{ W-s/°C} + 0.96 \text{ W-s/°C}$$
$$+ 0.96 \text{ W-s/°C} + 0.96 \text{ W-s/°C}) + (78.2 \text{ s})$$

$$= 94.5 \text{ s}$$

$$\Delta T_{47} = \left[1 - \exp\left(\frac{-t}{RC}\right)\right](\Delta T) \times (\% \text{ of power})$$

$$= \left[1 - \exp\left(\frac{-30\text{ s}}{94.5\text{ s}}\right)\right](10.2°\text{C}) \times (0.013)$$

$$= 0.04°\text{C}$$

$$\text{Percent of power} = \frac{\Delta T_{47}}{\Delta T_{47f}} = \frac{0.04°\text{C}}{10.2°\text{C}} = 0.004$$

$$RC_{48} = R_{48}(C_1 + C_2 + C_3 + C_{41} + C_{42} + C_{43} + C_{44}$$
$$+ C_{45} + C_{46} + C_{47} + C_{48}) + RC_{47}$$

15.28 Transient and Cyclic Calculations

$$= 1.7°\text{C/W} \,(1.313 \text{ W-s/}°\text{C} + 0.012 \text{ W-s/}°\text{C}$$
$$+ 1.55 \text{ W-s/}°\text{C} + 0.96 \text{ W-s/}°\text{C} + 0.96 \text{ W-s/}°\text{C}$$
$$+ 0.96 \text{ W-s/}°\text{C} + 0.96 \text{ W-s/}°\text{C} + 0.96 \text{ W-s/}°\text{C}$$
$$+ 0.96 \text{ W-s/}°\text{C} + 0.96 \text{ W-s/}°\text{C} + 0.96 \text{ W-s/}°\text{C}) + (78.2 \text{ s})$$
$$= 112.4 \text{ s}$$

$$\Delta T_{48} = \left[1 - \exp\left(\frac{-t}{RC}\right)\right] (\Delta T) \times (\% \text{ of power})$$

$$= \left[1 - \exp\left(\frac{-30 \text{ s}}{112.4 \text{ s}}\right)\right] (10.2°\text{C}) \times (0.004)$$

$$= 0.01°\text{C}$$

$$\text{Percent of power} = \frac{\Delta T_{48}}{\Delta T_{48f}} = \frac{0.01°\text{C}}{10.2°\text{C}} = 0.001$$

$$RC_5 = R_5(C_1 + C_2 + C_3 + C_{41} + C_{42} + C_{43} + C_{44} + C_{45}$$
$$+ C_{46} + C_{47} + C_{48} + C_5) + RC_{48}$$

$$= 1.28°\text{C/W} \,(1.313 \text{ W-s/}°\text{C} + 0.012 \text{ W-s/}°\text{C}$$
$$+ 1.55 \text{ W-s/}°\text{C} + 0.96 \text{ W-s/}°\text{C} + 0.96 \text{ W-s/}°\text{C}$$
$$+ 0.96 \text{ W-s/}°\text{C} + 0.96 \text{ W-s/}°\text{C} + 0.96 \text{ W-s/}°\text{C}$$
$$+ 0.96 \text{ W-s/}°\text{C} + 0.96 \text{ W-s/}°\text{C} + 0.96 \text{ W-s/}°\text{C}$$
$$+ 8 \times 10^{-6} \text{ W-s/}°\text{C}) + (112.4 \text{ s})$$

$$= 125.9 \text{ s}$$

$$\Delta T_5 = \left[1 - \exp\left(\frac{-t}{RC}\right)\right] (\Delta T) \times (\% \text{ of power})$$

$$= \left[1 - \exp\left(\frac{-30 \text{ s}}{125.9 \text{ s}}\right)\right] (10.2°\text{C}) \times (0.001)$$

$$= 0.002°\text{C}$$

$$\text{Percent of power} = \frac{\Delta T_5}{\Delta T_{5f}} = \frac{0.002°\text{C}}{10.2°\text{C}} = 0.0002$$

$$RC_6 = R_6(C_1 + C_2 + C_3 + C_{41} + C_{42} + C_{43} + C_{44}$$
$$+ C_{45} + C_{46} + C_{47} + C_{48} + C_5 + C_6) + RC_5$$

$$= 0.28°\text{C/W } (1.313 \text{ W-s/°C} + 0.012 \text{ W-s/°C}$$
$$+ 1.55 \text{ W-s/°C} + 0.96 \text{ W-s/°C} + 0.96 \text{ W-s/°C}$$
$$+ 0.96 \text{ W-s/°C} + 0.96 \text{ W-s/°C} + 0.96 \text{ W-s/°C}$$
$$+ 0.96 \text{ W-s/°C} + 0.96 \text{ W-s/°C} + 0.96 \text{ W-s/°C}$$
$$+ 8 \times 10^{-6} \text{ W-s/°C} + 9.45 \text{ W-s/°C}) + (125.9 \text{ s})$$
$$= 131.5 \text{ s}$$

$$\Delta T_6 = \left[1 - \exp\left(\frac{-t}{RC}\right)\right] (\Delta T) \times (\% \text{ of power})$$

$$= \left[1 - \exp\left(\frac{-30 \text{ s}}{131.5 \text{ s}}\right)\right] (1.7°\text{C}) \times (0.0002)$$

$$\Delta T_6 < 0.0001°\text{C}$$

$$\text{Percent of power} = \frac{\Delta T_6}{\Delta T_{6f}} = \frac{0.0001°\text{C}}{10.2°\text{C}} < 0.0002$$

$$RC_7 = R_7(C_1 + C_2 + C_3 + C_{41} + C_{42} + C_{43} + C_{44}$$
$$+ C_{45} + C_{46} + C_{47} + C_{48} + C_5 + C_6 + C_7) + RC_6$$
$$= 4.54°\text{C/W } (1.313 \text{ W-s/°C} + 0.012 \text{ W-s/°C}$$
$$+ 1.55 \text{ W-s/°C} + 0.96 \text{ W-s/°C} + 0.96 \text{ W-s/°C}$$
$$+ 0.96 \text{ W-s/°C} + 0.96 \text{ W-s/°C} + 0.96 \text{ W-s/°C}$$
$$+ 0.96 \text{ W-s/°C} + 0.96 \text{ W-s/°C} + 0.96 \text{ W-s/°C}$$
$$+ 8 \times 10^{-6} \text{ W-s/°C} + 9.45 \text{ W-s/°C} + 0.019 \text{ W-s/°C})$$
$$+ (131.5 \text{ s})$$
$$= 222.4 \text{ s}$$

$$\Delta T_7 = \left[1 - \exp\left(\frac{-t}{RC}\right)\right] (\Delta T) \times (\% \text{ of power})$$

$$= \left[1 - \exp\left(\frac{-30 \text{ s}}{222.4 \text{ s}}\right)\right] (27.2°\text{C}) \times (0.0002)$$

$$\Delta T_7 < 0.0001°\text{C}$$

The temperature gradients for all the sections have been calculated at a time of 30 s. During this 30-s transient only the parts close to the

15.30 Transient and Cyclic Calculations

TABLE 15.2 Approximate Solution Comparison

Time, s	Approximation temperature, °C	Detailed model temperature, °C	Percent difference
30	83.1	76.4	8.8
60	103.9	98.9	5.1
120	131.1	127.7	2.7
180	149.2	146.4	1.9
240	161.9	159.2	1.7
360	176.4	174.3	1.2
480	182.9	181.6	0.7
600	185.7	185.1	0.3

heat source respond. Adding the temperature gradients through each layer and adding them to the boundary node temperature of 25°C, we can calculate the temperature of the component at 30 s after power is applied to the component. This calculation is shown below:

$$T_1 = \Delta T_1 + \Delta T_2 + \Delta T_3 + \Delta T_{41} + \Delta T_{42} + \Delta T_{43} + \Delta T_{44} + \Delta T_{45}$$
$$+ \Delta T_{46} + \Delta T_{47} + \Delta T_{48} + \Delta T_5 + \Delta T_6 + \Delta T_7 + T_8$$
$$= 12°C + 15°C + 16.1°C + 6.7°C + 4.2°C + 2.2°C + 1.2°C$$
$$+ 0.46°C + 0.15°C + 0.04°C + 0.01°C + 0.002°C + 0°C$$
$$+ 0°C + 25°C$$
$$= 83.1°C$$

Using this approximation, we calculated a component temperature of 83.1°C at the end of the 30-s transient. At the same time period the detailed model using finite-difference software yields a temperature of 76.4°C, which is very close to this calculated temperature of 83.1°C. This approximation may be used for any time period. Because this calculation is time-consuming, a spreadsheet could be used to calculate the temperature at any point in time very easily. Table 15.2 lists the resulting temperatures in a spreadsheet form at different times to show the temperature response and the exact solution compared to this simple approximation calculation. In most cases the approximation will be within 10 percent of the exact solution.

System Response

The response to a thermal event depends on the duration and frequency of the event and the distribution of thermal mass within the thermal system. In general only the adjacent parts will respond to a short transient event. An iterative finite-difference solver was used to evaluate the system's response to these transients.

To a short transient

Short transients significantly affect only the adjacent parts of the system. The rapid responses of the junction and case in these transients indicate that more resolution might be needed to accurately simulate their response. As with meshing in general, it is good engineering practice to further divide a model to establish mesh sensitivity. In transient modeling the thermal mass distribution as well as the resistances are mesh-sensitive. The more remote parts, on the other hand, have little to no response and need no further resolution for this transient.

Single event. Figure 15.10 shows the transient response from a 25°C soak to 6 W applied to the junction (T_1). In this transient the parts near T_1 change quickly at first while the chassis barely changes at all. It should be noted that the response of T_1 slows from an exponential response as the remote parts begin to respond. Figure 15.11 presents the same data as Fig. 15.10 except that the temperature difference between adjacent parts is plotted as a fraction of its final steady-state value. This figure helps illustrate how the parts near T_1 tend to form relatively stable temperature gradients quickly. The remaining temperature fluctuations near T_1 are due to that fact that the gradients in the remote parts form slowly as the net heat load finally conducts to them.

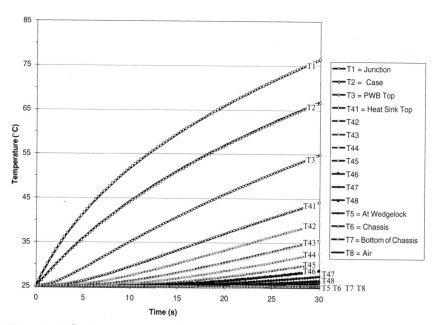

Figure 15.10 System response to 6 W over 30 s.

15.32   Transient and Cyclic Calculations

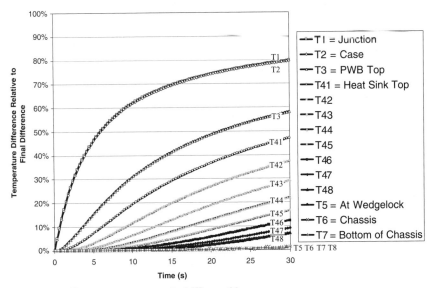

Figure 15.11 Percentage response to 6 W over 30 s.

Repeated event. Figure 15.12 shows the transient response of repeated applications of 6 W for 10 s to the junction (T_1) 25 percent of the time (10 s on and 30 s off). This analysis shows more clearly how only the nearby parts respond to short transients while the remote parts stay at relatively constant temperatures. The temperature gradients in the remote parts are equivalent to those resulting from the average power flowing through the parts, in this case 6 W × 25 percent = 1.5 W.

To a long transient

During longer transients the adjacent parts form stable temperature gradients while distant parts continue to respond.

Single event. Figure 15.13 shows the transient response from a 25°C soak to 6 W applied to the junction (T_1). Notice in this plot that the response of the remote parts (T_{45}) is not at all exponential over the entire transient. The fact that the nearby gradients form almost completely indicates that no further resolution of the component should be needed for this transient. The more remote parts, on the other hand, respond significantly, indicating that more resolution might be needed to accurately simulate their response. As with meshing in general, it is good engineering practice to further divide a model to establish mesh sensitivity. In transient modeling the thermal mass distribution as well as the resistances are sensitive to meshing.

Thermal System Transient Response 15.33

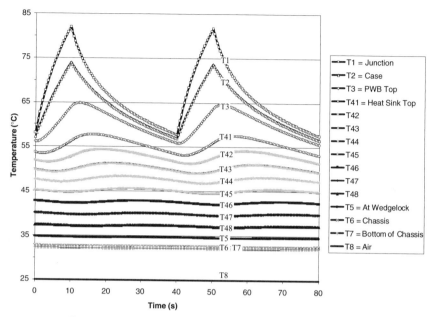

Figure 15.12 System response to 6 W pulsed for 10 s every 40 s.

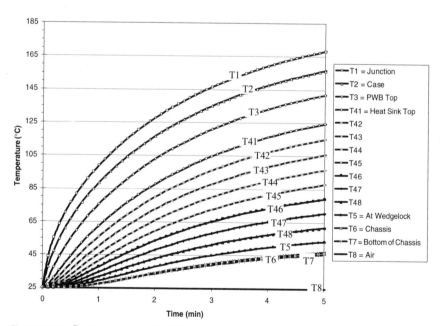

Figure 15.13 System response to 6 W over 5 min.

15.34 **Transient and Cyclic Calculations**

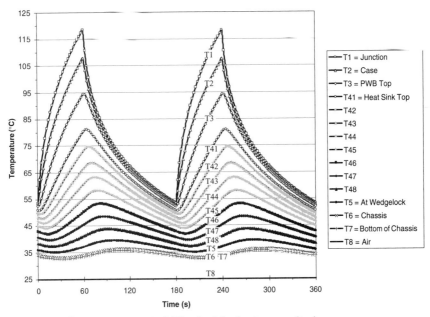

Figure 15.14 System response to 6 W pulsed for 1 min every 3 min.

Repeated event. Figure 15.14 shows the transient response of repeated applications of 6 W for 1 min to the junction (T_1) one-third of the time (1 min on and 2 min off). In this case the power lasts long enough for the remote parts to respond. Notice how their response is progressively smaller and more delayed.

Figure 15.15 shows the transient response to repeated 25°C steps in air temperature for 1 min out of every 3 min with a constant 3 W applied to T_1. This analysis shows that it is not the inherent nature of the junction to respond quickly or the chassis to respond slowly. What is inherent in a distributed mass thermal system is that the parts near the event respond quickly while the remote parts, in this case T_1, respond slowly and to a lesser degree.

Conclusion

The transient response of a thermal system with internal temperature gradients is the combined response of the individual exponential responses of its parts. In general the temperature of any part changes exponentially in the short term but more slowly in the long term because of the effects of more remote parts. Over longer transients a distributed system responds more slowly than does a single mass system. The parts closer to the thermal event respond quickly while the remote

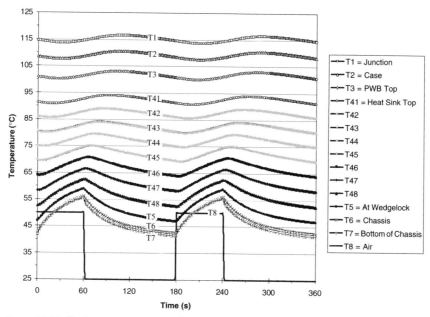

Figure 15.15 System response to 25°C spike for 1 min every 3 min with 3 W.

parts have a delayed, slower response. An iterative solution is possible using a solver. A noniterative approximation is also quite practical. However, a single-time-constant solution does *not* accurately represent the true transient response of a real thermal system.

Acknowledgments

The authors wish to acknowledge the support of Richard Rinick and Richard Porter for their expeditious review of this section. The authors would also be remiss in not acknowledging the steadfast support of their wives.

Nomenclature

A	Area of thermal path, cm²
C	Lumped thermal capacitance of solid, W-s/°C
C_p	Specific heat, W-s/kg-°C
D_h	Hydraulic diameter, cm
h	Heat-transfer coefficient, W/cm²-°C
k	Conductivity of material, W/cm-°C
L	Thermal path length, cm
$\dot{m}$	Mass flow rate, kg/s

Nu Nusselt number
P   Wetted perimeter of flow passage, cm
Pr Prandtl number
$\dot{Q}$ Heat-transfer rate, W
R   Resistance to conduction heat transfer, °C/W
Re Reynolds number
T   Temperature, °C
V   Volume, cm^3
V_a Flow velocity, cm/s

Greek

ΔT Temperature difference, °C
ν Kinematic viscosity of air, cm^2/s
ρ Density, kg/cm^3
τ Time constant, s

Reference

1. From Frank P. Incropera and David P. DeWitt, *Fundamentals of Heat and Mass Transfer*, Wiley, New York, 1985, pp. 5, 8, 64, 65, 68, 175–177, 369, 386, 397, 404. This material used by permission of John Wiley & Sons, Inc.

Chapter 16

Thermal Response of Laminates to Cyclic Heat Input from the Edge

Wataru Nakayama
ThermTech International
Kanagawa, Japan

Introductory Note

Laminates of dissimilar materials are commonly found in industrial products. Heat conduction in laminates is a well-studied subject, so there is a sizable volume of literature on it. However, actual calculation of temperature and heat flow in laminates is by no means simple, particularly when one deals with transient heat conduction problems. The difficulty amplifies where the laminate structure lacks regularity in the geometry and dimensions of component materials. For instance, printed-circuit boards (PCBs) in electronic equipment have diverse layouts of copper strips embedded in resin matrix. To solve heat conduction problems in such actual laminates, we resort to numerical analysis using the finite-difference or finite-element method. In more recent years, the art of numerical analysis (code) and the computer on which the code is run have been made powerful enough to deal with complex heat conduction problems. However, in many instances, it is still prohibitively time-consuming to find the full details of transient heat conduction in laminate structures. We still need to exercise the art of reducing the factors and variables involved in the problem in order to perform the analysis in a reasonable timeframe.

16.2 Transient and Cyclic Calculations

In this section we are concerned with one of the issues involved in the art of modeling. Suppose that a physical volume in that heat is generated intermittently from embedded heat sources, as is the case in almost all electronic devices and equipment. Transient heat generation produces a temperature distribution in the volume that is basically nonsteady (transient). However, when the heat generation occurs at a very high frequency, as on the microprocessor chip, the temporal variation of temperature is restricted to a very narrow region around the heat source, and the rest of the zone in the volume has a quasi-steady temperature distribution. So, for practical purposes, the assumption of steady-state temperature is applied, obviating the need to perform transient analysis. Consider another extreme end where the frequency of heat generation is very low. Again, the steady-state assumption makes sense, and the analysis for a situation of peak heat generation suffices to find whether the heat-source temperature can be held below a tolerable level in a given thermal environment. In between these extremes, very high and very low frequency ends, there are intermediate situations where the transient analysis cannot be bypassed altogether. But still, the transient analysis may be applied to a region near the heat source, and a quasi-steady temperature field may be assumed in the rest of the region. We suppose a boundary around the heat source that separates the zone of transient temperature and that of quasi-steady temperature. If we know such boundaries beforehand, we can program our task in a time-saving way, that is, the time-consuming transient analysis is performed only where the temperature varies perceptively with time. But how do we know the possible location of boundaries before embarking on full-scale numerical analysis? The size of the region of large temperature variation around the heat source depends on the heat generation frequency, the material composition and dimensions of the laminate, and the heat transfer from the surface(s) of the laminate to the environment. We need a guide that suggests a spatial extent of temporal variation of the temperature, and such a guide should involve relatively simple calculations. A heat conduction problem is devised here to develop a guide.

Problem Definition and Formal Solutions

Figure 16.1 explains the problem to be solved. A laminate is composed of low- and high-thermal-conductivity layers, alternately stacked and perfectly bonded rendering interfacial thermal resistance to zero. The laminate is semi-infinite in the direction of the layers (x), and cyclic heat input is applied to the edge ($x = 0$) of one of the high-conductivity layers. The edge of the laminate is adiabatic except for the edge of the heat source layer. The upper and lower surfaces of the laminate are convectively cooled.

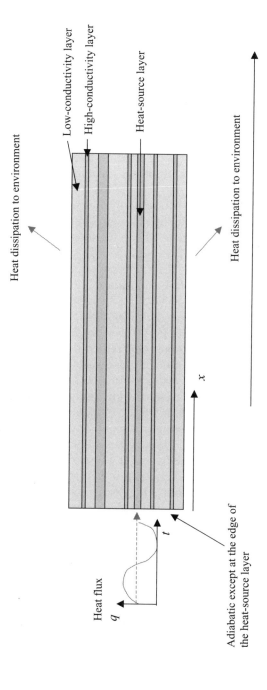

Figure 16.1 Laminate of alternate low- and high-conductivity layers with cyclic heat input at the edge of one of the high-conductivity layers.

16.4 Transient and Cyclic Calculations

In terms of the heat flux, q (W/m^2), the heat input at the edge ($x = 0$) of the heat source layer is given as

$$q = q_0[1 + \cos(2\pi f t)] \qquad (16.1)$$

where q_0 is the amplitude, f is the frequency in hertz, and t is the time in seconds. Note that the heat flux becomes maximum ($2q_0$) at $t = (n-1)/f$, and zero at $t = (2n-1)/2f$, where n is an integer. We assume that $t = 0$ is taken at an instant long after the start of cyclic heat application. Then, the thermal response of the laminate to the heat input includes the steady component corresponding to the first term (q_0) on the right-hand side of Eq. (16.1) and the cyclic component corresponding to the second term [$q_0\cos(2\pi f t)$].

Figure 16.2 shows conceptual sketches of the responses of the heat-source layer temperature to cyclic heat input. The temperature here is that averaged across the thickness of the heat-source layer. Also, the environment temperature is set as zero, so θ is the temperature rise from the environment. When the frequency is low, the cyclic component dominates the temperature distribution along x (Fig. 16.2a). For this quasi-steady situation the analysis can be focused on the temperature distribution produced by the peak heat input $2q_0$. When the frequency is high, the transient component is diffused in a short distance from the edge, and the rest of the region is covered by the steady component (Fig. 16.2b). We introduce the notion of penetration distance. The *penetration distance* of the transient component (l_t) is the distance from the edge to a point where the amplitude of cyclic temperature variation is reduced to $(\theta_{\max} - \theta_{\min}) \times 1/e = 0.368(\theta_{\max} - \theta_{\min})$, where $\theta_{\max}$ and $\theta_{\min}$ are the peak and the minimum temperatures at $x = 0$, respectively. The penetration distance of the steady component (l_s) is the distance from the edge to a point where the time-averaged temperature θ_{mid} is reduced to $\theta_{\text{mid}} \times 1/e = 0.368\theta_{\text{mid}}$. These penetration distances are indicative of the extent of the respective regions. Once we know the order of magnitude of penetration distances, we apply the knowledge to the numerical analysis on actual laminates that may have more complex planar patterns of high-conductivity layers, but retain essential features such as the number of layers, the layer thicknesses, the component materials, and the heat-transfer coefficients on the laminate surfaces. In other words, we define the region of analysis in a distance around the heat source by $l_s \times$ (a multiplication factor of the analyst's choice), and also program to perform transient heat conduction analysis only in a region around the heat source whose expanse is set as $l_t \times$ (a multiplication factor of the analyst's choice). Such an approach will produce considerable saving in computer memory usage and computing time compared to all-out analysis tracking temporal temperature variations in the whole spatial domain.

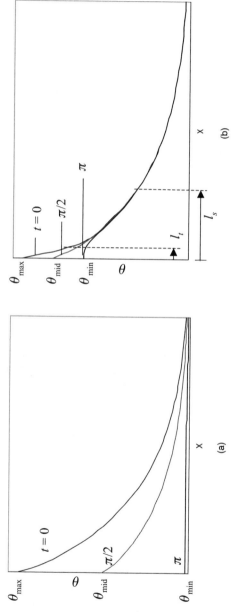

Figure 16.2 Responses of heat-source layer temperature to cyclic heat input from the edge $q = q_0(1 + \cos(2\pi f))$: (a) low-frequency input; (b) high-frequency input. In (b), the penetration distances are illustrated; l_s is the penetration distance of the steady component, and l_t, that of the transient component.

16.6 Transient and Cyclic Calculations

Besides the penetration distances, knowledge of the amplitude of temperature swing at the heat source, as indicated by θ_{max} and θ_{min} in Fig. 16.2, is useful, as the heat-source temperature is of primary interest in thermal design. So, our ultimate interest is in the penetration distances and the temperature swing at the heat source. Knowledge of these characteristic quantities is derived from the solution of the set of equations governing heat conduction in the laminate. However, as the literature testifies, for example, in Refs. 1 and 2, the rigorous analysis involves complex mathematics that deter practicing engineers from even trying to understand it. Also, the rigor of mathematics makes the analysis practically tractable only in those cases where the laminate has a regular composition, that is, a pair of low-conductivity and high-conductivity layers, each having a respectively fixed thickness, repeated in the direction of layer stack. In what follows we introduce an approximation that relaxes the constraint on the laminate composition.

The approximation is based on the observation of laminates commonly found in electronic equipment. Thus, thin metals (high conductivity layer) are embedded in resin matrix (low conductivity layer), and the following disparity relations hold:

$$k_M >> k \ (\alpha_M >> \alpha) \qquad d_M << d$$

where k is the thermal conductivity, α is the thermal diffusivity, d is the layer thickness, the subscript M is attached to the properties of the high-conductivity layer, and the properties of the low-conductivity layer come without subscript. These disparate relations suggest the systematic approximation based on the series expansion of variables in terms of small parameters such as the thermal diffusivity ratio (α/α_M). Here, we note only the first-order approximate equations as follows:

High-k layer

$$\omega_{jM} \frac{\partial \theta_j}{\partial t} - \frac{\partial^2 \theta_j}{\partial x^2} + \frac{1}{d_{kM}}(q_{Lj} - q_{Uj}) = 0 \qquad (16.2)$$

$$\text{At } x = 0: \quad j = j^* \quad \frac{\partial \theta_{j*}}{\partial x} = -e^{-it} \qquad (16.3)$$

$$j \neq j^* \quad \frac{\partial \theta_j}{\partial x} = 0 \qquad (16.4)$$

$$\text{At } x \to \infty: \quad \theta_j \to 0 \qquad (16.5)$$

$$q_{Uj} = -\lambda_j \left(\frac{\partial T_j}{\partial y_j} \right)_{y_j = d_j} \qquad (16.6)$$

$$q_{Lj} = -\lambda_{j+1} \left(\frac{\partial T_{j+1}}{\partial y_{j+1}} \right)_{y_{j+1} = 0} \qquad (16.7)$$

Low-k layer

$$\omega_j \frac{\partial T_j}{\partial t} - \frac{\partial^2 T_j}{\partial y_j^2} = 0 \qquad (16.8)$$

$$y_j = 0 \qquad T_j = \theta_{j-1} \qquad (16.9)$$

$$y_j = d_j \qquad T_j = \theta_j \qquad (16.10)$$

$$j = 1 \qquad \left(\frac{\partial T_1}{\partial y_1}\right)_{y_1=0} = B_U(T_1)_{y_1=0} \qquad (16.11)$$

$$j = J \qquad -\left(\frac{\partial T_J}{\partial y_J}\right)_{y_J=d_J} = B_L(T_J)_{y_J=d_J} \qquad (16.12)$$

Equations (16.2) to (16.12) are written in terms of nondimensional quantities. The definitions of nondimensional quantities are listed in Table 16.1. The formulation is intended for general situations; there are $J \times$ (low-conductivity layers) and $(J - 1) \times$ (high-conductivity layers).

TABLE 16.1 Symbol Definitions

	Nondimensional	Definition in dimensional symbols
Temperature	θ_j	$\theta_j/\Delta T$
	T_j	$T_j/\Delta T$
Time	t	ωt
Length	x	x/d
	y_j	y_j/d
	d_j	d_j/d
	d_{jM}	d_{jM}/d
Heat flux	q_{Uj}	q_{Uj}/q_0
	q_{Lj}	q_{Lj}/q_0
Thermal conductivity[†]	k_j	k_j/k_{jM*}
	k_{jM}	k_{jM}/k_{jM*}
Angular frequency	ω_j	$\omega d^2/\alpha_j$
	ω_{jM}	$\omega d^2/\alpha_{jM}$
Biot number[‡]	B_U	$h_U d/k_1$
	B_L	$h_L d/k_J$

Reference scales used in nondimensionalization:

Length	d, m ($=$ the total laminate thickness)
Time	$1/\omega$, s ($\omega = 2\pi f$)
Heat flux	q_0, W/m^2
Temperature	$\Delta T \equiv q_0 d/k_{jM*}$ K

[†]$k_{jM}^* =$ thermal conductivity of heat-source layer.
[‡]$h_U =$ heat-transfer coefficient on upper surface of laminate, W/m^2 K; $h_L =$ heat-transfer coefficient on lower surface of the laminate, W/m^2 K.

16.8 Transient and Cyclic Calculations

The subscript j is the index for a pair of low- and high-conductivity layers, $j = 1$ and J are the top and the bottom layers, respectively, cooled by convection heat transfer. The subscript j^* denotes the high-conductivity layer bonded to the heat source at its edge.

The symbol θ is used for the thickness-averaged temperature of the high-conductivity layer. Equation (16.2) is an integral equation expressing heat balance on an infinitesimal segment of the high-conductivity layer. Equations (16.3) to (16.7) define the boundary conditions for Eq. (16.2). Equation (16.3) is the boundary condition at the edge ($x = 0$) of the heat-source layer, where the heat input is expressed in complex form. The edges of other high-conductivity layers are thermally insulated as defined by Eq. (16.4). Equations (16.6) and (16.7) express the continuity condition for heat fluxes; q_{Uj} is the heat flux from the jth low-conductivity layer to the jth high-conductivity layer, and q_{Lj} is that from the jth high-conductivity layer to the $(j+1)$th low-conductivity layer. The coordinate y_j is defined in the jth low-conductivity layer, originating from the interface with the $(j-1)$th high-conductivity layer and directing normal to the interface. The y_j spans the thickness of the jth low-conductivity layer (d_j); hence, $0 \leq y_j \leq d_j$.

The temperature in the jth low-conductivity layer is denoted as T_j. Although T_j is a function of x and y_j, it is assumed to be governed by Eq. (16.8), which involves only the term for heat conduction in the y direction. (Formal series expansion of the original equation in terms of the thermal diffusivity ratio provides a way to take into account the heat conduction in the x direction. However, it is lengthy and will not be attempted here.) Equations (16.9) and (16.10) express the continuity condition for temperatures. Equation (16.11) is applied to the upper surface of the laminate where the convection heat-transfer condition is specified in terms of the Biot number (Table 16.1). Likewise, Eq. (16.12) is applied to the bottom surface of the laminate.

Although the present approximation renders the analysis as one involving only trigonometric and hyperbolic functions, a full description of the solution procedure is too lengthy to be included here, so only the key steps are described. The temperatures are written as

$$\theta_j = U_j(x) e^{-it} \tag{16.13}$$

$$T_j = V_j(\eta_j) e^{-it} \tag{16.14}$$

where $\eta_j \equiv y_j \sqrt{\omega_j/2}$. Equation (16.2) is reduced to the ordinary differential equation for the amplitude function $U_j(x)$ and Eq. (16.8), to that for $V_j(\eta_j)$. First, the equation for $V_j(\eta_j)$ is solved, and the solution for $V_j(\eta_j)$ is written in terms of $U_j(x)$, $U_{j-1}(x)$, η_j, and $\delta_j \equiv d_j \sqrt{\omega_j/2}$. Then, using the formal solution of $V_j(\eta_j)$ and $V_{j+1}(\eta_j)$, the simultaneous equations for high-conductivity layers are written as

$$\frac{d^2 U_j}{dx^2} + i\omega_{jM} U_j = \frac{1}{d_{jM}}(F_j U_{j-1} + G_j U_j + H_j U_{j+1}) \quad \text{for} \quad 1 \le j \le J-1$$

(16.15)

where F_j, G_j, and H_j are the functions of k_j, k_{j+1}, ω_j, ω_{j+1}, δ_j, δ_{j+1}. For $j = 1$ and $J-1$, these functions also include B_U and B_L. The $U_j(x)$ is further decomposed to the real part, U_j^R, and the imaginary part, U_j^I, and they are written in the form

$$U_j^R = a_j \cdot \exp(\beta x) \cdot \cos(\gamma x) \tag{16.16}$$

$$U_j^I = b_j \cdot \exp(\beta x) \cdot \sin(\gamma x) \tag{16.17}$$

Our task is now reduced to determining a_j, b_j, β, and γ, where $1 \le j \le J-1$. Substituting Eqs. (16.16) and (16.17) into Eq. (16.15), and equating the like terms of $\exp(\beta x) \cdot \cos(\gamma x)$, we have the simultaneous algebraic equations. The equations are in matrix form:

$$\mathbf{C}^R \cdot \mathbf{A} = (\beta^2 - \gamma^2) \mathbf{I} \cdot \mathbf{A} \tag{16.18}$$

$$\mathbf{C}^I \cdot \mathbf{A} = 2\beta\gamma \mathbf{I} \cdot \mathbf{B} \tag{16.19}$$

where $\mathbf{A} = [a_1 \ a_2 \ \cdots \ a_j \ \cdots \ a_{J-1}]^T$, $\mathbf{B} = [b_1 \ b_2 \ \cdots \ b_j \ \cdots \ b_{J-1}]^T$, $\mathbf{C}^R$ and $\mathbf{C}^I$ are the coefficient matrixes, and $\mathbf{I}$ is the unity matrix. When the like terms of $\exp(\beta x) \cdot \sin(\gamma x)$ are equated, $\mathbf{A}$ and $\mathbf{B}$ in Eqs. (16.18) and (16.19) are interchanged. To obtain finite values for the elements of $\mathbf{A}$ and $\mathbf{B}$, the following condition has to be satisfied:

$$\left| \mathbf{C}^R - (\beta^2 - \gamma^2) \mathbf{I} \right| = 0 \tag{16.20}$$

$$\left| (\mathbf{C}^I + 2\beta\gamma \mathbf{I})(\mathbf{C}^I - 2\beta\gamma \mathbf{I}) \right| = 0 \tag{16.21}$$

Equation (16.21) is derived from Eq. (16.18) and the one having $\mathbf{A}$ and $\mathbf{B}$ interchanged. Equations (16.20) and (16.21) are used to determine β and γ; there are at most $(J-1)$ values of β and also an equal number of γ. From among the solutions we select those that meet the condition in Eq. (16.5); that is, the temperature diminishes to zero for $x \to \infty$.

The elements of $\mathbf{A}$ and $\mathbf{B}$ take different values for different β and γ [written as $\beta^{(j')}$ and $\gamma^{(j')}$; $1 \le j' \le J-1$]. Hence, they are now given additional subscripts j', $a_{j,j'}$, and $b_{j,j'}$, where $1 \le j \le J-1$ and $1 \le j' \le J-1$. There are now $2 \times (J-1) \times (J-1)$ unknown quantities. Equations (16.20) and (16.21) guarantee that some rows in $\mathbf{C}^R - (\beta^2 - \gamma^2)\mathbf{I}$ and $\mathbf{C}^I - 2\beta\gamma \mathbf{I}$ (or $\mathbf{C}^I + 2\beta\gamma \mathbf{I}$) are identical. Hence

16.10 Transient and Cyclic Calculations

at most $2 \times (J-1) \times (J-2)$ equations among Eqs. (16.18) and (16.19) are independent, which provide a part of the equations used to determine $a_{j,j'}$ and $b_{j,j'}$. Other $2 \times (J-1)$ equations are obtained from the real and imaginary parts of the boundary condition expressed in Eqs. (16.3) and (16.4).

The formal solution for the temperature in the jth high-conductivity layer is written as

$$U_j^R = \sum_{j'=1}^{J-1} a_{j,j'} \cdot \exp\left(\beta^{(j')} x\right) \cdot \cos\left(\gamma^{(j')} x\right) \qquad (16.22)$$

$$U_j^I = \sum_{j'=1}^{J-1} b_{j,j'} \cdot \exp\left(\beta^{(j')} x\right) \cdot \sin\left(\gamma^{(j')} x\right) \qquad (16.23)$$

Amplitude: $\qquad A_{\theta j} = \sqrt{\left(U_j^R\right)^2 + \left(U_j^I\right)^2} \qquad (16.24)$

Phase angle: $\qquad \phi_{\theta j} = \tan^{-1}\left(\dfrac{U_j^I}{U_j^R}\right) \qquad (16.25)$

Temperature: $\qquad \theta_j^{tr} = A_{\theta j} \cos(t - \phi_{\theta j}) \qquad (16.26)$

This derivation is for the transient component; hence superscript tr appears on θ_j in Eq. (16.26). The steady component θ_j^{st} can be obtained by setting $\omega_{jM} \to 0$ in Eq. (16.15). Since the imaginary part of the solution vanishes, as suggested from omission of $i\omega_{jM}$ from Eq. (16.15), $\mathbf{C}^I$ in Eq. (16.19) becomes an all-zero element matrix; hence $\gamma = 0$. The $\beta^{(j')}$ and $a_{j,j'}$ for the steady component, denoted as $\beta_{st}^{(j')}$ and $a_{j,j'}^{st}$, can be determined following the steps described above for the transient component. The steady component and the sum of steady and transient components are, respectively

$$\theta_j^{st} = \sum_{j'=1}^{J-1} a_{j,j'}^{st} \exp\left(\beta_{st}^{(j')} x\right) \qquad (16.27)$$

$$\theta_j = \theta_j^{st} + \theta_j^{tr} \qquad (16.28)$$

The penetration distances are estimated as follows. The boundary condition at $x = 0$ [Eqs. (16.3) and (16.4)] suggests that the product $a_{j,j'} \times \beta^{(j')}$ appears on the left-hand side of these equations. Hence, generally, a large $a_{j,j'}$ is associated with a small $\beta^{(j')}$. This means that the largest terms in the series expression for θ_j^{st} and θ_j^{tr} penetrate deep in the x direction. Using the values of $\beta^{(j)}$ for the largest term (denoted as $\beta_{\text{pnt}}^{\text{st}}$ and $\beta_{\text{pnt}}^{\text{tr}}$), we calculate the penetration distances for the steady and transient components, respectively:

$$l_s = \frac{1}{\beta_{\text{pnt}}^{\text{st}}} \qquad (16.29)$$

$$l_t = \frac{1}{\beta_{\text{pnt}}^{\text{tr}}} \qquad (16.30)$$

Case Studies

Printed-circuit board

In a printed-circuit board the embedded copper layers have significant influence on heat spreading. In particular, the power supply and ground planes are continuous sheets of copper; hence, they are effective heat spreaders. We consider here a laminate having two high-conductivity layers and three low-conductivity layers; hence, $J = 3$. The high-conductivity layer $j = 1$ is the heat-source layer. The details of analytical solutions are given in Table 16.2.

The actual dimensional data and the Biot numbers are assumed to be as follows:

Thermal properties of copper
- $k_{1M} = k_{2M} = 372$ W/m K
- $\alpha_{1M} = \alpha_{2M} = 1.063 \times 10^{-4}$ m^2/s

Thermal properties of insulation (resin)
- $k_1 = k_2 = k_3 = 0.3$ W/m K
- $\alpha_1 = \alpha_2 = \alpha_3 = 1.50 \times 10^{-7}$ m^2/s

Thicknesses
- $d_{1M} = d_{2M} = 30$ μm $= 3 \times 10^{-5}$ m
- $d_1 = d_2 = d_3 = 300$ μm $= 3 \times 10^{-4}$ m
- $d = 0.96$ mm $= 9.6 \times 10^{-4}$ m

Biot numbers
- $B_U = B_L = 0.03$ ($h_U = h_L = 30$ W/m^2 K)

Figure 16.3 shows the maximum and minimum temperatures of the copper layer $j = 1$ at $x = 0$ versus the heat input frequency. For the frequency below approximately 0.01 Hz, that is, for the cycle time longer than 1.7 min, quasi-steady state is attained. Figure 16.3 also indicates that the amplitude of temperature variation is suppressed at high frequencies.

16.12 Transient and Cyclic Calculations

TABLE 16.2 Equations to Calculate Thermal Performance of Heat-Source Layer

Case of two high-conductivity layers laminated with three low-conductivity layers. All symbols used in (3) to (5) are nondimensional.

(1) Dimensional data of component layers

Layer no.	Thickness, m	Thermal conductivity, W/m K	Thermal diffusivity, m²/s
1	d_1	k_1	α_1
1M (heat source)	d_{1M}	k_{1M}	α_{1M}
2	d_2	k_2	α_2
2M	d_{2M}	k_{2M}	α_{2M}
3	d_3	k_3	α_3

(2) Scales used in nondimensionalization
Length $\quad d = d_1 + d_{1M} + d_2 + d_{2M} + d_3$
Time $\quad 1/\omega$
Temperature $\quad q_0 d / k_{1M}$

(3) Nondimensional parameters and variables; see Table 16.1
Additional parameters:

$$\delta_1 \equiv d_1 \sqrt{\frac{\omega_1}{2}} \qquad \delta_2 \equiv d_2 \sqrt{\frac{\omega_2}{2}} \qquad \delta_3 \equiv d_3 \sqrt{\frac{\omega_3}{2}}$$

where d_j and ω_j ($j = 1 \sim 3$) are the nondimensional thickness and the nondimensional angular frequency, respectively, as defined in Table 16.1

(4) Steps in calculation of steady component

$$\chi_+ \equiv \frac{1}{d_{1M}} \left(\frac{B_U}{B_U d_1 + 1} k_1 + \frac{k_2}{d_2} \right) + \frac{1}{d_{2M}} \left(\frac{B_L}{B_L d_3 + 1} k_3 + \frac{k_2}{d_2} \right)$$

$$\chi_- \equiv \frac{1}{d_{1M}} \left(\frac{B_U}{B_U d_1 + 1} k_1 + \frac{k_2}{d_2} \right) - \frac{1}{d_{2M}} \left(\frac{B_L}{B_L d_3 + 1} k_3 + \frac{k_2}{d_2} \right)$$

$$\psi \equiv \frac{2k_2}{d_2 \sqrt{d_{1M} d_{2M}}}$$

$$\beta_{st}^{(1)} = -\frac{1}{\sqrt{2}} \left[\chi_+ + \sqrt{\chi_-^2 + \psi^2} \right]^{1/2}$$

$$\beta_{st}^{(2)} = -\frac{1}{\sqrt{2}} \left[\chi_+ - \sqrt{\chi_-^2 + \psi^2} \right]^{1/2} \equiv \beta_{pnt}^{st}$$

$$a_{1,1}^{st} = -\frac{\chi_- + \sqrt{\chi_-^2 + \psi^2}}{2\beta_{st}^{(1)} \sqrt{\chi_-^2 + \psi^2}}$$

$$a_{1,2}^{st} = -\frac{-\chi_- + \sqrt{\chi_-^2 + \psi^2}}{2\beta_{st}^{(2)} \sqrt{\chi_-^2 + \psi^2}}$$

Heat-source temperature $\quad (\theta_1^{st})_{x=0} = a_{1,1}^{st} + a_{1,2}^{st}$

Penetration distance $\quad l_s = 1/\beta_{pnt}^{st}$

TABLE 16.2 Equations to Calculate Thermal Performance of Heat-Source Layer (Continued)

(5) Steps in calculation of transient component

$$D_1 = B_U^2\{\cosh(2\delta_1) - \cos(2\delta_1)\} + B_U\sqrt{2\omega_1}\{\sinh(2\delta_1) + \sin(2\delta_1)\}$$
$$\quad + \omega_1\{\cosh(2\delta_1) + \cos(2\delta_1)\}$$

$$D_2 = \cosh(2\delta_2) - \cos(2\delta_2)$$

$$D_3 = B_L^2\{\cosh(2\delta_3) - \cos(2\delta_3)\} + B_L\sqrt{2\omega_3}\{\sinh(2\delta_3) + \sin(2\delta_3)\}$$
$$\quad + \omega_3\{\cosh(2\delta_3) + \cos(2\delta_3)\}$$

Real parts of coefficients

$$G_{1R} = \frac{k_1}{D_1}\sqrt{\frac{\omega_1}{2}}\Big[B_U^2\{\sinh(2\delta_1) + \sin(2\delta_1)\} + B_U\sqrt{2\omega_1}\{\cosh(2\delta_1) + \cos(2\delta_1)\}$$
$$\quad + \omega_1\{\sinh(2\delta_1) - \sin(2\delta_1)\}\Big] + \frac{k_2}{D_2}\sqrt{\frac{\omega_2}{2}}[\sinh(2\delta_2) + \sin(2\delta_2)]$$

$$H_{1R} = \frac{k_2}{D_2}\sqrt{\frac{\omega_2}{2}}[-2\sinh(\delta_2)\cos(\delta_2) - 2\cosh(\delta_2)\sin(\delta_2)]$$

$$F_{2R} = \frac{k_2}{D_2}\sqrt{\frac{\omega_2}{2}}[-2\sinh(\delta_2)\cos(\delta_2) - 2\cosh(\delta_2)\sin(\delta_2)]$$

$$G_{2R} = \frac{k_2}{D_2}\sqrt{\frac{\omega_2}{2}}[\sinh(2\delta_2) + \sin(2\delta_2)]$$
$$\quad + \frac{k_3}{D_3}\sqrt{\frac{\omega_3}{2}}\Big[B_L^2\{\sinh(2\delta_3) + \sin(2\delta_3)\} + B_L\sqrt{2\omega_3}\{\cosh(2\delta_3) + \cos(2\delta_3)\}$$
$$\quad + \omega_3\{\sinh(2\delta_3) - \sin(2\delta_3)\}\Big]$$

$$c_{1,1}^R = \frac{G_{1R}}{d_{1M}} \qquad c_{1,2}^R = \frac{H_{1R}}{d_{1M}}$$

$$c_{2,1}^R = \frac{F_{2R}}{d_{2M}} \qquad c_{2,2}^R = \frac{G_{2R}}{d_{2M}}$$

$$X_R^{(1)} = \frac{1}{2}\left[c_{1,1}^R + c_{2,2}^R + \sqrt{\left(c_{1,1}^R + c_{2,2}^R\right)^2 - 4\left(c_{1,1}^R c_{2,2}^R - c_{1,2}^R c_{2,1}^R\right)}\right]$$

$$X_R^{(2)} = \frac{1}{2}\left[c_{1,1}^R + c_{2,2}^R - \sqrt{\left(c_{1,1}^R + c_{2,2}^R\right)^2 - 4\left(c_{1,1}^R c_{2,2}^R - c_{1,2}^R c_{2,1}^R\right)}\right]$$

Imaginary parts of coefficients

$$G_{1I} = \frac{k_1}{D_1}\sqrt{\frac{\omega_1}{2}}\Big[B_U^2\{-\sinh(2\delta_1) + \sin(2\delta_1)\} + B_U\sqrt{2\omega_1}\{-\cosh(2\delta_1) + \cos(2\delta_1)\}$$
$$\quad + \omega_1\{-\sinh(2\delta_1) - \sin(2\delta_1)\}\Big] + \frac{k_2}{D_2}\sqrt{\frac{\omega_2}{2}}[-\sinh(2\delta_2) + \sin(2\delta_2)]$$

16.14 Transient and Cyclic Calculations

TABLE 16.2 Equations to Calculate Thermal Performance of Heat-Source Layer (*Continued*)

$$H_{1I} = \frac{k_2}{D_2}\sqrt{\frac{\omega_2}{2}}[-2\cosh(\delta_2)\sin(\delta_2) + 2\sinh(\delta_2)\cos(\delta_2)]$$

$$F_{2I} = \frac{k_2}{D_2}\sqrt{\frac{\omega_2}{2}}[-2\cosh(\delta_2)\sin(\delta_2) + 2\sinh(\delta_2)\cos(\delta_2)]$$

$$G_{2I} = \frac{k_2}{D_2}\sqrt{\frac{\omega_2}{2}}[-\sinh(2\delta_2) + \sin(2\delta_2)]$$

$$+ \frac{k_3}{D_3}\sqrt{\frac{\omega_3}{2}}\Big[B_L^2\{-\sinh(2\delta_3) + \sin(2\delta_3)\} + B_L\sqrt{2\omega_3}\{-\cosh(2\delta_3) + \cos(2\delta_3)\}$$

$$+ \omega_3\{-\sinh(2\delta_3) - \sin(2\delta_3)\}\Big]$$

$$c_{1,1}^I = \frac{G_{1I}}{d_{1M}} - \omega_{1M} \qquad c_{1,2}^I = \frac{H_{1I}}{d_{1M}}$$

$$c_{2,1}^I = \frac{F_{2I}}{d_{2M}} \qquad c_{2,2}^I = \frac{G_{2I}}{d_{2M}} - \omega_{2M}$$

$$X_I^{(1)} = \frac{1}{2}\left[c_{1,1}^I + c_{2,2}^I + \sqrt{\left(c_{1,1}^I + c_{2,2}^I\right)^2 - 4\left(c_{1,1}^I c_{2,2}^I - c_{1,2}^I c_{2,1}^I\right)}\right]$$

$$X_I^{(2)} = \frac{1}{2}\left[c_{1,1}^I + c_{2,2}^I - \sqrt{\left(c_{1,1}^I + c_{2,2}^I\right)^2 - 4\left(c_{1,1}^I c_{2,2}^I - c_{1,2}^I c_{2,1}^I\right)}\right]$$

Exponents

$$\beta^{(1)} = -\frac{1}{\sqrt{2}}\left[X_R^{(1)} + \sqrt{X_R^{(1)2} + X_I^{(1)2}}\right]^{1/2}$$

$$\beta^{(2)} = -\frac{1}{\sqrt{2}}\left[X_R^{(2)} + \sqrt{X_R^{(2)2} + X_I^{(2)2}}\right]^{1/2} \equiv \beta_{pnt}^{tr}$$

Amplitude components

$$a_{1,1}^{tr} = -\frac{1}{\beta^{(1)}} \cdot \frac{c_{1,1}^R - X_R^{(2)}}{X_R^{(1)} - X_R^{(2)}}$$

$$a_{1,2}^{tr} = \frac{1}{\beta^{(2)}} \cdot \frac{c_{1,1}^R - X_R^{(1)}}{X_R^{(1)} - X_R^{(2)}}$$

Heat-source temperature $\qquad (\theta_1^{tr})_{x=0} = \left(a_{1,1}^{tr} + a_{1,2}^{tr}\right)\cos t$

Penetration distance $\qquad l_t = 1/\beta_{pnt}^{tr}$

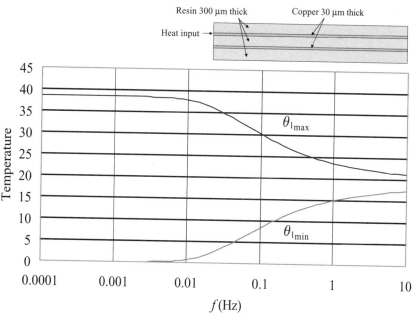

Figure 16.3 Maximum and minimum temperatures (nondimensional) of the heat source versus heat input frequency. The case of a printed-circuit board.

Figure 16.4 shows the penetration distances versus the heat input frequency. The penetration distance of the steady component l_s is about 36; in actual dimensions, this is $36 \times 0.96 = 34$ mm. In quasi-steady state below 0.01 Hz, the steady and transient components of temperature become indistinguishable. The penetration distance of transient component l_t decreases with increasing frequency. For example, at 1 Hz (1-s cycle), $l_t = 6.33$; in actual dimensions, this is 6 mm.

Graphite sheet heat spreader on substrates

Graphite sheets having high thermal conductivity have been developed. They are used to diffuse heat in compact electronic equipment such as handheld equipment. Such a graphite sheet, however, is less than approximately 0.5 mm thick because of constraints in manufacturing, so it has to be bonded to a host substrate that provides mechanical support. The substrate is often the equipment casing, and thus serves as intermediate heat conduction medium between the graphite sheet heat spreader and the environment. So, we consider here a bilayer (double layer) composed of a graphite sheet and the substrate, with the upper side of the substrate cooled by convection heat transfer and the lower side of the graphite sheet adiabatic. The heat source is attached to

16.16 Transient and Cyclic Calculations

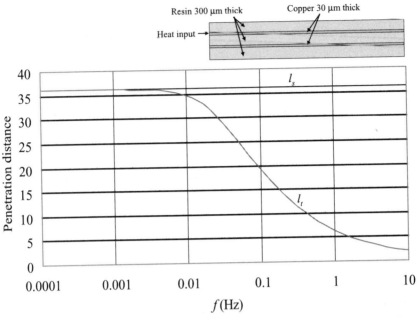

Figure 16.4 Penetration distances (nondimensional) along the printed-circuit board versus heat input frequency.

the edge ($x = 0$) of the graphite sheet. The equations used to calculate thermal responses of this bilayer system to cyclic heat input can be derived from those given in Table 16.2 without much elaboration, that is, by reducing the number of high- and low-conductivity layer pairs to one.

We also address here the issue of the effect of casing material on heat transfer by considering two materials for the substrate, one is plastic and the other, magnesium alloy. The dimensional data and the Biot numbers are assumed as follows:

Thermal properties of plastic
- $k_1 = 0.7$ W/m K
- $\alpha_1 = 3.5 \times 10^{-7}$ m²/s

Thermal properties of magnesium alloy
- $k_1 = 72$ W/m K
- $\alpha_1 = 4.04 \times 10^{-5}$ m²/s

Thermal properties of graphite
- $k_{1M} = 600$ W/m K
- $\alpha_{1M} = 8.46 \times 10^{-4}$ m²/s

Thermal Response of Laminates to Cyclic Heat Input

Thicknesses

- $d_1 = 1$ mm $= 1 \times 10^{-3}$ m
- $d_{1M} = 0.5$ mm $= 5 \times 10^{-4}$ m
- $d = 1.5$ mm $= 1.5 \times 10^{-3}$ m

Biot numbers

- Plastic substrate: $B_U = 0.02$ ($h_U = 9.3$ W/m² K)
- Magnesium alloy: $B_U = 0.0002$ ($h_U = 9.6$ W/m² K)
- For both substrates: $B_L = 0$

Figure 16.5 shows the maximum and minimum temperatures of the graphite sheet at $x = 0$ versus the heat input frequency. The solid curves denote the plastic substrate and the broken curves, the magnesium alloy substrate. Quasi-steady state is attained below 0.0001 Hz, that is, when the cycle time is longer than 2.8 min. Above 1 Hz (cycle time shorter than 1 s), the amplitude of temperature variation becomes so small that, again, quasi-steady state can be assumed for almost the whole expanse of this bilayer composite.

Figure 16.6 shows the penetration distances versus the heat input frequency. The solid curves represent the plastic substrate, and the

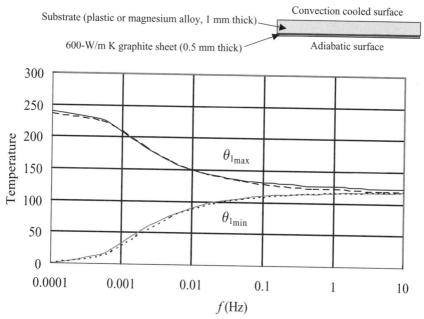

Figure 16.5 Maximum and minimum temperatures (nondimensional) of the heat source at the edge of a graphite sheet bonded to a convection-cooled substrate. The solid curves represent the plastic substrate; the broken curves, the magnesium alloy substrate.

16.18 Transient and Cyclic Calculations

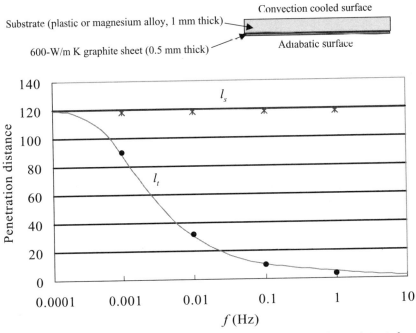

Figure 16.6 Penetration distances (nondimensional) along the graphite-substrate laminate. The solid curves denote the plastic substrate; the markers, the magnesium alloy substrate.

markers denote the magnesium alloy substrate. The penetration distance of the steady component l_s is about 120 for both substrates; in actual dimensions, this is $120 \times 1.5 = 180$ mm. In quasi-steady state below 0.0001 Hz, the steady and transient components of temperature become indistinguishable. The penetration distance of the transient component l_t decreases with increasing frequency; at 1 Hz (1-s cycle), $l_t = 5.3$ (8 mm).

Little difference in thermal response is seen between the graphite-plastic and the graphite-magnesium alloy substrates. This is attributed to the small Biot numbers assumed here, which correspond to natural convection cooling of the casing surface. The specific thermal resistance of the plastic substrate is $d_1/k_1 = 1.43 \times 10^{-3}$ m² K/W, that is, $1/75$th of the surface resistance, $1/h_U = 0.108$ m² K/W. The thermal resistance of the magnesium alloy substrate is even much smaller. Hence, the surface heat transfer has a dominant effect on the thermal performance of the bilayer composite. The role of the graphite is important, of course. Without the graphite sheet, longitudinal heat spreading is much slower in the plastic substrate than in the magnesium alloy substrate; hence, higher temperature results on the plastic substrate. Lining of a plastic casing by a graphite sheet is, therefore, a useful technique for cooling small electronic equipment that is most likely cooled by natural convection.

References

1. A. M. Manaker and G. Horvay, "Thermal Response in Laminated Composites," *ZAMM* **55**:503–513 (1975).
2. A. H. Nayfeh, "Continuum Modeling of Low Frequency Heat Conduction in Laminated Composites with Bonds," *ASME Journal of Heat Transfer* **102**:312–318 (1980).

Chapter 17

A Simple Calculation Procedure for Transient Heat and Mass Transfer with Phase Change: Moist Air Example

Pedro J. Mago
Department of Mechanical Engineering
Mississippi State University
Mississippi State, Mississippi

S. A. Sherif
Department of Mechanical and Aerospace
  Engineering
University of Florida
Gainesville, Florida

Introduction

Heat- and mass-transfer problems with phase change are encountered in a variety of applications. Examples include drying operations, the chemical process industry, air conditioning and refrigeration, and manufacturing operations, to name only a few. The calculation procedure described in this chapter will apply to a moist air heat- and mass-transfer example commonly found in coolers and freezers. This usually involves calculating heat and mass transfer to a surface whose temperature is below the dew-point temperature of water vapor in air. If the temperature of the surface in question is above the freezing point of water,

17.2 Transient and Cyclic Calculations

condensation will occur. If, on the other hand, the surface temperature is below the freezing point, frost will form. Typically, the frost thickness is highly time-dependent, thus making the problem computationally intensive because it involves a transient moving-boundary problem. Other complicating factors pertains to the fact that the rate at which water vapor is diffused into the frost layer is dependent on the rate at which heat is transferred from the moist air stream to the frost layer. Furthermore, the frost properties as well as the frost surface temperature change with time and position on the surface. The frost density is particularly difficult to model since it is not only time-dependent but also highly space-dependent, especially in the direction of frost growth. In higher-temperature and higher-humidity cases, the frost surface temperature is likely to undergo repeated cycles of melting and refreezing, with the melt generated seeping into the pores of the porous frost layer to the lower frost sublayers. This melt typically freezes into ice as it travels downward, resulting in an increase of density of the sublayers relative to the upper layers. Since the frost thermal conductivity is heavily dependent on the density, this spatial dependence also affects the thermal conductivity and the overall thermal resistance of the layer, as a consequence. Changes in thermal resistance affect the rate of heat transfer, which, in turn, affects the rate of mass transfer. This affects the rate of growth of the frost layer, which, in turn, affects the frost surface temperature. Thus, it is obvious that all these processes feed into each other, making the problem highly intractable.

In order to put this calculation procedure into context relative to the published work in the open literature, we will do a quick survey of some of that work. For plate coolers and freezers in free convection, we can cite the works of Whitehurst (1962a,b), Whitehurst et al. (1968), Goodman and Kennedy (1972), Kennedy and Goodman (1974), and Chuang (1976), among others. Forced convection to flat plates has been investigated by Coles (1954), Yonko (1965), Yonko and Sepsy (1967), Abdel-Wahed et al. (1984), O'Neal and Tree (1984), Padki et al. (1989), Sherif et al. (1990, 1993), Le Gall et al. (1997), Lee et al. (1997), Mao et al. (1999), and more recently Cheng and Wu (2003). For finned heat exchangers we can cite the works of Notestine (1966), Huffman and Sepsy (1967), Gatchilov and Ivanova (1979), Kondepudi (1988), Kondepudi and O'Neal (1987, 1989a,b,c, 1991), Ogawa et al. (1993), Tao et al. (1993), Besant (1999), Al-Mutawa et al. (1998a,b,c,d), and Al-Mutawa and Sherif (1998). A more comprehensive review can be found in Sherif et al. (2001).

Calculation Procedure

In this calculation procedure we focus on a rather simple, but general, method of analysis of simultaneous heat and mass transfer with phase

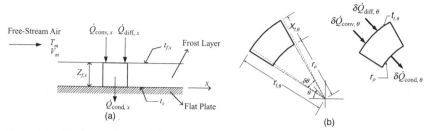

Figure 17.1 Energy and mass balance of a small segment of the frost layer: (a) flat-plate surface; (b) cylindrical surface.

change. The procedure is a quasi-steady-state one, but is accurate when it comes to modeling time-dependent problems. In quasi-steady-state analyses, time-dependent phenomena are treated as steady-state ones for sufficiently small time intervals. The overall solution is treated as a collection of a series of steady-state solutions at different time increments. So, while only steady-state equations are employed, the overall solution is time-dependent. The accuracy of this approach can be as good as the solution of the full-fledged time-dependent problem, provided the time increment chosen is sufficiently small. We will demonstrate the methodology and its accuracy for a problem involving the simultaneous heat and moisture transfer from a humid air stream to a cold surface.

The energy balance at any position on a small area on the surface (Fig. 17.1) can be written as follows:

$$\delta \dot{Q}_{\text{conv}} + \delta \dot{Q}_{\text{diff}} - \delta \dot{Q}_{\text{cond}} = 0 \quad (17.1)$$

where $\delta \dot{Q}_{\text{conv}}$, $\delta \dot{Q}_{\text{diff}}$, and $\delta \dot{Q}_{\text{cond}}$ are the heat-transfer rate by convection, due to diffusion, and by conduction through a finite segment of the surface, respectively.

The rate of heat convected can be determined from

$$\delta \dot{Q}_{\text{conv}} = h_0 A (t_m - t_f) \quad (17.2)$$

where t_m is the temperature of the free-stream air, t_f is the temperature of the frost, A is the area of the surface, and h_0 is the heat-transfer coefficient which depends on the geometry of the surface under study.

The heat-transfer rate due to mass diffusion is given by

$$\delta \dot{Q}_{\text{diff}} = L_s A h_{\text{mass}} (W_m - W_f) \quad (17.3)$$

where h_{mass} and W_f are the mass-transfer coefficient and the humidity ratio of air evaluated at t_f (assuming saturation conditions in the vicinity of the air-frost interface), respectively, and where L_s and W_m are the latent heat of sublimation of the frost and the humidity ratio of the free-stream air, respectively.

17.4 Transient and Cyclic Calculations

The humidity ratio of the free-stream air W_m can be determined from (ASHRAE, 2001)

$$W_m = \frac{m_w}{m_a} = 0.62198 \frac{\varepsilon(P_s\phi)}{P - \varepsilon(P_s\phi)} \qquad (17.4)$$

where m_w and m_a are the mass of water vapor and dry air, respectively; P_s is the saturation pressure of the water vapor; P is the total pressure; ϕ is the relative humidity of the air; and ε is the enhancement factor. The enhancement factor (ε) is a correction parameter that takes into account the effect of dissolved gases and pressure on the properties of the condensed phase, as well as the effect of intermolecular forces on the properties of moisture itself (ASHRAE, 2001). The enhancement factor can be determined from a polynomial equation which has been least-square curve-fit to the data as shown below:

$$\varepsilon = 1.00391 - 7.82205 \times 10^{-6} t + 6.94682 \times 10^{-7} t^2 + 3.04059 \\ \times 10^{-9} t^3 - 2.7852 \times 10^{-11} t^4 - 5.656 \times 10^{-13} t^5 \qquad (17.5)$$

The saturation pressure over ice appearing in Eq. (17.4) for the temperature range $173.15 \leq T \leq 273.15$ K is given by (ASHRAE, 2001)

$$\log_e(P_s) = \frac{a_0}{T} + a_1 + a_2 T + a_3 T^2 + a_4 T^3 + a_5 T^4 + a_6 \log_e(T) \qquad (17.6)$$

where $a_0 = -0.56745359 \times 10^4$
$a_1 = 6.3925247$
$a_2 = -0.9677843 \times 10^{-2}$
$a_3 = 0.62215701 \times 10^{-6}$
$a_4 = 0.20747825 \times 10^{-8}$
$a_5 = -0.9484024 \times 10^{-12}$
$a_6 = 4.1635019$

The saturation pressure over liquid water for the temperature range of 273.15 K $\leq T \leq 473.15$ K can be written as (ASHRAE, 2001)

$$\log_e(P_s) = \frac{b_0}{T} + b_1 + b_2 T + b_3 T^2 + b_4 T^3 + b_5 \log_e(T) \qquad (17.7)$$

where $b_0 = -0.58002206 \times 10^4$
$b_1 = 1.3914993$
$b_2 = -4.8640239 \times 10^{-2}$
$b_3 = 0.41764768 \times 10^{-4}$
$b_4 = -0.14452093 \times 10^{-7}$
$b_5 = 6.5459673$

In both Eqs. (17.6) and (17.7), P_s is in pascals.

The latent heat of sublimation (J/kg) for frost can be determined from the following equation [as modified from Parish (1970)]:

$$L_s = [-0.1083(1.8t_f + 32) + 2833] \qquad (17.8)$$

where L_s is expressed in J/kg.

The local mass-transfer coefficient can be determined using the Lewis analogy as follows:

$$h_{\text{mass}} = \frac{h_0}{C_p \text{Le}^{2/3}} \qquad (17.9)$$

where Le is the Lewis number, which can be expressed as

$$\text{Le} = \frac{\alpha}{D_{\text{diff}}} \qquad (17.10)$$

where α is the thermal diffusivity and D_{diff} is the mass diffusion coefficient. This can be determined in terms of the molecular weights and molar volumes of the dry air and water vapor (Holman, 1997):

$$D_{\text{diff}} = 0.04357(T_a)^{3/2} \frac{(1/M_a) + (1/M_w)}{P_{\text{atm}}\left(\forall_a^{1/3} + \forall_w^{1/3}\right)^2} \qquad (17.11)$$

where D_{diff} is in m²/s; P_{atm} is in pascals; M_a and M_w are the molecular weights of dry air (equal to 28.9) and water vapor (equal to 18), respectively; and $\forall_a$ and $\forall_w$ are the molar volumes of dry air (equal to 29.9) and water vapor (equal to 18.8), respectively.

The rate of heat transfer by conduction is determined using the Fourier heat conduction equation:

$$\delta \dot{Q}_{\text{cond}} = \begin{cases} -\dfrac{k_f A(t_f - t_P)}{dx} & \text{for plane wall surfaces} \\ -\dfrac{k_f A(t_f - t_P)}{dr} & \text{for cylindrical surfaces} \end{cases} \qquad (17.12)$$

where k_f is the thermal conductivity of frost.

Several equations can be used to determine the frost density and thermal conductivity, but for the purpose of this calculation procedure only two of them are used. For the frost density (ρ_f), an empirical expression proposed by Hayashi et al. (1977) is applied. This expression is a sole function of the frost surface temperature, as follows:

$$\rho_f = 650 \exp(0.227 t_f) \qquad (17.13)$$

where ρ_f is in kg/m³. This equation should be applied under the following conditions: frost surface temperature between -25 and $0°C$,

17.6 Transient and Cyclic Calculations

air-stream velocities between 2 and 6 m/s, and an air-stream humidity ratio equal to 0.0075 kg$_w$/kg$_a$.

For the frost thermal conductivity (k_f), an empirical correlation proposed by Yonko and Sepsy (1967) is applied. This correlation is given by the following equation:

$$k_f = 0.024248 + (0.00072311\rho_f) + (0.000001183\rho_f^2) \qquad (17.14)$$

where ρ_f is in kg/m^3 and k_f is expressed in W/m K. This equation is valid only for frost densities less than 573 kg/m^3.

The frost deposition rate per unit area can now be computed using

$$\dot{m}_f = h_{\text{mass}}(W_m - W_f) \qquad (17.15)$$

while the increase in frost thickness can be computed from

$$\Delta Z_f = \frac{\dot{m}_f \Delta \tau}{\rho_f} \qquad (17.16)$$

where $\Delta \tau$ is the time increment during which steady-state conditions were assumed to exist to make the quasi-steady-state assumption valid.

The new frost thickness at the new time is related to that at the previous time according to the equation

$$Z_{f,\text{new}} = Z_f + \Delta Z_f \qquad (17.17)$$

The frost surface temperature can now be obtained by substituting Eqs. (17.2), (17.3), and (17.12) into Eq. (17.1).

The calculation procedure can proceed as follows. The time and the frost thickness Z_f are first initialized to zero, and the frost surface temperature t_f is set equal to that of the surface (t_P). Air properties are evaluated at the air-frost boundary-layer film temperature. The local heat-transfer coefficient is computed depending on the geometry of the surface, and then this value is used in Eq. (17.9) to determine the mass-transfer coefficient. Equations (17.13), (17.14), and (17.15) are then used to determine the frost density, frost thermal conductivity, and the mass deposition rate per unit area, respectively. The increase in frost thickness and the new frost thickness can then be computed using Eqs. (17.16) and (17.17), respectively. Using the parameters determined above, a first approximation for the frost surface temperature is determined by solving Eq. (17.1). This approximate value of t_f is then used to determine the new boundary-layer film temperature in order to recompute the local heat- and mass-transfer coefficients, the local frost density and thermal conductivity, as well as a new approximate value for the frost surface temperature. This calculation procedure is repeated until the value of the new frost surface temperature computed is within a difference of 0.001°C (or any other reasonable difference) of the previous value.

Example

The procedure described in the previous section was applied to cold, flat plate surfaces in humid air under frosting conditions. However, before generating results from the model, we have to check the stability of the solution in terms of its dependence on the time step. This check is important as we do not want the results to depend on the size of the time step. Three time steps were tried: 1, 5, and 20 s, and the results revealed no discernable differences among the three cases. Therefore, for the purpose of this calculation, the time step used is one second. This means that steady-state conditions were assumed to exist for periods of one second at a time. Once the solution converged at a particular point in time, the generated results were used as initial conditions for the subsequent point in time.

Let's first examine how changing the surface temperature of the cold, flat plate affects the frost thickness, with all other things held constant. As can be seen from Fig. 17.2, decreasing the surface temperature of the cold plate increases the frost thickness. This can be explained since decreasing the surface temperature of the cold plate increases the temperature and humidity ratio driving forces, which causes the frost deposition rate over the surface to increase. To illustrate the validity of

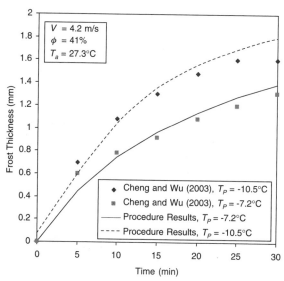

Figure 17.2 Variation of frost thickness with time for two different surface temperatures of the cold plate.

17.8 Transient and Cyclic Calculations

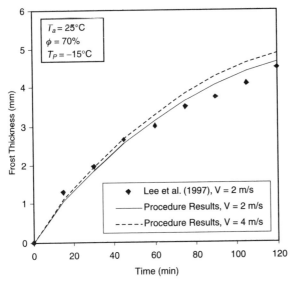

Figure 17.3 Effect of air velocity on frost thickness.

the calculation procedure, our results are compared with experimental results reported by Cheng and Wu (2003). As can be seen, results from the calculation procedure agree well with their results.

To demonstrate the effect of changing the air velocity on the frost thickness, we refer to Fig. 17.3. As expected, results show that the frost thickness grows with increasing air velocity. While this can be partially explained by the fact that increasing the air velocity increases the heat-transfer coefficient, which also increases the mass-transfer coefficient, thus causing the frost deposition rate to increase, increasing the air velocity also causes an increase in the frost-air interface temperature (see the next paragraph), which causes the temperature driving force between air and frost to decrease. A decrease in the temperature driving force is always associated with a decrease in the humidity ratio driving force, an effect that causes the frost deposition rate to decrease. Evidently, the former effect is more dominant than the latter, thus resulting in a net increase in the frost deposition rate. However, this effect tapers off as the frost thickness reaches a certain threshold after a relatively long period of time. Again, to validate the calculation procedure, our results are compared with experimental results obtained by Lee et al. (1997) for a case corresponding to an air velocity of 2 m/s. As can be seen, results from the calculation procedure agree well with the results presented by Lee et al. (1997).

Figure 17.4 illustrates the effect of changing the air velocity on the frost surface temperature. As can be seen, the frost surface temperature

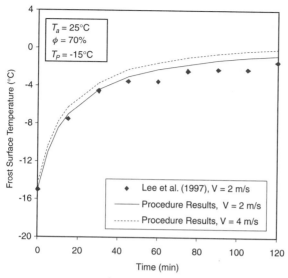

Figure 17.4 Effect of air velocity on frost surface temperature.

increases as the air velocity is increased as reported earlier. This can be explained using Fig. 17.3, since an increase in frost thickness increases the thermal resistance of the frost. Similarly to Fig. 17.3, the results of Fig. 17.4 are compared with those of Lee et al. (1997), indicating that the results from the calculation procedure agree well with their results.

Finally, let's examine how changing the Reynolds number affects the frost thickness, with all other things held constant. As can be seen from Fig. 17.5, increasing the Reynolds number increases the frost thickness, which validates the results presented in Fig. 17.3, since evaluating the effect of Reynolds number is the same as evaluating the effect of air velocity. It is important to point out here that the simulation time for Fig. 17.3 was only 2 h, while that for Fig. 17.5 was 6 h, which gives more confidence in the frost formation predictions of this calculation procedure for larger periods of time. The results obtained for this case are also compared with those of O'Neal and Tree (1984) for different Reynolds numbers. As can be seen, results from the calculation procedure agree well with their results.

Conclusions

A semiempirical, quasi-steady-state calculation procedure capable of studying heat and mass transfer with phase change on surfaces was presented. An example in which this procedure was employed is computation of the frost thickness and surface temperature along with the

17.10 Transient and Cyclic Calculations

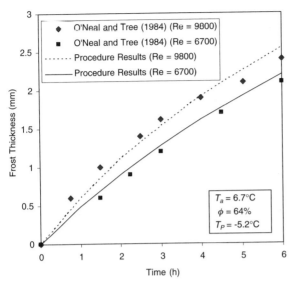

Figure 17.5 Effect of the Reynolds number on frost thickness.

mass of frost deposited on a cold, flat plate surface. Procedure results were compared with existing experimental data and were found to agree very well with available experimental data. This calculation procedure is aimed at describing a relatively simple but accurate technique to predict heat and mass transfer with phase change on different surfaces.

Nomenclature

A	Area, m^2
a_0–a_6	Constants defined by Eq. (17.6)
b_0–b_5	Constants defined by Eq. (17.7)
C_p	Specific heat at constant pressure, kJ/kg K
D_{diff}	Mass diffusion coefficient of water vapor in air, m^2/s
h_0	Local convective heat-transfer coefficient, W/m^2 K
h_{mass}	Local convective mass-transfer coefficient, kg/m^2 s
k	Thermal conductivity, W/m K
Le	Lewis number, dimensionless
L_s	Latent heat of sublimation of frost, J/kg
$\dot{m}_f$	Frost deposition rate per unit area, kg/m^2 s
M_a	Molecular weight of dry air = 28.9
M_w	Molecular weight of water vapor = 18
P	Total pressure, Pa

P_{atm}	Ambient pressure = 101,325 Pa
P_s	Saturation pressure, Pa
Re	Reynolds number, dimensionless
t	Temperature, °C
T	Absolute temperature, K
V	Velocity, m/s
$\forall_a$	Molar volume of dry air = 29.9
$\forall_w$	Molar volume of water vapor = 18.8
W	Humidity ratio, kg$_w$/kg$_a$
Z_f	Frost thickness, m

Greek

α	Thermal diffusivity of air, m²/s
$\Delta\tau$	Time increment, s
ΔZ_f	Increment in frost thickness, m
$\delta\dot{Q}_{cond}$	Heat-transfer rate by conduction through a finite segment of the surface, kW
$\delta\dot{Q}_{conv}$	Heat-transfer rate by convection to a finite segment of the surface, kW
$\delta\dot{Q}_{diff}$	Heat-transfer rate due to diffusion to a finite segment of the surface, kW
ε	Enhancement factor defined by Eq. (17.5)
ρ	Density, kg/m³
ϕ	Relative humidity, %

Subscripts

a	Air
f	Frost
m	Moist
P	Surface

References

Abdel-Wahed, R. M., Hifni, M. A., and Sherif, S. A., 1984, "Heat and Mass Transfer from a Laminar Humid Air Stream to a Plate at Subfreezing Temperature," *International Journal of Refrigeration*, vol. 7, no. 1, January, pp. 49–55.

Al-Mutawa, N. K., Sherif, S. A., Mathur, G. D., West, J., Tiedeman, J. S., and Urlaub, J., 1998a, "Determination of Coil Defrosting Loads: Part I—Experimental Facility Description (RP-622)," *ASHRAE Transactions*, vol. 104, pt. 1A, paper no. 4120, January, pp. 268–288.

Al-Mutawa, N. K., Sherif, S. A., Mathur, G. D., Steadham, J. M., West, J., Harker, R. A., and Tiedeman, J. S., 1998b, "Determination of Coil Defrosting Loads: Part II—Instrumentation and Data Acquisition Systems (RP-622)," *ASHRAE Transactions*, vol. 104, pt. 1A, paper no. 4121, January, pp. 289–302.

17.12 Transient and Cyclic Calculations

Al-Mutawa, N. K., Sherif, S. A., and Mathur, G. D., 1998c, "Determination of Coil Defrosting Loads: Part III—Testing Procedures and Data Reduction (RP-622)," *ASHRAE Transactions*, vol. 104, pt. 1A, paper no. 4122, January, pp. 303–312.

Al-Mutawa, N. K., Sherif, S. A., and Steadham, J. M., 1998d, "Determination of Coil Defrosting Loads: Part IV—Refrigeration/Defrost Cycle Dynamics (RP-622)," *ASHRAE Transactions*, vol. 104, pt. 1A, paper no. 4123, January, pp. 313–343.

Al-Mutawa, N. K., and Sherif, S. A., 1998, "Determination of Coil Defrosting Loads: Part V—Analysis of Loads (RP-622)," *ASHRAE Transactions*, vol. 104, pt. 1A, paper no. 4124, January, pp. 344–355.

ASHRAE, 2001, *2001 ASHRAE Handbook—Fundamentals*, the American Society of Heating, Refrigerating and Air-Conditioning Engineers, Inc., Atlanta, Ga.

Barron, R. F., and Han, L. S., 1965, "Heat and Mass Transfer to a Cryo-Surface in Free Convection," *ASME Journal of Heat Transfer*, vol. 87, no. 4, pp. 499–506.

Besant, R. W., 1999, *Characterization of Frost Growth and Heat Transfer at Low Temperatures*, final report, ASHRAE RP-824, The American Society of Heating, Refrigerating and Air-Conditioning Engineers, Atlanta, Ga., January.

Cheng, C.-H., and Wu, K.-H., 2003, "Observations of Early-Stage Frost Formation on a Cold Plate in Atmospheric Air Flow," *ASME Journal of Heat Transfer*, vol. 125, no. 1, February, pp. 95–102.

Chuang, M. C., 1976, "Frost Formation on Parallel Plates at Very Low Temperature in a Humid Stream," ASME paper 76-WA/HT-60, *ASME Winter Annual Meeting*, the American Society of Mechanical Engineers, New York.

Coles, W. D., 1954, *Experimental Determination of Thermal Conductivity of Low Density Ice*, NACA Technical Notes, no. 3143, National Advisory Committee for Aeronautics, March.

Gatchilov, T. S., and Ivanova, V. S., 1979, "Characteristics of the Frost Formed on the Surface of Finned Air Coolers," *15th International Congress of Refrigeration*, Venice, Italy, paper B2-71, pp. 997–1003.

Goodman, J., and Kennedy, L. A., 1972, "Free Convection Frost Formation on Cool Surfaces," *Proceedings of the 1972 Heat Transfer and Fluid Mechanics Institute*, Stanford University Press, Stanford, Calif., pp. 338–352.

Hayashi, Y., Aoki, K., and Yuhara, H., 1977, "Study of Frost Formation Based on a Theoretical Model of the Frost Layer," *Heat Transfer—Japanese Research*, vol. 6, no. 3, July–September, pp. 79–94.

Holman, J. P., 1997, *Heat Transfer*, 8th ed., McGraw-Hill, New York.

Huffman, G. D., and Sepsy, C. F., 1967, "Heat Transfer and Pressure Loss in Extended Surface Heat Exchangers Operating under Frosting Conditions—Part II: Data Analysis and Correlations," *ASHRAE Transactions*, vol. 73, pt. 2, June, pp. I.3.1–I.3.16.

Kennedy, L. A., and Goodman, J., 1974, "Free Convection Heat and Mass Transfer under Conditions of Frost Deposition," *International Journal of Heat and Mass Transfer*, vol. 17, pp. 477–484.

Kondepudi, S. N., 1988, *The Effects of Frost Growth on Finned Tube Heat Exchangers under Laminar Flow*, Ph.D. dissertation, Texas A&M University, College Station, Tex., December.

Kondepudi, S. N., and O'Neal, D. L., 1987, "The Effects of Frost Growth on Extended Surface Heat Exchanger Performance: A Review," *ASHRAE Transactions*, vol. 93, pt. 2, paper no. 3070, pp. 258–274.

Kondepudi, S. N., and O'Neal, D. L., 1989a, "Effect of Frost Growth on the Performance of Louvered Finned Tube Heat Exchangers," *International Journal of Refrigeration*, vol. 12, no. 3, pp. 151–158.

Kondepudi, S. N., and O'Neal, D. L., 1989b, "The Effects of Frost Formation on the Thermal Performance of Finned Tube Heat Exchangers," *AIAA 24th Thermophysics Conference*, AIAA paper no. 89-1741, Buffalo, N.Y., June.

Kondepudi, S. N., and O'Neal, D. L., 1989c, "The Performance of Finned Tube Heat Exchangers under Frosting Conditions," *Collected Papers in Heat Transfer—1989*, HTD vol. 123, ASME, New York, December, pp. 193–200.

Kondepudi, S. N., and O'Neal, D. L., 1991, "Modeling Tube-Fin Heat Exchangers under Frosting Conditions," *18th International Congress of Refrigeration*, Montreal, Quebec, Canada, August, paper no. 242.

Le Gall, R., Grillot, J. M., and Jallut, C., 1997, "Modelling of Frost Growth and Densification," *International Journal of Heat and Mass Transfer*, vol. 40, No. 13, pp. 3177–3187.

Lee, K, Kim, W., and Lee T., 1997, "A One-Dimensional Model for Frost Formation on a Cold Flat Surface," *International Journal of Heat and Mass Transfer*, vol. 40, no. 18, pp. 4359–4365.

Mao, Y., Besant, R. W., and Chen, H., 1999, "Frost Characteristics and Heat Transfer on a Flat Plate under Freezer Operating Conditions: Part I, Experimentation and Correlations," *ASHRAE Transactions*, vol. 105, pt. 2, June, pp. 231–251.

Notestine, H. E., 1966, *The Design, Fabrication, and Testing of an Apparatus to Study the Formation of Frost from Humid Air to an Extended Surface in Forced Convection*, M.S. thesis, The Ohio State University, Columbus, Ohio.

Ogawa, K., Tanaka, N., and Takeshita, M., 1993, "Performance Improvement of Plate Fin-and-Tube Heat Exchanger under Frosting Conditions," *ASHRAE Transactions*, vol. 99, pt. 1, January, pp. 762–771.

O'Neal, D. L., and Tree, D. R., 1984, "Measurements of Frost Growth and Density in Parallel Plate Geometry," *ASHRAE Transactions*, vol. 90, pt. 2.

Padki, M. M., Sherif, S. A., and Nelson, R. M., 1989, "A Simple Method for Modeling Frost Formation in Different Geometries," *ASHRAE Transactions*, vol. 95, pt. 2, June, pp. 1127–1137.

Parish, H. C., 1970, *A Numerical Analysis of Frost Formation under Forced Convection*, Master's thesis, The Ohio State University, Columbus, Ohio.

Sherif, S. A., Abdel-Wahed, R. M., and Hifni, M. A., 1990, "A Mathematical Model for the Heat and Mass Transfer on a Flat Plate under Frosting Conditions," *Chemical Engineering Communications*, vol. 92, June, pp. 65–80.

Sherif, S. A., Raju, S. P., Padki, M. M., and Chan, A. B., 1993, "A Semi-Empirical Transient Method for Modelling Frost Formation on a Flat Plate," *International Journal of Refrigeration*, vol. 16, no. 5, September, pp. 321–329.

Sherif, S. A., Mago, P. J., Al-Mutawa, N. K., Theen, R. S., and Bilen, K., 2001, "Psychrometrics in the Supersaturated Frost Zone," *ASHRAE Transactions*, vol. 107, pt. 2, June, pp. 753–767.

Tao, Y. X., Besant, R. W., and Mao, Y., 1993, "Characteristics of Frost Growth on a Flat Plate During the Early Growth Period," *ASHRAE Transactions*, vol. 99, pt. 1, January, pp. 739–745.

Whitehurst, C. A., 1962a, "Heat and Mass Transfer by Free Convection from Humid Air to a Metal Plate under Frosting Conditions," *ASHRAE Journal*, vol. 4, no. 5, May, pp. 58–69.

Whitehurst, C. A., 1962b, *An Investigation of Heat and Mass Transfer by Free Convection from Humid Air to a Metal Plate under Frosting Conditions*, Ph.D. dissertation, Department of Mechanical Engineering, Agricultural and Mechanical College of Texas.

Whitehurst, C. A., McGregor, O. W., and Fontenot, J. E., Jr., 1968, "An Analysis of the Turbulent Free-Convection Boundary Layer over a Frosted Surface," *Cryogenic Technology*, January/February.

Yonko, J. D., 1965, *An Investigation of the Thermal Conductivity of Frost While Forming on a Flat Horizontal Plate*, M.S. thesis, The Ohio State University, Columbus, Ohio.

Yonko, J. D. and Sepsy, C. F., 1967, "An Investigation of the Thermal Conductivity of Frost While Forming on a Flat Horizontal Plate," *ASHRAE Transactions*, vol. 73, pt. 2, paper no. 2043, June, pp. I.1.1–I.1.11.

Chapter 18

A Calculation Procedure for First- and Second-Law Efficiency Optimization of Refrigeration Cycles

J. S. Tiedeman
BKV Group
Minneapolis, Minnesota

S. A. Sherif
Department of Mechanical and Aerospace
  Engineering
University of Florida
Gainesville, Florida

Introduction

Thermodynamic performance optimization of any thermal system can be done by either maximizing the first-law or the second-law efficiency. The former process is equivalent to finding the maximum heat removal rate ($\dot{Q}_e$) for the minimum power input ($\dot{W}$). The latter process is equivalent to finding the parameters that minimize the destruction of available energy. The calculation procedure described here addresses both methods of optimization. It will be demonstrated for a two-stage vapor compression cycle, which is presented in Figs. 18.1 and 18.2, where IC/FT stands for the intercooler–flash tank.

When designing a system of this type, the designer must know the following before being able to start the calculation and sizing process: (1) the evaporator and condenser temperatures T_e and T_c and (2) the

18.2 Transient and Cyclic Calculations

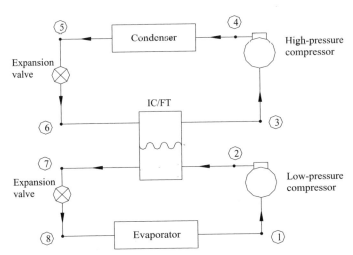

Figure 18.1 Schematic of a two-stage refrigeration system with an intercooler–flash tank.

intermediate-stage conditions. In general, the designer will know the condenser temperature from ambient conditions (T_0) in which the system is designed to operate, while the evaporator temperature will be set by the refrigerated space temperature (T_{Ref}). For the purposes of this calculation procedure, it is assumed that $T_0 = T_c - 5 \text{ K}$ and $T_{\text{Ref}} = T_e + 5 \text{ K}$.

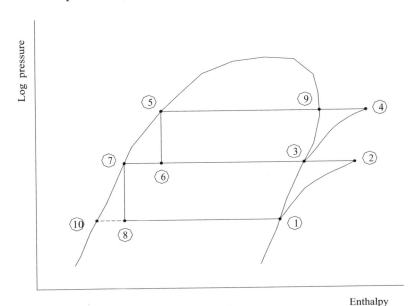

Figure 18.2 Pressure-enthalpy representation of a two-stage vapor compression refrigeration system.

The interstage conditions, however, are not as easily determined. According to Threlkeld (1962), "It is not possible to determine mathematically an optimum pressure for a vapor refrigerant, as would be the case for a perfect gas." Since this determination is not possible, Threlkeld proceeds to suggest a common approximation for two-stage systems: the use of the geometric mean of the condenser and evaporator pressures $p_{int} = (p_c p_e)^{0.5}$. The use of this approximation, however, leads to an interstage pressure (and thus temperature) that is noticeably lower than optimum. This can be attributed to the fact that the estimation does not account for the difference in the amount of work required to perform compression in the high- and low-pressure stages. The use of the geometric mean also fails to take into account the isentropic compressor efficiencies. This calculation will quantify the error associated with these common approximations and suggest more accurate mathematical approximations.

In order to put this calculation procedure into context relative to the published work in the open literature, we will do a quick survey of some of that work. Gupta and Prasad (1984, 1985) investigated optimization of the coefficient of performance (COP) of a two-stage system and found generalized equations for the optimum interstage conditions based on the interstage temperature. Their method is a good first step; however, the use of an approximation for gases in the superheated region as well as the determination of the optimum conditions based on interstage temperature leaves room for improvement. A preferable way to maximize the COP would be to determine the optimum conditions with respect to the interstage pressure. This is preferable because in an ideal cycle, the pressures will stay constant through the condenser, evaporator, and intercooler–flash tank, respectively, and the use of the pressure does not disregard the "superheat horn," as optimizing using the interstage temperature would, thus allowing a more accurate optimization. Gupta and Prasad's calculations were performed assuming isentropic compression in the compressors. As will be shown in the calculation procedure presented here, the COP of a system is a significant function of the compressor efficiency, and an analysis without considering this variable may leave a bit of room for improvement.

The general process of optimizing the refrigeration system with regard to the second law of thermodynamics involves determining the conditions that lead to the lowest destruction of exergy (or generation of entropy) for a process. For any refrigeration system, the process involves removing the most heat while generating the least entropy. Bejan et al. (1996) define the exergetic efficiency (η_{ex}), also known as the *second-law efficiency*, as the ratio of the "exergy of the product" to the "exergy of the fuel." As with the first-law analysis, the condenser and evaporator pressures are known, and from these, the optimum interstage conditions

18.4 Transient and Cyclic Calculations

can be determined. The determination of exergy destruction for the components of a somewhat similar system has been performed by Swers et al. (1972) and is contained in ASHRAE (1997). This work divided the refrigeration system into its components and found the exergy destruction contributed by each. Swers et al. (1972), however, did not find the maximum second-law efficiency for the system, instead focusing on the analysis of the individual components. Zubair et al. (1996) performed analyses on several types of two-stage systems revealing the rate of irreversibility production for the entire system.

Calculation Procedure

In this calculation procedure we focus on a method of optimizing vapor compression refrigeration cycles. The procedure will be demonstrated for a two-stage cycle based on the coefficient of performance (COP) and exergetic efficiency. The calculation procedure will indicate that the use of the common approximation of the geometric mean to find the optimum interstage pressure can lead to significant errors in interstage pressure. However, an optimum COP or exergetic efficiency based on the same interstage pressure will be shown to have relatively little error. Second-law optimization will reveal that the optimum data curves themselves have a maximum for each set of conditions tested. This leads to the conclusion that for a given system there is an optimum set of conditions that lead to the lowest amount of available energy (exergy) destruction for that system. Polynomial equations will be fitted to the resultant optimum data for the interstage pressure, COP, and second-law (exergetic) efficiency. These polynomial expressions will allow for the reproduction of optimum points based on high- and low-pressure compressor efficiencies and condenser and evaporator pressures.

While the procedure described here is demonstrated for a two-stage cycle, it is sufficiently general for application to more complex cycles and systems. For the chosen two-stage system, the COP is given by an energy balance of the evaporator and the two compressors where the ratio of the high- and low-pressure mass flow rates is found through an energy balance of the intercooler–flash tank (IC/FT). This then leads to the following COP equation for an ideal two-stage system:

$$\text{COP} = \frac{h_1 - h_8}{(h_2 - h_1) + \left(\dfrac{h_2 - h_7}{h_3 - h_6}\right)(h_4 - h_3)} \quad (18.1)$$

Optimization of the COP is based on the interstage pressure. Thus, the first derivative of the COP with respect to the interstage pressure can be set equal to zero to find the optimum condition.

The general process of development for the second law proceeds similarly to that of the first law. The exergetic efficiency, as described by Bejan et al. (1996), is defined as "useful exergy gained (ε_G)" divided by "exergy supplied (ε_S)":

$$\eta_{ex} = \frac{\varepsilon_G}{\varepsilon_S} \tag{18.2}$$

The *useful exergy gained* incorporates the exchange of energy and resultant destruction of exergy that occurs when heat is transferred from the air stream to the refrigerant at the evaporator. The *exergy supplied* is defined as the work provided to the system.

If one envisions a control volume that has entering and exiting streams of a working fluid, entering and exiting work, and entering and exiting heat interacting with constant-temperature thermal reservoirs, the exergetic efficiency can be derived. Realizing that whatever enters a control volume either exits or accumulates, the following equation is derived:

$$-T_0 S_{gen} = \Delta\varepsilon_{fluid} + \Delta\varepsilon_{heat} + \Delta\varepsilon_{work} \tag{18.3}$$

This equation expresses the exergy change of a process in terms of the exergy changes due to heat, work, and fluid interactions. By definition, the entropy generation term at the far right of Eq. (18.3) is equal to the negative of the exergy change for a process (as entropy is generated, exergy is destroyed). Substituting the definitions of exergy for a fluid stream, heat interaction, and work as defined by Sonntag et al. (1998), the following results:

$$\Delta\varepsilon = \Sigma(m\Psi)_{out} - \Sigma(m\Psi)_{in} + \Sigma\left[Q_{out}\left(1 - \frac{T_0}{T_{out}}\right)\right]$$

$$-\Sigma\left[Q_{in}\left(1 - \frac{T_0}{T_{in}}\right)\right] + \Sigma W_{out} - \Sigma W_{in} \tag{18.4}$$

For a two-stage refrigeration system, the exergy change of the airflow through the evaporator is the *useful exergy gained* by the system and the total work input is the *exergy supplied*. The following describes the exergy gained and exergy supplied, respectively:

$$\varepsilon_G = \Delta\varepsilon_{air} = \dot{m}_{air}\Psi_{air,out} - \dot{m}_{air}\Psi_{air,in}$$

$$= \dot{m}_{air}[(h_{air,out} - h_{air,in}) - T_0(s_{air,out} - s_{air,in})] \tag{18.5}$$

$$\varepsilon_S = \dot{W}_{total} = \dot{m}_1(h_2 - h_1) + \dot{m}_3(h_4 - h_3) \tag{18.6}$$

An energy balance for the IC/FT will produce the ratio of the mass flow rates in terms of enthalpy differences, while an energy balance of the

18.6   Transient and Cyclic Calculations

evaporator will produce the ratio of the high-pressure refrigerant mass flow rate to the air mass flow rate. Simplifying, the exergetic efficiency can now be derived as

$$\eta_{\text{ex}} = \frac{(h_{\text{air,out}} - h_{\text{air,in}}) - T_0(s_{\text{air,out}} - s_{\text{air,in}})}{\left(\dfrac{\dot{m}_1}{\dot{m}_{\text{air}}}\right)\left[(h_2 - h_1) + \left(\dfrac{\dot{m}_3}{\dot{m}_1}\right)(h_4 - h_3)\right]} \quad (18.7)$$

As with the first-law section, the optimum exergetic efficiency can be found by finding the point at which the derivative of the exergetic efficiency with respect to the interstage pressure is zero.

A code has been developed in house to calculate property data for selected refrigerants as well as to perform the differentiation of the COP and exergetic efficiency with respect to the interstage pressure. The error associated with property data calculation in the superheated region has been compared to empirical data obtained from Sonntag et al. (1998). The maximum error encountered was no greater than 1.2 percent. The error associated with property data calculation in the two-phase region was compared to empirical data obtained from Sonntag et al. (1998), which led to a maximum error of 1.6 percent. The calculation of the total error due to both data calculation and regression analysis will be presented in the next section. The reader can use any of several commercially available codes for calculating refrigerant properties for the purpose of their calculations.

Examples

In this part of the calculation procedure we will show some of the expected results of the optimization procedure described in the previous section. For demonstration purposes, we have selected ranges for the evaporator temperature from 213 to 273 K and for the condenser temperature from 298 to 313 K. The example refrigerants selected for this demonstration were R22 and ammonia. The isentropic efficiencies for the high- and low-pressure compressors were selected to range between 100 and 40 percent. For the purposes of this calculation procedure, both the high- and low-pressure compression efficiencies have been set equal. Varying the compressor efficiencies so that they are not equal would allow for a more realistic scenario; however, it is more convenient computationally not to vary them independently.

Let's first examine how changing the interstage pressure of the refrigeration system affects the COP, with all other things held constant. As can be seen from Fig. 18.3, there is a point at which the COP reaches a maximum value for a given evaporator temperature. This maximum COP occurs at higher interstage pressures as the evaporator temperature is raised. The locus of maxima can then be used to formulate an

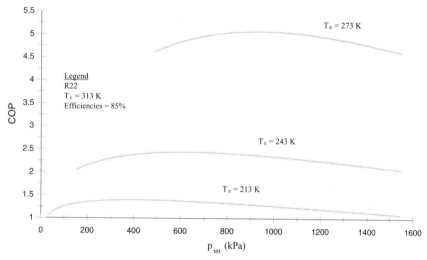

Figure 18.3 Variation of multiple COP data sets as a function of the interstage pressure.

equation that represents the optimum interstage pressure and COP for a system based on the condenser and evaporator pressures and the compression efficiencies.

To demonstrate how the interstage pressure changes as a function of the evaporator pressure, we refer to Fig. 18.4, which shows this trend employing three sets of data. The upper set of data is that obtained by using the geometric mean of the evaporator and condenser pressures. The middle and lower sets of data correspond to outputs of the computer code for ammonia and R22, respectively, using the suggested calculation

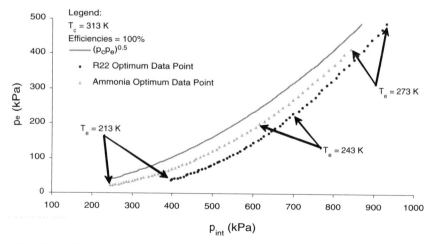

Figure 18.4 Effect of changing evaporator pressure on the optimum interstage pressure.

18.8 Transient and Cyclic Calculations

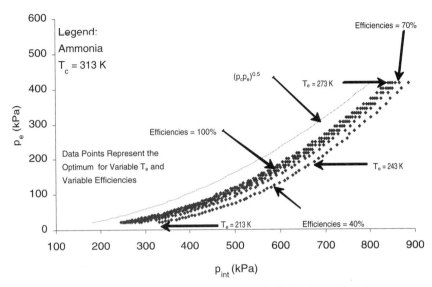

Figure 18.5 Effect of changing evaporator pressure and efficiencies on the interstage pressure for ammonia.

procedure. As can be seen, for the given condenser pressure and varying evaporator pressure, the geometric mean method will give a substantially lower optimum interstage pressure than the actual optimum interstage pressure, as determined by the suggested calculation procedure, for either refrigerant. This trend proves to be true for all operating ranges selected for the demonstration.

If the compressor efficiencies are now varied from 100 to 40 percent, as shown in Fig. 18.5, a substantial dependence on compressor efficiency can be observed. Although there is slightly increasing data scatter as the compression efficiencies are lowered, it can be seen that the interstage pressure is heavily dependent on the evaporator pressure and compressor efficiencies.

The general method of creating an equation for multiple sets of data, such as the optimum interstage pressure based on the condenser and evaporator pressures and the compressor efficiencies, is described by Stoecker and Jones (1982). The technique involves the formulation of several equations that are to be solved simultaneously. This formulation has been extended to a third-order equation

$$p_{\text{int}} = \sum_{k=0}^{3}\sum_{j=0}^{3}\sum_{i=0}^{3}\left[c_{(i+4j+16k+1)}\eta^i p_e^j p_c^k\right] \tag{18.8}$$

where the subscripts for the constant c resulting from the equation $i + 4j + 16k + 1$ correspond to those contained in Table 18.1. The error

TABLE 18.1 Polynomial Constants* for Optimum p_{int}

	R22	Ammonia		R22	Ammonia
c_1	−6.26e + 03	−3.51e + 04	c_{33}	−8.95e − 03	−6.04e − 02
c_2	4.20e + 04	1.50e + 05	c_{34}	6.40e − 02	2.56e − 01
c_3	−6.43e + 04	−1.99e + 05	c_{35}	−9.62e − 02	−3.38e − 01
c_4	2.85e + 04	8.43e + 04	c_{36}	4.08e − 02	1.42e − 01
c_5	−1.26e + 02	3.00e + 01	c_{37}	−2.78e − 04	−2.05e − 05
c_6	3.76e + 02	1.75e + 02	c_{38}	8.95e − 04	6.61e − 04
c_7	−5.30e + 02	−6.90e + 02	c_{39}	−1.30e − 03	−1.77e − 03
c_8	2.86e + 02	4.79e + 02	c_{40}	6.92e − 04	1.11e − 03
c_9	7.43e − 02	1.51e + 00	c_{41}	2.85e − 07	3.21e − 06
c_{10}	9.99e − 01	−8.78e + 00	c_{42}	1.21e − 06	−1.82e − 05
c_{11}	−2.07e + 00	1.51e + 01	c_{43}	−2.83e − 06	3.08e − 05
c_{12}	9.55e − 01	−7.80e + 00	c_{44}	1.25e − 06	−1.57e − 05
c_{13}	3.11e − 04	−3.64e − 03	c_{45}	4.34e − 10	−7.43e − 09
c_{14}	−3.47e − 03	1.96e − 02	c_{46}	−5.86e − 09	3.95e − 08
c_{15}	6.31e − 03	−3.21e − 02	c_{47}	1.08e − 08	−6.39e − 08
c_{16}	−3.09e − 03	1.59e − 02	c_{48}	−5.25e − 09	3.16e − 08
c_{17}	1.49e + 01	8.09e + 01	c_{49}	1.59e − 06	1.51e − 05
c_{18}	−9.73e + 01	−3.44e + 02	c_{50}	−1.31e − 05	−6.39e − 05
c_{19}	1.47e + 02	4.55e + 02	c_{51}	1.95e − 05	8.40e − 05
c_{20}	−6.38e + 01	−1.92e + 02	c_{52}	−7.88e − 06	−3.52e − 05
c_{21}	3.19e − 01	−2.43e − 02	c_{53}	8.09e − 08	1.54e − 08
c_{22}	−9.68e − 01	−6.10e − 01	c_{54}	−2.75e − 07	−2.16e − 07
c_{23}	1.38e + 00	1.93e + 00	c_{55}	4.03e − 07	5.20e − 07
c_{24}	−7.48e − 01	−1.27e + 00	c_{56}	−2.13e − 07	−3.15e − 07
c_{25}	−2.35e − 04	−3.83e − 03	c_{57}	−1.08e − 10	−8.81e − 10
c_{26}	−2.15e − 03	2.20e − 02	c_{58}	−1.50e − 10	4.93e − 09
c_{27}	4.55e − 03	−3.76e − 02	c_{59}	4.76e − 10	−8.29e − 09
c_{28}	−2.07e − 03	1.93e − 02	c_{60}	−1.95e − 10	4.20e − 09
c_{29}	−6.90e − 07	9.06e − 06	c_{61}	−7.45e − 14	2.00e − 12
c_{30}	8.10e − 06	−4.85e − 05	c_{62}	1.34e − 12	−1.06e − 11
c_{31}	−1.47e − 05	7.89e − 05	c_{63}	−2.51e − 12	1.70e − 11
c_{32}	7.20e − 06	−3.90e − 05	c_{64}	1.22e − 12	−8.37e − 12

*Standard (scientific) notation (e.g., −6.26e + 03 instead of −6.26 × 10^3) is retained here and in Tables 18.3 and 18.4.

in the approximation of Eq. (18.8) is no greater than 2.6 percent for either refrigerant, with a total error of 3.3 percent after accounting for additional approximations in the property data.

Table 18.2 is a comparison of the optimum interstage pressure at 100 percent isentropic compressor efficiencies [as defined by Eq. (18.8)] with that obtained using the geometric mean method. Any error contributed by using the geometric mean method will be even greater for efficiencies less than 100 percent.

Similar to the trends observed for the intermediate pressure p_{int}, the COP is also a function of the condenser and evaporator pressures and the high- and low-pressure compressor efficiencies. Figure 18.6, which is a plot of the optimum COP data points for varying compressor

18.10 Transient and Cyclic Calculations

TABLE 18.2 Deviation in Optimum Pressure of the Geometric Mean Approximation as Compared to the Optimized Data

Refrigerant	T_c, K	Maximum deviation	Minimum deviation
R22	298	30.65	2.80
R22	303	34.00	3.85
R22	308	36.56	5.13
R22	313	40.10	6.56
Ammonia	298	15.94	2.79
Ammonia	303	19.88	3.82
Ammonia	308	23.85	4.03
Ammonia	313	26.48	4.84

efficiencies at a constant condenser temperature, demonstrates this dependence. It can be seen that the COP is also heavily dependent on the high- and low-pressure compressor efficiencies. As with the interstage pressure, the optimum COP equation can be represented by a polynomial expression as described by Stoecker and Jones (1982). This polynomial can again be made third-order accurate and may be expressed as

$$\text{COP} = \sum_{k=0}^{3}\sum_{j=0}^{3}\sum_{i=0}^{3}\left[d_{(i+4j+16k+1)}\eta^i p_e^j p_c^k\right] \qquad (18.9)$$

Where the subscripts for the constant d resulting from the equation $i + 4j + 16k + 1$ correspond to those contained in Table 18.3. The error

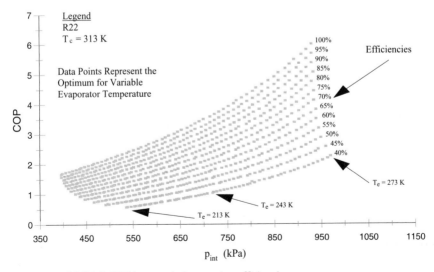

Figure 18.6 Multiple R22 locus sets for varying efficiencies.

TABLE 18.3 Polynomial Constants for Optimum COP

	R22	Ammonia		R22	Ammonia
d_1	5.63e − 01	−1.44e + 02	d_{33}	1.19e − 06	−2.74e − 04
d_2	−1.21e + 00	6.39e + 02	d_{34}	−5.77e − 06	1.22e − 03
d_3	5.57e + 00	−9.04e + 02	d_{35}	9.03e − 06	−1.73e − 03
d_4	−2.61e + 00	4.11e + 02	d_{36}	−4.27e − 06	7.84e − 04
d_5	−2.24e − 02	1.88e + 00	d_{37}	−3.87e − 08	3.59e − 06
d_6	1.67e − 01	−8.27e + 00	d_{38}	2.34e − 07	−1.59e − 05
d_7	−1.63e − 01	1.18e + 01	d_{39}	−2.88e − 07	2.26e − 05
d_8	7.69e − 02	−5.38e + 00	d_{40}	1.36e − 07	−1.03e − 05
d_9	1.69e − 04	−7.62e − 03	d_{41}	3.02e − 10	−1.45e − 08
d_{10}	−9.34e − 04	3.36e − 02	d_{42}	−1.61e − 09	6.43e − 08
d_{11}	1.26e − 03	−4.79e − 02	d_{43}	2.25e − 09	−9.14e − 08
d_{12}	−5.95e − 04	2.18e − 02	d_{44}	−1.07e − 09	4.15e − 08
d_{13}	−3.68e − 07	9.52e − 06	d_{45}	−6.62e − 13	1.82e − 11
d_{14}	2.26e − 06	−4.16e − 05	d_{46}	3.80e − 12	−7.98e − 11
d_{15}	−2.74e − 06	5.98e − 05	d_{47}	−4.94e − 12	1.14e − 10
d_{16}	1.30e − 06	−2.72e − 05	d_{48}	2.35e − 12	−5.19e − 11
d_{17}	−1.53e − 03	3.47e − 01	d_{49}	−3.03e − 10	7.12e − 08
d_{18}	6.84e − 03	−1.54e + 00	d_{50}	1.50e − 09	−3.15e − 07
d_{19}	−1.19e − 02	2.18e + 00	d_{51}	−2.27e − 09	4.47e − 07
d_{20}	5.63e − 03	−9.92e − 01	d_{52}	1.08e − 09	−2.03e − 07
d_{21}	5.05e − 05	−4.55e − 03	d_{53}	9.79e − 12	−9.32e − 10
d_{22}	−3.24e − 04	2.01e − 02	d_{54}	−5.71e − 11	4.12e − 09
d_{23}	3.76e − 04	−2.86e − 02	d_{55}	7.27e − 11	−5.86e − 09
d_{24}	−1.78e − 04	1.30e − 02	d_{56}	−3.45e − 11	2.66e − 09
d_{25}	−3.93e − 07	1.84e − 05	d_{57}	−7.66e − 14	3.77e − 12
d_{26}	2.11e − 06	−8.13e − 05	d_{58}	4.05e − 13	−1.67e − 11
d_{27}	−2.93e − 06	1.16e − 04	d_{59}	−5.71e − 13	2.37e − 11
d_{28}	1.39e − 06	−5.26e − 05	d_{60}	2.71e − 13	−1.08e − 11
d_{29}	8.59e − 10	−2.30e − 08	d_{61}	1.68e − 16	−4.71e − 15
d_{30}	−5.06e − 09	1.01e − 07	d_{62}	−9.47e − 16	2.07e − 14
d_{31}	6.42e − 09	−1.45e − 07	d_{63}	1.25e − 15	−2.96e − 14
d_{32}	−3.05e − 09	6.57e − 08	d_{64}	−5.95e − 16	1.35e − 14

in the approximation of Eq. (18.9) is no greater than 0.45 percent for either refrigerant, with a total error of 2.1 percent after accounting for additional errors in the property data.

To illustrate the validity of the use of Eq. (18.9), we offer Fig. 18.7, which is a plot of the same three sets of data as contained in Fig. 18.3 plus curves that represent the optimum interstage pressure using both the geometric mean method and Eq. (18.9). As can be seen, the use of Eq. (18.9) shows a slight increase in the optimum COP for the given conditions. However, the interstage pressures corresponding to these optimum COPs are noticeably larger than their counterparts based on the geometric mean method. This trend is consistent for all chosen operating conditions.

To illustrate second-law optimization, we examine how the exergetic efficiency is affected by changing the interstage pressure. Figure 18.8

18.12 Transient and Cyclic Calculations

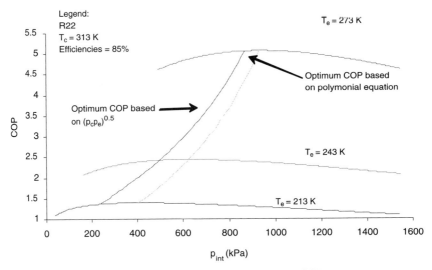

Figure 18.7 Comparison of optimum method for determining COP.

serves that purpose and, as can be seen, is similar in shape to the COP plot. Translated into specific trends, the optimum exergetic efficiency occurs at a corresponding optimum interstage pressure. This optimum pressure is the same for all previous optimizations and is provided in Eq. (18.8). This similarity results from the fact that the optimum

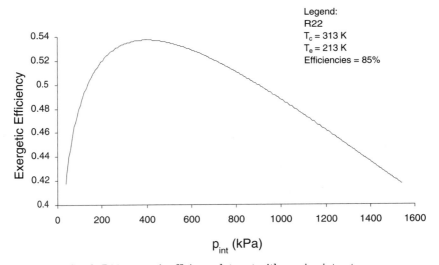

Figure 18.8 Single R22 exergetic efficiency data set with varying interstage pressure.

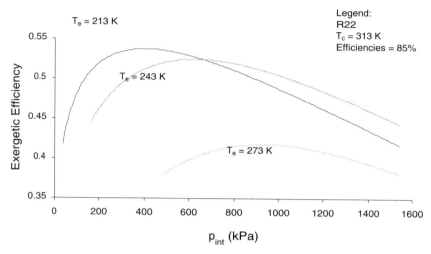

Figure 18.9 Multiple R22 exergetic efficiency data sets with varying interstage pressure.

interstage pressure occurs when the least amount of work is used by the system.

To demonstrate the effect of changing the evaporator temperature on the exergetic efficiency, we refer to Fig. 18.9. As can be seen, as the evaporator temperature increases, the maximum value of each curve decreases. This trend can also be seen in Fig. 18.10, which is a plot of the maximum values for evaporator temperatures ranging from 213 to 273 K. The optimum points for the exergetic efficiency all increase in value with decreasing condenser temperature (see Fig. 18.11). This is

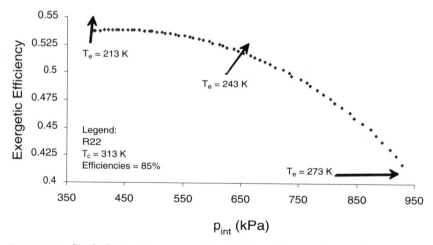

Figure 18.10 Single R22 optimum exergetic efficiency data set with varying interstage pressure.

18.14 Transient and Cyclic Calculations

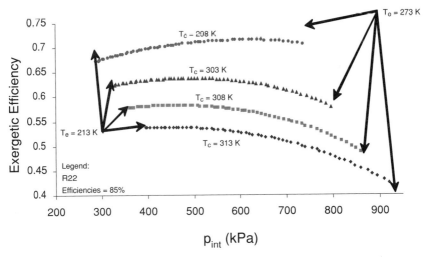

Figure 18.11 Refrigerant R22 optimum exergetic efficiency data sets with varying interstage pressure.

due to the fact that as more work is done, we are creating less entropy per unit of work, thus increasing our exergetic efficiency. Also, as the condenser temperature is decreased, the optimum data curves have maxima that increase with increasing interstage pressure indicating that the more "quality" work that is done, the higher the exergetic efficiency.

To describe the data contained in Fig. 18.11, a polynomial expression can be fitted. The following equation is a third-order polynomial expression:

$$\eta_{ex} = \sum_{k=0}^{3}\sum_{j=0}^{3}\sum_{i=0}^{3}\left[g_{(i+4j+16k+1)}\eta^i p_e^j p_c^k\right] \quad (18.10)$$

The coefficients g_1 through g_{64} for both refrigerants are contained in Table 18.4. The error in the approximation of Eq. (18.10) is no greater than 0.5 percent for either refrigerant, with a total error of 2.1 percent after accounting for errors in the property data representations.

Formulations of the exergetic efficiency given in this calculation procedure take a more system approach as opposed to the component approach employed by Swers et al. (1972) and Zubair et al. (1996).

Conclusions

In this chapter we have described a calculation procedure to optimize two-stage vapor compression refrigeration cycles from both COP and exergetic efficiency vantage points. We have shown that the interstage

TABLE 18.4 Polynomial Constants for Optimum Exergetic Efficiency

	R22	Ammonia		R22	Ammonia
g_1	$-7.57e-02$	$-6.74e+00$	g_{33}	$-4.61e-08$	$-1.27e-05$
g_2	$1.50e+00$	$3.10e+01$	g_{34}	$6.57e-07$	$5.69e-05$
g_3	$7.87e-02$	$-4.19e+01$	g_{35}	$-2.79e-07$	$-8.02e-05$
g_4	$-2.27e-02$	$1.91e+01$	g_{36}	$1.36e-07$	$3.65e-05$
g_5	$2.89e-04$	$8.78e-02$	g_{37}	$4.88e-10$	$1.67e-07$
g_6	$5.89e-03$	$-3.82e-01$	g_{38}	$3.35e-09$	$-7.37e-07$
g_7	$-4.66e-05$	$5.52e-01$	g_{39}	$3.58e-09$	$1.06e-06$
g_8	$-1.45e-04$	$-2.51e-01$	g_{40}	$-1.80e-09$	$-4.81e-07$
g_9	$-8.03e-07$	$-3.56e-04$	g_{41}	$-1.80e-12$	$-6.79e-10$
g_{10}	$-5.51e-06$	$1.57e-03$	g_{42}	$4.33e-12$	$3.01e-09$
g_{11}	$-1.62e-06$	$-2.24e-03$	g_{43}	$-1.42e-11$	$-4.29e-09$
g_{12}	$1.40e-06$	$1.02e-03$	g_{44}	$7.32e-12$	$1.95e-09$
g_{13}	$8.96e-10$	$4.45e-07$	g_{45}	$2.07e-15$	$8.49e-13$
g_{14}	$1.63e-08$	$-1.95e-06$	g_{46}	$1.18e-14$	$-3.75e-12$
g_{15}	$2.96e-09$	$2.80e-06$	g_{47}	$1.72e-14$	$5.37e-12$
g_{16}	$-2.16e-09$	$-1.28e-06$	g_{48}	$-8.95e-15$	$-2.45e-12$
g_{17}	$6.86e-05$	$1.61e-02$	g_{49}	$1.15e-11$	$3.30e-09$
g_{18}	$-1.29e-03$	$-7.24e-02$	g_{50}	$-1.36e-10$	$-1.47e-08$
g_{19}	$3.07e-04$	$1.01e-01$	g_{51}	$7.65e-11$	$2.08e-08$
g_{20}	$-1.47e-04$	$-4.62e-02$	g_{52}	$-3.74e-11$	$-9.46e-09$
g_{21}	$-5.97e-07$	$-2.12e-04$	g_{53}	$-1.30e-13$	$-4.34e-11$
g_{22}	$-7.37e-06$	$9.29e-04$	g_{54}	$-5.19e-13$	$1.91e-10$
g_{23}	$-3.82e-06$	$-1.34e-03$	g_{55}	$-9.93e-13$	$-2.74e-10$
g_{24}	$1.94e-06$	$6.08e-04$	g_{56}	$4.99e-13$	$1.25e-10$
g_{25}	$2.14e-09$	$8.58e-07$	g_{57}	$4.84e-16$	$1.76e-13$
g_{26}	$-1.24e-09$	$-3.80e-06$	g_{58}	$-1.51e-15$	$-7.81e-13$
g_{27}	$1.56e-08$	$5.42e-06$	g_{59}	$3.93e-15$	$1.11e-12$
g_{28}	$-8.11e-09$	$-2.47e-06$	g_{60}	$-2.01e-15$	$-5.07e-13$
g_{29}	$-2.46e-12$	$-1.07e-09$	g_{61}	$-5.57e-19$	$-2.20e-16$
g_{30}	$-2.21e-11$	$4.73e-09$	g_{62}	$-2.19e-18$	$9.74e-16$
g_{31}	$-1.91e-11$	$-6.79e-09$	g_{63}	$-4.73e-18$	$-1.39e-15$
g_{32}	$1.01e-11$	$3.09e-09$	g_{64}	$2.46e-18$	$6.36e-16$

pressure, COP, and the exergetic efficiency can all be expressed in terms of the condenser and evaporator pressures and the high- and low-pressure compression efficiencies for both ammonia and R22. This has been accomplished with a minimum amount of error for each case, resulting in polynomial expressions that faithfully reproduce the calculated data. The calculated optimum interstage pressure has been shown to be significantly different from the more widely used geometric mean method. Using the geometric mean method to calculate the optimum interstage pressure introduces deviations as large as 40 percent for R22 and 26 percent for ammonia. It has also been shown that the optimum interstage pressure equation produces the optimum interstage pressure for the COP as well as for the exergetic efficiency. An interesting result of this calculation procedure pertains to the fact that the exergetic efficiencies have optimum data points that fall on parabolically shaped curves, thus resulting in a maximum value for their respective optimum equations. Applying the procedure to more complex cycles and systems is relatively straightforward.

Acknowledgment

The authors are grateful for the support received from the Department of Mechanical and Aerospace Engineering at the University of Florida.

Nomenclature

a	Regression constant
COP	Coefficient of performance
d	Regression constant
g	Regression constant
h	Specific enthalpy, kJ/kg
IC/FT	Intercooler/flash tank
$\dot{m}$	Mass flow rate, kg/s
p	Pressure, kPa
Q	Amount of heat transfer, kJ
$\dot{Q}$	Heat-transfer rate, kW
s	Specific entropy, kJ/kg-K
T	Temperature, K
v	Specific volume, m^3/kg
W	Work, kJ
$\dot{W}$	Power, kW
Z	Compressibility factor

Greek

ε	Total exergy, kJ
Δ	Change in a quantity
Σ	Summation
η	Efficiency
Ψ	Specific availability, kJ/kg

Subscripts

0	Ambient conditions
c	Condenser
e	Evaporator
ex	Exergy

f	Fluid
G	Gained
gen	Generated
HP	High pressure
int	Interstage condition
LP	Low pressure
Ref	Refrigerated space condition
S	Supplied
s	Saturated condition

References

ASHRAE, 1997, *1997 ASHRAE Handbook—Fundamentals*, American Society of Heating, Refrigerating and Air-Conditioning Engineers, Inc., Atlanta, Georgia.

Bejan, A., Tsatsaronis, G., and Moran, M., 1996, *Thermal Design and Optimization*, Wiley, New York.

Gupta, V. K., and Prasad, M., 1984, "Graphic Estimation of Design Parameters for Two-Stage Ammonia Refrigerating Systems Parametrically Optimized," *Mechanical Engineering Bulletin*, vol. 14, no. 4, pp. 100–104.

Gupta, V. K., and Prasad, M., 1985, "On a Two-Stage R-22 Refrigerating System," *Israel Journal of Technology*, vol. 22, pp. 257–262.

Sonntag, R. E., Borgnakke, C., and Van Wylen, G. J., 1998, *Fundamentals of Thermodynamics*, Wiley, New York.

Stoecker, W. F., and Jones, J. W., 1982, *Refrigeration and Air Conditioning*, McGraw-Hill, New York.

Swers, R., Patel, Y. P., and Stewart, R. B., 1972, "Thermodynamic Analysis of Compression Refrigeration Systems," *ASHRAE Transactions*, vol. 78, no. 1, pp. 143–152.

Threlkeld, J. L., 1962, *Thermal Environmental Engineering*, Prentice-Hall, Englewood Cliffs, N. J.

Zubair, S. M., Yaqub, M., and Khan, S. H., 1996, "Second-Law-Based Thermodynamic Analysis of Two-Stage and Mechanical-Subcooling Refrigeration Cycles," *International Journal of Refrigeration*, vol. 19, no. 8, pp. 506–516.

Chapter 19

Transient Analysis of Low Temperature, Low Energy Carrier (LoTEC©)

Francis C. Wessling
Department of Mechanical and Aerospace Engineering
University of Alabama in Huntsville
Huntsville, Alabama

A *Low Temperature, Low Energy Carrier* (LoTEC©; see Fig. 19.1) is a passive thermal carrier designed to permit payload transport to and from the International Space Station at approximately constant temperature without external power [1]. The only power that LoTEC requires is for recording its interior temperatures. This power is at the milliwatt level and is provided by a temperature recorder housed with the payload inside LoTEC. LoTEC relies on a combination of high-resistance aerogel insulation and high-energy-density storage phase-change materials (PCMs). The capability of LoTEC is dependent on the type and amount of PCM chosen and on the ambient temperature. LoTEC has successfully operated at temperatures as low as −80°C and in a complete vacuum. It can hold an internal temperature of 2°C for approximately 2 weeks depending on the amount of PCM used and the ambient temperature.

LoTEC is a rectangular parallelepiped approximately 516 × 440 × 253 mm. The four 516-mm-long edges of LoTEC are chamfered at a 45° angle to facilitate installation into a storage facility on the Space Transportation System (STS), more commonly known as the *space shuttle*. LoTEC's composite shell protects the thermal insulation from

19.2 Transient and Cyclic Calculations

Figure 19.1 Low Temperature, Low Energy Carrier (LoTEC).

mechanical damage, and is a carrying case for transport on the ground, in the STS and in the ISS.

The insulation used in LoTEC is approximately 38 mm thick between the inside of the composite outer shell and an inner liner assembly. The inner liner assembly allows an interior open volume of approximately 158 × 343 × 413 mm to accommodate samples and PCMs. A LoTEC inner door on the inner liner assembly secures the inner volume, and another door (LoTEC outer door) on the front of LoTEC insulates the inner door. Each door has its own seal. This double-door technique prevents water condensation from forming on the outside of LoTEC when payloads that are cooler than the STS or ISS dew point are used.

An analysis of the transient response of LoTEC is needed to predict thermal performance. Although we have performed finite-element analyses, a quick analytical solution is necessary. Thus, several assumptions are made about the analytical model.

LoTEC can be considered as a well-insulated box with a small amount of structure contributing to heat gain to or loss from the interior. This can be simplified to a one-dimensional analysis (see Fig. 19.2). Ignore the chamfered corners of the aerogel insulation; this can be adjusted later on the basis of experimental data to give a proper area of the aerogel. Thus, the area of the aerogel A_i is originally taken to be that of the sides, top, bottom, front, and back of the inner liner. The thickness of the aerogel L equals 1.5 in. (0.0381 m). The temperature of the aerogel insulation T_i is a function of space and time. The thermal conductivity of the insulation is k_i, where i refers to the insulation. The inner liner is taken as a lumped mass $\rho_I V_I$, where I refers to the inner liner. The mass of the inner liner was actually determined by weighing. The specific heat of the liner is taken to be $C_{p,I}$. The temperature of the inner liner T_I is a function of time only; time is denoted as t. The distance perpendicular to the larger surfaces of the aerogel is x. The value of the effective

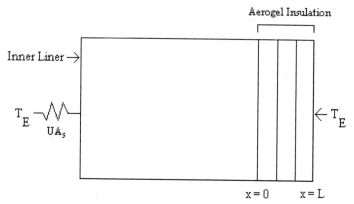

Figure 19.2 One-dimensional representation of LoTEC.

conductance of the structural heat gain or loss to the interior not due to aerogel is UA_s.

The temperature of the environment T_E varies in time as a temperature ramp that starts at the initial temperature of LoTEC and rises to a final temperature T_c of an environmental chamber in a time period t_0. All the temperatures are referenced to the initial temperature T_0 of LoTEC. The temperature ramp can be seen in Fig. 19.3. The time expression for this temperature as shown in Fig. 19.3, T_E, is given as

$$T_E = \left[\left(1 - \frac{t}{t_0}\right) \cdot u(t - t_0) + \frac{t}{t_0}\right] \cdot T_c \qquad (19.1)$$

where u is the unit step function or Helmholtz function. The transient solution for the temperatures requires the use of the Laplace

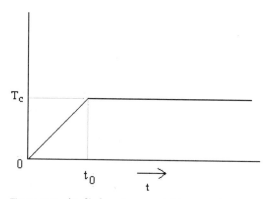

Figure 19.3 Applied environmental temperature ramp.

19.4 Transient and Cyclic Calculations

transform. Consequently, the Laplace transform of the temperature ramp is given as

$$L(T_E) = \frac{T_c}{t_0} \cdot \left(\frac{1}{s^2} - \frac{e^{-s \cdot t_0}}{s^2} \right) \qquad (19.2)$$

where s is the Laplace transform variable.

A partial differential equation (19.3) describes the time rate of change of the temperature of the inner liner to the energy transferred through the aerogel and through the structure directly:

$$\rho_I \cdot C_{p,I} \cdot V_I \cdot \frac{dT_I(t)}{dt} = \left(k_i \cdot A_i \cdot \frac{\partial T_i(0, t)}{\partial x} \right) + UA_s \cdot (T_E(t) - T_I) \qquad (19.3)$$

The temperature of the inner liner is set equal to the temperature at $x = 0$ of the aerogel that surrounds it. The temperature of the aerogel at $x = L$ is set equal to the environment temperature. Thus, perfect conductance between the aerogel and the surrounding air in the chamber is assumed. Measurements show this to be a good assumption because of the large resistance of the aerogel compared to the convective coefficient on the surface. Thus

$$T_I(t) = T_i(0, t) \qquad (19.4)$$

$$T_i(L, t) = T_E(t) \qquad (19.5)$$

The partial differential equation that describes the energy transfer through the aerogel ($x = 0$ to $x = L$) is given as

$$\frac{1}{\alpha} \cdot \frac{\partial T_i}{\partial t} = \frac{\partial^2 T_i}{\partial x^2} \qquad (19.6)$$

where α is the thermal diffusivity of the aerogel. The initial condition of the aerogel is set equal to the initial temperature T_0. Likewise, the initial temperature of the inner liner is set equal to T_0. These are set equal to 0.0 because every temperature is referenced to the initial temperature.

The initial conditions are $T_i(x, 0) = T_0$ and $T_I(0) = T_0 = 0$. Let

$$\alpha = \frac{k_i}{\rho_i \cdot C_{p,i}} \qquad (19.7)$$

$$\beta = \frac{k_i \cdot A_i}{\rho_I \cdot C_{p,I} \cdot V_I} \qquad (19.8)$$

$$\gamma = \frac{UA_s}{\rho_I \cdot C_{p,I} \cdot V_I} \qquad (19.9)$$

Let ϕ be the Laplace transform of the temperatures. The Laplace transform of the equations is given as follows:

From Eq. (19.3)

$$s \cdot \phi_I = \beta \cdot \frac{d\phi_i(0)}{dx} + \left[\gamma \cdot \frac{T_c}{t_0} \cdot \left(\frac{1}{s^2} - \frac{e^{-s \cdot t_0}}{s^2}\right)\right] - \gamma \phi_I \quad (19.10)$$

From Eq. (19.4)

$$\phi_I = \phi_i(0) \quad (19.11)$$

From Eq. (19.5)

$$\phi_i(L) = \frac{T_c}{t_0} \cdot \left(\frac{1}{s^2} - \frac{e^{-s \cdot t_0}}{s^2}\right) \quad (19.12)$$

or

$$\phi_i(L) = \frac{T_c}{t_0} \cdot \frac{1 - e^{-s \cdot t_0}}{s^2} \quad (19.13)$$

From Eq. (19.6)

$$\frac{s}{\alpha} \cdot \phi_i = \frac{d^2 \phi_i}{dx^2} \quad 0 \le x \le L \quad (19.14)$$

Let $q^2 = s/\alpha$ in order to solve the differential equation for ϕ_i. Then

$$q^2 = \frac{s}{\alpha} \quad (19.15)$$

Take the positive root in the solution because ϕ_I, as will subsequently be shown, is

$$\phi_I = \frac{1 + \dfrac{\gamma}{\beta \cdot q} \cdot \sinh(q \cdot L)}{\cosh(q \cdot L) + \dfrac{(s + \gamma) \cdot \sinh(q \cdot L)}{\beta \cdot q}} \cdot \left[\frac{T_E}{s^2 \cdot t_0} \cdot \left(1 - e^{-s \cdot t_0}\right)\right] \quad (19.16)$$

If $q > 0$
$$\frac{\sinh(q \cdot L)}{q} = \frac{\sinh(|q| \cdot L)}{|q|}$$

If $q < 0$
$$\frac{\sinh(q \cdot L)}{q} = \frac{-1 \cdot \sinh(|q| \cdot L)}{-|q|}$$

So, use the positive root in the expressions:

$$q = \text{positive} \sqrt{\frac{s}{\alpha}}$$

19.6 Transient and Cyclic Calculations

Substituting for q into Eq. (19.14) yields

$$\frac{d^2}{dx^2}\phi_i - q^2 \cdot \phi_i = 0 \qquad (19.17)$$

which has the following solution:

$$\phi_i = A \cdot \cosh(q \cdot x) + B \cdot \sinh(q \cdot x) \qquad (19.18)$$

$$\frac{d\phi_i}{dx} = A \cdot q \cdot \sinh(q \cdot x) + B \cdot q \cdot \cosh(q \cdot x) \qquad (19.19)$$

Applying the boundary condition at $x = 0$ yields

$$\phi_i = A = \phi_I \qquad (19.20)$$

and the differential equation [Eq. (19.10)] for ϕ_I gives the following equation for B:

$$s\phi_I = \beta \cdot B \cdot q + \left[\gamma \cdot \frac{T_c}{t_0 \cdot s^2} \cdot (1 - e^{-s \cdot t_0})\right] - \gamma \cdot \phi_I \qquad (19.21)$$

Thus,

$$B = \frac{(s+\gamma) \cdot \phi_I - \dfrac{\gamma \cdot T_c}{t_0 \cdot s^2} \cdot (1 - e^{-s \cdot t_0})}{\beta \cdot q} \qquad (19.22)$$

So

$$\phi_i = \phi_I \cdot \cosh(q \cdot x) + \sinh(q \cdot x)$$

$$\cdot \left[\frac{(s+\gamma) \cdot \phi_I}{\beta \cdot q} - \gamma \cdot \frac{T_c}{t_0 \cdot s^2 \cdot \beta \cdot q} \cdot (1 - e^{-s \cdot t_0})\right] \qquad (19.23)$$

Applying the boundary condition at $x = L$ yields

$$\frac{T_c}{t_0 \cdot s^2} \cdot (1 - e^{-s \cdot t_0}) = \phi_I \cdot \cosh(q \cdot L)$$

$$+ \left[\frac{s+\gamma}{\beta \cdot q} \cdot \phi_I - \frac{\gamma \cdot T_c}{t_0 \cdot s^2 \cdot \beta \cdot q} \cdot (1 - e^{-s \cdot t_0})\right] \cdot \sinh(qL)$$

$$(19.24)$$

or

$$\frac{T_c}{t_0 \cdot s^2} \cdot (1 - e^{-s \cdot t_0}) \cdot \left(1 + \frac{\gamma}{\beta \cdot q} \cdot \sinh(q \cdot L)\right)$$

$$= \phi_I \cdot \left(\cosh(q \cdot L) + \frac{s+\gamma}{\beta \cdot q} \cdot \sinh(q \cdot L)\right) \qquad (19.25)$$

Solving (19.25) for ϕ_m yields

$$\phi_I = \frac{\frac{T_c}{t_0 \cdot s^2} \cdot (1 - e^{-s \cdot t_0}) \cdot \left(1 + \frac{\gamma}{\beta \cdot q} \cdot \sinh(q \cdot L)\right)}{\cosh(q \cdot L) + \frac{s + \gamma}{\beta \cdot q} \cdot \sinh(q \cdot L)} \quad (19.26)$$

Now, return to the time domain. Use the Laplace transform inversion theorem:

$$T_I = \int_{R - i \cdot \infty}^{R + i \cdot \infty} \phi_I \cdot e^{s \cdot t} dt \quad (19.27)$$

Because the integral is in complex space, let $s = z$, then $q = \sqrt{z/\alpha}$ and

$$\frac{dq}{dz} = \frac{1}{2\sqrt{\alpha \cdot z}} = \frac{\sqrt{\alpha}}{2\alpha\sqrt{z}} = \frac{1}{2\alpha \cdot q}$$

Then

$$T_I = \int_{R - i \cdot \infty}^{R + i \cdot \infty} \frac{T_c}{t_0 \cdot z^2} \cdot \frac{(1 - e^{-z \cdot t_0}) \cdot \left(1 + \frac{\gamma}{\beta \cdot q} \cdot \sinh(q \cdot L)\right)}{\cosh(q \cdot L) + \left(\frac{z + \gamma}{\beta \cdot q}\right) \cdot \sinh(q \cdot L)} \cdot e^{z \cdot t} dt \quad (19.28)$$

Now let

$$g(z) = \frac{T_c}{t_0 \cdot z^2} \cdot (1 - e^{-z \cdot t_0}) \cdot \left(1 + \frac{\gamma}{\beta \cdot q} \cdot \sinh(q \cdot L)\right) \cdot e^{z \cdot t} \quad (19.29)$$

$$h(z) = \cosh(q \cdot L) + \left(\frac{z + \gamma}{\beta \cdot q}\right) \cdot \sinh(q \cdot L) \quad (19.30)$$

There is a pole of order 2 at $z = 0$ in Eq. (19.28), so take the derivative to get the residue; then take the limit as z approaches zero. The limit turns out to be T_c. For the many zeros of Eq. (19.30), the residue is evaluated as

$$\text{Residue} = \lim_{z \to 0} \frac{d}{dz} \left[\frac{\frac{T_c}{t_0} \cdot (1 - e^{-z \cdot t_0}) \cdot \left(1 + \frac{\gamma}{\beta \cdot q} \cdot \sinh(q \cdot L)\right) \cdot e^{z \cdot t}}{\cosh(q \cdot L) + \frac{z + \gamma}{\beta \cdot q} \cdot \sinh(q \cdot L)} \right]$$

$$(19.31)$$

19.8 Transient and Cyclic Calculations

There are many poles of order 1 for

$$\cosh(q \cdot L) + \frac{z+\gamma}{\beta \cdot q} \cdot \sinh(q \cdot L) = 0 \tag{19.32}$$

So

$$\text{Residue} = \frac{g(z)}{h'(z)} \tag{19.33}$$

This leads to an infinite series solution because many values of z satisfy Eq. (19.32). Thus, the solution for the temperature is

$$T_I(t) = T_c + \sum_{j=0}^{\infty} T_m(\alpha, \beta, \gamma, z_j, L, t) \tag{19.34}$$

where

$T_m(\alpha, \beta, \gamma, L, z, t)$

$$= \frac{\frac{T_c - T_0}{t_0 \cdot z^2} \cdot \exp(z \cdot t) \cdot (1 - \exp(-t_0 \cdot z)) \cdot \left(1 + \frac{\gamma}{\beta \cdot z} \cdot \sqrt{\alpha \cdot z} \cdot \sinh\left(\sqrt{\frac{z}{\alpha}} \cdot L\right)\right)}{\frac{1}{2} \cdot \sinh[L \cdot \sqrt{z/\alpha}] \cdot \frac{L}{\sqrt{z/\alpha} \cdot \alpha} + \frac{1}{\beta \cdot \sqrt{z/\alpha}} \cdot \sinh[L \cdot \sqrt{z/\alpha}] - \frac{1}{2} \cdot \frac{(z+\gamma)/z}{\beta \cdot \sqrt{z/\alpha}} \cdot \sinh[L \cdot \sqrt{z/\alpha}] + \frac{1}{2} \cdot \frac{z+\gamma}{\beta \cdot z} \cdot \cosh[L \cdot \sqrt{z/\alpha}] \cdot L}$$

$$\tag{19.35}$$

The results of this equation compare favorably with measurements of the transient temperature (see Fig. 19.4). The values of variables α, β, γ, and L are 4.80×10^{-8} m²/s, 3.15×10^{-6} m/s, 5.25×10^{-6}/s, and 0.0381 m. The sum-of-squares error is $0.0764 K^2$. The analytical temperatures are calculated using only five terms of the infinite series. The measured values are determined from inserting LoTEC into an environmental chamber and applying the ramp temperature of Fig. 19.3. The time of the ramp t_0 is 1 h. The temperature T_c is 20°C, and the temperature T_E is 60°C. The temperatures measured are an average of six temperatures on the inner liner walls. The measured temperatures differ from each other by less than 1°C, with an uncertainty of 0.2°C. Note that the square root of the sum-of-squares error is 0.276°C, which is slightly greater than the uncertainty in the measurement.

The agreement between the measurements and the analytical results shows that even though the LoTEC has an intricate geometry, the one-dimensional simplification is amazingly accurate.

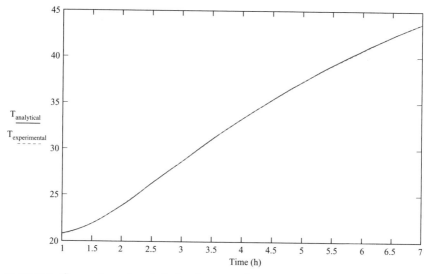

Figure 19.4 Comparison of analytical and measured temperatures using five terms of the summation. Times shorter than 1 h require more than five terms in the infinite series with these values of properties. The sum-of-squares error between the two temperatures is $0.0764 K^2$.

Reference

1. F. C. Wessling, L. S. Stodleck, A. Hoehn, S. Woodard, S. O'Brien, and S. Thomas, "Low Temperature, Low Energy Carrier (LoTEC©) and Phase Change Materials (PCMs) for Biological Samples," 30th *Proceedings of the International Conference on Environmental Systems,* paper no. 2000-01-2280, Toulouse, France, July 2000.

Chapter 20

Parameter Estimation of Low Temperature, Low Energy Carrier (LoTEC©)

Francis C. Wessling
Department of Mechanical and Aerospace Engineering
University of Alabama in Huntsville
Huntsville, Alabama

James J. Swain
Department of Industrial and Systems Engineering and Engineering Management
University of Alabama in Huntsville
Huntsville, Alabama

A *Low Temperature, Low Energy Carrier* (LoTEC©; see Fig. 20.1) is a passive thermal carrier designed to permit payload transport to and from the International Space Station at approximately constant temperature without external power [1]. Further details on LoTEC are given in a companion calculation in Chap. 19.

An analysis of the transient response of LoTEC with one-dimensional assumptions (see Fig. 20.2) is presented in the companion calculation. This solution is for the case of a ramp temperature as shown in Fig. 20.3. As shown in the figure, the temperature of the environment T_E varies over time, starting at the initial temperature of the LoTEC and rising to a final temperature of T_c over a period t_0.

The solution of the transient temperature response $T_I(t)$ for the inner liner of the LoTEC is given in Eq. (19.35) of Chap. 19 (above) and is

20.2 Transient and Cyclic Calculations

Figure 20.1 Low Temperature, Low Energy Carrier (LoTEC).

reproduced here as Eq. (20.1):

$$T_I(t) = T_c + \sum_{i=1}^{\infty} T_m(\alpha, \beta, \gamma, z_i, L, t) \qquad (20.1)$$

where

$$T_m(\alpha, \beta, \gamma, L, z, t) = \frac{\frac{T_c - T_0}{t_0 z^2} e^{(zt)}(1 - \exp(-t_0 z))\left(1 + \frac{\gamma}{\beta z}\sqrt{\alpha z}\sinh[L\sqrt{z/\alpha}]\right)}{\frac{L}{2\alpha\sqrt{z/\alpha}}\sinh[L\sqrt{z/\alpha}] + \frac{1}{\beta L\sqrt{z/\alpha}}\sinh[L\sqrt{z/\alpha}] - \frac{(z+\gamma)/z}{2\beta L\sqrt{z/\alpha}}\sinh[L\sqrt{z/\alpha}] + \frac{L(z+\gamma)}{2\alpha}\cosh[L\sqrt{z/\alpha}]}$$

The solution is given in terms of five variables: α, β, γ, L, and t. The roots $z_i, i = 1, \ldots$, are dependent on the other variables through a

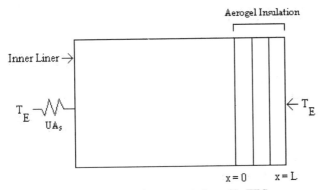

Figure 20.2 One-dimensional representation of LoTEC.

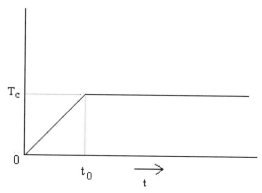

Figure 20.3 Applied environmental temperature ramp.

transcendental equation [Eq. (20.2)], and are independent of time t:

$$F(\alpha, \beta, \gamma, L, z) = \cosh[L\sqrt{z/\alpha}] + \frac{z+\gamma}{\beta\sqrt{z/\alpha}} \sinh\left[L\sqrt{\frac{z}{\alpha}}\right] \qquad (20.2)$$

The variables α, β, and γ contain eight primitive variables: $\alpha = k_i/\rho_i C_{p,i}$, $\beta = k_i A_i/\rho_m V_m C_{p,m}$, and $\gamma = UA_s/\rho_m V_m C_{p,m}$. Measurements exist for some of these primitive variables, but not all of them. For example, the effective area of the aerogel is unknown because of the chamfered corners of LoTEC. The area of the aerogel A_i is originally taken to be that of the sides, top, bottom, front, and back of the inner liner. This is adjusted later according to experimental data to give a proper area of the aerogel. The thickness of the aerogel is designated by L and equal to 1.5 in. (0.0381 m). The thermal conductivity of the insulation is k_i. Although it is measured separately in the laboratory, its actual value in this application is unknown. The inner liner is taken as a lumped mass $\rho_m V_m$ and is obtained directly by weighing. The specific heat of the liner is taken to be $C_{p,m}$, and this value is not well known. The effective conductance of the structural heat gain or loss to the interior not due to aerogel is given by UA_s.

Each of the three variables α, β, and γ have at least one unknown. These are k_i, A_i, $C_{p,m}$, and UA_s. A nonlinear least-squares parameter estimation technique using Gauss linearization [2] is used to determine the variables α, β, and γ rather than the primitive variables. The variables A_i, $C_{p,m}$, and UA_s should be repeatable from one LoTEC to another. Their values are dependent on the geometric construction of LoTEC. The LoTEC are assembled using jigs to hold the dimensions to close tolerance. The approach used here is to test two LoTECs to

20.4 Transient and Cyclic Calculations

estimate unique values of α for each LoTEC and estimate the common values of β and γ from both LoTECs.

The two unique values of α (α_1 and α_2), β, and γ can be estimated together using the results of both experiments in combination as will be described. The temperature for each LoTEC is measured every 6 min from the end of the ramp time (t_0 is one hour) to 7 h for a total of 61 data points per LoTEC and 122 data points overall. The first 61 points are from LoTEC 1 and the second 61 are from LoTEC 2.

The nonlinear estimation problem is used to solve for the values of $\hat{\alpha}_1$, $\hat{\alpha}_2$, $\hat{\beta}$, and $\hat{\gamma}$ that most closely match the 122 observed temperatures to the values predicted in Eq. (20.1). A Gauss approximation algorithm is used to iteratively solve for the values of the parameters that achieve this best fit. The Gauss algorithm is based on a linearization of the response $T_I(t)$ using the derivatives of $T_I(t)$ with respect to the parameters α_1, α_2, β, and γ. These derivatives are contained in the sensitivity matrix **X** consisting of 122 rows and 4 columns, in which each row of the matrix represents the derivative of the $T_I(t)$ for a particular observation of t and the value of the parameters at that particular iteration of the algorithm. Note that for the first 61 rows, $T_I(t)$ depends on α_1 and the final 61 depend on α_2 as the only terms that differ in the two experiments. The derivatives are evaluated next.

As noted, the variable z are the roots of the transcendental Eq. (20.2). Derivatives must be taken with respect to α, β, γ, and L. These are needed in the formulation of the derivatives of $T_m(t)$, which are also functions of z and are given, for example, by $DT_m/D\alpha = \partial T_m/\partial \alpha + \partial T_m/\partial z \, \partial z/\partial \alpha$, where $DT_m/D\alpha$ is the derivative of T_m with respect to α considering both the explicit derivative $\partial T_m/\partial \alpha$ and the derivative $\partial T_m/\partial z$ from the variable z in Eq. (20.2). Similar derivatives exist for the other three variables. Because z is an implicit function of α, β, γ, and L as defined by F of Eq. (20.2), its derivative with respect to α is given by $\partial z/\partial \alpha = -(\partial F/\partial \alpha)/(\partial F/\partial z)$. Similar expressions obtain for the other derivatives. The actual equations follow.

Derivatives of F

The transcendental function F is given in Eq. (20.2). The partial derivatives of F with respect to the variables α, β, γ, L, and z are now developed in the following equations using the shorthand notation $dF\,d\alpha$ to represent $\partial F/\partial \alpha$:

$$dF\,d\alpha(\alpha, \beta, \gamma, L, z) = \frac{-Lz}{2\alpha^2 \sqrt{z/\alpha}} \sinh\left[L\sqrt{\frac{z}{\alpha}}\right] + \frac{z(z+\gamma)}{2\alpha^2 \beta (z/\alpha)^{3/2}}$$

$$\times \sinh\left[L\sqrt{\frac{z}{\alpha}}\right] + \frac{(z+\gamma)L}{2\alpha\beta} \cosh\left[L\sqrt{\frac{z}{\alpha}}\right] \quad (20.3)$$

$$dF\,d\beta(\alpha, \beta, \gamma, L, z) = \frac{-(z+\gamma)}{\beta^2 \sqrt{z/\alpha}} \sinh\left[L\sqrt{\frac{z}{\alpha}}\right] \qquad (20.4)$$

$$dF\,d\gamma(\alpha, \beta, \gamma, L, z) = \frac{1}{\beta \sqrt{z/\alpha}} \sinh\left[L\sqrt{\frac{z}{\alpha}}\right] \qquad (20.5)$$

$$dF\,dL(\alpha, \beta, \gamma, L, z) = \sqrt{\frac{z}{\alpha}} \sinh\left(L\sqrt{\frac{z}{\alpha}}\right) + \left[\frac{z+\gamma}{\beta}\cosh\left(L\sqrt{\frac{z}{\alpha}}\right)\right] \qquad (20.6)$$

$$dF\,dz(\alpha, \beta, \gamma, L, z) = \frac{L}{2\alpha\sqrt{z/\alpha}} \sinh\left[L\sqrt{\frac{z}{\alpha}}\right] + \frac{1}{\beta\sqrt{z/\alpha}} \sinh\left[L\sqrt{\frac{z}{\alpha}}\right]$$
$$- \frac{z+\gamma}{2\beta z\sqrt{z/\alpha}} \sinh\left[L\sqrt{\frac{z}{\alpha}}\right] + \frac{(z+\gamma)L}{2\beta z} \cosh\left[L\sqrt{\frac{z}{\alpha}}\right] \qquad (20.7)$$

The derivatives of z with respect to the four variables are given in the following equations:

$$dz\,d\alpha(\alpha, \beta, \gamma, L, z) = \frac{-dF\,d\alpha(\alpha, \beta, \gamma, L, z)}{dF\,dz(\alpha, \beta, \gamma, L, z)} \qquad (20.8)$$

$$dz\,d\beta(\alpha, \beta, \gamma, L, z) = \frac{-dF\,d\beta(\alpha, \beta, \gamma, L, z)}{dF\,dz(\alpha, \beta, \gamma, L, z)} \qquad (20.9)$$

$$dz\,d\gamma(\alpha, \beta, \gamma, L, z) = \frac{-dF\,d\gamma(\alpha, \beta, \gamma, L, z)}{dF\,dz(\alpha, \beta, \gamma, L, z)} \qquad (20.10)$$

$$dz\,dL(\alpha, \beta, \gamma, L, z) = \frac{-dF\,dL(\alpha, \beta, \gamma, L, z)}{dF\,dz(\alpha, \beta, \gamma, L, z)} \qquad (20.11)$$

Derivatives of Temperature

We now develop derivatives of temperature. In order to keep the equations within a reasonable length, the temperature terms are separated into differentiable pieces that are then reassembled.

The numerator of Eq. (20.1) is evaluated first. Call it $N(\alpha, \beta, \gamma, L, z, t)$. The partial derivatives will be designated in the form $dN\,d\alpha(\alpha, \beta, \gamma, L, z, t)$. Similar nomenclature is used for the other variables. Note that since observations are taken after $t \geq t_0$, T_E is a constant equal to $T_c - T_0$

20.6 Transient and Cyclic Calculations

as used in Eq. (20.1):

$$N(\alpha, \beta, \gamma, L, z, t) = \frac{T_E}{t_0 z^2} e^{(zt)}(1 - \exp(-t_0 z))\left(1 + \frac{\gamma}{\beta z}\sqrt{\alpha z}\sinh\left[L\sqrt{\frac{z}{\alpha}}\right]\right) \quad (20.12)$$

$$dNd\alpha(\alpha, \beta, \gamma, L, z, t) = \frac{T_E}{t_0 z^2} e^{zt}(1 - \exp(-t_0 z))\left[\frac{\gamma}{2\beta\sqrt{\alpha z}}\sinh\left[L\sqrt{\frac{z}{\alpha}}\right]\right.$$
$$\left. -\frac{L\gamma}{2\alpha\beta}\cosh\left[L\sqrt{\frac{z}{\alpha}}\right]\right] \quad (20.13)$$

$$dNd\beta(\alpha, \beta, \gamma, L, z, t) = \frac{-\gamma T_E\sqrt{\alpha z}}{t_0 \beta^2 z^3} e^{(zt)}(1 - \exp(-t_0 z))\sinh\left[L\sqrt{\frac{z}{\alpha}}\right] \quad (20.14)$$

$$dNd\gamma(\alpha, \beta, \gamma, L, z, t) = \frac{T_E\sqrt{\alpha z}}{\beta t_0 z^3} e^{(zt)}(1 - \exp(-t_0 z))\sinh\left[L\sqrt{\frac{z}{\alpha}}\right] \quad (20.15)$$

$$dNdL(\alpha, \beta, \gamma, L, z, t) = \frac{T_E\gamma}{t_0 \beta z^2} e^{(zt)}(1 - \exp(-t_0 z))\cosh\left[L\sqrt{\frac{z}{\alpha}}\right] \quad (20.16)$$

The derivative of the numerator with respect to z is considered next. The numerator is separated into two parts, N_1 and N_2, to simplify the expressions:

$$N_1(\alpha, \beta, \gamma, L, z, t) = \frac{T_E}{t_0 z^2} e^{(zt)}(1 - \exp(-t_0 z)) \quad (20.17)$$

$$N_2(\alpha, \beta, \gamma, L, z, t) = 1 + \frac{\gamma}{\beta z}\sqrt{\alpha z}\sinh\left[L\sqrt{\frac{z}{\alpha}}\right] \quad (20.18)$$

Taking derivatives of these expressions yields

$$dN_1\,dz(\alpha, \beta, \gamma, L, z, t) = \frac{-2T_E}{t_0 z^3} e^{(zt)}(1 - \exp(-t_0 z)) + \frac{T_E t}{t_0 z^2} e^{(zt)}(1 - \exp(-t_0 z))$$
$$+ \frac{T_E}{z^2} e^{(zt)}\exp(-t_0 z) \quad (20.19)$$

$$dN_2\,dz(\alpha, \beta, \gamma, L, z, t) = \frac{-\gamma\sqrt{\alpha z}}{\beta z^2}\sinh\left[L\sqrt{\frac{z}{\alpha}}\right] + \frac{\alpha\gamma}{2\beta z\sqrt{\alpha z}}\sinh\left[L\sqrt{\frac{z}{\alpha}}\right]$$
$$+ \frac{\gamma L}{2\beta z}\cosh\left[L\sqrt{\frac{z}{\alpha}}\right] \quad (20.20)$$

These terms are then combined to yield the expression

$$dN\,dz(\alpha, \beta, \gamma, L, z, t) = N_1(\alpha, \beta, \gamma, L, z, t)dN_2\,dz(\alpha, \beta, \gamma, L, z, t)$$
$$+ N_2(\alpha, \beta, \gamma, L, z, t)dN_1\,dz(\alpha, \beta, \gamma, L, z, t) \quad (20.21)$$

The denominator $D(\alpha, \beta, \gamma, L, z, t)$ is approached in the same manner as the numerator by separation into four parts $D_i(\alpha, \beta, \gamma, L, z, t)$, for $i = 1, 2, 3, 4$:

$$D(\alpha, \beta, \gamma, L, z, t) = \frac{L}{2\alpha\sqrt{z/\alpha}}\sinh\left[L\sqrt{\frac{z}{\alpha}}\right] + \frac{1}{\beta\sqrt{z/\alpha}}\sinh\left[L\sqrt{\frac{z}{\alpha}}\right]$$
$$- \frac{z+\gamma}{2\beta z\sqrt{z/\alpha}}\sinh\left[L\sqrt{\frac{z}{\alpha}}\right] + \frac{(z+\gamma)L}{2\beta z}\cosh\left[L\sqrt{\frac{z}{\alpha}}\right]$$
$$(20.22)$$

Only the nonzero denominator derivatives are presented below (e.g., $dD_1/d\beta = 0$):

$$dD_1\,d\alpha\,(\alpha, \beta, \gamma, L, z, t) = \frac{-L^2}{4\alpha^2}\cosh\left[L\sqrt{\frac{z}{\alpha}}\right] + \frac{L}{4\alpha\sqrt{\alpha z}}\sinh\left[L\sqrt{\frac{z}{\alpha}}\right]$$
$$- \frac{L}{2\alpha\sqrt{\alpha z}}\sinh\left[L\sqrt{\frac{z}{\alpha}}\right] \quad (20.23)$$

$$dD_1\,dL\,(\alpha, \beta, \gamma, L, z, t) = \frac{L}{2\alpha}\cosh\left[L\sqrt{\frac{z}{\alpha}}\right] + \frac{1}{2\sqrt{\alpha z}}\sinh\left[L\sqrt{\frac{z}{\alpha}}\right]$$
$$(20.24)$$

$$dD_1\,dz\,(\alpha, \beta, \gamma, L, z, t) = \frac{L^2}{4\alpha z}\cosh\left[L\sqrt{\frac{z}{\alpha}}\right] - \frac{L}{4z\sqrt{\alpha z}}\sinh\left[L\sqrt{\frac{z}{\alpha}}\right]$$
$$(20.25)$$

The derivatives for the second denominator term $D_2(\alpha, \beta, \gamma, L, z, t)$ are provided next:

$$dD_2\,d\alpha\,(\alpha, \beta, \gamma, L, z, t) = \frac{1}{2\beta\sqrt{\alpha z}}\cosh\left[L\sqrt{\frac{z}{\alpha}}\right] + \frac{L}{2\alpha\beta}\sinh\left[L\sqrt{\frac{z}{\alpha}}\right]$$
$$(20.26)$$

$$dD_2\,d\beta\,(\alpha, \beta, \gamma, L, z, t) = \frac{-1}{2\beta^2\sqrt{z/\alpha}}\sinh\left[L\sqrt{\frac{z}{\alpha}}\right] \quad (20.27)$$

20.8 Transient and Cyclic Calculations

$$dD_2\, dL(\alpha, \beta, \gamma, L, z, t) = \frac{1}{\beta} \cosh\left[L\sqrt{\frac{z}{\alpha}}\right] \tag{20.28}$$

$$dD_2\, dz(\alpha, \beta, \gamma, L, z, t) = \frac{-1}{2\beta z \sqrt{z/\alpha}} \sinh\left[L\sqrt{\frac{z}{\alpha}}\right] + \frac{L}{2\beta z} \cosh\left[L\sqrt{\frac{z}{\alpha}}\right] \tag{20.29}$$

The derivatives for the third denominator term $D_3(\alpha, \beta, \gamma, L, z, t)$ are provided next:

$$dD_3\, d\alpha(\alpha, \beta, \gamma, L, z, t) = \frac{-(z+\gamma)}{4\beta z \sqrt{\alpha z}} \sinh\left[L\sqrt{\frac{z}{\alpha}}\right] + \frac{L(z+\gamma)}{4\alpha\beta z} \cosh\left[L\sqrt{\frac{z}{\alpha}}\right] \tag{20.30}$$

$$dD_3\, d\beta(\alpha, \beta, \gamma, L, z, t) = \frac{z+\gamma}{2\beta^2 z \sqrt{z/\alpha}} \sinh\left[L\sqrt{\frac{z}{\alpha}}\right] \tag{20.31}$$

$$dD_3\, d\gamma(\alpha, \beta, \gamma, L, z, t) = \frac{1}{2\beta z \sqrt{z/\alpha}} \sinh\left[L\sqrt{\frac{z}{\alpha}}\right] \tag{20.32}$$

$$dD_3\, dL(\alpha, \beta, \gamma, L, z, t) = \frac{-(z+\gamma)}{2\beta z} \cosh\left[L\sqrt{\frac{z}{\alpha}}\right] \tag{20.33}$$

$$dD_3\, dz(\alpha, \beta, \gamma, L, z, t) = \frac{-1}{2\beta z \sqrt{z/\alpha}} \sinh\left[L\sqrt{\frac{z}{\alpha}}\right] + \frac{z+\gamma}{2\beta z^2 \sqrt{z/\alpha}}$$

$$\times \sinh\left[L\sqrt{\frac{z}{\alpha}}\right] + \frac{z+\gamma}{4\beta z^2 \sqrt{z/\alpha}} \sinh\left[L\sqrt{\frac{z}{\alpha}}\right]$$

$$- \frac{L(z+\gamma)}{4\beta z^2} \cosh\left[L\sqrt{\frac{z}{\alpha}}\right] \tag{20.34}$$

The final denominator term D_4 has the following derivatives:

$$dD_4\, d\alpha(\alpha, \beta, \gamma, L, z, t) = \frac{-L^2(z+\gamma)}{4\alpha^2 \beta \sqrt{z/\alpha}} \sinh\left[L\sqrt{\frac{z}{\alpha}}\right] \tag{20.35}$$

$$dD_4\, d\beta(\alpha, \beta, \gamma, L, z, t) = \frac{-L(z+\gamma)}{2\beta^2 z} \cosh\left[L\sqrt{\frac{z}{\alpha}}\right] \tag{20.36}$$

$$dD_4\, d\gamma(\alpha, \beta, \gamma, L, z, t) = \frac{L}{2\beta z} \cosh\left[L\sqrt{\frac{z}{\alpha}}\right] \tag{20.37}$$

$$dD_4\,dL(\alpha,\beta,\gamma,L,z,t) = \frac{z+\gamma}{2\beta\sqrt{\alpha z}}\sinh\left[L\sqrt{\frac{z}{\alpha}}\right]$$
$$+\frac{z+\gamma}{2\beta z}\cosh\left[L\sqrt{\frac{z}{\alpha}}\right] \qquad (20.38)$$

$$dD_4\,dz(\alpha,\beta,\gamma,L,z,t) = \frac{1}{2\beta z}\cosh\left[L\sqrt{\frac{z}{\alpha}}\right] - \frac{L(z+\gamma)}{2\beta z^2}\cosh\left[L\sqrt{\frac{z}{\alpha}}\right]$$
$$+\frac{L^2(z+\gamma)}{4\beta z\sqrt{\alpha z}}\sinh\left[L\sqrt{\frac{z}{\alpha}}\right] \qquad (20.39)$$

The derivatives of the denominator can be obtained by summation over the four terms:

$$dD\,d\alpha(\alpha,\beta,\gamma,L,z,t) = \sum_{i=1}^{4} dD_i\,d\alpha(\alpha,\beta,\gamma,L,z,t) \qquad (20.40)$$

$$dD\,d\beta(\alpha,\beta,\gamma,L,z,t) = \sum_{i=1}^{4} dD_i\,d\beta(\alpha,\beta,\gamma,L,z,t) \qquad (20.41)$$

$$dD\,d\gamma(\alpha,\beta,\gamma,L,z,t) = \sum_{i=1}^{4} dD_i\,d\gamma(\alpha,\beta,\gamma,L,z,t) \qquad (20.42)$$

$$dD\,dL(\alpha,\beta,\gamma,L,z,t) = \sum_{i=1}^{4} dD_i\,dL(\alpha,\beta,\gamma,L,z,t) \qquad (20.43)$$

$$dD\,dz(\alpha,\beta,\gamma,L,z,t) = \sum_{i=1}^{4} dD_i\,dz(\alpha,\beta,\gamma,L,z,t) \qquad (20.44)$$

The partial derivatives of the temperature expression T_m with respect to the various variables can be expressed in terms of the following numerator and denominator terms:

$dT_m\,d\alpha(\alpha,\beta,\gamma,L,z,t)$
$$= \frac{D(\alpha,\beta,\gamma,L,z,t)dN\,d\alpha(\alpha,\beta,\gamma,L,z,t) - N(\alpha,\beta,\gamma,L,z,t)dD\,d\alpha(\alpha,\beta,\gamma,L,z,t)}{D^2(\alpha,\beta,\gamma,L,z,t)}$$
$$(20.45)$$

$dT_m\,d\beta(\alpha,\beta,\gamma,L,z,t)$
$$= \frac{D(\alpha,\beta,\gamma,L,z,t)dN\,d\beta(\alpha,\beta,\gamma,L,z,t) - N(\alpha,\beta,\gamma,L,z,t)dD\,d\beta(\alpha,\beta,\gamma,L,z,t)}{D^2(\alpha,\beta,\gamma,L,z,t)}$$
$$(20.46)$$

20.10 Transient and Cyclic Calculations

$$dT_m \, d\gamma(\alpha, \beta, \gamma, L, z, t)$$
$$= \frac{D(\alpha, \beta, \gamma, L, z, t) dN \, d\gamma(\alpha, \beta, \gamma, L, z, t) - N(\alpha, \beta, \gamma, L, z, t) dD \, d\gamma(\alpha, \beta, \gamma, L, z, t)}{D^2(\alpha, \beta, \gamma, L, z, t)}$$
(20.47)

$$dT_m \, dL(\alpha, \beta, \gamma, L, z, t)$$
$$= \frac{D(\alpha, \beta, \gamma, L, z, t) dN \, dL(\alpha, \beta, \gamma, L, z, t) - N(\alpha, \beta, \gamma, L, z, t) dD \, dL(\alpha, \beta, \gamma, L, z, t)}{D^2(\alpha, \beta, \gamma, L, z, t)}$$
(20.48)

$$dT_m \, dz(\alpha, \beta, \gamma, L, z, t)$$
$$= \frac{D(\alpha, \beta, \gamma, L, z, t) dN \, dz(\alpha, \beta, \gamma, L, z, t) - N(\alpha, \beta, \gamma, L, z, t) dD \, dz(\alpha, \beta, \gamma, L, z, t)}{D^2(\alpha, \beta, \gamma, L, z, t)}$$
(20.49)

The total derivatives of the temperature T_m with respect to the variables can be found with an implicit z as a function of the variables:

$$DT_m \, D\alpha(\alpha, \beta, \gamma, L, z, t) = dT_m \, d\alpha(\alpha, \beta, \gamma, L, z, t) + dT_m \, dz(\alpha, \beta, \gamma, L, z, t)$$
$$\times dz \, d\alpha(\alpha, \beta, \gamma, L, z) \quad (20.50)$$

$$DT_m \, D\beta(\alpha, \beta, \gamma, L, z, t) = dT_m \, d\beta(\alpha, \beta, \gamma, L, z, t) + dT_m \, dz(\alpha, \beta, \gamma, L, z, t)$$
$$\times dz \, d\beta(\alpha, \beta, \gamma, L, z) \quad (20.51)$$

$$DT_m \, D\gamma(\alpha, \beta, \gamma, L, z, t) = dT_m \, d\gamma(\alpha, \beta, \gamma, L, z, t) + dT_m \, dz(\alpha, \beta, \gamma, L, z, t)$$
$$\times dz \, d\gamma(\alpha, \beta, \gamma, L, z) \quad (20.52)$$

$$DT_m \, DL(\alpha, \beta, \gamma, L, z, t) = dT_m \, dL(\alpha, \beta, \gamma, L, z, t) + dT_m \, dz(\alpha, \beta, \gamma, L, z, t)$$
$$\times dz \, dL(\alpha, \beta, \gamma, L, z) \quad (20.53)$$

Having determined the form of the derivative of $T_m(\alpha, \beta, \gamma, L, z, t)$, we can obtain the derivative of $T_I(t)$ through summation over the roots z_i from Eq. (20.2):

$$DT_I \, D\alpha(\alpha, \beta, \gamma, L, z, t) = \sum_{i=1}^{\infty} dT_m \, d\alpha(\alpha, \beta, \gamma, L, z_i, t) \quad (20.54)$$

$$DT_I \, D\beta(\alpha, \beta, \gamma, L, z, t) = \sum_{i=1}^{\infty} dT_m \, d\beta(\alpha, \beta, \gamma, L, z_i, t) \quad (20.55)$$

$$DT_I \, D\gamma(\alpha, \beta, \gamma, L, z, t) = \sum_{i=1}^{\infty} dT_m \, d\gamma(\alpha, \beta, \gamma, L, z_i, t) \quad (20.56)$$

$$DT_I \, DL(\alpha, \beta, \gamma, L, z, t) = \sum_{i=1}^{\infty} dT_m \, dL(\alpha, \beta, \gamma, L, z_i, t) \quad (20.57)$$

Parametric Estimation Using Gauss Linearization

The Gauss algorithm is used to solve the nonlinear estimation problem, using the derivatives $DT_I\,D\alpha$, $DT_I\,D\beta$, and $DT_I\,D\gamma$ in Eqs. (20.54) through (20.56). These expressions are also known as the "sensitivity coefficients." To simplify the description of the algorithm, let the parameters be summarized using the column vector θ as the transpose of $(\alpha_1, \alpha_2, \beta, \gamma)$. The initial value of the parameters are denoted by $\hat{\theta}_0$, values of the parameters during iterations $u = 1, 2, \ldots$ are denoted by $\hat{\theta}_u$, and the final estimate will be denoted $\hat{\theta}$. The sensitivity matrix $\mathbf{X}$ contains $DT_I(t_j)/D\theta_k$, for row $j = 1, \ldots, 122$ and column $k = 1, 2, 3, 4$. The sensitivity matrix $\mathbf{X}_u$ is computed using $\hat{\theta}_u$.

The sensitivity coefficients can be plotted as functions of time t and compared. The independence of these functions is a condition for identifiability in the estimation problem. To reduce the variability in the magnitudes among the sensitivity coefficients, normalized sensitivity coefficients are plotted: $V\alpha = \alpha DT_I\,D\alpha$, $V\beta = \beta DT_I\,D\beta$, $V\gamma = \gamma DT_I\,D\gamma$, $VL = LDT_I\,DL$. As shown in Fig. 20.4, the sensitivity functions are independent over the timeframe of the experiments, from 1 to 7 h, so this timeframe is sufficient for determining the parameters α, β, and γ.

Define the actual temperature observations $\mathbf{y} = (y_1, y_2, \ldots, y_{122})^T$, where the first 61 observations are from the first LoTEC and the remaining 61 observations are from the second LoTEC. The residual error between observations and the predicted temperatures is given by $\mathbf{e}_u = (e_{1u}, e_{2u}, \ldots, e_{122u})^T$, where $e_{ju} = y_j - T_I(t_j)$, where $T_I(t_j)$ is evaluated

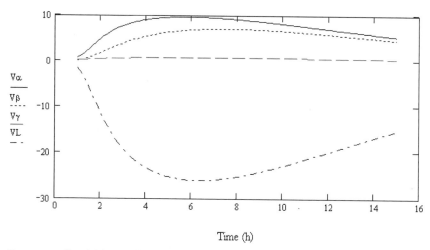

Figure 20.4 Sensitivity coefficients for calculated temperatures.

20.12 Transient and Cyclic Calculations

TABLE 20.1 Summary of Parameter Estimation by Gauss Method

Trial no.	$\hat{\alpha}_1 \times 10^8$	$\hat{\alpha}_1 \times 10^8$	$\hat{\beta} \times 10^6$	$\hat{\gamma} \times 10^6$	SS_e
0	5.0339	5.0004	3.0805	3.905	1.311
1	5.109	5.0754	3.1905	1.6825	0.4299
2	5.1061	5.0725	3.1943	1.6428	0.4268
3	5.1061	5.0725	3.1943	1.6425	0.4268
4	5.1061	5.0725	3.1943	1.6425	0.4268
Uncertainty	0.1168	0.1157	0.0438	0.9117	

using $\hat{\theta}_u$. The estimates are chosen to minimize the sum of squared residuals:

$$SS_e = \mathbf{e}^T\mathbf{e} = \sum_{j=1}^{122} e_j^2 \qquad (20.58)$$

In each iteration an improved estimate of the parameters can be obtained using

$$\hat{\theta}_{u+1} = \hat{\theta}_u + (\mathbf{X}_u^T\mathbf{X}_u)^{-1}\mathbf{X}_u\mathbf{e}_u \qquad (20.59)$$

The initial estimates $\hat{\theta}_0$ were calculated from measurements and/or estimates of the eight primitive variables. Poor initial estimates result in imaginary roots to Eq. (20.2).

Approximately four iterations were required to give the minimum least squares of the errors. As shown in Table 20.1, convergence occurred rapidly using these starting values. For the experimental data used, the values of $\hat{\alpha}_1, \hat{\alpha}_2, \hat{\beta}_2,$ and $\hat{\gamma}_1$ are 5.1061×10^{-8} m²/s, 5.0725×10^{-8} m²/s, and 3.1943×10^{-6} m/s, 1.6425×10^{-6} Hz. The total sum of squared residuals SS_e at this solution is $0.4268K^2$. This is a good fit to the data, as shown in Fig. 20.5.

An estimate of the uncertainties in the estimated parameters can be obtained using the SS_e and the sensitivity matrix $\mathbf{X}_u$ using $\hat{\theta}$. An estimate of the error variance is given by $s^2 = SS_e/(n-p)$, where n is the number of observations (122) and p is the number of estimated parameters (4). Then

$$\text{Var}[\hat{\theta}] = s^2(\mathbf{X}_u^T\mathbf{X}_u)^{-1} \qquad (20.60)$$

The square roots of the diagonal terms of this expression are the standard deviations of the individual estimated coefficients. Using twice the standard deviations provides an uncertainty amount for each parameter. The uncertainties for $\hat{\alpha}_1, \hat{\alpha}_2, \hat{\beta}_2,$ and $\hat{\gamma}_1$ are, respectively, 2.1, 2.3, 1.4,

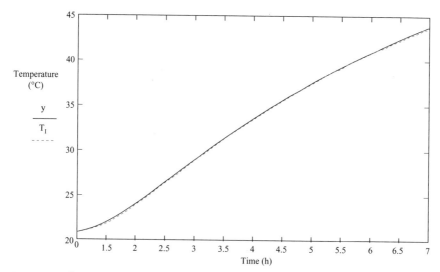

Figure 20.5 Comparison of measured and calculated temperatures for two LoTECs.

and 55 percent. Fortunately, the temperature is only weakly dependent on the value of γ.

References

1. F. C. Wessling, L. S. Stoieck, A. Hoehn, S. Woodard, S. O'Brien, and S. Thomas, "Low Temperature, Low Energy Carrier (LoTEC©) and Phase Change Materials (PCMs) for Biological Samples," 30th Proceedings of the International Conference on Environmental Systems, paper no. 2000-01-2280, Toulouse, France, July 2000.
2. J. V. Beck and K. J. Arnold, *Parameter Estimation in Engineering and Science*, Wiley, 1977.

Part 4

Heat-Transfer-Coefficient Determination

As can be seen in a number of calculations presented earlier in this handbook, heat-transfer coefficients play a key role. Their determination can be difficult in instances in which surface geometries are complex or surfaces are fouled or corroded. This brief part consists of two chapters, both by academics. The first confronts geometrically complex surfaces exposed to convection by using liquid crystal or infrared thermography in conjunction with a semi-infinite solid assumption. Where there is buildup of scale or contamination around the tubes of a heat exchanger, or corrosion of internal baffles alters the individual film coefficient for heat transfer on the side affected, determining individual film coefficients experimentally is simplified by a method called the Wilson analysis; this is described in the second chapter.

Chapter 21

Calculation of Convective Heat-Transfer Coefficient Using Semi-infinite Solid Assumption

Srinath V. Ekkad
Mechanical Engineering Department
Louisiana State University
Baton Rouge, Louisiana

Introduction

Two-dimensional surface heat-transfer distributions on convective surfaces provide a complete picture for complex flow-solid interactions in heat transfer. The semi-infinite solid assumption provides an opportunity to obtain detailed surface heat-transfer behavior using thermography. Typically, liquid crystal or infrared thermography is used in conjunction with the semi-infinite solid assumption.

Calculation of Heat-Transfer Coefficient Using Semi-infinite Solid Assumption

The local heat-transfer coefficient h over a surface coated with liquid crystals can be obtained by using a one-dimensional (1D) semi-infinite solid assumption for the test surface. The 1D transient conduction

21.4 Heat-Transfer-Coefficient Determination

equation, the initial condition, and the convective boundary condition on the liquid crystal coated surface are

$$k\frac{\partial^2 T}{\partial x^2} = \rho C_p \frac{\partial T}{\partial t} \qquad (21.1)$$

Boundary conditions:

At $t = 0, T = T_i$

At $x = 0, -k\dfrac{\partial T}{\partial x} = h(T_w - T_m)$

As $x \to \infty, T = T_i$

Solving Eq. (21.1) with prescribed initial and boundary conditions, the solution thus obtained is

$$\frac{T(x,t) - T_i}{T_m - T_i} = \text{erfc}\left(\frac{x}{2\sqrt{\alpha t}}\right) - \left[\exp\left(\frac{h^2 \alpha t}{k^2} + \frac{hx}{k}\right)\right]$$

$$\times \left[\text{erfc}\left(\frac{h\sqrt{\alpha t}}{k} + \frac{x}{2\sqrt{\alpha t}}\right)\right] \quad (21.2)$$

Since the surface heat-transfer coefficient is required, one obtains the nondimensional temperature at the convective boundary surface (at $x = 0$) [1]:

$$\frac{T_w - T_i}{T_m - T_i} = 1 - \exp\left(\frac{h^2 \alpha t}{k^2}\right)\text{erfc}\left(\frac{h\sqrt{\alpha t}}{k}\right) \qquad (21.3)$$

The color-change temperature or the prescribed wall temperature T_w is obtained either using a liquid crystal coating or through infrared thermography. The initial temperature T_i of the test surface and the mainstream temperature T_m are measured before and during the test, respectively. The time of the wall temperature change from the initial transient test temperature is determined. The local heat-transfer coefficient h can be calculated from Eq. (21.3). Use of a low-conductivity material such as plexiglass or polycarbonate and maintaining test conditions such that the time of temperature imposition on the surface is short enables one to maintain the validity of the semi-infinite solid assumption on the test surface as the test duration does not allow for temperature penetration into the test surface. The theory allows for a short-duration experiment, and complex geometries can be tested for obtaining heat-transfer coefficients.

Other variations of these experiments involve the use of a secondary flow that is at a temperature different from that of the primary flow. In that situation, the heat-transfer coefficient is dependent on the mixing temperature of both flows. There are two unknowns in Eq. (21.3).

Film cooling over a surface is a three-temperature problem involving the mainstream temperature T_m, the coolant temperature T_c, and the wall temperature T_w. In film-cooling situations, the mainstream temperature T_m in Eq. (21.3) must be replaced by a film temperature T_f, which is a mixed temperature between the mainstream and coolant temperatures that governs the convection from the liquid-crystal-coated surface. To find the unknown T_f in terms of known quantities T_m and T_c, a nondimensional temperature is defined as the film cooling effectiveness (η):

$$\eta = \frac{T_f - T_m}{T_c - T_m} \quad \text{or} \quad T_f = \eta(T_c - T_m) + T_m = \eta T_c + (1 - \eta) T_m \quad (21.4)$$

Replacing T_m in Eq. (21.3) by T_f from Eq. (21.4), we get the following equation with two unknowns, h and η:

$$T_w - T_i = \left[1 - \exp\left(\frac{h^2 \alpha t}{k^2}\right) \operatorname{erfc}\left(\frac{h\sqrt{\alpha t}}{k}\right)\right] [\eta T_c + (1 - \eta) T_m - T_i] \quad (21.5)$$

To obtain both the heat-transfer coefficient h and the film effectiveness η, it is necessary to obtain two equations with two unknowns (h and η) and solve for h and η. Therefore, two similar transient tests are run to obtain two different sets of conditions.

Concluding Remarks

A method for calculating heat-transfer coefficients on complex geometries exposed to convection is explained. The semi-infinite solid assumption is used to determine the heat-transfer coefficient in a short-duration transient test. The advantage of obtaining full surface results for complex geometries is extremely significant in improving quality of heat transfer experiment. Several results are presented in Ref. 2.

References

1. Incropera, F. P., and Dewitt, D. P., *Introduction to Heat Transfer*, 4th ed., Wiley, New York, 2002.
2. Ekkad, S. V., and Han, J. C., "A Transient Liquid Crystal Thermography Technique for Gas Turbine Heat Transfer Measurements," *Measurement Science & Technology*, special edition on gas turbine measurements, vol. 11, pp. 957–968, July 2000.

Chapter 22

Determination of Heat-Transfer Film Coefficients by the Wilson Analysis

Ronald J. Willey
Department of Chemical Engineering
Northeastern University
Boston, Massachusetts

Introduction

Over time, the performance of shell-and-tube heat exchangers has changed. The common explanation is "fouling," the buildup of scale or contamination around the tubes. Another possibility is the corrosion of internal baffles or related components that alter the individual film coefficient for heat transfer on the side affected. Determination of individual film coefficients experimentally is simplified by a method called the *Wilson analysis*. This method is most successful for single-phase (with no boiling or condensation on the side of interest) heat exchangers. It requires the ability to vary the flow rate of the fluid on the side of interest, while holding the flow rate of the opposite fluid constant. The overall heat-transfer coefficient U is determined for each flow rate. By plotting the inverse of U versus the inverse of the velocity or the flow rate, normally raised to a fractional power between 0.6 and 0.8, the inverse of the film coefficient h_i for the side of interest can be experimentally determined. The film coefficient can be compared to values predicted by correlations in the literature. If the experimentally determined

22.2 Heat-Transfer-Coefficient Determination

film coefficient is 25 percent or more below that predicted, then most likely the exchanger internals are not functioning as designed. The results assist in troubleshooting heat exchangers, and pinpointing what service may be needed to repair (e.g., cleaning versus replacement).

My personal experience is in the academic laboratory. Students analyze several types of heat exchangers. They determine the experimental film coefficients by the Wilson analysis, and then compare the results to the empirical correlations as offered in textbooks and handbooks. Amazingly, the tube side matches predictions by the Dittus-Boelter correlation (or related tube correlations) within ± 20 percent. For shell-and-tube heat exchangers, the shell-side film coefficient often varies from that predicted by correlations. The closest empirical correlation is the Donohue equation [1]. It predicts results well for new shell-and-tube (S&T) exchangers that have no internal leakage between baffles. For plate heat exchangers, experimental work at Northeastern University in the early 1960s provided some of the first empirical correlations for film coefficients with these exchangers. The Wilson analysis method helped the researchers obtain film coefficients that later led to Nusselt number correlations for plate heat exchangers operating in the turbulent flow regime [2].

Elemental Tubular Heat-Exchanger Design Equations

Heat exchangers are sized and designed by using an overall heat-transfer coefficient U (for further background about heat transfer, see, for example, the text by McCabe et al. [1]). The inverse of the overall heat-transfer coefficient, in its simplest form, is a sum of resistances due to heat transfer attributed to convection resistance on both sides of the wall that separates the two fluids, conduction resistance across the wall, and fouling resistances:

$$\frac{1}{U_i} = \frac{1}{h_i} + \Sigma \text{ other resistances} \qquad (22.1)$$

where U_i is the overall heat-transfer coefficient for the design equation

$$Q = U_i A_i \Delta T_{\text{lm}} \qquad (22.2)$$

where Q = heating or cooling duty required
A = area required for heat transfer
ΔT_{lm} = log-mean driving force temperature between fluids (definition varies depending on flow patterns and geometry)
h_i = [in Eq. (22.1)] film coefficient for convective heat transfer on the fluid side of interest

Individual Film Coefficients Estimated by Empirical Equations

Film coefficients are defined by Newton's law of cooling

$$h_i = \frac{Q}{A}(T_b - T_w) \qquad (22.3)$$

where Q/A = heat flux flowing from the hot fluid to the cold fluid
T_b = local temperature of the bulk fluid (the local hot stream temperature as expressed above)
T_w = local temperature at the heat-exchanger wall

Over the years, engineers and scientists have developed empirical equations that allow the estimation of film coefficients depending on the flow regime and whether a phase change is involved. For a single-phase fluid flowing in the turbulent flow regime, a classical empirical equation is the Dittus-Boelter relation [3]

$$N_{\text{Nu}} = 0.023 N_{\text{Re}}^{0.8} N_{\text{Pr}}^{n} \qquad (22.4)$$

where $n = 0.4$ for heating and 0.3 for cooling, and

$$N_{\text{Nu}} = \frac{h}{kD_i} \quad \text{(Nusselt number for tube-side fluid)} \qquad (22.5)$$

$$N_{\text{Re}} = \frac{vD_i\rho}{\mu} \quad \text{(Reynolds number for tube-side fluid)} \qquad (22.6)$$

$$N_{\text{Pr}} = \frac{c_p\mu}{k} \quad \text{(Prandtl number for tube-side fluid)} \qquad (22.7)$$

where c_p = fluid's heat capacity
D_i = inside diameter in tube
k = fluid's thermal conductivity
v = average velocity of fluid inside tube
μ = fluid's viscosity
ρ = fluid's density

For shell-side correlations, see chap. 11 in Ref. 4 or the Donohue correlation as presented in Ref. 1, chap. 15.

Individual Film Coefficients by Experimental Means

The Wilson analysis (originally described by Wilson in 1915 [5]), provides researchers and engineers with an experimental method for determining individual heat-transfer film coefficients in convective heat

22.4 Heat-Transfer-Coefficient Determination

transfer in which no phase change occurs. The objective is to maintain constant flow rate on one side of the heat exchanger while varying the flow rate on the opposite side (the side of interest). The overall heat-transfer coefficient U is determined for each point on the basis of the flow rates and temperatures acquired for the exchanger under investigation. The area used in the calculation is the area normal to the heat transfer on the side that flow rate is being varied. Since flow rate for only one side is varied, the change in the overall heat transfer coefficient is directly related to a change in the individual convective film coefficient h_i for the side on which the flow rate is being varied. By plotting $1/U$ versus $1/\text{flow rate}^x$ using the proper exponent x (detailed below), a straight line results. This plot is sometimes called a *Wilson plot*. The intercept is equaled to the sum of all resistances to heat transfer *except* the side that is being varied. The resistance on the variable flow side, $1/h_i$ for each point, is given as the difference of $1/U_i$ minus the intercept found from in the Wilson plot. The exponent x depends on the flow regime, and is exchanger-dependent. Typically, 0.8 is used for turbulent flow inside tubes, 0.33 for laminar flow inside tubes, and 0.6 for crossflow for tube bundles.

Problem Statement

Determine the individual tube-side film coefficients and corresponding Nusselt (N_{Nu}) numbers for the five data points listed in Table 22.1 for a 12-ft-long, six-tube (i.d. = 0.425 in.), single-pass shell-and-tube heat exchanger. Water is the working fluid used in the experiment on both sides. The cold fluid is on the tube side. The streams are flowing countercurrently to each other. Compare the results to the N_{Nu} predicted by an empirical formula available in the literature.

Solution Outline

1. Determination of A_i, area normal to heat transfer for the heat exchanger.

TABLE 22.1 Data Acquired in a Shell-and-Tube Heat-Exchanger Experiment

	Flow rate	Stream temperatures			
Point	Cold flow, L/min	Cold in, °C	Cold out, °C	Hot in, °C	Hot out, °C
1	4.7	12.2	40.6	45.0	40.6
2	8.5	11.7	37.8	45.0	37.6
3	16.3	11.1	33.9	45.0	32.7
4	20.2	11.1	32.2	45.0	30.9
5	35.7	10.6	27.1	45.0	25.4

2. Determination of the heating load for each point acquired.
3. Determination of ΔT_{lm} for each data point acquired.
4. Determination of U_i for each data point.
5. Determination of v_i for each point acquired.
6. Formation of Wilson plot by plotting $1/U_i$ against $1/v_i^x$ for each point; see "Introduction" above for suggested values of x to use.
7. Determination of the underlying resistances to heat transfer for all resistances *except* the film coefficient of interest. This is the intercept given on the Wilson plot.
8. Determination of the film coefficient for each point ($1/U_i$—the intercept found above).
9. Comparison of each film coefficient to its counterpart empirical values. Huge differences indicate fouling or other problems.

Determination of area

$$A_i = \pi D_i L N_t$$

where $D_i = 0.425$ in. $= 0.425$ in. $\cdot$ 0.0254 m/in. $= 0.0108$ m
$L = 12$ ft $= 12$ ft $\cdot$ 0.3049 m/ft $= 3.66$ m
$A_i = \pi \cdot 0.0108$ m $\cdot$ 3.66 m $\cdot$ $6 = 0.745$ m^2

Determination of Q_c for point 1

Given a cold stream average temperature of $(40.6 + 12.2)/2 = 26.4°$C from the tables, find the following properties of water:

$$\rho = 996 \text{ kg/m}^3$$
$$C_p = 4176 \text{ J/kg C}$$
$$\mu = 0.866 \cdot 10^{-3} \text{ kg/m s}$$
$$k = 0.622 \text{ W/m C}$$
$$N_{Pr} = 5.81$$
$$Q_c = m\, C_p (T_{c\,out} - T_{c\,in})$$

where m is the mass flow rate, kg/s. Then

$m = $ flow rate times ρ

$= 4.7$ L/min $\cdot$ 1 m^3/1000 L $\cdot$ 1 min/60 s $\cdot$ 996 kg/m^3

$= 0.078$ kg/s

$Q_c = 0.078$ kg/s $\cdot$ $(40.6°$C $- 12.2°$C$) \cdot 4176$ J/kg C $= 9250$ W

22.6 Heat-Transfer-Coefficient Determination

Few details are provided for the hot stream for comparison of the heat balance; therefore, Q_h cannot be computed and compared. Normally, this value should be within 5 percent of Q_c.

Determination of ΔT_{lm} for point 1

For countercurrent flow, ΔT_{lm} is as follows:

$$\Delta T_{lm} = \frac{(T_{h\,in} - T_{c\,out}) - (T_{h\,out} - T_{c\,in})}{\ln\left(\dfrac{T_{h\,in} - T_{c\,out}}{T_{h\,out} - T_{c\,in}}\right)}$$

$$= \frac{(45 - 40.6) - (40.6 - 12.2)}{\ln\left(\dfrac{45 - 40.6}{40.6 - 12.2}\right)} = 12.9°C$$

Determination of U_i for point 1

$$U_i = \frac{Q/A}{\Delta T_{lm}} = 9250\ \text{W}/0.745\ \text{m}^2/12.9°C = 962\ \text{W/m}^2\ \text{C}$$

Determination of v_i for point 1

v_i = flow rate/total cross-sectional area

Flow rate = 4.7 L/min · 1 m³/1000 L · 1 min/60 s = 7.83×10^{-5} m³/s

Cross-sectional area for flow = $N_t \cdot \pi \cdot D_i^2/4$ = 6 tubes · π · $(0.0108\ \text{m})^2/4 = 5.5 \times 10^{-4}$ m²

$v_i = (7.83 \times 10^{-5}\ \text{m}^3/\text{s})/(5.5 \times 10^{-4}\ \text{m}^2) = 0.142$ m/s

Formation of Wilson plot

See Table 22.2 and Fig. 22.1 for the completed calculations. The plot is of $1/U_i$ versus $1/v_i$ for each point.

Determination of the underlying resistances

From Fig. 22.1, the data are linear with the exponent on the velocity equal to 0.8 (turbulent flow throughout). In this case linear regression can be used to determine the intercept of 1.16×10^{-4} m² °C/W.

Determination of film coefficient for each point

From Fig. 22.1

$$\frac{1}{U_1} = \frac{1}{h_1} + \text{intercept} \tag{22.8}$$

TABLE 22.2 Completed Calculations for Experimental h_i and Resultant Nusselt Number Comparisons

Point	Flow rate Cold flow, L/min	Stream temperatures Cold in, °C	Cold out, °C	Hot in, °C	Hot out, °C	U_i, W/m²°C	Experimental h_i, W/m²°C	Empirical h_i, W/m²°C
1	4.7	12.2	40.6	45.0	40.6	970	1100	1060
2	8.5	11.7	37.8	45.0	37.6	1410	1690	1680
3	16.3	11.1	33.9	45.0	32.7	2200	2970	2760
4	20.2	11.1	32.2	45.0	30.9	2480	3510	3240
5	35.7	10.6	27.1	45.0	25.4	3370	5600	4960

22.8 Heat-Transfer-Coefficient Determination

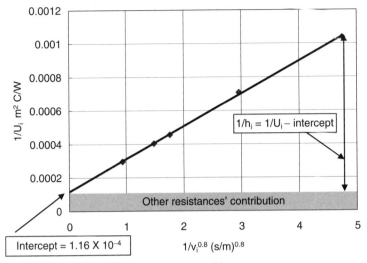

Figure 22.1 Wilson plot for data given in problem.

$$\frac{1}{h_1} = \frac{1}{U_1} - \text{intercept} = 1.038 \times 10^{-3} - 0.116 \times 10^{-3} = 0.922 \, \text{m}^2\,\text{C/W}$$

$$h_1 = \frac{1}{h_1} = \frac{1}{0.922}(\text{m}^2\,\text{C/W})^{-1} = 1090 \, \text{W/m}^2\,\text{C}$$

$$N_{\text{Nu}_1} = \frac{h_1 D}{k} = \frac{1090 \, \text{W/m}^2\,\text{C} \cdot 0.0108 \, \text{m}}{0.622 \, \text{W m C}} = 18.8$$

Comparison of each film coefficient to its counterpart empirical values

$$N_{\text{Nu}} = 0.023 N_{\text{Re}}^{0.8} N_{\text{Pr}}^{0.4} \tag{22.4}$$

$$N_{\text{Re}} = \frac{v D_i \rho}{\mu} = \frac{0.142 \, \text{m/s} \cdot 0.0108 \, \text{m} \cdot 996 \, \text{kg/m}^3}{0.866 \cdot 10^{-3} \, \text{kg/m s}} = 1770$$

$$N_{\text{Pr}} = \frac{c_p \mu}{k} = 5.81$$

$$N_{\text{Nu empir}} = 0.023 \cdot 1770^{0.8} \cdot 5.81^{0.4} = 18.4$$

$$h_i = \text{Nu} \cdot \frac{k}{D} = \frac{18.4 \cdot 0.622 \, \text{W/m C}}{0.0108 \, \text{m}} = 1060 \, \text{W/m}^2\,\text{C}$$

The percentage difference is $(18.8 - 18.4)/18.4 \cdot 100\% = 2$ percent, which is considered excellent agreement in this case; the exchanger is functioning as expected on the tube side.

References

1. W. McCabe, J. Smith, and P. Harriott, *Unit Operations of Chemical Engineering*, McGraw-Hill, New York, 2001, chap. 12.
2. R. A. Troupe, J. C. Morgan, and J. Prifti, *Chemical Engineering Progress*, **56** (1):124–128 (1960).
3. F. W. Dittus and L. M. K. Boelter, *Publication on Engineering* (Univ. Calif. Berkeley), **2**:443 (1930).
4. R. H. Perry and D. W. Green, *Perry's Chemical Engineers' Handbook*, McGraw-Hill, New York, 1997.
5. R. E. Wilson, *Transactions ASME*, **37**:47 (1915).

Part

5

Tubes, Pipes, and Ducts

Even when a fluid flows through a circular pipe, heat-transfer calculations are not as simple as a practicing engineer would like them to be. The following calculations, which were contributed by academics, a consultant, and an engineer working in industry, are useful for dealing with real-life situations:

- *Assessment of performance of heat-exchanging equipment and development of heat-transfer correlations*
- *The case of a pumped fluid entering a heated pipe as a slightly subcooled or saturated liquid, being heated sufficiently to cause vaporization, and leaving the pipe as liquid-vapor mixture or totally vapor; or the inverse, which is related, in which the vessel pressure and the exit pressure fix the driving force, and the flow rate from the vessel is required*
- *When a "waxy" crude oil containing high-molecular-weight alkanes or paraffin waxes flowing through a pipeline is exposed to a cold environment that is below its freezing-point temperature (or solubility temperature), and solids deposition on the pipe wall is likely to occur*
- *The combustor of a hypersonic vehicle over a critical portion of its flight trajectory*

Chapter

23

Calculation of Local Inside-Wall Convective Heat-Transfer Parameters from Measurements of Local Outside-Wall Temperatures along an Electrically Heated Circular Tube

Afshin J. Ghajar and Jae-yong Kim
School of Mechanical and Aerospace Engineering
Oklahoma State University
Stillwater, Oklahoma

Introduction

Heat-transfer measurements in pipe flows are essential for assessment of performance of heat-exchanging equipment and development of heat-transfer correlations. Usually, the experimental procedure for a uniform wall heat flux boundary condition consists of measuring the tube outside-wall temperatures at discrete locations and the inlet and outlet bulk temperatures in addition to other measurements such as the flow rate, voltage drop across the test section, and current carried by the test section. Calculations of the local peripheral heat-transfer coefficients and local Nusselt numbers thereafter are based on knowledge

23.4 Tubes, Pipes, and Ducts

of the tube inside-wall temperatures. Although measurement of the inside-wall temperature is difficult, it can be accurately calculated from the measurements of the outside-wall temperature, the heat generation within the tube, and the thermophysical properties of the pipe material (electrical resistivity and thermal conductivity).

This chapter presents the general finite-difference formulations used for this type of heat-transfer experiment, provides specific applications of the formulations to single- and two-phase convective heat transfer experiments, gives details of implementation of the calculation procedure (finite-difference method) in a computer program, and shows representative reduced heat-transfer results.

Finite-Difference Formulations

The numerical solution of the conduction equation with internal heat generation and non-uniform thermal conductivity and electrical resistivity was originally developed by Farukhi (1973) and introduced by Ghajar and Zurigat (1991) in detail. The numerical solution is based on the following assumptions:

1. Steady-state conditions exist.
2. Peripheral and radial wall conduction exists.
3. Axial conduction is negligible.
4. The electrical resistivity and thermal conductivity of the tube wall are functions of temperature.

On the basis of these assumptions, expressions for calculation of local inside-wall temperature and heat flux and local and average peripheral heat-transfer coefficients will be presented next.

The heat balance on a control volume of the tube at a node P (refer to Fig. 23.1) is given by

$$\dot{q}_g = \dot{q}_n + \dot{q}_e + \dot{q}_s + \dot{q}_w \tag{23.1}$$

From Fourier's law of heat conduction in a given direction n, we have

$$\dot{q} = -kA\frac{dT}{dn} \tag{23.2}$$

Now, substituting Fourier's law and applying the finite-difference formulation for a control volume on a segment (slice) of the tube with nonuniform thermal conductivity in Eq. (23.1), we obtain

$$\dot{q}_n = \left[\frac{\delta_{n^-}}{k_P} + \frac{\delta_{n^+}}{k_N}\right]^{-1} A_n (T_P - T_N) \tag{23.3}$$

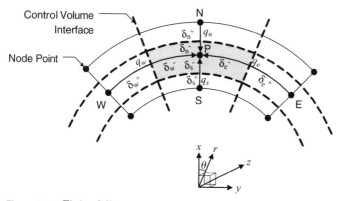

Figure 23.1 Finite-difference node arrangement on a segment (slice).

$$\dot{q}_e = \left[\frac{\delta_{e^-}}{k_P} + \frac{\delta_{e^+}}{k_E}\right]^{-1} A_e \left(T_P - T_E\right) \quad (23.4)$$

$$\dot{q}_s = \left[\frac{\delta_{s^-}}{k_P} + \frac{\delta_{s^+}}{k_S}\right]^{-1} A_s \left(T_P - T_S\right) \quad (23.5)$$

$$\dot{q}_w = \left[\frac{\delta_{w^-}}{k_P} + \frac{\delta_{w^+}}{k_W}\right]^{-1} A_w \left(T_P - T_W\right) \quad (23.6)$$

Note that, in order to deal with the nonuniform thermal conductivity, the thermal conductivity at each control volume interface is evaluated as the sum of the thermal conductivities of the neighboring node points based on the concept that the thermal conductance is the reciprocal of the resistance (Patankar, 1991).

The heat generated at the control volume is given by

$$\dot{q}_g = I^2 R \quad (23.7)$$

Substituting $R = \gamma l / A_c$ into Eq. (23.7) gives

$$\dot{q}_g = I^2 \gamma \frac{l}{A_c} \quad (23.8)$$

Substituting Eqs. (23.3) to (23.6) and (23.8) into Eq. (23.1) and solving for T_S gives

$$T_S = T_P - \left(\dot{q}_g - \dot{q}_n - \dot{q}_e - \dot{q}_w\right) \bigg/ \left\{\left[\frac{\delta_{s^-}}{k_P} + \frac{\delta_{s^+}}{k_S}\right]^{-1} A_s\right\} \quad (23.9)$$

Equation (23.9) is used to calculate the temperature of the interior nodes. Once the local inside-wall temperatures are calculated from Eq. (23.9), the local peripheral inside-wall heat flux can be calculated from the heat-balance equation, Eq. (23.1).

From the local inside-wall temperature, the local peripheral inside-wall heat flux, and the local bulk fluid temperature, the local peripheral heat-transfer coefficient can be calculated as follows:

$$h = \frac{\dot{q}''}{T_w - T_b} \qquad (23.10)$$

Note that, in these analyses, it is assumed that the bulk fluid temperature increases linearly from the inlet to the outlet according to the following equation:

$$T_b = T_{in} + \frac{(T_{out} - T_{in})z}{L} \qquad (23.11)$$

The local average heat-transfer coefficient at each segment can be calculated by the following equation:

$$\bar{h} = \frac{\dot{q}''}{\bar{T}_w - \bar{T}_b} \qquad (23.12)$$

In this section, we have presented the basic formulations of the local inside-wall temperature, the local peripheral heat-transfer coefficient, the local average heat-transfer coefficient, and the overall heat-transfer coefficient from the given local outside-wall temperature at a particular segment (slice) of an electrically heated circular tube. Next, we will show specific applications of the formulations to single- and two-phase convective heat-transfer experiments.

Application of the Finite-Difference Formulations

In this section, the finite-difference formulations developed in the previous section are applied to actual heat-transfer experiments. The experiments were performed to study single- and two-phase heat transfer in an electrically heated tube under a variety of flow conditions. The obtained data were then used to develop robust single- and two-phase heat-transfer correlations. For this purpose, accurate heat-transfer measurement is critical; therefore, finite-difference formulations were used as a key tool in obtaining accurate heat-transfer coefficients from the measured outside-wall temperatures.

Experimental setup

A brief description of the experimental setup is presented to help the readers understand how the finite-difference formulations are used in actual experimental work.

A schematic diagram of the overall experimental setup is shown in Fig. 23.2. The experimental setup shown in the figure is designed to systemically collect pressure drop and heat-transfer data for single- and two-phase flows for various flow conditions and flow patterns (in case of two-phase flow) and different inclination angles.

The test section is a 27.9-mm i.d. (inside diameter) straight standard stainless-steel 316 schedule 10S pipe with a length : diameter ratio of 100. The uniform wall heat flux boundary condition is maintained by a welder which is a power supply to the test section. The entire length of the test section is insulated using fiberglass pipe wrap insulation, which provides the adiabatic boundary condition for the outside wall. T-type thermocouples are cemented with an epoxy adhesive having high thermal conductivity and electrical resistivity to the outside wall of the test section at uniform intervals of 254 mm (refer to Fig. 23.3). There are 10 thermocouple stations with four thermocouples in 90° peripheral intervals at each station in the test section. The inlet and exit bulk temperatures are measured by T-type thermocouple probes. To ensure a uniform fluid bulk temperature at the inlet and exit of the test section, a mixing well is utilized. More details of the experimental setup may be found in Ghajar et al. (2004).

Heat-transfer measurements at uniform wall heat flux boundary condition are carried out by measuring the outside wall temperatures at the 10 thermocouple stations along the test section and the inlet and outlet bulk temperatures in addition to other measurements such as the flow rates of test fluids, system pressure, voltage drop across the test section, and current carried by the test section.

A National Instruments data acquisition system is used to acquire the data measured during the experiments. The computer interface used to monitor and record the data is a LabVIEW Virtual Instrument written for this specific application.

Finite-difference formulations for the experimental setup

In order to apply the finite-difference formulations developed in the previous section to the experimental setup, the grid configuration of the test section is designed as shown in Fig. 23.4 according to the basic node configuration shown in Fig. 23.1. The computation domain is divided into control volumes of uniform thickness except at the inside

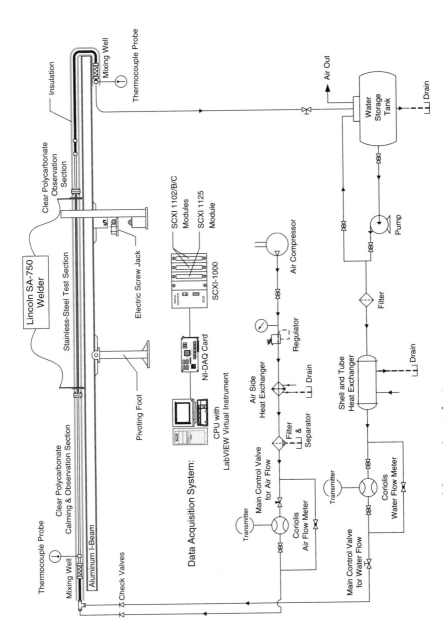

Figure 23.2 Schematic of the experimental setup.

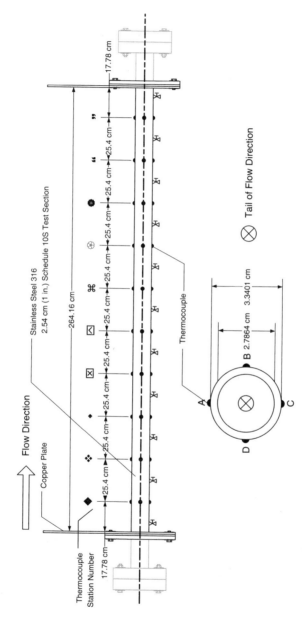

Figure 23.3 Test section.

23.10 Tubes, Pipes, and Ducts

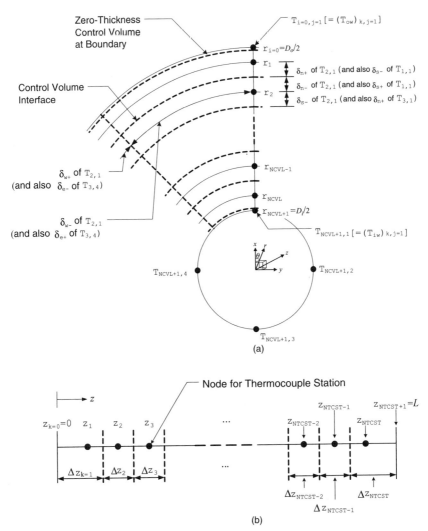

Figure 23.4 Grid configuration for the experimental setup: (a) at a segment (slice); (b) at axial direction.

and outside walls of the tube in the radial direction, where the control volumes have a zero thickness. The node points are placed in the centerpoint of the corresponding control volumes, and the node points along the circumferential and axial directions are placed in accordance with the actual thermocouple locations at the test section.

According to the assumption that the axial conduction is negligible, each thermocouple station (segment) is solved independently of the others. The calculation in a segment is marched from the outside-wall node points, the temperatures of these nodes are experimentally measured, to the inside-wall node points based on the heat balance at each control volume.

First, determination of the geometric variables at a node point (i, j) is as follows. The distances from a grid point to the interface of the control volume are

$$\delta_{n^+} = \delta_{n^-} = \delta_{s^+} = \delta_{s^-} = \frac{\Delta r}{2} = \frac{(D_o - D_i)/N_{\text{CVL}}}{2} \tag{23.13}$$

for the radial direction and

$$(\delta_{e^+})_i = (\delta_{e^-})_i = (\delta_{w^+})_i = (\delta_{w^-})_i = \frac{2\pi r_i / N_{\text{TC@ST}}}{2} \tag{23.14}$$

for the circumferential direction.

Because of the zero-thickness control volume layer at the boundaries ($i = 0$ and $N_{\text{CVL}} + 1$), we have $\delta_{n^+} = 0$ at $i = 1$ and $\delta_{s^+} = 0$ at $i = N_{\text{CVL}}$. The areas of all faces for a control volume are

$$(A_n)_i = \frac{2\pi \left[r_i + (\delta_{n^-})_i\right] \Delta z_k}{N_{\text{TC@ST}}} \tag{23.15}$$

$$(A_e)_i = (A_w)_i = \left[(\delta_{n^-})_i + (\delta_{s^-})_i\right] \Delta z_k \tag{23.16}$$

$$(A_s)_i = \frac{2\pi \left[r_i - (\delta_{s^-})_i\right] \Delta z_k}{N_{\text{TC@ST}}} \tag{23.17}$$

The length and the cross-sectional area of the control volume in the axial direction are set as

$$l = \Delta z_k \tag{23.18}$$

$$(A_c)_i = \frac{2\pi r_i \Delta r}{N_{\text{TC@ST}}} \tag{23.19}$$

Now we can calculate the heat generation and the heat flux at a control volume by applying these geometric variables.

Note that because of the zero-thickness control volume layer at the boundaries ($i = 0$ and $N_{\text{CVL}} + 1$), all heat fluxes (q_n, q_e, q_s, and q_w) and the heat generation (q_g) terms are set to zero at $i = 0$ and $N_{\text{CVL}} + 1$.

Therefore, Eqs. (23.3), (23.4), (23.5), (23.6), and (23.8) for $i = 1, 2, \ldots, N_{\text{CVL}}$ and $j = 1, 2, \ldots, N_{\text{TC@ST}}$ become as follows:

$$(\dot{q}_n)_{i,j} = \begin{cases} \dfrac{k_{i,j}}{\delta_{n^-}} (A_n)_i \left(T_{i,j} - T_{i-1,j}\right) & \text{for } i = 1 \\[2ex] \left[\dfrac{\delta_{n^-}}{k_{i,j}} + \dfrac{\delta_{n^+}}{k_{i-1,j}}\right]^{-1} (A_n)_i \left(T_{i,j} - T_{i-1,j}\right) & \text{for } i = 2, 3, \ldots, N_{\text{CVL}} \end{cases}$$

$$\tag{23.20}$$

23.12 Tubes, Pipes, and Ducts

$$(\dot{q}_e)_{i,j} = \left[\frac{(\delta_{e-})_i}{k_{i,j}} + \frac{(\delta_{e+})_i}{k_{i,j+1}}\right]^{-1} (A_e)_i \left(T_{i,j} - T_{i,j+1}\right) \quad \text{for} \quad i = 1, 2, \ldots, N_{\text{CVL}}$$

(23.21)

$$(\dot{q}_s)_{i,j} = \begin{cases} \left[\dfrac{\delta_{s-}}{k_{i,j}} + \dfrac{\delta_{s+}}{k_{i+1,j}}\right]^{-1} (A_s)_i \left(T_{i,j} - T_{i+1,j}\right) & \text{for} \quad i = 1, 2, \ldots, N_{\text{CVL}} - 1 \\[2mm] \dfrac{k_{i,j}}{\delta_{s-}} (A_s)_i \left(T_{i,j} - T_{i+1,j}\right) & \text{for} \quad i = N_{\text{CVL}} \end{cases}$$

(23.22)

$$(\dot{q}_w)_{i,j} = \left[\frac{(\delta_{w-})_i}{k_{i,j}} + \frac{(\delta_{w+})_i}{k_{i,j-1}}\right]^{-1} (A_w)_i \left(T_{i,j} - T_{i,j-1}\right) \quad \text{for} \quad i = 1, 2, \ldots, N_{\text{CVL}}$$

(23.23)

The heat generated at the control volume is given by

$$(\dot{q}_g)_{i,j} = I_{i,j}^2 \, \gamma_{i,j} \, \frac{\Delta z_k}{(A_C)_i} \quad \text{for} \quad i = 1, 2, \ldots, N_{\text{CVL}}; \, j = 1, 2, \ldots, N_{\text{TC@ST}}$$

(23.24)

Substituting Eqs. (23.20) to (23.23) into Eq. (23.1) and solving for $T_{i+1,j}$ gives

$$T_{i+1,j} = \frac{T_{i,j} - \left\{(\dot{q}_g)_{i,j} - (\dot{q}_n)_{i,j} - (\dot{q}_e)_{i,j} - (\dot{q}_w)_{i,j}\right\}}{\left[\dfrac{\delta_{s-}}{k_{i,j}} + \dfrac{\delta_{s+}}{k_{i+1,j}}\right]^{-1} (A_s)_i}$$

(23.25)

$$\text{for} \quad i = 0, 1, 2, \ldots, N_{\text{CVL}}; \, j = 1, 2, \ldots, N_{\text{TC@ST}}$$

With $T_{i=0,j}\, [=(T_{\text{ow}})_{k,j}]$ given, Eq. (23.25) for the kth thermocouple station will be forced to converge after some iterations since the equation is developed on the basis of heat balance in a control volume.

Then, the local inside-wall temperature at the kth thermocouple station is

$$(T_{\text{iw}})_{k,j} = T_{N_{\text{CVL}}+1,j} \quad \text{for} \quad j = 1, 2, \ldots, N_{\text{TC@ST}}$$

(23.26)

The local inside-wall heat flux at the kth thermocouple station is

$$(\dot{q}_{\text{in}}'')_{k,j} = k_{N_{\text{CVL}},j} \, \frac{T_{N_{\text{CVL}},j} - T_{N_{\text{CVL}}+1,j}}{\Delta r/2} \quad \text{for} \quad j = 1, 2, \ldots, N_{\text{TC@ST}}$$

(23.27)

The bulk temperature at the kth thermocouple station is

$$(T_b)_k = T_{\text{in}} + \frac{(T_{\text{out}} - T_{\text{in}})z_k}{L} \tag{23.28}$$

The local heat-transfer coefficient at the kth thermocouple station is

$$h_{k,j} = \frac{(\dot{q}''_{\text{in}})_{k,j}}{(T_{\text{iw}})_{k,j} - (T_b)_k} \quad \text{for} \quad j = 1, 2, \ldots, N_{\text{TC@ST}} \tag{23.29}$$

Finally, the local average heat-transfer coefficient at the kth thermocouple station is

$$\bar{h}_k = \frac{\dfrac{1}{N_{\text{TC@ST}}} \sum_{j=1}^{N_{\text{TC@ST}}} (\dot{q}''_{\text{in}})_{k,j}}{\dfrac{1}{N_{\text{TC@ST}}} \sum_{j=1}^{N_{\text{TC@ST}}} (T_{\text{iw}})_{k,j} - (T_b)_k} \tag{23.30}$$

At times it is necessary to calculate the overall heat-transfer coefficient for certain test runs. In these cases, the overall heat-transfer coefficient can be calculated as follows:

$$\bar{\bar{h}} = \frac{1}{L} \int \bar{h} \, dz = \frac{1}{L} \sum_{k=1}^{N_{\text{TCST}}} \bar{h}_k \, \Delta z_k \tag{23.31}$$

Note that, to execute these calculations, the thermophysical properties of the pipe material (thermal conductivity and electrical resistivity) are required. These and other thermophysical properties needed for these type of calculations will be presented next.

Thermophysical properties

In order to implement the formulations above in a computer program and represent the heat transfer and flow data in a dimensionless form, knowledge of thermophysical properties of pipe material and the working fluids are required. For the heat-transfer experiments presented in this chapter to demonstrate the applications of the finite-difference formulations, the pipe material was made of stainless steel 316 and the working fluids were air, water, ethylene glycol, and mixtures of ethylene glycol and water. The equations used in Tables 23.1 to 23.4 present the necessary equations for the calculation of the pertinent thermophysical properties. The equations presented were mostly curve fitted from the available data in the literature for the range of experimental temperatures.

23.14 Tubes, Pipes, and Ducts

TABLE 23.1 Equations for Thermophysical Properties of Stainless Steel 316

Fitted equation	Range and accuracy	Data source
Thermal conductivity: $k_{ss} = 13.0 + 1.6966 \times 10^{-2} T - 2.1768 \times 10^{-6} T^2$, with T in °C and k_{ss} in W/m-K	300–1000 K; $R^2 = 0.99985$	Incropera and Dewitt (2002)
Electrical resistivity: $\gamma_{ss} = 73.152 + 6.7682 \times 10^{-2} T - 2.6091 \times 10^{-6} T^2 - 2.2713 \times 10^{-8} T^3$, with T in °C and γ_{ss} in $\mu\Omega$-cm	−196 to 600°C; $R^2 = 0.99955$	Davis (1994)

TABLE 23.2 Equations for Thermophysical Properties of Air

Fitted equation	Range and accuracy	Data source
Density: $\rho_{air} = p/[T(R/M_{air})]$, where T is in kelvins, ρ_{air} in kg/m³, p is absolute pressure in Pa, R is the universal gas constant (= 8314.34 J/kmol·K), and M_{air} is the molecular weight of air (= 28.966 kg/kmol)		
Viscosity: $\mu_{air} = 1.7211 \times 10^{-5} + 4.8837 \times 10^{-8} T - 2.9967 \times 10^{-11} T^2$, with T in °C and μ_{air} in Pa·s	−10 to 120°C; $R^2 = 0.99994$	Kays and Crawford (1993)
Thermal conductivity: $k_{air} = 2.4095 \times 10^{-2} + 7.6997 \times 10^{-5} T - 5.189 \times 10^{-8} T^2$, with T in °C and k_{air} in W/m·K	−10 to 120°C; $R^2 = 0.99996$	Kays and Crawford (1993)
Specific heat: $(c_p)_{air} = 1003.6 + 3.1088^{-2} T + 3.4967 \times 10x^{-4} T^2$, with T in °C and $(c_p)_{air}$ in J/kg·K	−10 to 330°C; $R^2 = 0.99956$	Kays and Crawford (1993)

TABLE 23.3 Equations for Thermophysical Properties of Water

Fitted equation	Range and accuracy	Data source
Density: $\rho_{water} = 999.96 + 1.7158 \times 10^{-2} T - 5.8699 \times 10^{-3} T^2 + 1.5487 \times 10^{-5} T^3$, with T in °C and ρ_{water} in kg/m³	0–100°C; $R^2 = 0.99997$	Linstrom and Mallard (2003)
Viscosity: $\mu_{water} = 1.7888 \times 10^{-3} - 5.9458 \times 10^{-5} T + 1.3096 \times 10^{-6} T^2 - 1.8035 \times 10^{-8} T^3 + 1.3446 \times 10^{-10} T^4 - 4.0698 \times 10^{-13} T^5$, with T in °C and μ_{water} in Pa·s	0–100°C; $R^2 = 0.99998$	Linstrom and Mallard (2003)
Thermal conductivity: $k_{water} = 5.6026 \times 10^{-1} - 2.1056 \times 10^{-3} T - 8.6806 \times 10^{-6} T^2 - 5.4451 \times 10^{-9} T^3$, with T in °C and k_{water} in W/m·K	0–100°C; $R^2 = 0.99991$	Linstrom and Mallard (2003)
Specific heat: $(c_p)_{water} = 4219.8728 - 3.3863 T + 0.11411 T^2 - 2.1013 \times 10^{-3} T^3 + 2.3529 \times 10^{-5} T^4 - 1.4167 \times 10^{-7} T^5 + 3.58520 \times 10^{-10} T^6$, with T in °C and $(c_p)_{water}$ in J/kg·K	0–100°C; $R^2 = 0.99992$	Linstrom and Mallard (2003)

TABLE 23.4 Equations for Thermophysical Properties of Mixture of Ethylene Glycol and Water

Fitted equation	Range and accuracy	Data source
Density: $\rho_{mix} = \sum_{i=1}^{3}\sum_{j=1}^{3} C_{ij} X^{(j-1)} T^{(i-1)}$ where $C = \begin{pmatrix} 1.0004 & 0.17659 & -0.049214 \\ -1.2379 \times 10^{-4} & -9.9189 \times 10^{-4} & 4.1024 \times 10^{-4} \\ -2.9837 \times 10^{-6} & 2.4264 \times 10^{-6} & -9.5278 \times 10^{-8} \end{pmatrix}$ with T in °C and ρ_{mix} in g/cm³	0–150°C within ±0.25%	Ghajar and Zurigat (1991)
Viscosity: $\ln \mu_{mix} = \sum_{i=1}^{2}\sum_{j=1}^{3} C_{ij} X^{(j-1)} T^{(i-1)} + \left[\sum_{j=1}^{3} C_{3j} X^{(j-1)}\right]^{1/4} T^2$ where $C = \begin{pmatrix} 0.55164 & 2.6492 & 0.82935 \\ -2.7633 \times 10^{-2} & -3.1496 \times 10^{-2} & 4.8136 \times 10^{-3} \\ -6.0629 \times 10^{-17} & 2.2389 \times 10^{-15} & -5.8790 \times 10^{-16} \end{pmatrix}$ with T in °C and μ_{mix} in mPa·s	−10 to 100°C within ±5%	Ghajar and Zurigat (1991)

TABLE 23.4 Equations for Thermophysical Properties of Mixture of Ethylene Glycol and Water (*Continued*)

Fitted equation	Range and accuracy	Data source
Prandtl number: $$\ln \Pr_{\text{mix}} = \sum_{i=1}^{2}\sum_{j=1}^{3} C_{ij} X^{(j-1)} T^{(i-1)} + \left[\sum_{j=1}^{3} C_{3j} X^{(j-1)}\right]^{1/4} T^2$$ where $$C = \begin{pmatrix} 2.5735 & 3.0411 & 0.60237 \\ -3.1169 \times 10^2 & -2.5424 \times 10^{-2} & 3.7454 \times 10^{-3} \\ 1.1605 \times 10^{-16} & 2.5283 \times 10^{-15} & 2.3777 \times 10^{-17} \end{pmatrix}$$ with T in °C	−10 to 100°C within ±5%	Ghajar and Zurigat (1991)
Thermal conductivity: $k_{\text{mix}} = (1-X) k_{\text{water}} + X k_{\text{eg}} - k_F (k_{\text{water}} - k_{\text{eg}})(1-X) X$ where $k_{\text{eg}} = 0.24511 + 1.755 \times 10^{-4} T - 8.52 \times 10^{-7} T^2$, $k_F = 0.6635 - 0.3698 X - 8.85 \times 10^{-4} T$, with T in °C and k_{water}, k_{eg}, k_F, and k_{mix} in W/m·K	0–150°C within ±1%	Ghajar and Zurigat (1991)
Thermal expansion coefficient: $$\beta_{\text{mix}} = -\frac{1}{\rho_{\text{mix}}} \left\{ \begin{array}{l} -1.2379 \times 10^{-4} - 9.9189 \times 10^{-4} X + 4.1024 \times 10^{-4} X^2 \\ +2 \left(-2.9837 \times 10^{-6} + 2.4614 \times 10^{-6} X - 9.5278 \times 10^{-8} X^2\right) T \end{array} \right\}$$ with T in °C, ρ_{mix} in g/cm³, and β_{mix} in 1/°C	0–150°C within ±0.25%	Ghajar and Zurigat (1991)

Stainless steel 316. In the finite-difference equations, the thermal conductivity and electrical resistivity of each node are determined as a function of temperature from the equations shown in Table 23.1 for stainless steel 316.

Test fluids. The test fluids used in the experiments were air, distilled water, ethylene glycol, and different mixtures of distilled water and ethylene glycol. The equations for thermophysical properties of these fluids are presented in Tables 23.2 to 23.4.

Computer program

The finite-difference formulations presented in the previous section are implemented into a computer program written for the experimental works described in this chapter. The computer program consists of five parts: reading and reducing input data, executing the finite-difference formulations, providing thermophysical properties, evaluating all the heat-transfer and flow parameters, and printing outputs. Since the finite-difference formulations and thermophysical properties have already been discussed in detail, the inputs and outputs from the computer program will be presented next.

Input data. All the necessary inputs for the computer program are provided by a database file managing a data set and each raw data file for the data set. The raw data file is produced through recordings from the experimental data acquisition process. A sample of the input files is given as Fig. 23.5a and b.

The database file shown in Fig. 23.5a includes the preset information for an experimental run:

GroupNo	Data group number (e.g., the first digit indicates 1→single-phase, 2→two-phase)
RunNo	Test run number
Phase	1→single-phase flow test, 2→two-phase flow test
Liquid	Test liquid (W→water, G→ethylene glycol)
MC_EG	Mass fraction of ethylene glycol in the liquid mixture
Pattern	Flow pattern information of the two-phase flow test
IncDeg	Inclination angle of the test section

With the information provided in the database file, the computer program opens and reads data from each individual raw data file for an experimental run shown in Fig. 23.5b and proceeds to all the required data reductions and calculations and then saves the results in an output file for each test run.

Output. A sample of the outputs from the computer program is shown in Figs. 23.6 and 23.7 for a single-phase flow experimental run and a

GroupNo	RunNo	Phase	Liquid	MC_EG	Pattern	IncDeg
1010	RN0608	1	W	0.00	NONE	0.00
1020	RN0709	1	G	0.50	NONE	0.00
2010	RN4438	2	W	0.00	A	0.00
2210	RN4658	2	W	0.00	ABS	2.00
2510	RN8750	2	W	0.00	BS	5.00
2710	RN4860	2	W	0.00	S	7.00
...						

(a)

Date	Time	TC01-1	TC01-2	TC01-3	TC01-4	...	TP_IN	TP_OUT	...	FR_w	FR_a	Ref.P	Current	Voltage
MM-DD-YYYY	HH:MM:SS	C	C	C	C		C	C		lb/min	lb/min	psi	A	V
11-29-2003	06:47:14	24.876	21.110	16.341	21.391	...	15.988	21.039	...	2.659	1.662	2.459	299.746	2.299
11-29-2003	06:47:23	24.861	21.069	16.341	21.399	...	15.988	21.038	...	2.770	1.658	2.437	299.732	2.299
11-29-2003	06:47:31	24.860	21.045	16.347	21.353	...	15.994	21.048	...	2.656	1.669	2.460	299.746	2.300
11-29-2003	06:47:40	24.864	21.021	16.343	21.367	...	15.986	21.049	...	2.659	1.668	2.430	299.553	2.299
11-29-2003	06:47:49	24.862	20.978	16.342	21.316	...	15.988	21.055	...	2.657	1.659	2.439	299.679	2.299
11-29-2003	06:47:57	24.852	21.008	16.342	21.320	...	15.991	21.063	...	2.658	1.663	2.472	299.865	2.300
11-29-2003	06:48:06	24.856	21.086	16.343	21.331	...	16.007	21.050	...	2.663	1.667	2.457	299.877	2.300
...														

(b)

Figure 23.5 Input data formats. (*a*) Database format for experimental run; (*b*) raw data format for experimental run.

Inside-Wall Parameters from Outside-Wall Temperatures   23.19

```
=================================================
               RUN NUMBER 0608
               SINGLE PHASE TEST
               TEST FLUID IS WATER
               Test Date: 04-01-2003
               FPS UNIT VERSION
=================================================

      MASS FLOW RATE          =    1308.3   LBM/HR
      MASS FLUX               =  199333     LBM/(SQ.FT-HR)
      FLUID VELOCITY          =        .88  FT/S
      ROOM TEMPERATURE        =      73.32  F
      INLET TEMPERATURE       =      80.74  F
      OUTLET TEMPERATURE      =      87.07  F
      AVERAGE RE NUMBER       =    9213
      AVERAGE PR NUMBER       =       5.59
      CURRENT TO TUBE         =     535.0   AMPS
      VOLTAGE DROP IN TUBE    =       4.29  VOLTS
      AVERAGE HEAT FLUX       =    3161     BTU/(SQ.FT-HR)
      Q=AMP*VOLT              =    7831     BTU/HR
      Q=M*C*(T2-T1)           =    8267     BTU/HR
      HEAT BALANCE ERROR      =      -5.57  %
```
(a)

```
              OUTSIDE SURFACE TEMPERATURES [F]
        1       2       3       4       5       6       7       8       9       10
  1   93.99   95.63   96.36   97.01   97.01   97.88   98.40   98.97   99.78  100.26
  2   94.21   96.01   96.72   96.94   97.56   98.30   98.89   99.53  100.20  100.90
  3   93.67   95.70   96.86   97.19   97.87   98.23   99.10   99.25  100.59  101.33
  4   93.24   95.59   96.49   97.23   97.37   98.22   98.53   99.30  100.04  100.34

              INSIDE SURFACE TEMPERATURES [F]
        1       2       3       4       5       6       7       8       9       10
  1   92.31   93.94   94.67   95.32   95.31   96.18   96.71   97.27   98.08   98.56
  2   92.53   94.33   95.03   95.25   95.87   96.61   97.20   97.85   98.51   99.21
  3   91.98   94.01   95.18   95.50   96.19   96.54   97.42   97.56   98.91   99.65
  4   91.54   93.90   94.80   95.54   95.68   96.53   96.84   97.61   98.35   98.64

              REYNOLDS NUMBER AT THE INSIDE TUBE WALL
        1       2       3       4       5       6       7       8       9       10
  1   10161   10349   10434   10509   10509   10610   10671   10738   10833   10889
  2   10186   10394   10476   10501   10574   10660   10729   10805   10883   10966
  3   10123   10357   10493   10531   10610   10652   10754   10771   10930   11018
  4   10073   10345   10449   10535   10551   10651   10687   10778   10864   10899

              INSIDE SURFACE HEAT FLUXES [BTU/HR/FT^2]
        1       2       3       4       5       6       7       8       9       10
  1   2923    2938    2942    2939    2950    2949    2949    2953    2953    2954
  2   2920    2925    2933    2941    2934    2933    2936    2931    2943    2942
  3   2932    2937    2929    2934    2927    2940    2930    2946    2931    2926
  4   2946    2936    2939    2933    2940    2935    2946    2937    2947    2957

              PERIPHERAL HEAT TRANSFER COEFFICIENT BTU/(SQ.FT-HR-F)
        1       2       3       4       5       6       7       8       9       10
  1    262    241     239     238     251     246     248     249     245     247
  2    257    233     231     240     239     236     237     235     235     234
  3    271    240     229     234     232     238     232     242     227     224
  4    283    242     236     233     243     238     245     241     239     246
```
(b)

Figure 23.6 Output for single-phase flow heat-transfer test run. (*a*) Part 1: summary of a test run. (*b*) Part 2: detailed information at each thermocouple location. (*c*) Part 3: detailed information at each thermocouple station.

```
=====================================================
                    RUN NUMBER 0608 CONTINUED
                       SINGLE PHASE TEST
                      TEST FLUID IS WATER
                     Test Date: 04-01-2003
                        FPS UNIT VERSION
=====================================================
```

ST	RE	PR	X/D	MUB	MUW	TB	TW	DENS	NU
1	8911.67	5.80	6.38	2.045	1.798	81.17	92.09	62.20	69.71
2	8978.30	5.75	15.50	2.030	1.759	81.77	94.04	62.20	62.10
3	9045.13	5.71	24.61	2.015	1.742	82.38	94.92	62.19	60.77
4	9112.15	5.66	33.73	2.000	1.732	82.99	95.40	62.18	61.35
5	9179.36	5.61	42.84	1.985	1.725	83.60	95.76	62.18	62.56
6	9246.76	5.57	52.96	1.971	1.712	84.21	96.47	62.17	62.05
7	9314.36	5.52	61.08	1.956	1.701	84.82	97.04	62.17	62.21
8	9382.14	5.48	70.19	1.942	1.691	85.43	97.57	62.16	62.58
9	9450.11	5.43	79.31	1.928	1.675	86.04	98.46	62.15	61.15
10	9518.27	5.39	88.42	1.914	1.665	86.64	99.02	62.15	61.39

NOTE: TBULK IS GIVEN IN DEGREES FAHRENHEIT
MUB AND MUW ARE GIVEN IN LBM/(FT*HR)

(c)

Figure 23.6 (*Continued*)

two-phase flow experimental run, respectively. As shown in the figures [Fig. 23.6 is in English units and Fig. 23.7 is in SI (metric) units], the user has the option of specifying SI or English units for the output.

The output file starts with a summary of some of the important information about the experimental run for a quick reference as shown in Figs. 23.6a and 23.7a. The program then lists (see Figs. 23.6b and 23.7b) the measured outside-wall temperatures, the calculated inside-wall temperatures, the calculated Reynolds numbers (superficial Reynolds numbers for each phase of the two-phase flow run) of the flow based on the inside-wall temperature, the calculated heat flux, and the calculated peripheral heat-transfer coefficients for each thermocouple location. In the last part of the output (refer to Figs. 23.6c and 23.7c), the tabulated summary of the local averaged heat-transfer results at each thermocouple station, such as its location from the tube entrance, bulk Reynolds numbers, bulk Prandtl numbers, viscosities at the wall and bulk, local average heat-transfer coefficient, and Nusselt numbers are displayed. In the case of two-phase flow runs, some auxiliary information for the two-phase flow parameters is listed as shown in Fig. 23.7d. More details about the information that appears in Fig. 23.7d may be found in Ghajar (2004).

Utilization of the finite-difference formulations

The outputs presented in Figs. 23.6 and 23.7 contain all the necessary information for an in-depth analysis of the experiments. One way

```
=====================================================
                    RUN NUMBER 4897
                     TWO-PHASE TEST
                   FLOW PATTERN: ABS
                 Test Date: 04-01-2004
                     SI UNIT VERSION
=====================================================
     LIQUID                          :     WATER
     GAS                             :     AIR
     LIQUID  MASS FLOW RATE          :     1370.36   [kg/hr]
     GAS     MASS FLOW RATE          :     33.902    [kg/hr]
     LIQUID  V_SL                    :     0.625     [m/s]
     GAS     V_SG                    :     6.482     [m/s]
     ROOM    TEMPERATURE             :     14.89     [C]
     INLET   TEMPERATURE             :     13.67     [C]
     OUTLET  TEMPERATURE             :     14.70     [C]
     AVG REFERENCE GAGE PRESSURE     :     95165.30  [Pa]
     AVG LIQUID  RE_SL               :     14962
     AVG GAS     RE_SG               :     24043
     AVG LIQUID  PR_L                :     8.278
     AVG GAS     PR_G                :     0.714
     AVG LIQUID  DENSITY             :     999.1     [kg/m^3]
     AVG GAS     DENSITY             :     2.382     [kg/m^3]
     AVG LIQUID  SPECIFIC HEAT       :     4.190     [kJ/kg-K]
     AVG GAS     SPECIFIC HEAT       :     1.004     [kJ/kg-K]
     AVG LIQUID  VISCOSITY           :     116.25e-05 [Pa-s]
     AVG GAS     VISCOSITY           :     17.90e-06  [Pa-s]
     AVG LIQUID  CONDUCTIVITY        :     0.588      [W/m-K]
     AVG GAS     CONDUCTIVITY        :     25.18e-03  [W/m-K]
     CURRENT TO TUBE                 :     466.13    [A]
     VOLTAGE DROP IN TUBE            :     3.54      [V]
     AVG HEAT FLUX                   :     7138.39   [W/m^2]
     Q = AMP*VOLT                    :     1650.66   [W]
     QGEN CALCULATED                 :     1600.13   [W]
     QGEN CALCULATION ERROR          :     -3.06     [%]
     Q = M*C*(T2 -T1)                :     1652.02   [W]
     HEAT BALANCE ERROR              :     0.08      [%]
                                  (a)
```

				OUTSIDE SURFACE TEMPERATURE OF TUBE [C]					
	1	2	3	4	5	6	7	8	9	10
1	16.71	17.09	17.35	17.64	17.83	18.11	18.11	18.31	18.12	18.25
2	16.17	16.28	16.65	16.69	16.97	17.14	17.26	17.03	17.33	17.33
3	15.55	15.71	15.75	15.89	16.18	16.20	16.35	16.23	16.33	16.40
4	16.23	16.38	16.41	16.70	16.85	17.01	17.18	17.26	17.13	17.30

				INSIDE SURFACE TEMPERATURES [C]					
	1	2	3	4	5	6	7	8	9	10
1	16.02	16.41	16.66	16.95	17.15	17.43	17.43	17.63	17.43	17.57
2	15.48	15.58	15.95	16.00	16.27	16.44	16.56	16.33	16.63	16.64
3	14.84	15.00	15.04	15.18	15.47	15.49	15.64	15.52	15.62	15.69
4	15.54	15.68	15.71	16.00	16.15	16.31	16.48	16.57	16.43	16.60

			SUPERFICIAL REYNOLDS NUMBER OF GAS AT THE INSIDE TUBE WALL						
	1	2	3	4	5	6	7	8	9	10
1	23925	23901	23885	23866	23854	23836	23836	23824	23836	23828
2	23960	23954	23930	23927	23910	23899	23891	23906	23887	23886
3	24001	23991	23988	23979	23961	23959	23950	23957	23951	23947
4	23956	23947	23945	23927	23918	23907	23896	23891	23900	23888

			SUPERFICAL REYNOLDS NUMBER OF LIQUID AT THE INSIDE TUBE WALL						
	1	2	3	4	5	6	7	8	9	10
1	15718	15878	15985	16107	16188	16309	16307	16391	16310	16366
2	15492	15533	15689	15707	15823	15893	15943	15848	15974	15975
3	15230	15297	15311	15369	15488	15498	15558	15511	15550	15580
4	15517	15577	15589	15710	15770	15837	15911	15945	15887	15961

				INSIDE SURFACE HEAT FLUXES [W/m^2]					
	1	2	3	4	5	6	7	8	9	10
1	6840	6805	6797	6781	6785	6768	6791	6752	6791	6785
2	6909	6935	6903	6929	6925	6924	6918	6956	6906	6922
3	7011	7007	7032	7038	7028	7049	7050	7057	7055	7058
4	6900	6920	6938	6928	6943	6944	6929	6923	6936	6926

			PERIPHERAL HEAT TRANSFER COEFFICIENT [W/m^2-K]						
	1	2	3	4	5	6	7	8	9	10
1	3002	2655	2499	2329	2258	2119	2198	2117	2343	2313
2	3988	3998	3436	3545	3245	3144	3111	3667	3291	3458
3	6386	6029	6407	6180	5288	5622	5425	6502	6529	6691
4	3846	3761	3923	3530	3458	3357	3229	3252	3667	3516

(b)

Figure 23.7 Output for two-phase flow heat-transfer experimental run. (*a*) Part 1: summary of a test run. (*b*) Part 2: detailed information at each thermocouple location. (*c*) Part 3: detailed information at each thermocouple station. (*d*) Part 4: auxiliary information.

23.22 Tubes, Pipes, and Ducts

```
================================================================
                      RUN NUMBER 4897 CONTINUED
                           TWO-PHASE TEST
                         FLOW PATTERN: ABS
                       Test Date: 04-01-2004
                           SI UNIT VERSION
================================================================
```

ST	X/D	RESL	RESG	PRL	PRG	MUB/W(L)	MUB/W(G)	HT/HB	HFLUX	TB[C]	TW[C]	HCOEFF	NU_L
1	6.38	14781	24072	8.39	0.714	1.048	0.995	0.470	6915	13.74	15.47	4006.7	190.01
2	15.50	14821	24065	8.37	0.714	1.051	0.995	0.440	6917	13.84	15.67	3790.2	179.69
3	24.61	14861	24059	8.34	0.714	1.053	0.995	0.390	6918	13.94	15.84	3643.0	172.66
4	33.73	14902	24053	8.32	0.714	1.055	0.995	0.377	6919	14.04	16.03	3473.7	164.58
5	42.84	14942	24046	8.29	0.714	1.059	0.994	0.427	6920	14.14	16.26	3265.9	154.69
6	51.96	14983	24040	8.27	0.714	1.060	0.994	0.377	6921	14.24	16.42	3175.8	150.37
7	61.08	15023	24033	8.24	0.714	1.060	0.994	0.405	6922	14.34	16.53	3161.4	149.65
8	70.19	15064	24027	8.22	0.714	1.057	0.994	0.326	6922	14.44	16.51	3335.6	157.84
9	79.31	15104	24021	8.19	0.714	1.055	0.995	0.359	6922	14.54	16.53	3474.5	164.37
10	88.42	15145	24014	8.17	0.714	1.054	0.995	0.346	6923	14.63	16.62	3478.9	164.52

(c)

```
================================================================
                      RUN NUMBER 4897 CONTINUED
                           TWO-PHASE TEST
                         FLOW PATTERN: ABS
                       Test Date: 04-01-2004
                           SI UNIT VERSION
================================================================
         INCLINATION ANGLE      :    7.000   [DEG]
         TOTAL MASS FLUX(Gt)    :  639.699   [kg/m^2-s]
         QUALITY(x)             :    0.024
         SLIP RATIO(K)          :    3.332
         VOID FRACTION(alpha)   :    0.757
         V_SL                   :    0.625   [m/s]
         V_SG                   :    6.482   [m/s]
         RE_SL                  :   14962
         RE_SG                  :   24043
         RE_TP                  :   39005
         X(Taitel & Dukler)     :    2.070
         T(Taitel & Dukler)     :    0.139
         Y(Taitel & Dukler)     :   27.098
         F(Taitel & Dukler)     :    0.609
         K(Taitel & Dukler)     :   74.440
```

(d)

Figure 23.7 (*Continued*)

to verify the reliability of the finite-difference formulations is to compare the experimental heat-transfer coefficients with the predictions from the well-known empirical heat-transfer correlations. Figure 23.8 shows the comparison of the experimental Nusselt numbers (dimensionless heat-transfer coefficients) obtained from the computer program with those calculated from selected well-known single-phase heat-transfer correlations of Colburn (1933), Sieder and Tate (1936), Gnielinski (1976), and Ghajar and Tam (1994). As shown in the figure, the majority of the experimental data is excellently matched with the calculated values from the correlations, and the maximum deviation between the calculated and the experimental values is within ±20 percent. Note that in Fig. 23.8 Gnielinski's (1976) correlations, labeled as [1] and [3], refer to the first and the third correlations proposed in his work.

The finite-difference formulations presented here have been successfully applied to develop correlation in single- and two-phase flow convective heat-transfer studies. Ghajar and Tam (1994) utilized the

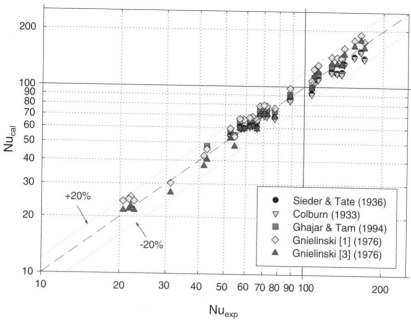

Figure 23.8 Comparison of experimental Nusselt numbers with predictions from selected single-phase heat-transfer correlations at thermocouple station 6.

finite-difference formulations for heat-transfer measurements in a horizontal circular straight tube with three different inlet configurations (reentrant, square-edged, and bell-mouth inlets) under uniform wall heat flux boundary condition. From the measurements, they successfully developed a correlation for prediction of the developing and fully developed forced and mixed single-phase convective heat-transfer coefficients in the transition region for each inlet. They also proposed heat-transfer correlations for laminar, transition, and turbulent regions for the three inlets.

In the case of the gas-liquid two-phase flow heat transfer, Kim and Ghajar (2002) applied the finite-difference formulations in order to measure heat-transfer coefficients and developed correlations for the overall heat-transfer coefficients with different flow patterns in a horizontal tube. A total of 150 two-phase heat-transfer experimental data were used to develop the heat-transfer correlations. Ghajar et al. (2004) extended the work of Kim and Ghajar (2002) and studied the effect of slightly upward inclination ($2°$, $5°$, and $7°$) on heat transfer in two-phase flow.

In the research works mentioned here, a computer program based on the presented finite-difference formulations was used as the key tool to analyze the experimental data. As demonstrated, the computational procedure (computer program) presented in this chapter can be utilized as an effective design tool to perform parametric studies or to develop heat-transfer correlations. It can also be used as a demonstration tool for the basic principles of heat conduction and convection.

Summary

A detailed computational procedure has been developed to calculate the local inside-wall temperatures and the local peripheral convective heat-transfer coefficients from the local outside-wall temperatures measured at different axial locations along an electrically heated circular tube (uniform wall heat flux boundary condition). The computational procedure is based on finite-difference formulation and the knowledge of heat generation within the pipe wall and the thermophysical properties of the pipe material and the working fluids. The method has applications in a variety industrial heat-exchanging equipment and can be used to reduce the heat-transfer experimental data to a form suitable for development of forced- and mixed-convection flow heat-transfer correlations in an electrically heated circular tube for different flow regimes.

Nomenclature

A	Area, m^2
A_c	Cross-sectional area of a control volume in the axial direction, m^2
C	Coefficient; refer to Table 23.4
c_p	Specific heat at constant pressure, J/(kg · K)
D_i	Circular tube inside diameter, m
D_o	Circular tube outside diameter, m
E	Neighboring node point to a given node point in the eastern direction; refer to Fig. 23.1
h	Heat-transfer coefficient, W/(m^2 · K)
$\bar{h}$	Local average heat-transfer coefficient, W/(m^2 · K); refer to Eqs. (23.12) and (23.30)
$\check{h}$	Overall heat-transfer coefficient, W/(m^2 · K); refer to Eq. (23.31)
I	Electrical current, A
i	Index in the radial direction
j	Index in the circumferential direction
k	Thermal conductivity, W/(m · K), or index in the axial direction
L	Length of the test section, m

l	Length of a control volume in the axial direction, m
N	Neighboring node point to a given node point in the northern direction; refer to Fig. 23.1
N_{CVL}	Number of control volume layers in the radial direction except the boundary layers
$N_{TC@ST}$	Number of thermocouples at a thermocouple station
N_{TCST}	Number of thermocouple stations
Nu	Nusselt number, hD_i/k, dimensionless
n	Normal distance, m
P	A given node point P; refer to Fig. 23.1
Pr	Prandtl number, $\mu c_p/k$, dimensionless
p	Pressure, Pa
$\dot{q}$	Heat-transfer rate, W
$\dot{q}_g$	Heat generation rate, W
$\dot{q}''$	Heat flux, W/m²
R	Electrical resistance, Ω or the universal gas constant (= 8314.34 J/kmol · K)
R^2	Correlation coefficient, dimensionless
r	Radius, m
T	Temperature, °C or K
S	Neighboring node point to a given node point in the southern direction; refer to Fig. 23.1
W	Neighboring node point to a given node point in the western direction; refer to Fig. 23.1
X	Mass fraction of ethylene glycol in the mixture of ethylene glycol and water, dimensionless
x	A spatial coordinate in a cartesian system, m
y	A spatial coordinate in a cartesian system, m
z	A spatial coordinate in a cartesian system; axial direction, m

Greek

β	Thermal expansion coefficient, 1/°C
γ	Electrical resistivity, $\Omega \cdot m$
δ	Distance from a node point to a control volume face at a given direction, m
Δ	Designates a difference
θ	Angular coordinate, °
μ	Dynamic viscosity, Pa · s
ρ	Density, kg/m³

Subscripts

b	Bulk
cal	Calculated
E	Evaluated at neighboring node point to a given node point in eastern direction; refer to Fig. 23.1
e	Evaluated at eastern control volume interface of a given node point; refer to Fig. 23.1
e^+	Evaluated at eastern control volume interface of a given node point in side of neighboring node point; refer to Fig. 23.1
e^-	Evaluated at eastern control volume interface of a given node point in side of given node point; refer to Fig. 23.1
eg	Ethylene glycol
exp	Experimental
i	Index in radial direction
in	Evaluated at inlet
iw	Evaluated at inside wall of a tube
j	Index in circumferential direction
k	Index in axial direction
mix	Mixture of ethylene glycol and water
N	Evaluated at neighboring node point to a given node point in northern direction; refer to Fig. 23.1
n	Evaluated at northern control volume interface of a given node point; refer to Fig. 23.1
n^+	Evaluated at northern control volume interface of a given node point in side of neighboring node point; refer to Fig. 23.1
n^-	Evaluated at northern control volume interface of a given node point in side of given node point; refer to Fig. 23.1
out	Evaluated at outlet
ow	Evaluated at outside wall of a tube
P	Evaluated at a given node; refer to Fig. 23.1
S	Evaluated at neighboring node point to a given node point in southern direction; refer to Fig. 23.1
s	Evaluated at southern control volume interface of a given node point; refer to Fig. 23.1
s^+	Evaluated at southern control volume interface of a given node point in side of neighboring node point; refer to Fig. 23.1
s^-	Evaluated at southern control volume interface of a given node point in side of given node point; refer to Fig. 23.1
ss	Stainless steel

W	Evaluated at neighboring node point to a given node point in western direction; refer to Fig. 23.1
w	Evaluated at western control volume interface of a given node point (refer to Fig. 23.1); evaluated at inside wall of a tube
w^+	Evaluated at western control volume interface of a given node point in side of neighboring node point; refer to Fig. 23.1
w^-	Evaluated at western control volume interface of a given node point in side of given node point; refer to Fig. 23.1

Superscript

$-$	Average

References

Colburn, A. P. (1933), "A Method of Correlating Forced Convective Heat Transfer Data and a Comparison with Liquid Friction," *Trans. Am. Inst. Chem. Eng.*, vol. 29, pp. 174–210.

Davis J. R. (ed.) (1994), *Stainless Steels*, in series of ASM Specialty Handbook, ASM International, Materials Park, Ohio.

Farukhi M. N. (1973), *An Experimental Investigation of Forced Convective Boiling at High Qualities inside Tubes Preceded by 180 Degree Bend*, Ph.D. thesis, Oklahoma State University, Stillwater.

Ghajar, A. J. (2004), "Non-Boiling Heat Transfer in Gas-Liquid Flow in Pipes—a Tutorial," invited tutorial, *Proceedings of the 10th Brazilian Congress of Thermal Sciences and Engineering—ENCIT 2004*, Rio de Janeiro, Brazil, Nov. 29–Dec. 3.

Ghajar, A. J., Kim, J., Malhotra, K., and Trimble, S. A. (2004), "Systematic Heat Transfer Measurements for Air-Water Two-Phase Flow in a Horizontal and Slightly Upward Inclined Pipe," *Proceedings of the 10th Brazilian Congress of Thermal Sciences and Engineering—ENCIT 2004*, Rio de Janeiro, Brazil, Nov. 29–Dec. 3.

Ghajar, A. J., and Tam, T. M. (1994), "Heat Transfer Measurements and Correlations in the Transitional Region for a Circular Tube with Three Different Inlet Configurations," *Experimental Thermal and Fluid Science*, vol. 8, no. 1, pp. 79–90.

Ghajar, A. J., and Zurigat, Y. H. (1991), "Microcomputer-Assisted Heat Transfer Measurement/Analysis in a Circular Tube," *Int. J. Appl. Eng. Educ.*, vol. 7, no. 2, pp. 125–134.

Gnielinski, V. (1976), "New Equations for Heat and Mass Transfer in Turbulent Pipe and Channel Flow," *Int. Chem. Eng.*, vol. 16, no. 2, pp. 359–368.

Incropera, F. P., and DeWitt, D. P. (2002), *Introduction to Heat Transfer*, 4th ed., Wiley, New York.

Kays, W. M., and Crawford, M. E. (1993), *Convective Heat and Mass Transfer*, 3d ed., McGraw-Hill, New York.

Kim, D., and Ghajar, A. J. (2002), "Heat Transfer Measurements and Correlations for Air-Water Flow of Different Flow Patterns in a Horizontal Pipe," *Experimental Thermal and Fluid Science*, vol. 25, no. 8, pp. 659–676.

Linstrom, P. J., and Mallard, W. G. (eds.) (2003), *NIST Chemistry WebBook*, NIST Standard Reference Database no. 69, National Institute of Standards and Technology, http://webbook.nist.gov.

Patankar, S. V. (1991), *Computation of Conduction and Duct Flow Heat Transfer*, Innovative Research, Inc., Minn.

Sieder, E. N., and Tate, G. E. (1936), "Heat Transfer and Pressure Drop in Liquids in Tube," *Ind. Eng. Chem.*, vol. 29, pp. 1429–1435.

Chapter 24

Two-Phase Pressure Drop in Pipes

S. Dyer Harris
Equipment Engineering Services, P.A.
Wilmington, Delaware

Problem

Calculating the head pressure required to force a given flow rate through a length of pipe is a common problem in engineered systems, and is straightforward when the fluid is a liquid far from its saturation state. A more complex problem is the case of the pumped fluid entering a heated pipe as a slightly subcooled or saturated liquid, being heated sufficiently to cause vaporization, and leaving the pipe as liquid-vapor mixture or totally vapor. This is called a *single-component, two-phase flow* (an example of a multicomponent two-phase flow would be a mixture of air and water). The inverse case is related, in which the vessel pressure and the exit pressure fix the driving force, and the flow rate from the vessel is required. In either case the change in specific volume requires a higher local velocity for a given mass flow, and consequently there is an increased frictional pressure drop. Additional losses from liquid-vapor interfaces, elevation changes, and other factors can combine to make the two-phase pressure drop considerably higher than that of the single-phase.

To properly design a flowing system, derive an algorithm to calculate the pressure drop expected when a single-phase liquid at saturation conditions is pumped through a heated tube, considering two-phase flow effects.

24.2 Tubes, Pipes, and Ducts

Calculate conditions over a range of pressures, heater powers, and tube diameters to illustrate the effect of each parameter on pressure change.

Given information

Geometry
- Heated horizontal tube
- Single-phase liquid water at inlet
- Smooth internal pipe surface

Conditions
- Tube length (heated) = 3 m
- Tube inside diameter = 19 mm, 25 mm
- Inlet pressure = 10 bar, 35 bar, 70 bar
- Mass flow = 1 kg/s
- Heater power = 15 kW, 18 kW

Assumptions

Saturated liquid conditions at pipe inlet

No noncondensible gases present

Steady-state operation

Uniform wall heat flux over heated length

Neglect inlet/outlet-fitting effects

Coolant state may be represented by mean bulk properties (this important assumption is called the *homogeneous model*, and although commonly used, there are others; the homogeneous model is used here for illustration, and alternatives are discussed)

Nomenclature

A	Flow area
D	Tube diameter
f	Friction factor
G	Mass flux (mass flow rate per unit cross section); acceleration due to gravity
g_c	Unit system factor
h_f	Enthalpy of liquid phase at saturation
h_{fg}	Enthalpy change from liquid to vapor (latent heat)
L	Heated length of tube
M	Mass flow rate
N	Number of subdivisions of heated tube for analysis

n	Subdivision number
P	Pressure (absolute)
x	Thermodynamic quality (vapor mass fraction)
V	Fluid velocity
v	Fluid specific volume
v_f	Specific volume of saturated liquid
v_{fg}	Specific volume change from liquid to vapor
W_f	Mass of liquid
W_g	Mass of vapor
z	Distance along tube from inlet
θ	Angle of elevation, such that increase in height $= z \sin \theta$
μ	Dynamic viscosity
ρ	Fluid density

Use of a consistent set of units, such as SI or English, is always required.

Two-Phase Flow Pressure Drop Fundamentals

With incompressible single-phase flow, coupling between flow rate and pressure loss is usually through property variation with temperature and is relatively weak. Bulk temperatures and averages suffice. In contrast, for heated two-phase flow the heat transfer and the pressure drop are very closely coupled. Heat addition causes a phase change and therefore significant changes in bulk properties, vapor bubbles, and flow patterns along the tube length. In turn these affect the heat-transfer rates and the pressure drop. In many cases the two phases may not be in thermodynamic equilibrium at a particular point. Furthermore, particularly at high heat-transfer rates, the combination of proposed flow, heat transfer, and phase change may not converge to a stable operating point, a situation called *flow instability*.

There is significant literature on this subject. Much of this literature treats special cases meaningful to a particular author or employer. For instance, there is a lot of data and discussion of conditions at high pressure and temperature, driven primarily by the nuclear power industry of the 1970s and 1980s. If your need is for a process design with steam and water in tube bundles above 70 bar (at 1000 psia), someone probably has solved your problem (e.g., see Refs. 1 and 5).

In general, the pressure drop through a tube will be higher for a two-phase flow than for the same mass flow of the liquid. Much of the research has focused on finding or computing a multiplier for the single-phase calculation to determine how much higher. Ideally, it would be a

24.4 Tubes, Pipes, and Ducts

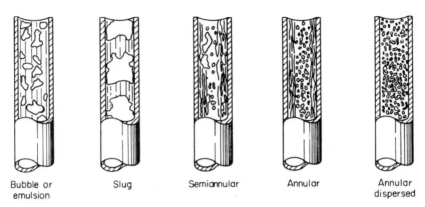

Bubble or emulsion | Slug | Semiannular | Annular | Annular dispersed

Figure 24.1 Flow patterns in boiling two-phase flow in a heated tube.

simple number, like 2.5. A little reflection immediately squashes that idea because there are so many variables. Among these are

- Flow regime changes with level of boiling (Fig. 24.1).
- There is an acceleration term, often significant, because of the major density changes. (Lowering density requires a higher velocity to maintain constant total mass flow.)
- There is tight coupling of the flow regimes to the heat transfer.
- There are more fluids than just water used in two-phase flow systems. For many of them detailed flow and thermodynamic property data are not available.

Two-Phase Forced-Convection Flow Patterns

It is useful to visualize the flow patterns in a heated pipe with two-phase flow (please refer to Fig. 24.1). Conceptually, it shows consecutive sections of a heated tube, with wall temperature above T_{sat}. Subcooled liquid enters at the bottom, is heated to saturation over some length, then vapor bubbles appear and are swept into the flow (bubble pattern). With more heating the pattern changes as shown from left to right. The sequence of patterns illustrated is fuzzily defined and tends to move from one to another gradually. There is usually no sharp demarcation between them. A simple way to visualize the process is of a liquid flow, initially containing trace vapor, transitioning over the length of the tube to a vapor containing trace liquid. Many applications will not completely vaporize the liquid, and only one or another of the regimes will be present over most of the length.

The Momentum Equation for Two-Phase Flow

A basic equation will be derived for modeling flow suitable for practical calculation, albeit with the aid of a computer. This expression forms the starting point for many analyses. Increased accuracy for specific situations is obtained by substitution of appropriate friction term models and assumptions. It should also not be surprising that the many correlations and methods are rooted in empirical data, not developed from theory. The general method will be shown, and examples of the simplest given.

A control volume momentum analysis of a fluid in many fluid mechanics texts [2,3] produces an expression similar to the following by summing the forces on an element of fluid in a stream tube:

$$v\,dP + \left(\frac{V\,dV}{g_c}\right) + \left(\frac{g}{g_c}\sin\theta\right)dz + \frac{f}{D}\left(\frac{V^2}{2g_c}\right)dz = 0$$

Applying this equation in one dimension to an internal pipe flow implies that bulk or mean values of the variables are used. Boundary-layer and velocity profile effects are not explicit. A boiling, two-phase flow is compressible, and local mean velocity varies with density and phase. It is simpler to replace the velocity by the mass flow rate per unit cross-sectional area, using a form of the continuity equation:

$$M = \rho AV = \frac{VA}{v}$$

$$V = \left(\frac{M}{A}\right)v = Gv$$

where $G = (M/A)$ is the mass flux (kg/s-m^2). Substituting for V and dV, and dividing by v, we obtain

$$dP + \frac{G^2\,dv}{g_c} + \left(\frac{g}{g_c}\frac{\sin\theta}{v}\right)dz + \frac{f}{D}\frac{G^2 v}{2g_c}dz = 0$$

The first term is the pressure differential, the second term describes acceleration from volume change, the third term is static head, and the fourth is a model for pressure losses due to friction.

24.6 Tubes, Pipes, and Ducts

For flow between two points along the streamtube or pipe, we obtain

$$\int_{P_1}^{P_2} dP + \frac{G^2}{g_c}\int_{v_1}^{v_2} dv + \frac{g}{g_c}\int_{z_1}^{z_2}\frac{\sin\theta}{v}dz + \frac{1}{2g_c}\frac{G^2}{D}\int_{z_1}^{z_2} fv\,dz = 0$$

$$(P_1 - P_2) + \frac{G^2}{g_c}(v_2 - v_1) + \frac{g}{g_c}\left(\frac{z_2 - z_1}{\bar{v}}\right)\sin\theta + \frac{1}{2g_c}\frac{G^2}{D}\int_{z_1}^{z_2} fv\,dz = 0$$

For a single-phase and incompressible flow, both specific volume and the friction factor f are constants. The second term becomes zero, the third term is static pressure change from elevation, and the fourth term becomes the familiar expression for friction loss in a pipe.

In this problem, the fluid is initially a liquid at or near saturation. Increasing elevation (flow upward) and friction pressure drop along the wall will cause flashing of the liquid to vapor. If there is also heat addition from the walls fluid, enthalpy increases and vaporization is intensified. Because the quality varies along the length of the pipe, the effective friction factor also must vary. The two-phase challenge is evaluation of the integrals by suitable definitions or models for the friction factor–specific volume relationship. An iteration scheme is required. The value of the friction factor depends on fluid properties, and the fluid properties depend on the pressure and enthalpy at each point along the pipe, which in turn depend on the friction loss. An added complication is the variation in the flow patterns as shown in Fig. 24.1. Intuitively, each pattern should require a different treatment of the relation between liquid and vapor phases, and the resulting fluid state.

A simplifying solution scheme is to divide the line into small segments over which values of these properties might be assumed nearly constant, compute the segment ΔP, and then sum the whole. Iteration is required until the changes in state and the property values are in balance. Alternatively, if the overall pressure difference is given and G is to be found (as in a relief line), the scheme might involve a double iteration wherein a G is assumed, then the ΔP calculated, tested against the available ΔP, and an adjustment made to G.

The example problem illustrates use of the equation with the same common assumptions, and illustrates the solution procedure and the general nature of a two-phase flow.

Procedure

The following series of steps may be programmed in Basic, FORTRAN, or one of the math packages, according to one's preference:

1. Obtain tables or correlations of properties for the fluid used.
2. Divide the pipe into a number of segments and assign the appropriate heat addition to each segment.
3. For the given pipe geometry and flow rate, calculate the single-phase, isothermal pressure profile as the first approximation.
4. Beginning with the single-phase isothermal profile, calculate the fluid state properties in each segment, then revise the pressure in each using the equation derived above. The change in enthalpy from the heat flux over the segment is added to the saturation enthalpy at each step. This produces a new pressure profile.
5. Iterate the pressure-property solutions until the calculated pressure profile converges.

Each of these steps will now be discussed and illustrated by calculated results for the problem posed.

Solution Details

1. Obtaining fluid property values can be the most vexing part of the computation. Tabulations with detail equivalent to the steam tables for water are rare for most other fluids of interest. The two-phase flow phenomena being modeled depend on incremental changes in properties from segment to segment. Whatever the source, the property data should be fitted with polynomial or power-law regressions for use by the computer, so that property change is continuous computationally.

2. When dividing the pipe into segments, the guiding principle is that segments should be short enough that pressure and hence the properties can be reasonably represented with average values in the segment. However, the segments should not be so small that coarseness in the property tables or accuracy of regression fit will affect the solution. While the computer is willing work with a large number of small subdivisions, this may give a false optimism about the accuracy of results.

3. A reasonable initial estimate of the pressure profile in the tube is the single-phase isothermal pressure profile, calculated in the traditional way. Various correlations for the friction factor have been proposed. Any will do that fit the physical circumstances. In this example the Blasius [6] smooth pipe correlation is used. Note that this and any other friction factor correlation will contain a Reynolds number as an independent variable. Calculating the Reynolds number in turn requires the fluid properties density and viscosity.

4. This section will comment on evaluation of each term in the equation for each segment. The key parameters defining fluid state are the pressure and the thermodynamic quality. The pressure at beginning and end of a segment is calculated in the previous iterative step. From classsical thermodynamics, *quality* is defined as the mass of vapor divided by the total mass of fluid:

$$x = \frac{W_g}{W_f + W_g}$$

This definition is strictly only true for thermodynamic equilibrium and in addition for a flowing system in which the liquid and vapor velocities are equal. These assumptions are tacit in the homogeneous model, but may not be for other models. The classical thermodynamics definition of quality should be noted when evaluating or using other models.

In this example, the mass quality will be evaluated from the enthalpy in the segment. Initially, the enthalpy is the saturated liquid enthalpy $h_f(1)$ commensurate with the inlet pressure $P(1)$. The enthalpy, or energy content, will be increased by addition of heat Δh over the segment. At any segment $n\, (n = 1, 2, 3, \ldots, N)$ the energy content will be

$$h(n) = h(1) + (n-1)\Delta h$$

(If the heat flux were not uniform, then a Δh defined for each segment could be used.) Because the local pressure has changed from friction losses, elevation changes, and vapor acceleration, the thermodynamic state values of h_f and h_{fg} have changed slightly, so that a new quality can be calculated from

$$x = \frac{h(n) - h_f}{h_{fg}}$$

All the other thermodynamic properties can be obtained from this revised segment quality:

- The *acceleration term* represents the change in pressure related to the change in velocity between two points, in this case the beginning and the end of each segment. Evaluation is straightforward, requiring calculation of the specific volume at beginning and end using the quality.
- The *elevation term* is essentially the vertical component of the weight per unit mass of fluid in the segment. For a reasonably short segment in which conditions are not changing rapidly, a mean value of the specific volume in a segment is convenient to use.

- The *friction loss term* is the most difficult term to evaluate and the subject of many studies and models. Clearly there should be differences in irreversible losses depending on the flow pattern extant. The homogeneous model is reasonable for a liquid flow with trace vapor (low quality, bubbly), or a vapor flow with trace liquid (high quality, annular dispersed). A pattern such as semiannular or annular, with a liquid film on the wall surface and a mostly vapor core, does not fit the homogeneous concept at all. Alternatives are referenced and mentioned in the concluding section. Nearly all analyses begin with the concept of a multiplier, generally defined by the ratio of the two-phase pressure gradient at a point to the single-phase gradient with liquid at the same total mass flow [1]:

$$\varphi^2 = \frac{(dP/dz)_{tp}}{(dP/dz)_{sp}}$$

Under the assumptions of homogeneous flow, this is effectively defining a two-phase friction factor as the single-phase factor multiplied by the specific volume. However, recall that the friction factor correlates as a function not only of the density but also as the viscosity through the Reynolds number. The viscosity of the two-phase mixture therefore needs better definition. One possibility is to evaluate viscosity in the same manner as the formal thermodynamic properties as a linear function of quality:

$$\mu = \mu_f + (1-x)\mu_{fg}$$

Another viscosity model proposed [1] is

$$\frac{1}{\mu} = \frac{x}{\mu_g} + \frac{1-x}{\mu_f}$$

The latter gives more weight to the vapor phase at increasing quality, reflecting the observation that the vapor void fraction, especially at lower pressures, is physically dominant in the tube. Pick one or the other depending on knowledge of or belief about the dominating pattern in the system being analyzed. In this example the latter definition is used. Collier and Thome [1] present an expression for the friction multiplier that combines this definition with the Blasius form for smooth pipe friction factor:

$$\phi^2 = \left[1 + x\left(\frac{v_{fg}}{v_f}\right)\right]\left[1 + x\left(\frac{\mu_{fg}}{\mu_g}\right)\right]^{-1/4}$$

Figure 24.2 is a plot of this expression for steam and water over a range of pressures. Note that it can be a very large number at low pressures

24.10 Tubes, Pipes, and Ducts

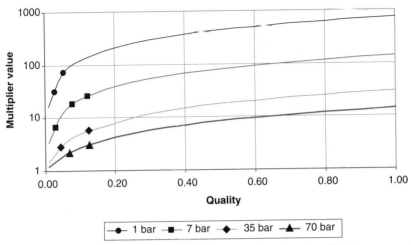

Figure 24.2 Homogeneous multiplier for water as a function of quality for a range of pressures.

and quality. This expression is effectively the integrand of the friction term in the equation. The procedure used here for evaluation of each segment is effectively the trapezoidal rule for numerical integration, using the inlet and outlet property values for a segment.

Results

Calculations made for the problem conditions specified and the assumptions and models discussed above are exhibited in graphical form in Figs. 24.3 through 24.6.

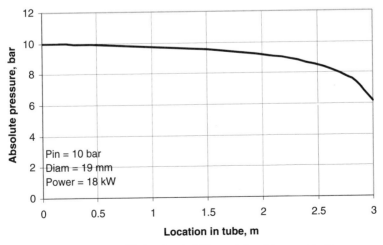

Figure 24.3 Pressure profile in a heated 19-mm tube at low pressure.

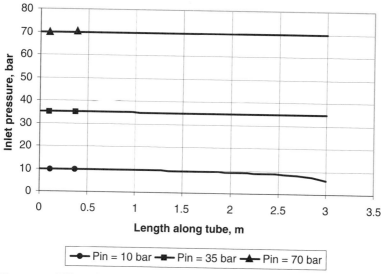

Figure 24.4 Effect of increasing system pressure on the pressure profile.

Figure 24.3 shows the calculated pressure profile in the 19-mm tube at 18 kW total heat at 10-bar inlet pressure. As energy is added, quality increases and the ΔP per segment increases. The vapor specific volume of water at this pressure is relatively large for a given quality. The effect is reminiscent of compounding interest.

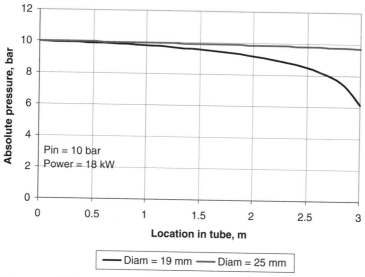

Figure 24.5 Effect of increasing tube diameter on the pressure profile.

24.12 Tubes, Pipes, and Ducts

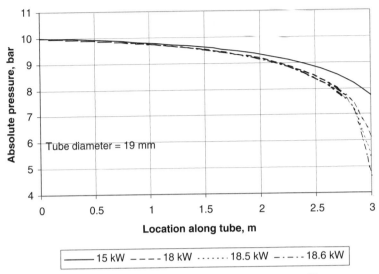

Figure 24.6 Effect of increasing tube power on the pressure profile.

Figure 24.4 shows the same tube and power, but compares inlet pressure. Consistent with intuition, at the higher pressures, 35 and 70 bar (500 and 1000 psia), the total ΔP values are much less because the qualities and hence void fractions are much lower. Of course, this also means that to increase permissible heat load for a given available pumping head, increasing overall system pressure is an option.

Figure 24.5 shows ΔP profiles for the same flow rate, total power, and lower inlet pressure, but an increased tube diameter. Increasing the tube size for the same power decreases the wall heat flux just from geometry. It also decreases the mean velocity (or equivalently the mass flux), so that frictional and acceleration pressure losses are less for any quality. This means that excessive losses can be avoided by increasing the pipe size, just as in single-phase flow. As before, the additional size would allow higher power loads for the same losses.

Figure 24.6 is a study of limiting flows. Two-phase flow is compressible, and the effects are nonlinear as quality and void fraction increase. In single-phase compressible (gas) flow, phenomena such as supersonic flow, shock waves, and choking (mass flow limit) occur, all related to the speed of sound in the gas. Two-phase flow does not exhibit limits that are closely related to its sonic velocity, but there are still limits that are analogous to choking. Looking at the 18-kW line in Fig. 24.3, one can surmise that increasing the power added will continue to increase the pressure drop. The practical limit of the pressure drop is the difference between the supply pressure and ambient pressure to which it is discharged. Obviously, absolute outlet pressure cannot be a negative number, even if discharging into a vacuum. Leung [7] has derived expressions for the limiting rate of discharge of an adiabatic (the tube is not heated) two-phase flow from a relief line. The algorithm just illustrated can be used to find such limits by repeatedly running the problem at incrementally higher input powers as shown in Fig. 24.6. If modeling an adiabatic flow, one can estimate the discharge flow rate by incrementing flow rate until the calculated outlet pressure equals the available discharge pressure.

Other Models

The homogeneous model, while simple in concept, does not reflect the true physical characteristics of all two-phase flows encountered. For example, situations typified by an annular flow (Fig. 24.1) are not modeled well. Therefore, a lot of effort has been made to improve the flow model for this regime, often at the expense of increasing complexity. The most frequently referenced model originated with Martinelli and Nelson [4]. Their approach was to treat the liquid film and the core vapor separately, with some recognition for the fluid vapor interface as a source of friction loss. Part of the method included empirical fits to available data. Since their initial paper, many improvements have been made both in the formulation and in the data supporting it [1].

The more recent trend toward miniaturization of electronics has increased interest and development of cooling methods in small channels with high heat flux. Intentionally or unintentionally, boiling of the coolant occurs. However, the dimensions of the flow channels drastically affect the phenomena. A nucleating bubble, small relative to a 15-mm tube, may completely span a 0.01-mm passage. High mass flux levels may still be a laminar flow because of the small dimension. Some of the methods and models traditionally used, including the ones discussed here, may not apply. A paper by Kandlikar [8] reviews these issues and suggests possible approaches to analysis.

The analysis and example results shown used the homogeneous model. Two-phase models that treat each phase separately with some form of coupling between them to account for energy and momentum exchange can be found and substituted in the basic equation. Here is a brief summary of the major properties of each approach.

Homogeneous model
- Easiest to use, but simplistic in assumptions
- Expect to work best at
    - Very low-quality (low-void-fraction) flows
    - High-pressure systems where liquid and vapor-specific volumes are converging
    - High mass flow rates, implying turbulent mixing of the phases

- Adiabatic flow where the vapor phase appears uniformly from reduction in pressure

Annular models
- Harder to use, but takes flow patterns into better account
- Basis has roots in empirical data
- Expect to work best at
  - Annular or separated flow occurs over most of tube length
  - Lower mass flow rates

References

1. Collier, J. G., and J. R. Thome, 1996, *Convective Boiling and Condensation*, 3d ed., Oxford University Press.
2. Rohsenhow, W. M., and J. P. Hartnett, 1973, *Handbook of Heat Transfer*, McGraw-Hill.
3. *ASHRAE Handbook—Fundamentals*, 2001.
4. Martinelli, R. C., and D. B. Nelson, 1948, "Prediction of Pressure Drop during Forced Circulation Boiling of Water," *Trans. ASME* **70**:165.
5. Tong, L. S., 1975, *Boiling Heat Transfer and Two-Phase Flow,* Kriegler Publishing.
6. Pitts, Donald R., and L. E. Sissom, 1998, *Theory and Problems of Heat Transfer*, 2d ed., Schaum's Outline Series.
7. Leung, J. C., 1986, "Simplified Vent Sizing Equations for Emergency Relief Requirements in Reactors and Storage Vessels," *J. AIChE*, **32**(10):1622.
8. Kandlikar, S. G., 2004, "Heat Transfer Mechanisms during Flow Boiling in Microchannels," *ASME J. Heat Transfer*, **126**:8.

Chapter 25

Heat-Transfer Calculations for Predicting Solids Deposition in Pipeline Transportation of "Waxy" Crude Oils

Anil K. Mehrotra and Hamid O. Bidmus
Department of Chemical and Petroleum Engineering
University of Calgary
Calgary, Alberta, Canada

When a liquid flowing through a pipeline is exposed to a cold environment that is below its freezing-point temperature (or solubility temperature), solids deposition on the pipe wall is likely to occur. This phenomenon takes place frequently during the transportation of "waxy" crude oils that contain high-molecular-weight alkanes or paraffin waxes. Paraffin waxes have a reduced solubility in crude oils at lower temperatures, causing their crystallization and deposition on cooler surfaces.

The adverse effects of wax deposition are encountered in all sectors of the petroleum industry, ranging from oil reservoir formations to blockage of pipelines and process equipment. The deposited wax impedes the flow of oil through the pipeline, causing an increase in the pumping power. If not controlled adequately, the deposited solids may block the pipeline, resulting in high costs for its cleaning. Remediation can be carried out using mechanical cleaning methods, chemical cleaning methods, or by heating to melt the deposit.

25.2 Tubes, Pipes, and Ducts

The highest temperature at which the first crystals of paraffin wax start to appear on cooling of paraffinic mixtures, such as waxy crude oils, is called the *wax appearance temperature* (WAT) or the cloud-point temperature (CPT). Wax molecules start to crystallize out of the liquid mixture when the crude-oil temperature falls below the WAT, which leads eventually to solid deposition on the pipe wall.

Holder and Winkler [1] observed solid wax deposits using cross-polarized microscopy and found that the wax crystallites have structures of platelets that overlap and interlock. A solid network structure is formed when sufficient quantities of solid paraffin crystals are formed, which leads to the formation of a gel-like structure with entrapped liquid oil. The wax deposit is thus not entirely solid but rather consists of two phases: liquid oil and solid wax. The wax-oil gel is formed as a result of the flocculation of orthorhombic wax crystallites that appear in the solution during cooling [2]. It has been reported that as little as 2 percent of precipitated wax can be sufficient to form a gel-like deposit [3].

Solids deposition from waxy crude oils is a complex engineering problem, which involves the consideration of thermodynamics, solid-liquid multiphase equilibria, crystallization kinetics, fluid dynamics, heat transfer, mass transfer, rheology, and thermophysical and transport properties. Modeling of solids deposition has been attempted via molecular diffusion and mass transfer, shear dispersion, Brownian motion, and heat transfer. In this chapter, however, we will treat the solids deposition from waxy crude oils mainly as a heat-transfer problem; that is, solids deposition at steady state is assumed to occur primarily because of the presence of a thermal driving force, and momentum and mass-transfer effects are neglected.

Energy-Balance and Heat-Transfer Equations for Solids Deposition at Steady State

Consider a long pipeline carrying a "waxy" mixture that is completely immersed in water at a constant temperature T_c. This is similar to the transportation of waxy crude oils through sub-sea pipelines from offshore wells. Heat transfer to the surrounding "colder" seawater would decrease the crude-oil temperature T_h, which could lead to the precipitation and deposition of paraffin solids on the pipe wall. As mentioned above, wax crystals would form a deposit layer comprising a solid (wax) phase with immobile liquid oil trapped within it. With time, the deposit layer would grow in thickness until such time that its growth does not occur any more. At this point, the rates of heat transfer across the flowing oil, the wax deposit layer, and the pipe wall would be the same and remain constant with time. When this happens, the thickness

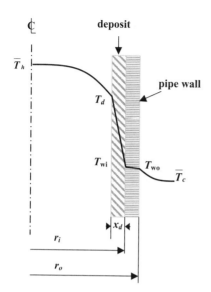

Figure 25.1 Radial temperature profile through the various thermal resistances.

of the deposit as well as the oil and wall temperature would each attain a constant value. Once the deposit layer achieves a constant layer thickness, the rate of heat transfer can be assumed to be under a thermal steady state because of the stable temperatures.

As shown schematically with radial temperature profiles in Fig. 25.1, the transfer of thermal energy from the "hot" fluid (the flowing crude oil) to the "cold" fluid (the surrounding seawater) at steady state involves four thermal resistances in series: two convective resistances due to the flowing crude oil and the seawater and two conductive resistances offered by the pipe wall and the deposited layer.

Assuming one-dimensional heat transfer in the radial direction of the pipe, the rate of heat transfer could be equated to the rate of thermal energy lost by the hot waxy crude oil, the rate of thermal energy gained by the cold seawater, and the rate of heat exchange between hot and cold fluids, as follows:

$$q = \dot{m}_h C_h (T_{h\,in} - T_{h\,out}) = \dot{m}_c C_c (T_{c\,out} - T_{c\,in}) = U_i A_i (\overline{T}_h - \overline{T}_c) \quad (25.1)$$

where
q = rate of heat transfer
$\dot{m}_h, \dot{m}_c$ = mass flow rates of "hot" and "cold" streams
C_h, C_c = average specific-heat capacities of "hot" and "cold" streams
$T_{h\,in}, T_{h\,out}$ = inlet and outlet "hot" stream temperatures
$T_{c\,out}, T_{c\,in}$ = outlet and inlet "cold" stream temperatures

25.4 Tubes, Pipes, and Ducts

U_i = overall heat-transfer coefficient based on inside pipe surface area A_i
$\overline{T}_h, \overline{T}_c$ = average temperatures of "hot" and "cold" streams, respectively

The rate of heat transfer can also be equated to the heat flow across the two convective thermal resistances, as follows:

$$q = h_c A_c (T_{wo} - \overline{T}_c) = h_h A_h (\overline{T}_h - T_d) \tag{25.2}$$

where h_c, h_h = convective heat-transfer coefficients on outside and inside of the pipe, respectively
A_c = pipe outside surface area in contact with cold stream
A_h = inside surface area at deposit-oil interface
T_d = temperature at this interface

Note that, for a clean pipe with no deposit, A_h would be replaced by A_i and T_d would become equal to T_{wi}.

Next, the overall or combined thermal resistance is expressed as the sum of four individual thermal resistances:

$$\frac{1}{U_i A_i} = \frac{1}{h_h A_h} + \frac{\ln(r_o/r_i)}{2\pi k_m L} + \frac{\ln(r_i/(r_i - x_d))}{2\pi k_d L} + \frac{1}{h_c A_c} \tag{25.3}$$

At steady state, the heat flux (i.e., the rate of heat transfer per unit inside pipe wall area) through each of the four thermal resistances included in U_i is as follows:

$$\frac{q}{A_i} = \frac{h_h(\overline{T}_h - T_d)}{r_i/(r_i - x_d)} = \frac{k_d(T_d - T_{wi})}{r_i \ln(r_i/(r_i - x_d))} = \frac{k_m(T_{wi} - T_{wo})}{r_i \ln(r_o/r_i)} = \frac{h_c(T_{wo} - \overline{T}_c)}{r_i/r_o} \tag{25.4}$$

where x_d = deposit thickness (assumed to be uniform along pipe length)
T_{wi}, T_{wo} = average inside and outside pipe wall temperatures
k_d, k_m = thermal conductivities of deposit layer and pipe wall, respectively

When dealing with the solidification or melting of a pure substance, the liquid-deposit interface temperature T_d would be the melting or freezing temperature; however, it would be the liquidus or saturation temperature for a multicomponent mixture. From a modeling study, Singh et al. [4] estimated T_d to approach the WAT of waxy mixtures when the deposit layer thickness stops growing, while Bidmus and Mehrotra [5] verified experimentally that T_d and WAT are the same temperature at pseudo-steady-state conditions.

Example 1: Determination of Deposit Thickness x_d

A "waxy" crude oil is produced and transported via a pipeline from an offshore oil production platform to an onshore refinery for processing. The crude oil has a WAT of 26°C. At the beginning of the operation, about 10,000 barrels of oil per day (bopd) at a temperature of 40°C leaves the offshore platform into the pipeline. After a few months of pipeline operation, it is observed that the pressure drop across the pipeline has increased considerably and that the crude oil arrives at the refinery at a temperature of 28°C, which has been found to be more or less constant for several days. The engineer managing the operation believes that the increased pressure drop is a result of solids deposition, and the deposit thickness needs to be estimated before deciding on a suitable remedial action. It can be assumed that the seawater at an average temperature of 10°C flows across (or normal to) the pipeline at an average velocity of 0.1 m/s. How could the deposit thickness be estimated from heat-transfer considerations? Pertinent pipeline data as well as average crude-oil and seawater properties are listed in Table 25.1.

TABLE 25.1 Data and Average Properties Used in Example Calculations

Property	Value
Crude oil	
Flow rate F	10,000 bopd or 0.0184 m³/s
Specific-heat capacity C_h	2400 J/kg K
Thermal conductivity k_h	0.15 W/m K
Viscosity μ_h	10 cP or 0.010 Pa s
Density ρ_h	750 kg/m³
Seawater	
Specific-heat capacity C_c	4200 J/kg K
Thermal conductivity k_c	0.65 W/m K
Viscosity μ_c	1 cP or 0.001 Pa s
Density ρ_c	1020 kg/m³
Deposit	
Thermal conductivity k_d	0.24 W/m K
Pipeline	
Wall thermal conductivity k_m	24 W/m K
Inside diameter D_i	10 in. or 0.254 m
Wall thickness $(r_o - r_i)$	0.625 in. or 0.0159 m
Outside diameter D_o	11.25 in. or 0.286 m

25.6 Tubes, Pipes, and Ducts

Solution

Since the temperature of the crude oil arriving at the plant has been observed to be constant for some time, it can be assumed that the pipeline-seawater system has attained a thermal steady state. Under these conditions, the deposit layer would have a constant thickness and the temperature at the deposit-oil interface would be at the crude-oil WAT. Although the deposit thickness will most likely not be uniform along the pipeline, due to the varying temperature of the crude oil from the inlet to the outlet, it will be assumed to be uniform for these calculations. Equation (25.4) can thus be used to solve for the three unknowns: T_{wi}, T_{wo}, and x_d.

Estimation of inside heat-transfer coefficient h_h. The average velocity of crude oil in the pipeline is

$$\bar{u}_h = \frac{F}{A_i} = \frac{4F}{\pi D_i^2} = \frac{4 \times 0.0184}{\pi \times 0.254^2} = 0.36 \, \text{m/s}$$

The Reynolds number of crude oil in the pipeline is

$$\text{Re} = \frac{\rho_h \bar{u}_h D_i}{\mu_i} = \frac{750 \times 0.36 \times 0.254}{0.010} = 6915$$

The Prandtl number is

$$\text{Pr} = \frac{C_h \mu_h}{k_h} = \frac{2400 \times 0.010}{0.15} = 160$$

Given that the flow is turbulent (Re > 4000) and assuming L/D_i to be large, the Dittus-Boelter correlation [6] can be used to estimate the average heat-transfer coefficient in the pipeline h_h:

$$\text{Nu} = \frac{h_h D_i}{k_h} = 0.023 \, \text{Re}^{0.8} \, \text{Pr}^{0.3} \Rightarrow h_h = \frac{0.023 \times 6915^{0.8} \times 160^{0.3} \times 0.15}{0.254}$$

$$= 73 \, \text{W/m}^2 \, \text{K}$$

Estimation of outside heat-transfer coefficient h_c

$$\text{Re} = \frac{\rho_c \bar{u}_c D_o}{\mu_c} = \frac{1020 \times 0.1 \times 0.286}{0.001} = 29{,}150$$

$$\text{Pr} = \frac{C_c \mu_c}{k_c} = \frac{4200 \times 0.001}{0.65} = 6.46$$

The Churchill-Bernstein equation [7] can be used to obtain the average heat-transfer coefficient for crossflow across a cylindrical surface:

$$\text{Nu} = 0.3 + \frac{0.62\,\text{Re}^{1/2}\text{Pr}^{1/3}}{\left[1+(0.4/\text{Pr})^{2/3}\right]^{1/4}}\left[1+\left(\frac{\text{Re}}{282{,}000}\right)^{5/8}\right]^{4/5}$$

$$\Rightarrow h_c = \left[0.3 + \frac{0.62 \times 29{,}150^{1/2} \times 6.46^{1/3}}{\left[1+(0.4/6.46)^{2/3}\right]^{1/4}}\left[1+\left(\frac{29{,}150}{282{,}000}\right)^{5/8}\right]^{4/5}\right]$$

$$\times \frac{0.65}{0.286} = 515\,\text{W/m}^2\,\text{K}$$

Estimation of deposit thickness x_d. The average oil temperature is $\overline{T}_h = (40+28)/2 = 34°\text{C}$. As already explained, the deposit-oil interface temperature T_d will be assumed to be equal to the WAT of 26°C. Therefore, using Eq. (25.4), we obtain

$$\frac{73(34-26)}{0.127/(0.127-x_d)} = \frac{0.24(26-T_{\text{wi}})}{0.127\ln(0.127/(0.127-x_d))}$$

$$= \frac{24(T_{\text{wi}}-T_{\text{wo}})}{0.127\ln(0.143/0.127)} = \frac{502(T_{\text{wo}}-10)}{0.127/0.143}$$

Solving these equalities simultaneously for T_{wi}, T_{wo}, and x_d would yield $T_{\text{wi}} = 11.3°\text{C}$, $T_{\text{wo}} = 11.0°\text{C}$, and $x_d = 6.1 \times 10^{-3}$ m or 6.1 mm. Thus, the deposit layer in the pipeline is estimated to be approximately 6 mm thick, which is about 5 percent of the pipeline inner radius. It should be noted that the inside heat-transfer coefficient for the pipeline with deposit has been assumed to be the same as that for the clean pipeline. Including the effect of deposit thickness would increase the value of h_h from 73 to 77 W/m² K. Also, while a radial temperature gradient would exist across each section of the pipeline, the inlet and outlet crude-oil temperatures are bulk values and the calculated crude-oil temperatures are average values for the entire length of the pipeline.

Example 2: The Case of Insulated Pipeline

If the pipeline in Example 1 were insulated over its entire length with a waterproof plastic material having a thermal conductivity of 0.15 W/m K, what minimum thickness of insulation material would be required to prevent solids deposition?

25.8 Tubes, Pipes, and Ducts

Solution

With an additional thermal resistance (due to insulation), Eq. (25.4) becomes

$$\frac{h_h(\overline{T}_h - T_d)}{r_i/(r_i - x_d)} = \frac{k_d(T_d - T_{wi})}{r_i \ln(r_i/(r_i - x_d))} = \frac{k_m(T_{wi} - T_{wo})}{r_i \ln(r_o/r_i)} = \frac{k_{ins}(T_{wo} - T_{ins})}{r_i \ln((r_o + x_{ins})/r_o)}$$

$$= \frac{h_c(T_{ins} - \overline{T}_c)}{r_i/(r_o + x_{ins})} \tag{25.5}$$

where k_{ins} and x_{ins} are the thermal conductivity and the thickness of the insulation material, respectively, and T_{ins} is the temperature of the insulation material surface in contact with the seawater. Without any solids deposition (i.e., $x_d = 0$), $T_{wi} \geq T_d$. Equation (25.5), therefore, becomes

$$h_h(\overline{T}_h - T_{wi}) = \frac{k_m(T_{wi} - T_{wo})}{r_i \ln(r_o/r_i)} = \frac{k_{ins}(T_{wo} - T_{ins})}{r_i \ln((r_o + x_{ins})/r_o)} = \frac{h_c(T_{ins} - \overline{T}_c)}{r_i/(r_o + x_{ins})} \tag{25.6}$$

The minimum thickness of the insulation material required can be obtained for $T_{wi} = T_d = 26°C$:

$$73(34 - 26) = \frac{24(26 - T_{wo})}{0.127 \ln(0.143/0.127)} = \frac{0.15(T_{wo} - T_{ins})}{0.127 \ln((0.143 + x_{ins})/0.143)}$$

$$= \frac{502(T_{ins} - 10)}{0.127/(0.143 + x_{ins})}$$

Solving these equalities gives $T_{wo} = 25.6°C$, $T_{ins} = 11.0°C$, and $x_{ins} = 4.3 \times 10^{-3}$ m. The minimum insulation thickness required is, therefore, 4.3 mm. Note that this thickness is not much less than the deposit layer thickness obtained in Example 1, because the thermal conductivity of insulation material is not much lower than that of the deposit layer. During the solids deposition process, the deposit layer acts as an insulation to heat transfer, thereby preventing further solids deposition. The importance of the deposit thermal conductivity is illustrated in Example 5. Also note the much larger temperature difference across the insulation material ($25.6 - 11.0 = 14.6°C$) compared to that across the pipe wall ($26.0 - 25.6 = 0.4°C$).

Example 3: Effect of Crude-Oil Composition (or WAT)

A change in the crude-oil composition will affect its WAT. Basically, the waxier a crude oil is, the higher is its WAT. The effect of wax composition will, therefore, be investigated by varying the WAT. For the operating conditions and crude-oil properties used in Example 1, what would be the deposit thickness for the crude-oil WAT values of (a) 27°C and (b) 25°C?

Solution

Since crude-oil properties and other data (listed in Table 25.1) as well as operating conditions are the same, the individual heat-transfer coefficients obtained in Example 1 can be used. Equation (25.4) can be solved for x_d by substituting the given WAT values for the interface temperature T_d:

Part a

$$\frac{73\,(34-27)}{0.127/(0.127-x_d)} = \frac{0.24\,(27-T_{wi})}{0.127\,\ln(0.127/(0.127-x_d))}$$

$$= \frac{24(T_{wi}-T_{wo})}{0.127\,\ln(0.143/0.127)} = \frac{515(T_{wo}-15)}{0.127/0.143}$$

Solving these three equalities gives $x_d = 7.6 \times 10^{-3}$ m or 7.6 mm.

Part b

$$\frac{73(34-25)}{0.127/(0.127-x_d)} = \frac{0.24(25-T_{wi})}{0.127\,\ln(0.127/(0.127-x_d))}$$

$$= \frac{24(T_{wi}-T_{wo})}{0.127\,\ln(0.143/0.127)} = \frac{515(T_{wo}-15)}{0.127/0.143}$$

Again, solving these three equalities gives $x_d = 5.0 \times 10^{-3}$ m or 5.0 mm.

The higher WAT of 27°C gave a larger deposit thickness of 7.6 mm, and a lower WAT of 25°C gave a smaller deposit thickness of 5.0 mm. These results imply that an increase in the crude-oil wax concentration would lead to an increase in the amount of deposit in the pipeline. This effect is further illustrated in Fig. 25.2, where the deposit thickness has been obtained for WAT values ranging from 20 to 27°C. It should be noted that, in these calculations, it was assumed that all crude-oil properties remain constant even though the concentration of wax was varied. Properties such as the viscosity and density of the crude oil can be altered by an increase in wax concentration, which would alter the value of inside heat-transfer coefficient. However, even with such crude-oil property changes, an increase in wax concentration would lead to increased amount of deposit in the pipeline.

25.10 Tubes, Pipes, and Ducts

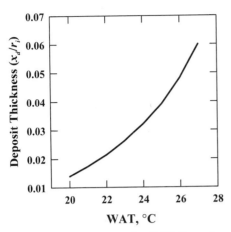

Figure 25.2 Effect of crude-oil WAT on deposit thickness under similar operating conditions.

Example 4: Effect of Crude-Oil Flow Rate

The pipeline throughput is increased from 10,000 to 15,000 bopd (0.0276 m³/s). What would be the thickness of the deposit layer in the pipeline at steady state?

Solution

Estimation of inside heat-transfer coefficient h_h. The inside heat-transfer coefficient would be altered by an increase in the production rate. The new velocity of crude oil in the pipeline would be $\bar{u}_h = F/A_i = 4F/(\pi D_i^2) = (4 \times 0.0276)/(3.142 \times 0.254^2) = 0.54$ m/s. Then

$$\text{Re} = \frac{\rho_h \bar{u}_h D_i}{\mu_h} = \frac{750 \times 0.54 \times 0.254}{0.01} = 10{,}370 \qquad \text{Pr} = 160$$

The inside heat-transfer coefficient is

$$h_h = \frac{0.023 \times 10{,}370^{0.8} \times 160^{0.3} \times 0.15}{0.254} = 102 \text{ W/m}^2 \text{ K}$$

Estimation of deposit thickness x_d. Using Eq. (25.4), we obtain

$$\frac{102(34-26)}{0.127/(0.143-x_d)} = \frac{0.24(26-T_{wi})}{0.127 \ln(0.127/(0.127-x_d))}$$

$$= \frac{24(T_{wi}-T_{wo})}{0.127 \ln(0.143/0.127)} = \frac{515(T_{wo}-10)}{0.127/0.143}$$

Solving these equations would give $x_d = 4.3 \times 10^{-3}$ m or 4.3 mm. Thus, when compared to Example 1, the deposit thickness is reduced as a result of an increase in the crude-oil flow rate. This is caused by the increased inside heat-transfer coefficient due to the increased flow rate. The resulting higher wall temperature then yields a reduction in the amount of deposition.

Example 5: Deposit Thermal Conductivity

As was shown in Example 2, when solids deposition occurs in a pipeline, the layer of deposit already formed in the pipeline acts as insulation to further reduce heat transfer between the crude oil and the seawater. The deposit thickness at which this happens would depend on the thermal conductivity of the deposit layer. Suppose that a section of the pipeline in Example 1 was removed for inspection. It was found that the actual deposit thickness was higher than the estimated value by 50 percent. An analysis suggested that this error was perhaps due to the use of an incorrect deposit thermal conductivity value in the calculations. What is a more appropriate estimate for the thermal conductivity of the deposit layer?

Solution

Once again, all other conditions will be assumed to be unchanged. Although Eq. (25.4) can still be applied, a different approach will be used to solve this problem.

Deposit thickness. The deposit thickness is 50 percent higher: $x_d = 1.5 \times 6.1 \times 10^{-3} = 9.2 \times 10^{-3}$ m or 9.2 mm.

Rate of heat transfer. With the estimated value of the inside heat-transfer coefficient, the rate of heat transfer can be obtained from Eq. (25.2):

$$q = h_h A_h (\overline{T}_h - T_d) = h_h \, 2\pi(r_i - x_d) L (\overline{T}_h - T_d)$$
$$\Rightarrow q = 73 \times 2 \times \pi L (0.127 - 9.2 \times 10^{-3})(34 - 26) = 435 L \text{ W}$$

Thermal conductivity. The revised rate of heat transfer is the same as that for the temperature difference $(T_d - \overline{T}_c)$

$$q = \frac{T_d - \overline{T}_c}{\frac{\ln(r_o/r_i)}{2\pi k_m L} + \frac{\ln(r_i/(r_i - x_d))}{2\pi k_d L} + \frac{1}{h_c 2\pi r_o L}} \tag{25.7}$$

25.12 Tubes, Pipes, and Ducts

where k_d is the only unknown, which can be obtained by solving

$$435L = \frac{26 - 15}{\frac{\ln(0.143/0.127)}{2 \times \pi \times 24 \times L} + \frac{\ln(0.143/(0.127 - 9.2 \times 10^{-3}))}{2 \times \pi \times k_d \times L} + \frac{1}{515 \times 2 \times \pi \times 0.143 \times L}}$$

$$\Rightarrow k_d = \frac{5.20}{14.72} = 0.353 \text{ W/m K}$$

So the actual thermal conductivity of the deposit layer in the pipeline is 0.353 W/m K. Considering that the wax deposit comprises a gel-like mixture of liquid oil entrapped in a solid wax matrix, it is expected that the thermal conductivity value of the deposit would be between the values of the crude oil and the solid wax. Since the thermal conductivities of crude oil and solid wax are in the range of 0.10 to 0.35 W/m K, the value of 0.353 W/m K for the deposit obtained above is somewhat higher. However, high thermal conductivity values for deposited solids have been reported experimentally [8,9]. When the deposit was removed and the thermal conductivity was measured, they were generally lower than those calculated from the heat-transfer consideration [8]. The actual conditions of the deposit during the deposition process are complex and difficult to reproduce in a thermal conductivity measurement. It is pointed out that the deposit consists of a network of solid wax with liquid oil entrapped in it, which is subjected to a temperature difference during the deposition process. Therefore, the liquid trapped in the deposit may provide convective effects due to the temperature difference across the deposit layer. While these conditions cannot be reproduced in actual measurements, the calculated thermal conductivity values will be the apparent or effective values for the whole deposit layer and will include any convective effects in the deposit.

Example 6: Effect of Crude-Oil and Seawater Temperatures

What will be the deposit thickness if (a) the seawater temperature in Example 1 is 15°C and (b) the pipeline inlet and outlet temperatures are 45 and 28°C, respectively?

Solution

Using Eq. (25.4) with the same properties and heat-transfer coefficients, we get the following.

Part a

Deposit thickness

$$\frac{73(34-26)}{0.127/(0.127-x_d)} = \frac{0.24(26-T_{wi})}{0.127 \ln(0.127/(0.127-x_d))}$$

$$= \frac{24(T_{wi}-T_{wo})}{0.127 \ln(0.143/0.127)} = \frac{515(T_{wo}-15)}{0.127/0.143}$$

Solving these equations gives $x_d = 5.2$ mm.

Part b

Deposit thickness. The average oil temperature $T_h = (45+28)/2 = 36.5°C$. Using this value in Eq. (25.4), we obtain

$$\frac{73(36.5-26)}{0.127/(0.127-x_d)} = \frac{0.24(26-T_{wi})}{0.127 \ln(0.127/(0.127-x_d))}$$

$$= \frac{24(T_{wi}-T_{wo})}{0.127 \ln(0.143/0.127)} = \frac{515(T_{wo}-10)}{0.127/0.143}$$

which gives $x_d = 3.8$ mm.

The calculations presented above were repeated for several values of the average crude oil and seawater temperatures. The predictions are shown in Figs. 25.3 and 25.4, which indicate that there is a reduction in the amount of deposited solids with an increase in either the

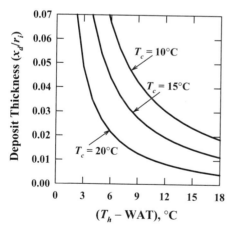

Figure 25.3 Predicted variations in deposit thickness with crude-oil temperature at different seawater temperatures.

25.14 Tubes, Pipes, and Ducts

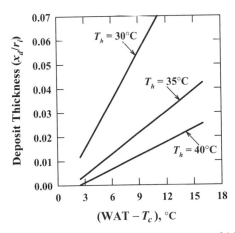

Figure 25.4 Predicted variations in deposit thickness with seawater temperature at different crude-oil temperatures.

crude-oil temperature or the seawater temperature. Conversely, a reduction in either the crude-oil or the seawater temperature would yield a larger amount of deposited solids. These predictions are in good agreement with experimental results reported by Bidmus and Mehrotra [5], who had performed wax deposition experiments over a range of operating conditions using a concentric draft tube assembly with a "hot" wax-solvent mixture flowing in the inner tube and cold water flowing through the annular region.

Example 7: Minimum Crude-Oil Temperature for Preventing Solids Deposition in Pipeline

As was shown in Example 6 as well as Figs. 25.3 and 25.4, an increase in either the crude-oil temperature or the seawater temperature would decrease the amount of deposited solids. The maximum seawater temperature above which solids deposition would not occur is the crude-oil WAT. What is the minimum average crude-oil temperature for which there would be no solids deposition in the pipeline?

Solution

Equating the two expressions for the rate of heat transfer given by Eqs. (25.2) and (25.7), we obtain

$$h_h A_h (\overline{T}_h - T_d) = \frac{T_d - \overline{T}_c}{\frac{\ln(r_o/r_i)}{2\pi k_m L} + \frac{\ln(r_i/(r_i - x_d))}{2\pi k_d L} + \frac{1}{h_c 2\pi r_o L}} \qquad (25.8)$$

Equation (25.8) is rearranged as follows:

$$\frac{\ln(r_i/(r_i - x_d))}{2\pi k_d L} = \frac{1}{h_h A_h}\left[\frac{T_d - \overline{T}_c}{\overline{T}_h - T_d}\right] - \left[\frac{\ln(r_o/r_i)}{2\pi k_m L} + \frac{1}{h_c 2\pi r_o L}\right] \quad (25.9)$$

Note that, when there is no deposit layer, $A_h = 2\pi r_i L$ and $x_d = 0$:

$$\frac{1}{h_h r_i}\left[\frac{T_d - \overline{T}_c}{\overline{T}_h - T_d}\right] = \left[\frac{\ln(r_o/r_i)}{k_m} + \frac{1}{h_c r_o}\right] \quad (25.10)$$

The minimum average crude-oil temperature to avoid any solids deposition can be expressed as

$$\overline{T}_h = T_d + \frac{T_d - \overline{T}_c}{h_h r_i}\left[\frac{\ln(r_o/r_i)}{k_m} + \frac{1}{h_c r_o}\right]^{-1} \quad (25.11)$$

Furthermore, because of the relatively high thermal conductivity of the pipeline material, its thermal resistance can be neglected, which would simplify Eq. (25.11) as follows:

$$\overline{T}_h = T_d + \frac{h_c r_o}{h_h r_i}(T_d - \overline{T}_c) \quad (25.12)$$

Accordingly, the average crude-oil temperature should be higher than the crude-oil WAT (or T_d) by $[(h_c r_o/h_h r_i)(T_d - \overline{T}_c)]$, in order to avoid any solids deposition in the pipeline. For the case considered in Example 1,

$$\overline{T}_h = 26 + \frac{515 \times 0.143}{73 \times 0.127}(26 - 10) = 152.2°C$$

that is, the average or bulk crude-oil temperature would have to be maintained at about 152°C to prevent any solids deposition in the pipeline. Since this value is the average temperature in the pipeline, the inlet crude-oil temperature would have to be higher than 152°C. The technical and economic feasibilities of preheating the crude oil (at the offshore platform) to a temperature higher than 152°C, prior to its pumping, would need to be investigated.

Importance of Fractional Thermal Resistance Offered by the Deposited Solids

As was shown in Example 6, when either the crude-oil temperature or the seawater temperature is increased, the amount of deposited solids decreases. The difference between these two temperatures is the driving force for the deposition process, and it can be related to the amount of solids deposited in the pipeline. This would mean that the larger the difference between the crude-oil and the seawater temperatures, the greater would be the amount of deposit in the pipeline. However, the results in Fig. 25.3 indicate that, at a constant seawater temperature, the amount of deposit actually would decrease when the crude-oil temperature is increased (thereby increasing the temperature differential). Whereas the results in Fig. 25.4 show that when the temperature difference is increased, while keeping the crude oil temperature constant and decreasing the seawater temperature, the amount of deposit would increase. This reasoning suggests that the overall temperature difference alone is not sufficient for determining the amount of deposited solids.

A more appropriate quantity for describing the amount of deposit is the ratio of the thermal resistance due to the deposit layer and the total thermal resistance; this ratio of thermal resistances is denoted by θ_d [5]:

$$\theta_d = \frac{R_d}{R_h + R_d + R_m + R_c} \quad (25.13)$$

The deposit fractional thermal resistance is a useful quantity for estimating the amount of deposited solids [5]. For the case considered in Example 1, a plot between the individual fractional thermal resistances and the amount of deposit for various conditions of temperature is shown in Fig. 25.5.

While all the other fractional thermal resistances decrease with an increase in the deposit thickness, the fractional thermal resistance for the deposit layer increases. However, the use of Eq. (25.13) would require a priori knowledge of the deposit thickness and thermal conductivity.

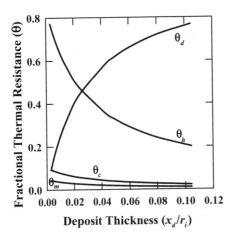

Figure 25.5 Relationship between fractional resistances and deposit thickness.

Bidmus and Mehrotra [5] showed that θ_d is also related to the ratio of the temperature difference across the deposit layer and the overall temperature difference at steady state:

$$\theta_d = \frac{T_d - T_{wi}}{\overline{T}_h - \overline{T}_c} \tag{25.14}$$

For a pipeline made of metal with a relatively high thermal conductivity and small wall thickness, the average wall temperature can be assumed to be equal to the seawater temperature; hence, it can be shown that

$$\theta_d \cong \frac{T_d - \overline{T}_c}{\overline{T}_h - \overline{T}_c} \tag{25.15}$$

This equation can be used to estimate the extent of solids deposition at steady state with the knowledge of only the bulk "hot" and "cold" fluid temperatures and the crude-oil WAT (or T_d).

Example 8: Application of θ_d

A waxy crude oil with an average temperature of 36°C and a WAT of 25°C is transported in a pipeline submerged in seawater at 10°C. If the same crude oil with a lower average temperature of 34°C is transported in the same pipeline submerged in warmer seawater at 15°C, which scenario is likely to result in a higher solids deposition? Assume that the pipeline is made of a metal having a high thermal conductivity and with a small wall thickness such that its thermal resistance can be neglected.

Solution

For both cases, the fractional thermal resistance of the deposit layer θ_d can be estimated from Eq. (25.15) as follows:

For scenario 1: $\theta_d = (25 - 10)/(36 - 10) = 0.58$

For scenario 2: $\theta_d = (25 - 15)/(34 - 15) = 0.53$

Scenario 1 has a slightly higher value of θ_d and is, therefore, likely to have more deposit under the same operating conditions. It should be noted that this type of analysis involving a comparison of θ_d is appropriate for similar values of WAT and hydrodynamic conditions in the pipeline. Even though further study is required to establish and validate a relationship between θ_d and other operating conditions, θ_d is an important parameter that could be useful for scaling up of solids deposition in pipelines [5].

Concluding Remarks

This chapter presented a mathematical framework for estimating solids deposition in a pipeline for transporting "waxy" crude oils under steady state. It is based entirely on heat-transfer considerations, involving relationships for energy balance and heat transfer. The mathematical model is relatively simple, and its use was illustrated through a number of example calculations for predicting the amount (or thickness) of deposited solids.

It is emphasized that this heat-transfer model does not deal with the effects of wax concentration on the WAT value as well as the composition of the deposited solids; these considerations would require appropriate thermodynamic calculations involving solid-liquid phase equilibria along with the estimation of thermophysical and transport properties. Similarly, fluid dynamics (or momentum transfer) considerations would determine the pressure drop across the pipeline with or without solids deposition. Moreover, waxy crude oils and wax-solvent mixtures display complex rheological (thixotropic) behavior, especially at temperatures below the WAT [10,11]. The liquid-solid phase ratio as well as the composition of deposited solids have been found to change somewhat with time through a process referred to as "deposit aging" [4,5,8,9], which may be explained in terms of diffusional mass transfer as well as the shear stress between the flowing crude oil and the deposited solid layer.

References

1. Holder, G. A., and Winkler, J., "Wax Crystallization from Distillate Fuels. I. Cloud and Pour Phenomena Exhibited by Solutions of Binary n-Paraffin Mixtures," *J. Inst. Petrol.* **51,** 228, 1965.
2. Dirand, M., Chevallier, V., Provost, E., Bouroukba, M., and Petitjean, D., "Multicomponent Paraffin Waxes and Petroleum Solid Deposits: Structural and Thermodynamic State," *Fuel* **77,** 1253, 1998.
3. Holder, G. A., and Winkler, J., "Wax Crystallization from Distillate Fuels. II. Mechanism of Pour Point Depression," *J. Inst. Petrol.* **51,** 235, 1965.
4. Singh, P., Venkatesan, R., Fogler, H. S., and Nagarajan, N., "Formation and Aging of Incipient Thin Film Wax-Oil Gels," *AIChE J.* **46,** 1059, 2000.
5. Bidmus, H. O., and Mehrotra, A. K., "Heat-Transfer Analogy for Wax Deposition from Paraffinic Mixtures," *Ind. Eng. Chem. Res.* **43,** 791, 2004.
6. Holman, J. P., *Heat Transfer*, 9th ed., McGraw-Hill, New York, 2002.
7. Churchill, S. W., and Bernstein, M. A., "Correlating Equation for Forced Convection from Gases and Liquids to a Circular Cylinder in Crossflow," *J. Heat Transfer* **99,** 300, 1977.
8. Bidmus, H. O., *A Thermal Study of Wax Deposition from Paraffinic Mixtures*, M.Sc. thesis, University of Calgary, Calgary, Canada, 2003.
9. Parthasarathi, P., *Deposition and Aging of Waxy Solids from Paraffinic Mixtures*, M.Sc. thesis, University of Calgary, Calgary, Canada, 2004.
10. Wardhaugh, L. T., and Boger, D. V., "Measurement of the Unique Flow Properties of Waxy Crude Oils," *Chem. Eng. Res. Des.* **65,** 74, 1987.
11. Tiwary, D., and Mehrotra, A. K., "Phase Transformation and Rheological Behaviour of Highly Paraffinic 'Waxy' Mixtures," *Can. J. Chem. Eng.* **82,** 162, 2004.

Chapter 26

Heat Transfer in a Circular Duct

Nicholas Tiliakos
Technology Development Dept.
ATK-GASL
Ronkonkoma, New York

Introduction

The following calculations were performed to assess the thermal loads (i.e., heat flux q, convective heat-transfer coefficient h_{conv}, etc.) for the combustor of a hypersonic vehicle over a critical portion of its flight trajectory. Calculation 1 provides the relevant information to describe the combustor thermal environment (i.e., convective and radiative heat flux, recovery temperature, convective heat-transfer coefficient, radiative equilibrium temperature), which is then used in calculation 3 to determine how a section of the combustor, namely, a test specimen, can be instrumented with thermocouples which are then used to determine the heat flux into the specimen, from experimental data obtained during aeropropulsion wind tunnel testing. Aeropropulsion tunnels are used to simulate the thermal and aerodynamic environment that the hypersonic vehicle will encounter at a critical point in its trajectory. Calculation 2 was required to assess the survivability of a fuel injector embedded inside a duct upstream of the combustor. Our concern was with the integrity of a brazed component internal to this fuel injector.

26.2 Tubes, Pipes, and Ducts

Calculation 1: Convective and Radiative Heat Transfer in a Circular Duct

Determine the convective and radiative heat transfer of a high-temperature combustible mixture flowing in a circular duct; then determine the shear stress acting on the duct wall.

Calculation procedure

The circular duct could represent the combustor section of an air-breathing engine, or the exhaust stack of a steam power plant; in the case presented below it shall represent the combustor of a hypersonic engine. The working fluid is a high-temperature combustible mixture whose properties can be determined using standard chemical equilibrium codes [1]. The conditions in the combustor, obtained from the ramjet performance analysis (RJPA) chemical equilibrium code [2], are presented in Table 26.1.

1. Determine the recovery temperature inside the combustor. The recovery temperature is given by

$$T_{\text{recovery}} = r\,[T_t - T_s] + T_s \tag{26.1}$$

where r is the recovery factor and T_t and T_s are the total and static temperatures of the flowfield, respectively. The recovery factor is defined as

$$r = (\text{Pr})^{1/n} \tag{26.2}$$

where $n = 1/2$ for laminar flow and $n = 1/3$ for turbulent flow.

1a. To assess whether the flow in the duct is laminar or turbulent, calculate the Reynolds number on the basis of the duct diameter. (*Note:* If this were

TABLE 26.1 Geometric, Thermodynamic, and Transport Properties of the Combustor (Calculation 1)

Combustor diameter D	5 in.
Combustor length L	36 in.
Static pressure P_s	32 psia
Total pressure P_t	90 psia
Static temperature T_s	4200°R
Total temperature T_t	4800°R
Density ρ	0.0211 lb$_m$/ft^3
Equilibrium specific heat C_p	0.444 Btu/lb$_m$°R
Equilibrium specific-heat ratio γ	1.21
Mach number M	1.3
Velocity V	3700 ft/s
Prandtl number Pr	0.68
Absolute viscosity μ	5.1×10^{-5} lb$_m$/ft·s
Thermal conductivity κ	1.02×10^{-4} Btu/ft·s·°R

not a circular duct, the Reynolds number would have to be calculated from the hydraulic diameter, defined as $D_h = 4A_c/P$, where A_c is the flow cross-sectional area and P is the wetted perimeter [3].)

The Reynolds number based on the combustor (duct) diameter is

$$\text{Re}_D = \frac{\rho V D}{\mu} \tag{26.3}$$

where ρ, V, and μ are the density (lb$_m$/ft^3), velocity (ft/s), and absolute viscosity (lb$_m$/ft · s) of the working fluid in the combustor, respectively, and D is the combustor (duct) diameter (ft). The viscosity is a transport property that can be obtained from standard chemical equilibrium codes, such as NASA's Chemical Equilibrium Analysis (CEA) code [1]. If the mixture is predominately nitrogen, then Sutherland's viscosity power law can be used to estimate the mixture viscosity [4]:

$$\frac{\mu}{\mu_0} \approx \left(\frac{T}{T_0}\right)^{3/2} \left[\frac{T_0 + S}{T + S}\right] \tag{26.4}$$

where S is an effective temperature, called the *Sutherland constant*, which is characteristic of the gas [table 1-2 of Ref. 4]; T_0 and μ_0 are a reference temperature and reference viscosity, respectively; μ_0 is evaluated at T_0.

Using Eq. (26.3), and the combustor (i.e., duct) properties in Table 26.1, the Reynolds number is calculated to be

$$\text{Re}_D = \frac{(0.0211 \text{ lb}_m/\text{ft}^3)(3700 \text{ ft/s})(5 \text{ in.} \cdot 1 \text{ ft}/12 \text{ in.})}{5.1 \times 10^{-5} \text{ lb}_m/\text{ft} \cdot \text{s}} = 6.378 \times 10^5$$

Since $\text{Re}_D > 2300$, the transition Reynolds number in a circular pipe [3], it follows that flow in the combustor is *turbulent*. Now that we know that the flow is turbulent, the recovery temperature is calculated using $n = 1/3$ in Eq. (26.1), to get

$$T_{\text{recovery}} = (0.68^{1/3})[4800°R - 4200°R] + 4200°R = 4728°R$$

2. Determine the hydrodynamic length in the circular duct. Internal flows are usually characterized as either developing or fully developed. For external flow there are no constraints to the flow, and the only consideration is whether the flow is laminar or turbulent. In contrast, for internal flow the fluid is confined by the surface, so that the boundary layer is unable to develop without eventually being constrained by the duct walls [3]. The fluid flowing over the internal duct wall will form a boundary layer, which, after a certain distance, will merge at the centerline. Beyond this merger, viscous effects dominate throughout the duct

26.4 Tubes, Pipes, and Ducts

cross section and the velocity profile no longer changes with increasing distance from the duct entrance. At this point the flow is then said to be fully developed, and the distance from the duct entrance to the merger point is called the *hydrodynamic entry length* $x_{\text{fd},h}$ [3].

For turbulent flow in a duct, the hydrodynamic length $x_{\text{fd},h}$ is approximately independent of Reynolds number, and is given, as a first approximation, as [3,5]

$$10 \leq \left(\frac{x_{\text{fd},h}}{D}\right)_{\text{turb}} \leq 60 \qquad (26.5)$$

For the combustor, $x_{\text{fd},h}$ is between $= 10 \cdot (D = 5 \text{ in.}) = 50$ in. and $60 \cdot (D = 5 \text{ in.}) = 300$ in. Since both the lower and upper limits of $x_{\text{fd},h} > L \, (= 36 \text{ in.})$, it follows that the flow in the combustor is *developing, turbulent* flow. Note that for laminar flow in a duct, the hydrodynamic length $x_{\text{fd},h}$ is defined as follows [3]:

$$\left(\frac{x_{\text{fd},h}}{D}\right)_{\text{laminar}} \approx 0.05 \text{Re}_D \qquad (26.6)$$

3. **Determine the thermal entry length in the circular duct.** For turbulent flow in a duct, the thermal entry length $x_{\text{fd},t}$ is defined similar to the hydrodynamic length, except that in this case beyond the merger of the wall's thermal boundary layer, viscous effects dominate throughout the duct cross section and the temperature profile no longer changes with increasing distance from the duct entrance. At this point the flow is then said to be thermally fully developed, and the distance from the duct entrance to the merger point is called the *thermal entry length* $x_{\text{fd},t}$ [3].

For turbulent flow in a duct, the thermal entry length $x_{\text{fd},t}$ is approximately independent of Reynolds number and Prandtl number, and is given, as a first approximation, as [3,5]

$$\left(\frac{x_{\text{fd},t}}{D}\right)_{\text{turb}} = 10 \qquad (26.7)$$

For the combustor, $x_{\text{fd},t} = 10 \cdot (D = 5 \text{ in.}) = 50$ in. and since $x_{\text{fd},t} > L$ ($= 36$ in.), it follows that the flow in the combustor is *thermally developing, turbulent* flow. Note that for laminar flow in a duct, the thermal entry length $x_{\text{fd},t}$ is defined as [3]

$$\left(\frac{x_{\text{fd},t}}{D}\right)_{\text{laminar}} \approx 0.05 \text{Re}_D \, \text{Pr} \qquad (26.8)$$

4. **Determine the Nusselt number in the circular duct.** We have determined that the flow in the combustor is still developing (both hydrodynamically and thermally), so that the bottom surface of the duct has no

"communication" with the top surface; therefore, for all practical purposes, flow over the bottom and top surfaces is similar to external flow over a flat plate. As a result we shall analyze the flow inside the combustor (duct) as external flow over a flat plate. For such a configuration, many closed-form correlations for determining the Nusselt number exist [3].

For a critical Reynolds number = $\mathrm{Re}_{x,c} = 5 \times 10^5$, $\mathrm{Re}_L \leq 10^8$, and $0.60 < \mathrm{Pr} < 60$, the average Nusselt number for external flow over a flat plate is given as [3]

$$(\mathrm{Nu})_L = \lfloor 0.037\,\mathrm{Re}_L^{0.8} - 871 \rfloor \mathrm{Pr}^{1/3} \qquad (26.9)$$

Substituting in the previously determined values, we obtain

$$(\mathrm{Nu})_L = \lfloor 0.037(4.592 \times 10^6)^{0.8} - 871 \rfloor (0.68)^{1/3} = 6184.3$$

Note that in the calculation above we used the Re number based on the duct length L, since, as mentioned above, the flow can be analyzed as if it were external flow over a flat plate. If, however, the flow were fully developed, then we could use the following Nusselt number correlation for turbulent, fully developed flow in a circular duct [3]:

$$(\mathrm{Nu})_D = \frac{(f/8)(\mathrm{Re}_D - 1000)\,\mathrm{Pr}}{1 + 12.7(f/8)^{1/2}(\mathrm{Pr}^{2/3} - 1)} \qquad (26.10)$$

where f is the friction factor, given as [3]

$$f = (0.790\,\ln \mathrm{Re}_D - 1.64)^{-2} \qquad (26.10\mathrm{a})$$

Note that Eq. (26.10) applies for turbulent, fully developed flow in a circular duct, for $0.5 < \mathrm{Pr} < 2000$, $3000 \leq \mathrm{Re}_D \leq 5 \times 10^6$, $(L/D) \geq 10$, and Eq. (26.10a) applies for turbulent, fully developed flow in a circular duct, for $3000 \leq \mathrm{Re}_D \leq 5 \times 10^6$.

5. Determine the convective heat-transfer coefficient in the circular duct. Since the flow in the combustor is still developing (both hydrodynamically and thermally), we can analyze the flow inside the combustor as external flow over a flat plate. Using the Nusselt number for external flow over a flat plate [Eq. (26.9) in step 4], the convective heat-transfer coefficient is determined as follows:

$$h = \frac{\mathrm{Nu}_L \kappa}{L} \qquad (26.11)$$

26.6 Tubes, Pipes, and Ducts

Substituting in numbers from above and from Table 26.1, the convective heat-transfer coefficient is

$$h = \frac{(6184.3)(1.02 \times 10^{-4} \text{ Btu/ft} \cdot \text{s} \cdot {}^\circ\text{R})}{3 \text{ ft}} = 0.21 \text{ Btu/ft}^2 \cdot \text{s} \cdot {}^\circ\text{R}$$

$$= 1.458 \times 10^{-3} \text{ Btu/in.}^2 \cdot \text{s} \cdot {}^\circ\text{R}$$

If, however, the flow were fully developed, then we could still use Eq. (26.11), but with the Nusselt number correlation for flow in a circular duct [Eqs. (26.10) and (26.10a)] [3], as presented at the end of step 4.

6. Determine the cold-wall convective heat transfer in the circular duct, for a duct cold-wall temperature equal to $T_{\text{wall}} = 520°\text{R}$. The convective heat flux is given by Newton's law of cooling [3]:

$$q_{\text{conv}} = h(T_{\text{recovery}} - T_{\text{wall}}) \tag{26.12}$$

where h was calculated in step 5, T_{recovery} in step 1, and T_{wall} is given as $520°\text{R}$. Substituting the appropriate numbers from steps 1 and 5, the cold-wall convective heat-transfer coefficient is

$$q_{\text{conv, cold}} = (0.21 \text{ Btu/ft}^2 \text{ s } {}^\circ\text{R})(4728°\text{R} - 520°\text{R})$$

$$= 883.7 \text{ Btu/ft}^2 \text{ s} = 6.14 \text{ Btu/in.}^2 \text{ s}$$

7. Determine the hot-wall convective heat transfer in the circular duct, for a duct hot-wall temperature equal to the radiation equilibrium temperature. To determine the duct wall temperature for radiation equilibrium, neglecting conduction into the wall as small compared to the other modes of heat transfer (especially at steady state, where temperature gradients in the wall go to zero), we balance the convective and radiative heat-transfer rates $q_{\text{conv}} = q_{\text{rad}}$, with q_{conv} given in Eq. (26.12) and q_{rad} given as

$$q_{\text{rad}} = \sigma \varepsilon F \left(T_{\text{wall}}^4 - T_{\text{surr}}^4 \right) \tag{26.13}$$

where σ is the Stefan-Boltzmann constant, 3.3063×10^{-15} Btu/in.$^2 \cdot$ s $\cdot {}^\circ\text{R}^4$; ε is the duct wall emissivity, which may be a function of the duct wall temperature; F is the view factor, which depends on geometry [3]; and T_{surr} is the temperature of the fluid surrounding the duct. We shall assume $\varepsilon = 1$, $F = 1$ for this calculation.

Setting Eq. (26.12) equal to Eq. (26.13), we obtain

$$F\sigma\varepsilon T_{\text{wall}}^4 - F\sigma\varepsilon T_{\text{surr}}^4 + hT_{\text{wall}} - hT_{\text{recovery}} = 0 \tag{26.14}$$

and noting that for radiation equilibrium $T_{wall} = T_{rad,equil}$, Eq. (26.14) becomes

$$F\sigma\varepsilon T_{rad,equil}^4 - F\sigma\varepsilon T_{surr}^4 + hT_{rad,equil} - hT_{recovery} = 0 \quad (26.15)$$

Substituting $F = 1$, $\sigma = 3.3063 \times 10^{-15}$ Btu/in.$^2 \cdot$s$\cdot$°R^4, $\varepsilon = 1$, $T_{surr} = 520$°R, we can guess a $T_{rad,equil}$ and iterate until it satisfies Eq. (26.15). Doing this, we obtain $T_{rad,equil} = 4092$°R.

Knowing the duct hot-wall temperature, $T_{wall,hot} = T_{rad,equil} = 4092$°R, we can calculate the hot-wall convective heat-transfer coefficient $q_{conv,hot}$ using Eq. (26.12) in step 6, with $T_{recovery} = 4728$°R, given in step 1, and $T_{wall} = T_{wall,\,hot} = T_{rad,\,equil} = 4092$°R, given above:

$$q_{conv,\,hot} = (1.458 \times 10^{-3} \text{ Btu/in.}^2 \text{ s °R})(4728 - 4092°\text{R}) = 0.93 \text{ Btu/in.}^2 \text{ s}$$

8. Determine the shear stress acting on the interior of the duct walls. The duct wall shear stress is given by

$$\tau_w = c_f Q_{dyn} = c_f \left(\tfrac{1}{2}\gamma P_s M^2\right) \quad (26.16)$$

where γ is the specific-heat ratio, P_s is the static pressure in the duct (combustor), M is the Mach number in the duct, and c_f is the skin friction coefficient, which can be found from the Chilton-Colburn analogy [3,6,7]:

$$\frac{c_f}{2} = \text{StPr}^{2/3} \quad (26.17)$$

where St is the Stanton number, St = Nu/(RePr), and Pr is the Prandtl number; Eq. (26.17) applies for $0.6 < \text{Pr} < 60$. For laminar flows, Eq. (26.17) is applicable only when $dP/dx \sim 0$; however, in turbulent flow, conditions are less sensitive to pressure gradients, so that Eq. (26.17) is approximately valid [3]. Substituting in the values for Re_L (step 4), Nu (step 4), and Pr (Table 26.1), $c_f = [2(6184.3)/(4.592 \times 10^6)(0.68)](0.68)^{2/3} = 0.0031$.

The values for γ, P_s, and M (see Table 26.1) should be local values (as close to the duct wall as possible). Substituting these values into Eq. (26.16), we find that the shear stress acting on the duct wall is $\tau_{wall} = (0.0031)(0.5)(1.21)(32 \text{ psia})(1.3)^2 = 14.6 \text{ lb}_f/\text{ft}^2$.

Calculation 2: Convective and Conductive Heat Transfer in a Circular Duct

A liquid fuel injector, made of stainless steel 440 (SS 440) shown in Fig. 26.1, is embedded inside a combustor wall, made of Poco Graphite®,

26.8 Tubes, Pipes, and Ducts

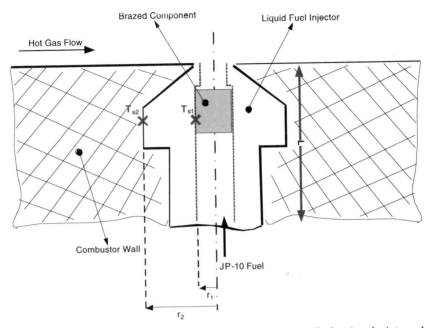

Figure 26.1 Liquid fuel injector embedded inside a combustor wall, showing the internal brazed component.

which reaches 2700°F (3160°R) at steady state. Type JP-10 fuel, initially at $T_{\text{JP-10}} = 70°F$, flows through the fuel injector, at 0.1 lb$_m$/s. Material properties for SS 440 and Poco Graphite, as well as JP-10 fuel properties, are presented in Table 26.2. Inside the fuel injector, there is a brazed component (assume same material properties as

TABLE 26.2 Material and JP-10 Fuel Properties for Liquid Fuel Injector (Calculation 2)

	JP-10 Fuel
Specific heat C_p	0.365 Btu/lb$_m$°R
Fuel Injector and Components: Stainless Steel 440 [8]	
Density ρ	0.28 lb$_m$/in.^{3}
Specific heat C_p	0.11 Btu/lb$_m$°F at 32–212°F
Thermal conductivity κ	14 Btu/h·ft·°F at 212°F
	$= 3.241 \times 10^{-4}$ Btu/in.·s·°R
Melting temperature T_{melt}	2500–2750°F
Combustor Wall: Poco Graphite [8,9]	
Density ρ	0.0641 lb$_m$/in.^{3}
Thermal conductivity κ	55 Btu/h·ft·°F
	$= 1.273 \times 10^{-3}$ Btu/in.·s·°R

SS 440). Note that the fuel injector is $L = 0.4$ in. long, with inner radius $r_1 = 0.0545$ in. and outer radius $r_2 = 0.0885$ in. At steady state

1. What temperature does the brazed component reach, and will it melt?
2. What is the temperature distribution inside the fuel injector wall (in the region $r_1 \leq r \leq r_2$?

Calculation procedure

1. Determine the steady-state temperature of the braze component T_{s1}. The fuel injector, shown in Fig. 26.1, can be modeled as a hollow cylinder, where the brazed component extends to radius r_1 at temperature T_{s1} and the outside wall of the fuel injector extends to radius r_2 at temperature T_{s2}. At steady-state conditions the combustor wall temperature, 2700°F, will equal the fuel injector outer-wall temperature T_{s2}. We shall assume one-dimensional (1D) radial heat transfer through a cylinder (i.e., the fuel injector) with an inner and outer radius of r_1 and r_2, respectively, as shown in Fig. 26.1. The fuel injector outer wall and the combustor wall will get to the steady-state temperature more quickly than will the fuel injector body since the surrounding combustor wall material has higher thermal conductivity than does the fuel injector body (compare SS 440 and Poco Graphite thermal conductivity values in Table 26.2). The fuel injector material will soak up this heat via radial heat conduction, which will be balanced by the heat transfer to the JP-10 fuel.

The JP-10 fuel, flowing at $\dot{m} = 0.1$ lb$_\text{m}$/s through the injector, will have heat transferred to it at the rate of

$$q_\text{conv} = \dot{m}_\text{JP-10} C_{p,\text{JP-10}} (T_\text{JP-10} - T_{s1}) \tag{26.18}$$

Radial heat conduction through a hollow cylinder is given as [3]

$$q_\text{cond} = \frac{2\pi L \kappa (T_{s1} - T_{s2})}{\ln(r_2/r_1)} \tag{26.19}$$

where T_{s1} and T_{s2} are the braze component and combustor wall steady-state temperatures, respectively (Fig. 26.1). Equating Eqs. (26.18) and (26.19) and solving for T_{s1}, we get

$$T_{s1} = \frac{\dot{m}_\text{JP-10} C_{p,\text{JP-10}} \ln(r_2/r_1) T_\text{JP-10} + 2\pi L \kappa T_{s2}}{\dot{m}_\text{JP-10} C_{p,\text{JP-10}} \ln(r_2/r_1) + 2\pi L \kappa} \tag{26.20}$$

where the thermal conductivity κ is for SS 440, presented in Table 26.2. Substituting in the appropriate numbers in Eq. (26.20) and solving for T_{s1}, we get $T_{s1} = 645°\text{R} = 185°\text{F}$; this temperature is less than the melting temperature of the SS 440 injector braze material, given as

26.10 Tubes, Pipes, and Ducts

2500°F in Table 26.2. Therefore for the combustor conditions outlined above, the fuel injector braze component will not melt.

2. Determine the temperature distribution in the cylindrical fuel injector. The temperature distribution inside the fuel injector, between r_2 and r_1 as shown in Fig. 26.1, is given by [3]

$$T(r) = \frac{T_{s1} - T_{s2}}{\ln(r_1/r_2)} \ln\left(\frac{r_1}{r_2}\right) + T_{s,2} \qquad (26.21)$$

Calculation 3: Conductive Heat Transfer inside a Duct Wall

A SS 440 material test piece, of thickness $t = 0.75$ in., is embedded inside a combustor engine wall, subject to the same flow conditions discussed in calculation 1 (presented in Table 26.1) and is to be tested inside an aeropropulsion wind tunnel, simulating Mach 1.3 hot-gas flow, as shown in Fig. 26.2. The material test piece will require instrumentation with type K thermocouples. You are asked to determine (1) whether the test piece can be modeled as a lumped capacitance; (2) the theoretical temperature distribution within the test piece, without the zirconia; (3) how many thermocouples should be embedded inside the material, and where they should be located; (4) the steady-state heat flux into the specimen deduced from experimental (i.e., thermocouple temperature) data; and (5) the SS 440 material wall temperature (below the zirconia coating).

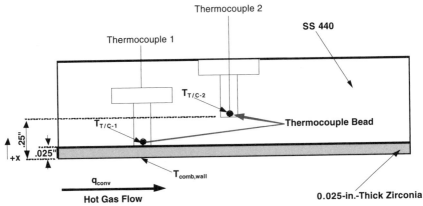

Figure 26.2 Thermocouples embedded in a test section made of SS 440 and covered with 0.025-in. zirconia, exposed to a hot-gas flow.

Calculation procedure

1. Determine whether the SS 440 material test piece can be modeled as a lumped capacitance. A material's temperature distribution can be modeled as a lumped capacitance if its Biot number Bi < 0.1 [3]. The *Biot number* is defined as

$$\text{Bi} = \frac{\text{resistance to conduction}}{\text{resistance to convection}} = \frac{hL_c}{\kappa} \quad (26.22)$$

where h is the convective heat-transfer coefficient and L_c is a characteristic length in the direction of heat transfer, defined as

$$L_c = \frac{\text{volume}}{\text{surface area}} = \frac{V}{A_s} \quad (26.23)$$

For this calculation, let's take L_c = material thickness, equal to $t = 0.75$ in. The thermal conductivity of SS 440 is 3.241×10^{-4} Btu/in.·s·°R (Table 26.2), and the convective heat transfer for these flow conditions was determined in calculation 1, step 5 and $h = 1.458 \times 10^{-3}$ Btu/in.²·s·°R. Substituting these numbers into Eq. (26.23), we determine that the Bi = 3.37. Since Bi ≫ 0.1, we cannot use the lumped-capacitance model for the SS 440 material test piece, and as such more than 1 thermocouple is required to assess the heat transfer into the test piece.

2. Determine the theoretical temperature distribution within the test specimen (without the zirconia). In order to calculate the temperature distribution inside the SS 440 test piece, we need to assume that the SS 440 specimen can be modeled as a 1D, semi-infinite solid, with a constant heat flux boundary condition and with constant thermal properties. Note that with respect to the thermocouple bead, the 0.75-in.-thick specimen does appear as a semi-infinite solid. The 1D, time-dependent heat-transfer equation is

$$\frac{\partial^2 T(x,t)}{\partial x^2} = \frac{1}{\alpha}\frac{\partial T(x,t)}{\partial t} \quad (26.24)$$

with the boundary conditions (BCs)

$$-\kappa \frac{\partial T(0,t)}{\partial x} = q = \text{constant} \quad \text{for} \quad x=0, t>0 \quad (26.25)$$

$$T(\infty, t) = T_i \quad \text{as} \quad x \to \infty, t > 0 \quad (26.26)$$

and the initial condition (IC)

$$T(x,0) = T_i \quad \text{at} \quad t = 0 \text{ for all } x \quad (26.27)$$

26.12 Tubes, Pipes, and Ducts

The boundary condition is a nonhomogenous BC [Eq. (26.26)]. To readily obtain an analytical solution to Eq. (26.24), we let $T^* = T - T_i$, which will transform the nonhomogenous BC [Eq. (26.26)] into a homogenous BC, as shown below, where Eqs. (26.24) to (26.27) have been recast for the "new, homogenous" problem

$$\frac{\partial^2 T^*(x,t)}{\partial x^2} = \frac{1}{\alpha}\frac{\partial T^*(x,t)}{\partial t} \tag{26.28}$$

with the BCs

$$-\kappa \frac{\partial T^*(0,t)}{\partial x} = q = \text{constant} \quad \text{for} \quad x = 0, t > 0 \tag{26.29}$$

$$T^*(\infty, t) = 0 \quad \text{as} \quad x \to \infty, t > 0 \tag{26.30}$$

and the IC

$$T^*(x, 0) = 0 \quad \text{at} \quad t = 0 \quad \text{for all } x \tag{26.31}$$

The integral method can be used to solve the homogenous set of Eqs. (26.28) to (26.31) [10].

The temperature distribution can be derived to give

$$T^*(x,t) = \frac{q\sqrt{\alpha t}}{\kappa} \exp\left[\frac{-x}{\sqrt{\alpha t}}\right] \tag{26.32}$$

or, noting that $T^* = T - T_i$

$$T(x,t) = T_i + \frac{q\sqrt{\alpha t}}{\kappa} \exp\left[\frac{-x}{\sqrt{\alpha t}}\right] \tag{26.33}$$

where T_i = initial temperature of the specimen at $t = 0$ (°R), q is the heat flux into the specimen (Btu/in.$^2 \cdot$ s), α is the specimen thermal diffusivity (in.^{2}/s), κ is the specimen thermal conductivity (Btu/in. $\cdot$ s $\cdot$ °R), and x is the location of the temperature $T(x,t)$ (provided by an embedded thermocouple) into the specimen, at depth x (in).

Equation (26.33) can be used to assess the temperature distribution into the test specimen (without zirconia) and allow us to determine where to embed the thermocouples into the test specimen. Assuming that $T_i = 530°$R, and, for steady-state conditions, the hot-wall convective heat flux into the combustor wall was calculated in calculation 1, step 7, as $q_{\text{conv,hot}} = 0.93$ Btu/in.^{2} s. Using the thermal diffusivity, $\alpha = \kappa/\rho C_p = 0.0105$ in.^{2}/s, for SS 440 (Table 26.2) and the thermal conductivity (Table 26.2), the theoretical temperature distribution inside the test specimen, for several steady-state times, is presented in Fig. 26.3. Note that when deciding where to locate thermocouples inside a test piece, Eq. (26.33) can be used to determine $T(x,t)$ for a range of q values [recall that Eq. (26.33) was derived for a semi-infinite solid].

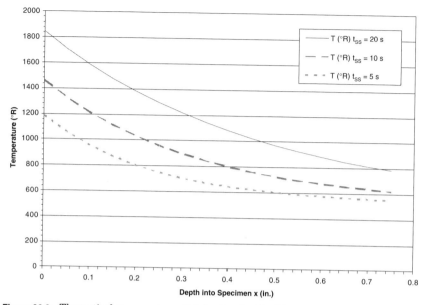

Figure 26.3 Theoretical temperature distribution inside SS 440 test specimen (without the zirconia coating), for several steady-state times, and for a specimen $q = 0.93$ Btu/in.^{2} s (calculation 1). Specimen initial temperature $T_i = 530°$R.

3. How many thermocouples should be embedded inside the material, and where should they be located? We saw in step 1 that the test specimen temperature distribution cannot be modeled as a lumped capacitance because its Biot number Bi $\gg 0.1$. In step 2 we determined the theoretical temperature distribution, also shown in Fig. 26.3, where we see that at least three thermocouples should be embedded inside the material thermocouples to adequately determine the temperature distribution inside the test specimen, as well as the heat flux to the test specimen. These locations could be at the specimen surface ($x = 0$ in.), at $x = 0.35$ in., and $x = 0.75$ in. (backside of test piece). If the temperature distribution calculation is repeated for a different material, such as Poco Graphite, which has a thermal diffusivity and thermal conductivity approximately 10 and 13 times that of SS 440, respectively, the thermal gradients are much less, thus requiring at most two thermocouples inside that material.

4. Determine the SS heat flux into the specimen from experimental temperature data. Equation (26.33), step 2, can be rearranged and solved for the steady state q, as shown below:

$$q(x,t)_{SS} = \frac{\kappa[T_{SS} - T_i]}{\sqrt{\alpha t_{SS}}} \exp\left[\frac{-x}{\sqrt{\alpha t_{SS}}}\right] \qquad (26.34)$$

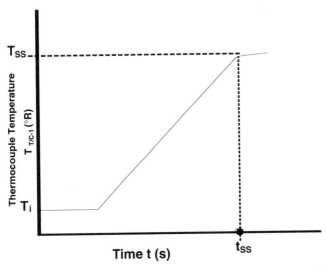

Figure 26.4 Schematic of experimental data from a thermocouple embedded inside a test specimen, described in calculation 3, and shown in Fig. 26.2.

If we use the experimental temperature data, shown in Fig. 26.4, from one of the thermocouples embedded to a known depth, $x = 0.25$ in. (Fig. 26.2), we can determine the steady-state heat flux into the SS 440 test specimen using Eq. (26.34). The experimental temperature data are $t_{SS} = 11$ s, $T_{SS} = 2530°$R, $T_i = 530°$R, $x = 0.25$ in., and thermal diffusivity α (SS 440) $= 0.0105$ in.^{2}/s; therefore $q_{SS} = 0.913$ Btu/in.^{2} s at $x = 0.25$ in. Recall that Eq. (26.34) was derived neglecting the zirconia layer and applies only for the homogenous SS 440 solid beneath.

5. **Determine the test specimen wall temperature $T_{comb,wall}$ as shown in Fig. 26.2.** The heat flux across the zirconia coating is given by Fourier's law of heat conduction, for steady-state conditions only [3]:

$$q_{SS} = \frac{\kappa_{zirc}[T_{comb,wall} - T_{T/C\text{-}1}]}{\Delta x} \qquad (26.35)$$

The thermal conductivity of the zirconia coating is $\kappa_{zirc} = 5.087 \times 10^{-4}$ Btu/in. $\cdot$ s $\cdot$ C, and $T_{comb,wall}$ and $T_{T/C\text{-}1}$ are the steady-state temperatures (shown labeled in Fig. 26.2); Δx is the separation distance, in the direction of heat flow, between $T_{T/C\text{-}1}$ and $T_{comb,wall}$, and in this case this is the thickness of the zirconia coating, 0.025 in. The heat flux into the test specimen was determined in step 4, Eq. (26.34); therefore,

Eq. (26.35) can be rearranged to solve for the steady-state $T_{\text{comb,wall}}$

$$T_{\text{comb,wall}} = \frac{q\,\Delta x}{\kappa_{\text{zirc}}} + T_{\text{T/C-1}} \qquad (26.36)$$

where $q = 0.913$ Btu/in.^{2} s is the steady-state heat flux into the test specimen, calculated in step 4 using Eq. (26.34); $T_{\text{T/C-1}}$ is the steady-state thermocouple measurement [$T_{\text{T/C-1}} = 2530°$R at $t_{\text{SS}} = 11$ s (shown in Figs. 26.2 and 26.4)]; and $\Delta x = 0.025$ in. Substituting these numbers into Eq. (26.36), $T_{\text{comb, wall}} = 3103°$R $= 2643°$F.

References

1. McBride, B. J., and Gordon, S., Chemical Equilibrium Analysis (CEA) Code, NASA Reference Publication 1311, June 1996, Computer Program for Calculation of Complex Chemical Equilibrium Compositions & Applications, vol. I, *Analysis*; vol II, *User's Manual & Program Description*.
2. Pandolfini, P. P., and Friedman, M. A., *Instructions for Using Ramjet Performance Analysis* (RJPA), IBM PC version 1.24, document JHU/APL AL-92-P175, June 1992.
3. Incropera, F. P., and DeWitt, D. P., *Fundamentals of Heat and Mass Transfer*, 5th ed., Wiley, New York, 2002.
4. White, F. M., *Viscous Fluid Flow*, McGraw-Hill, New York, 1974.
5. Kays, W. M., and Crawford, M. E., *Convective Heat and Mass Transfer*, McGraw-Hill, New York, 1980.
6. Colburn, A. P., *Trans. Am. Inst. Chem. Eng.*, **29**, 174, 1933.
7. Chilton, T. H., and Colburn, A. P., *Ind. Eng. Chem.*, **26**, 1183, 1934.
8. *Materials Selector*, published by Materials Engineering, a Penton publication, issued Dec. 1992, p. 45.
9. www.poco.com/graphite/igrade.asp.
10. Ozisik, M. N., *Heat Conduction*, Wiley, New York, 1993.

Part

6

Heat Exchangers

Heat-exchanger analyses and design calculations, including those involving compact heat exchangers, are a staple of many heat-transfer books. This handbook is no exception. The calculations and analyses included in this part are all by academics in the United States and Canada. Three of them deal with the real world:

- *Air cooling of a high-voltage power supply using a compact heat exchanger*
- *A heat recovery system for an industrial clothes dryer*
- *"Passive" condensers (heat exchangers driven by natural forces and not requiring any power supplies, moving parts, or operator actions) in nuclear reactors*

The remaining three chapters present general methodologies:

- *A rigorous, step-by-step methodology for calculating core dimensions of a compact heat exchanger with the most intricate basic flow arrangement situation in a single-pass configuration—a crossflow in which fluids do not mix orthogonal to the respective flow directions*
- *An analysis methodology for an air-cooled heat exchanger with single-phase natural circulation loops*
- *A design methodology based on optimization via an evolutionary algorithm that manipulates populations of solutions by mating the best members of the current population (natural selection) and by randomly introducing perturbations into the newly created population (mutation)*

Chapter 27

Air Cooling of a High-Voltage Power Supply Using a Compact Heat Exchanger

Kirk D. Hagen
Weber State University
Ogden, Utah

Introduction

A heat exchanger with a heat-transfer surface area-to-volume ratio of at least 700 m^2/m^3 is classified as a *compact heat exchanger*. Compact heat exchangers are typically used in applications that require a heat exchanger with a high heat-transfer capacity and a small volume and when one of the fluids is a gas. Consequently, this type of heat exchanger is usually found in aerospace and other applications where the size and weight of the cooling unit must be minimized. Of all the applications for which compact heat exchangers find effective use, the thermal management of electronics is probably the most common.

The following heat-transfer analysis pertains to the air cooling of a high-voltage power supply using a single stream compact heat exchanger, commonly called a "coldplate."[1] A coldplate typically consists of a layer of folded metal fins sandwiched between two metal baseplates. Coldplates are usually constructed by dip brazing the fins to the baseplates to ensure intimate thermal contact. The heat-producing

27.4 Heat Exchangers

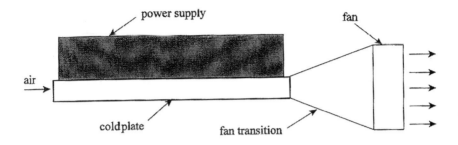

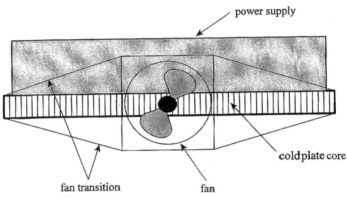

Figure 27.1 Air-cooling system.

component, a power supply in this case, is mounted to one of the baseplates. A fan forces ambient air through the passages formed by the folded fins and baseplates, which make up the coldplate core. The air receives the heat dissipated by the power supply and transports that heat to the surroundings, thereby maintaining the power supply temperature below a maximum allowable value specified by the manufacturer. This basic cooling system is illustrated in Fig. 27.1.

Statement of the Problem

The mounting plate of a high-voltage power supply in a ground-based communications system is to be held below a temperature of 70°C while operating in an environment where the air temperature is 27°C and the atmospheric pressure is 1 atm. Under normal operating conditions, the power supply dissipates 550 W. The mounting plate of the power supply measures 18.0 × 12.0 cm. Because the unit in which the power supply operates is a ground-based system, a 110-V alternating-current (ac) power source is available. The most convenient and cost-effective means of cooling the high-voltage power supply is to utilize a compact coldplate with a small ac axial fan. Because of packaging constraints, the coldplate cannot extend more than 3.0 cm beyond the perimeter of the power supply baseplate, and it cannot extend more than 3.5 cm normal to the power supply baseplate surface. Hence, the coldplate cannot exceed the overall size of 24.0 × 18.0 × 3.5 cm. There is no constraint, however, on a reasonably sized fan and a transition duct from the fan to the coldplate core. To meet these constraints, we choose a coldplate that measures 22.0 × 16.0 cm, with the smaller dimension corresponding to the direction of airflow. The third dimension is governed by the required fin height, and will be determined as part of the analysis. The analysis is performed using SI (International System; metric) units.

Analysis Assumptions

To simplify the analysis, the following assumptions are made:

1. Heat transfer and airflow rate are steady.
2. Conduction in the coldplate is one-dimensional (normal to the coldplate surface).
3. Radiation is neglected.
4. All the heat dissipated by the power supply is transferred into the coldplate.
5. Heat dissipated by the power supply is uniformly distributed over the coldplate baseplate.
6. The flow rate of air is equal in all ducts of the coldplate core.
7. The coldplate surface is isothermal.
8. The thermal resistance of the coldplate baseplate is negligible.
9. Thermal resistance at the power supply–coldplate interface is negligible.
10. The tips of the fins are adiabatic. Thus, even though only one baseplate receives heat, the other baseplate is not considered as an extension of the fins.
11. The heat-transfer coefficient is constant over the entire surface of the fins.
12. All thermal properties are constant, and are based on a temperature of 27°C (300 K).

27.6 Heat Exchangers

Thermal Properties

Aluminum 1100 (folded fins): $k = 222$ W/m · °C

Air at 1 atm: $\rho = 1.1614$ kg/m^3, $c_p = 1007$ J/kg · °C, $\nu = 15.89 \times 10^{-6}$ m^2/s, Pr $= 0.707$

Calculations

We begin by considering the dimensions of the coldplate and the folded fins. The overall length and width, respectively, of the coldplate are $L = 16.0$ cm and $W = 22.0$ cm, where the smaller dimension (L) is chosen to correspond to the direction of airflow to minimize the pressure drop. Plain folded fins are commercially available in a wide range of fin thicknesses, heights, and pitches. *Fin pitch* is a spacing parameter that defines the number of fins per inch. Folded fins are usually manufactured with either a rounded or a flat crest. The flat-crest type is used here. Illustrated in Fig. 27.2 is an end view of a portion of the coldplate core showing the fin thickness δ, fin height b, and fin pitch N_f.

At the beginning of the calculations, we do not know the heat-transfer surface area or airflow rate required to maintain the power supply temperature below its allowable value. Recognizing the potential iterative nature of the analysis, we choose a reasonable set of fin dimensions and a flow rate to initiate the calculations. For the fin thickness and pitch, we let

$$\delta = 0.010 \text{ in.} = 0.0254 \text{ cm}$$

$$N_f = 8 \text{ fins/in.} = 3.1496 \text{ fins/cm}$$

Using the duct geometry shown in Fig. 27.2, the duct aspect ratio α^* is given by the relation

$$\alpha^* = \frac{b - \delta}{1/N_f - \delta} \tag{27.1}$$

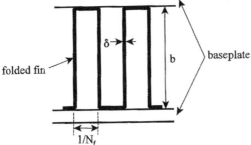

Figure 27.2 Coldplate fin configuration.

Based on the work of Kays and London,[2] a common duct aspect ratio for which heat transfer and friction data exist is $\alpha^* = 8$. Solving Eq. (27.1) for the fin height b, we obtain

$$b = \alpha^* \left(\frac{1}{N_f} - \delta \right) + \delta$$

$$= 8 \left(\frac{1}{3.1496} - 0.0254 \right) + 0.0254$$

$$= 2.362 \text{ cm} = 0.02362 \text{ m}$$

Recall that the overall thickness of the coldplate cannot exceed 3.5 cm. Using the calculated fin height and assuming that the baseplates have a thickness of 0.3 cm, the overall coldplate thickness t is $t = 2(0.3) + 2.362 = 2.962$ cm, which is below the allowable value. The wetted perimeter P of a duct is

$$P = 2 \left(b + \frac{1}{N_f} - 2\delta \right)$$

$$= 2 \left[2.362 + \frac{1}{3.1496} - 2(0.0254) \right]$$

$$= 5.257 \text{ cm} = 0.05257 \text{ m} \tag{27.2}$$

and the cross-sectional area for flow A of a single duct is

$$A = (b - \delta) \left(\frac{1}{N_f} - \delta \right)$$

$$= (2.362 - 0.0254) \left(\frac{1}{3.1496} - 0.0254 \right)$$

$$= 0.6825 \text{ cm}^2 = 6.825 \times 10^{-5} \text{ m}^2 \tag{27.3}$$

The hydraulic diameter D_h of a duct is

$$D_h = \frac{4A}{P}$$

$$= \frac{4(6.825 \times 10^{-5})}{0.05257}$$

$$= 5.193 \times 10^{-3} \text{ m} \tag{27.4}$$

27.8 Heat Exchangers

The number of ducts N_d in the coldplate is the product of the fin pitch and coldplate width

$$N_d = N_f W$$
$$= (3.1496)(22.0)$$
$$= 69.3 \qquad (27.5)$$

which we round down to 69 to avoid a fractional number of ducts. The total cross-sectional area for flow A_t is

$$A_t = N_d A$$
$$= (69)(6.825 \times 10^{-5})$$
$$= 4.709 \times 10^{-3} \text{ m}^2 \qquad (27.6)$$

Now that the coldplate dimensions have been determined, we need to estimate a flow rate. A reasonable volume flow rate for a typical axial fan suitable for this type of application is 65 cfm (ft³/min). Thus, we have

$$Q = 65 \text{ ft}^3/\text{min} \times 1 \text{ m}^3/35.3147 \text{ ft}^3 \times 1 \text{ min}/60 \text{ s}$$
$$= 0.03068 \text{ m}^3/\text{s}$$

The air velocity v based on this assumed flow rate is

$$v = \frac{Q}{A_t}$$
$$= \frac{0.03068}{4.709 \times 10^{-3}}$$
$$= 6.515 \text{ m/s} \qquad (27.7)$$

which yields a mass flow rate $\dot{m}$ of

$$\dot{m} = \rho A_t v$$
$$= (1.1614)(4.709 \times 10^{-3})(6.515)$$
$$= 0.0356 \text{ kg/s} \qquad (27.8)$$

and a mass velocity G of

$$G = \rho v$$
$$= (1.1614)(6.515)$$
$$= 7.567 \text{ kg/s} \cdot \text{m}^2 \qquad (27.9)$$

The Reynolds number Re is

$$\mathrm{Re} = \frac{vD_h}{\nu}$$

$$= \frac{(6.515)(5.193 \times 10^{-3})}{15.89 \times 10^{-6}}$$

$$= 2129 \qquad (27.10)$$

which indicates laminar flow. For laminar flow, the hydrodynamic entry length $x_{\mathrm{fd},h}$ may be obtained from the approximation

$$\left(\frac{x_{\mathrm{fd},h}}{D_h}\right)_{\mathrm{lam}} \approx 0.05\,\mathrm{Re}$$

$$= (0.05)(2129)$$

$$= 107 \qquad (27.11)$$

Similarly, the thermal entry length $x_{\mathrm{fd},t}$ may be obtained from the approximation

$$\left(\frac{x_{\mathrm{fd},t}}{D_h}\right)_{\mathrm{lam}} \approx 0.05\,\mathrm{Re}\,\mathrm{Pr}$$

$$= (0.05)(2129)(0.707)$$

$$= 75 \qquad (27.12)$$

The actual duct length-to-diameter ratio for our coldplate is

$$\frac{L}{D_h} = \frac{0.16}{5.193 \times 10^{-3}} = 31$$

which indicates that the velocity and thermal boundary layers are not fully developed. The graph of heat transfer and friction data from Kays and London[2] for a duct with an aspect ratio of $\alpha^* = 8$ includes curves for the friction factor f and the Colburn j factor j_H, for $L/D_h = 40$. These curves can be used in this analysis with little error. A version of this graph is shown in Fig. 27.3.

From Fig. 27.3, knowing that $\mathrm{Re} = 2129$, the friction factor and Colburn j factor are $f = 0.013$ and $j_H = 0.004$, respectively. The Colburn j factor is defined by the relation

$$j_H = \mathrm{St}\,\mathrm{Pr}^{2/3} \qquad (27.13)$$

where St is the Stanton number and Pr is the Prandtl number. The Stanton number may be expressed in terms of the heat-transfer

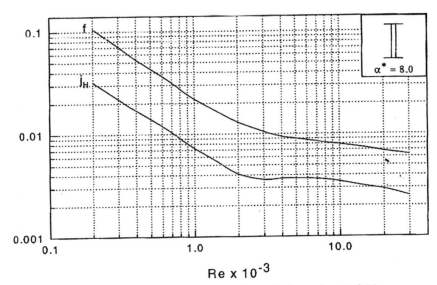

Figure 27.3 Friction factor and Colburn j factor data. (Redrawn from Ref. 2.)

coefficient h and the mass velocity G as

$$\text{St} = \frac{h}{Gc_p} \tag{27.14}$$

where c_p is the specific heat of air. Combining Eqs. (27.9), (27.13), and (27.14), and solving for the heat-transfer coefficient, we obtain

$$h = j_H \rho c_p v \, \text{Pr}^{-2/3}$$
$$= (0.004)(1.1614)(1007)(6.515)(0.707)^{-2/3}$$
$$= 38.4 \text{ W/m}^2 \cdot {}^\circ\text{C} \tag{27.15}$$

The next step in the analysis is to determine the fin efficiency. From Kern and Kraus,[3] the relation for fin efficiency η of a longitudinal fin is

$$\eta = \frac{\tanh(mb)}{mb} \tag{27.16}$$

where

$$m = \left(\frac{2h}{k\delta}\right)^{1/2}$$
$$= \left(\frac{2(38.4)}{(222)(2.54 \times 10^{-4})}\right)^{1/2}$$
$$= 36.905 \text{ m}^{-1} \tag{27.17}$$

Thus

$$\eta = \frac{\tanh[(36.905)(0.02362)]}{(36.905)(0.02362)}$$

$$= 0.7022/0.8717$$

$$= 0.8056$$

Now, we find the heat-exchanger surface efficiency. The fin heat-transfer surface area S_f for one duct is

$$S_f = 2(b - \delta)L$$

$$= 2(2.362 - 0.0254)(16.0)$$

$$= 74.771 \text{ cm}^2 = 7.477 \times 10^{-3} \text{ m}^2 \quad (27.18)$$

and the total heat-transfer surface area S_t for one duct (fin area plus interfin area) is

$$S_t = S_f + \left(\frac{1}{N_f} - \delta\right)L$$

$$= 74.771 + \left(\frac{1}{3.1496} - 0.0254\right)(16.0)$$

$$= 79.445 \text{ cm}^2 = 7.945 \times 10^{-3} \text{ m}^2 \quad (27.19)$$

Note that this result does not include the interfin area near the fin tips because, as stated in assumption 10 in the "Analysis Assumptions" section, the baseplate on the other side of the coldplate is not considered as an extension of the fins and therefore does not participate in the heat transfer. From Kraus and Bar-Cohen,[4] the heat-exchanger surface efficiency κ is

$$\kappa = 1 - \frac{S_f}{S_t}(1 - \eta)$$

$$= 1 - (74.771/79.445)(1 - 0.8056)$$

$$= 0.8170 \quad (27.20)$$

We are now ready to calculate the coldplate surface temperature. Neglecting changes in potential and kinetic energies, a simple energy balance on the air in the coldplate core yields the relation

$$q = \dot{m} c_p(T_o - T_i) \quad (27.21)$$

where q is the heat dissipated by the power supply and T_o and T_i are the outlet and inlet air temperatures, respectively. Solving Eq. (27.21) for T_o, we obtain

$$T_o = \frac{q}{\dot{m}\,c_p} + T_i$$

$$= \frac{550}{(0.0356)(1007)} + 27$$

$$= 42.3°C$$

For an isothermal coldplate surface, Incropera and DeWitt[5] derive a relation for the surface temperature T_s as

$$\frac{T_s - T_o}{T_s - T_i} = e^{\phi} \qquad (27.22)$$

where

$$\phi = -\frac{PL\kappa h}{\dot{m}\,c_p} \qquad (27.23)$$

Rearranging Eq. (27.22) and solving for T_s, we obtain

$$T_s = \frac{T_o - T_i\,e^{\phi}}{1 - e^{\phi}} \qquad (27.24)$$

Because the wetted perimeter P in Eq. (27.23) pertains to a single duct, the mass flow rate $\dot{m}$ must also pertain to a single duct. Thus, the value for ϕ is

$$\phi = \frac{-(0.05257)(0.16)(0.8170)(38.4)}{(0.0356/69)(1007)}$$

$$= -0.5079$$

Finally, solving for T_s from Eq. (27.24), we obtain

$$T_s = \frac{42.3 - 27e^{-0.5079}}{1 - e^{-0.5079}}$$

$$= 65.4°C$$

which is below the maximum allowable temperature of 70°C.

Recognizing that this result is based on some assumed fin dimensions and an estimated flow rate of air, we now need to do a pressure drop analysis for the coldplate in order to select a fan for this application. This is accomplished by generating the system curve for the coldplate and superimposing that curve on some candidate fan curves until the operating point coincides with the assumed flow rate in the foregoing heat-transfer analysis. If this process does not lead to an acceptable operating point, the coldplate dimensions and/or the fin dimensions will have to be modified. Using information given by Mott[6] and referring to Fig. 27.1, we see that there are three energy losses in the system. There is an entrance loss, a friction loss in the coldplate core, and a gradual enlargement loss in the fan transition. The total loss h_L is the sum of these three losses.

Hence, we have

$$h_L = \left(K_1 + f\frac{L}{D_h} + K_2\right)\frac{v^2}{2g} \tag{27.25}$$

where K_1 and K_2 are the loss or resistance coefficients for the entrance and fan transition, respectively, f is the friction factor obtained from Fig. 27.3, and v is the air velocity in the coldplate core. For the entrance loss coefficient, we let $K_1 = 0.5$. The gradual enlargement loss coefficient is a function of the cone angle and the ratio of the hydraulic diameter of the fan to the hydraulic diameter of the coldplate exit. To be conservative, we will assume an infinite hydraulic diameter ratio and a cone angle of 60°. These two assumptions lead to a loss coefficient of $K_2 = 0.72$. The pressure drop Δp corresponding to the total loss is

$$\Delta p = \gamma h_L = \left(K_1 + f\frac{L}{D_h} + K_2\right)\frac{\rho v^2}{2} \tag{27.26}$$

where γ is the specific weight of air. Because fan curves are expressed in terms of pressure as a function of volume flow rate, it is more convenient to express Eq. (27.26) in terms of Q rather than v. Substituting Eq. (27.7) into Eq. (27.26), we obtain

$$\Delta p = \left(K_1 + f\frac{L}{D_h} + K_2\right)\frac{\rho Q^2}{2A_t^2} \tag{27.27}$$

The units on Δp and Q in Eq. (27.27) are Pa and m³/s, respectively.

At this point in the analysis, we generate a system curve for the coldplate using Eq. (27.27). This step is best facilitated by creating a table of pressure drops and corresponding volume flow rates and recognizing that the friction factor f is a function of Reynolds number obtained from Fig. 27.3. The results are shown in Table 27.1. We then select a candidate axial fan that runs on 110 V ac power and that has a fan curve that appears to supply the required flow rate of air for our application. Shown in Fig. 27.4 is a graph of such a fan curve and the system curve

27.14 Heat Exchangers

TABLE 27.1 System Curve Data Generated from Eq. (27.27)

Q		Re	f	Δp	
m³/s	ft³/min			Pa	in. H₂O
0.00	0.0	0	—	0.0	0.0
0.01	21.9	694	0.033	5.86	0.0235
0.02	42.4	1388	0.018	18.6	0.0743
0.03	63.6	2082	0.014	38.9	0.156
0.03068	65.0	2129	0.013	39.9	0.160
0.04	84.8	2776	0.010	64.0	0.257

that was obtained from the data in Table 27.1. Note that one of the entries in Table 27.1 is for the estimated volume flow rate of 65 ft³/min. The point at which the fan and system curves cross is the operating point. The *operating point* defines the volume flow rate and pressure drop for the system when the fan in question is used. We see that the operating point is approximately 65 ft³/min and 0.16 in. H₂O. This volume flow rate matches the value that was estimated at the beginning of the analysis, so no modifications of coldplate or fin dimensions are necessary. However, if we desired a larger margin between the calculated baseplate temperature and the allowable temperature, we would have to increase the fin pitch and/or fin height and perhaps use a fan with a greater flow capacity.

As a final calculation, we determine the heat-transfer surface area-to-volume ratio to verify that our coldplate can be classified as a compact

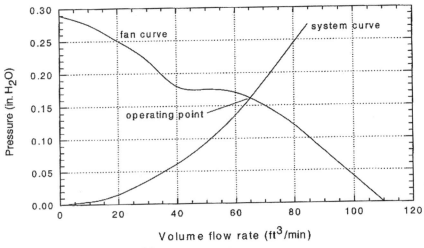

Figure 27.4 System curve and fan curve.

heat exchanger. The total heat-transfer surface area S_{total} is

$$S_{total} = 2\left[(b-\delta) + \left(\frac{1}{N_f} - \delta\right)\right]LN_d$$

$$= 2\left[(2.362 - 0.0254) + \left(\frac{1}{3.1496} - 0.0254\right)\right](16.0)(69)$$

$$= 5804 \text{ cm}^2 = 0.5804 \text{ m}^2 \tag{27.28}$$

For consistency with the standard definition of heat-transfer surface area in heat exchangers, Eq. (27.28) includes the interfin area near the fin tips. The volume of the coldplate V, excluding the baseplates, is

$$V = WLb$$

$$= (22.0)(16.0)(2.362)$$

$$= 831 \text{ cm}^3 = 8.31 \times 10^{-4} \text{ m}^3 \tag{27.29}$$

Hence, the heat-transfer surface area-to-volume ratio β is

$$\beta = \frac{S_{total}}{V}$$

$$= \frac{0.5804}{8.31 \times 10^{-4}}$$

$$= 698 \text{ m}^2/\text{m}^3 \tag{27.30}$$

Strictly speaking, β must be at least 700 m²/m³ to classify the heat exchanger as "compact," but since our result is extremely close to this value, we consider our coldplate to be a compact heat exchanger. A value of β much less than 700 m²/m³ would not have invalidated the heat-transfer analysis, however.

Concluding Remarks

This chapter presents a heat-transfer analysis of a coldplate for air cooling a high-voltage power supply. The coldplate consists of a standard configuration of folded aluminum fins brazed between two baseplates. A small axial fan, selected by means of a pressure drop analysis, draws ambient air through the fin passages, carrying the dissipated heat to the surroundings. The analysis shows that the power supply baseplate temperature is lower than the allowable value, thereby assuring reliable power supply operation with respect to thermal considerations.

While this analysis did not require modifications of coldplate or fin dimensions to find a suitable operating point, most analyses of this type involve an iterative process, which is most efficiently facilitated by the use of a computer. In order to meet the temperature specification, the thermal analyst or designer must iterate on the primary dimensional parameters while simultaneously selecting a fan that supplies sufficient airflow. This chapter outlines the basic analytical steps required to successfully do this.

References

1. Hagen, K. D., *Heat Transfer with Applications*, Prentice-Hall, Upper Saddle River, N.J., 1999, pp. 388, 410.
2. Kays, W. M., and A. L. London, *Compact Heat Exchangers*, 3d ed., McGraw-Hill, New York, 1984, p. 146.
3. Kern, D. Q., and A. D. Kraus, *Extended Surface Heat Transfer*, McGraw-Hill, New York, 1972, p. 90.
4. Kraus, A. D., and A. Bar-Cohen, *Thermal Analysis and Control of Electronic Equipment*, McGraw-Hill, New York, 1983, p. 234.
5. Incropera, F. P., and D. D. DeWitt, *Introduction to Heat Transfer*, 4th ed., Wiley, New York, 2002, pp. 449, 450.
6. Mott, R. L., *Applied Fluid Mechanics*, 5th ed., Prentice-Hall, Upper Saddle River, N.J., 2000, pp. 239–283.

Chapter 28

Energy Recovery from an Industrial Clothes Dryer Using a Condensing Heat Exchanger

David Naylor
Department of Mechanical and Industrial Engineering
Ryerson University
Toronto, Ontario, Canada

P. H. Oosthuizen
Department of Mechanical Engineering
Queen's University
Kingston, Ontario, Canada

Description of the Problem

A heat recovery system for an industrial clothes dryer is shown in Fig. 28.1. Warm moist air at a temperature of 45°C and relative humidity of 90 percent is exhausted from the clothes dryer at flow rate of 0.13 m³/s. The total pressure is assumed to be 100 kPa. The building cold-water supply enters a thin-walled copper tube ($D = 1.0$ cm) at 8°C and at a flow rate of 6 L/min. The water supply makes nine passes across the exhaust duct, before going to the building's hot-water heater. The total length of piping inside the duct is $L = 1.7$ m. The intent of the system is to recover latent heat by the formation of condensation on the outer surface of the tube. This will serve to preheat the cold water and reduce the energy demand at the water heater.

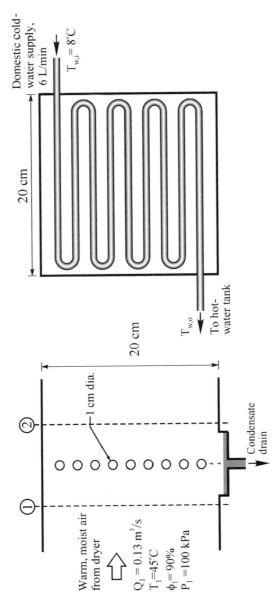

Figure 28.1 Side view and front view of the heat recovery system.

The following calculation illustrates a method to estimate the amount of energy that can be recovered with this design. The heat exchanger is analyzed as a combined heat/mass-transfer problem using the log mean temperature-difference (LMTD) method.

Solution Procedure

The thermal resistance diagram for this problem, shown in Fig. 28.2, has three resistances: R_i is the thermal resistance associated with the convection inside the pipe, R_{cond} is the thermal resistance associated with the condensate film, and R_{mass} is the effective resistance associated with the mass transfer of vapor to the pipe wall. The thermal resistance across the pipe wall is neglected.

The resistance R_{mass} requires further discussion. When noncondensable gases (air in the present case) are present in the flow, the concentration of these "noncondensables" builds up at the pipe surface. This effect reduces the partial pressure of vapor at the vapor-liquid interface. As a result, there can be a substantial temperature difference between the freestream dew-point temperature ($T_{1,dp}$) and the interface temperature (T_i). It should be kept in mind that R_{mass} is not actually a thermal resistance, but it will be treated as such for the purpose of this iterative method.

For a thin-walled pipe ($A_i \approx A_o$), the overall heat-transfer coefficient U can be calculated as

$$U = \frac{1}{A}(R_{mass} + R_{cond} + R_i)^{-1} = \left(AR_{mass} + \frac{1}{h_{cond}} + \frac{1}{h_i}\right)^{-1} \quad (28.1)$$

where h_{cond} and h_i are the average convective heat-transfer coefficients produced by condensation on the outside of the pipe and forced convection on inside of the pipe.

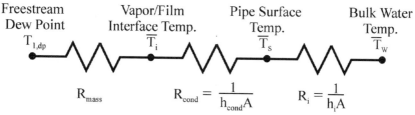

Figure 28.2 Thermal resistance diagram.

28.4 Heat Exchangers

Mass-transfer resistance

The mass transfer of water vapor to the surface of the pipe is driven by the vapor concentration gradient. This gradient can be estimated using a heat/mass-transfer analogy, combined with an empirical correlation for forced convection over a cylinder. Forced convection from a cylinder in crossflow can be calculated from the correlation by Hilpert [1]:

$$\text{Nu} = \frac{hD}{k} = C\text{Re}_D^m \text{Pr}^{1/3} \tag{28.2}$$

The Chilton-Colburn analogy [1] can be used to get an expression for the average Sherwood number. This analogy is applied by replacing the Nusselt number with the Sherwood number and the Prandtl number with the Schmidt number:

$$\text{Sh} = \frac{g_m D}{\rho \delta} = C\text{Re}_D^m \text{Sc}^{1/3} \tag{28.3}$$

where g_m is the mass-transfer conductance, ρ is the mean density, δ is the binary diffusion coefficient (water vapor diffusing into air), and Sc is the Schmidt number (Sc = ν/δ).

To initiate the iterative solution procedure, it is necessary to assume a value for the mean temperature on the outside of the condensate film. In the first iteration, the mean vapor-condensate interface temperature will be arbitrarily taken to be $\overline{T}_i = 15°C$. This is just an initial guess, which will be corrected as the calculation proceeds.

The air and mass-transfer properties are obtained (from Ref. 1) at an approximate film temperature $T_f = (T_1 + \overline{T}_i)/2 = 30°C$: $\delta = 2.667 \times 10^{-5}$ m²/s, $\mu = 1.861 \times 10^{-5}$ N s/m², Sc = 0.61. Since the concentration of vapor is low, the dynamic viscosity is approximated as that of dry air.

The density of the vapor and air can now be calculated at the free-stream conditions and at the interface conditions. For a relative humidity of $\phi_1 = 90$ percent at $T_1 = 45°C$, the partial pressure of water vapor at the inlet is

$$P_{v,1} = \phi_1 P_{\text{sat},1} = 0.9(9593 \text{ Pa}) = 8634 \text{ Pa} \tag{28.4}$$

where the saturation pressure is obtained from steam tables. Using this vapor pressure, the vapor density in the freestream can be calculated from the ideal-gas equation of state:

$$\rho_{v,1} = \frac{P_{v,1} M_v}{RT_1} = \frac{(8634)(18)}{(8314)(318)} = 0.0587 \text{ kg/m}^3 \tag{28.5}$$

The air density in the freestream is

$$\rho_{A,1} = \frac{P_{A,1} M_A}{RT_1} = \frac{(100{,}000 - 8634)(29)}{(8314)(318)} = 1.002 \text{ kg/m}^3 \tag{28.6}$$

At the vapor-condensate interface, the vapor pressure will correspond to the saturation pressure at the assumed interface temperature (for $\bar{T}_i = 15°\text{C}$, $P_{v,i} = 1705$ Pa). So, the vapor density at the interface is

$$\rho_{v,i} = \frac{P_{v,i} M_v}{R\bar{T}_i} = \frac{(1705)(18)}{(8314)(288)} = 0.0128 \text{ kg/m}^3 \tag{28.7}$$

Similarly, the air density at the interface is

$$\rho_{A,i} = \frac{P_{A,i} M_A}{R\bar{T}_i} = \frac{(100{,}000 - 1705)(29)}{(8314)(288)} = 1.191 \text{ kg/m}^3 \tag{28.8}$$

Now, the mass fraction of vapor can be calculated in the freestream

$$m_1 = \frac{\rho_{v,1}}{\rho_{v,1} + \rho_{A,1}} = \frac{0.0587}{0.0587 + 1.002} = 0.0553 \tag{28.9}$$

and at the interface

$$m_i = \frac{\rho_{v,i}}{\rho_{v,i} + \rho_{A,i}} = \frac{0.0128}{0.0128 + 1.191} = 0.0106 \tag{28.10}$$

In the present problem the mean freestream velocity is $Q/A_c = (0.13 \text{ m}^3/\text{s})/(0.2)^2 \text{ m}^2 = 3.25$ m/s. So, the Reynolds number for the external flow is

$$\text{Re}_D = \frac{\rho \bar{V} D}{\mu} = \frac{(1.13)(3.25)(0.01)}{1.861 \times 10^{-5}} = 1973 \tag{28.11}$$

where the density in the Reynolds number has been evaluated as the mean of the total density ($\rho_v + \rho_A$) in the freestream and at the interface. At this Reynolds number the constant and exponent for Hilpert's correlation are $C = 0.683$ and $m = 0.466$. Evaluating the average Sherwood number, we obtain

$$\text{Sh} = 0.683 \, (1973)^{0.466} (0.61)^{1/3} = 19.88 \tag{28.12}$$

Using this result, the convective mass-transfer conductance g_m is

$$g_m = \frac{\rho \, \text{Sh} \, \delta}{D} = \frac{(1.13)(19.88)(2.667 \times 10^{-5})}{0.01} = 0.0599 \text{ kg/m}^2 \text{ s} \tag{28.13}$$

The convective mass transfer is driven by the difference in the mass fraction of vapor in the freestream and at the vapor-condensate interface. The mass-transfer rate, which is equal to the condensation rate, is given by

$$\dot{m}_v = g_m A(m_{v,1} - m_{v,i}) \tag{28.14}$$

The total surface area of the pipe is $A = \pi DL = \pi(0.01)(1.7) = 0.0534 \text{ m}^2$. Applying Eq. (28.14) and the results from Eqs. (28.9), (28.10), and (28.13) gives the condensation rate:

$$\dot{m}_v = (0.0599)(0.0534)(0.0553 - 0.0106) = 1.43 \times 10^{-4} \text{ kg/s} \tag{28.15}$$

The latent heat-transfer rate required to condense vapor at this rate is

$$q = \dot{m}_v h_{fg} = (1.43 \times 10^{-4})(2{,}466{,}000) = 353 \text{ W} \tag{28.16}$$

where h_{fg} is the latent heat of vaporization, evaluated at the interface temperature.

The dew-point temperature corresponding to $P_{v,1} = 8.364$ kPa is $T_{1,\text{dp}} = 43.0°\text{C}$. For the assumed vapor-condensate interface temperature, the effective thermal resistance due to mass transfer can be calculated as

$$R_{\text{mass}} = \frac{T_{1,\text{dp}} - \overline{T}_i}{q} = \frac{43.0 - 15.0}{353} = 0.0793°\text{C/W} \tag{28.17}$$

Condensation heat transfer

The heat-transfer coefficient on the outer surface of the condenser coil (h_{cond}) can be estimated using relations for laminar film condensation on vertical tube bundles. For the first iteration it is necessary to make an initial guess for the temperature drop across the vapor film. Initially, it will be assumed that $\overline{T}_i - \overline{T}_S = 2°\text{C}$, where $\overline{T}_S$ is the mean surface temperature of the pipe. Again, an improved estimate will be made for this quantity, later in the calculation.

The outer heat-transfer coefficient can be estimated from a theoretical relation for laminar film condensation on a bank of horizontal cylinders [2]. For n cylinders of diameter D, which are stacked vertically, the average Nusselt number is

$$\text{Nu}_{\text{cond}} = \frac{h_{\text{cond}}(nD)}{k_\ell} = 0.729 \left[\frac{g h_{fg} \rho_\ell (\rho_\ell - \rho_v)(nD)^3}{\mu_\ell k_\ell (\overline{T}_i - \overline{T}_S)} \right]^{1/4} \tag{28.18}$$

In this equation, the properties of the liquid film are evaluated at mean temperature of the condensate film $(\overline{T}_S + \overline{T}_i)/2$. For the first iteration we will evaluate the liquid water properties at 14°C: $k_\ell = 0.592$ W/m K,

$\rho_\ell = 1000$ kg/m³, and $\mu_\ell = 1.18 \times 10^{-3}$ N s/m². The latent heat is evaluated at the interface temperature, $h_{fg} = 2466$ kJ/kg K.

The density of the air-vapor mixture is about 1000 times lower than the density of the liquid film. So, in Eq. (28.18), the term $\rho_\ell(\rho_\ell - \rho_v)$ will be approximated as ρ_ℓ^2. In the current design, there are nine stacked pipes ($n = 9$). Evaluating Eq. (28.18), we obtain

$$\mathrm{Nu}_{\mathrm{cond}} = 0.729 \left[\frac{(9.81)(2466 \times 10^3)(1000)^2(9 \times 0.01)^3}{(1.18 \times 10^{-3})(0.592)(2.0)} \right]^{1/4} = 1374 \tag{28.19}$$

The average condensation heat-transfer coefficient is then

$$h_{\mathrm{cond}} = \frac{\mathrm{Nu}_{\mathrm{cond}} k_\ell}{nD} = \frac{(1374)(0.592)}{9(0.01)} = 9037 \text{ W/m}^2 \text{ K} \tag{28.20}$$

Internal pipe flow (forced convection)

The properties of the water inside the pipe are evaluated at the mean bulk temperature. For the first iteration, the properties are evaluated at $T_{w,i} = 8°C$: $k_w = 0.584$ W/m K, $c_{p,w} = 4196$ J/kg K, $\rho_w = 1000$ kg/m³, $\mu_w = 1.377 \times 10^{-3}$ N s/m², and $\mathrm{Pr}_w = 9.93$. The mass flow rate in the pipe is 6 L/min or $\dot{m}_w = 0.1$ kg/s. Now the internal Reynolds number can be calculated:

$$\mathrm{Re}_D = \frac{\rho_w \bar{V} D}{\mu_w} = \frac{4\dot{m}_w}{\pi D \mu} = \frac{4(0.1)}{\pi(0.01)(1.377 \times 10^{-3})} = 9246 \text{ (turbulent)} \tag{28.21}$$

There are many correlations in the literature for turbulent fully developed forced convection in circular pipes. In a review of empirical correlations, the most recent edition of the *Handbook of Heat Transfer* [3] recommends the equation by Gnielinski [4]:

$$\mathrm{Nu}_i = 0.012(\mathrm{Re}_D^{0.87} - 280)\mathrm{Pr}^{0.4} \quad 1.5 \leq \mathrm{Pr} \leq 500 \quad 3 \times 10^3 \leq \mathrm{Re}_D \leq 10^6 \tag{28.22}$$

Evaluating Eq. (28.22), we obtain

$$\mathrm{Nu}_i = 0.012[(9246)^{0.87} - 280](9.93)^{0.4} = 76.4 \tag{28.23}$$

Evaluating the average heat-transfer coefficient, we obtain

$$h_i = \frac{\mathrm{Nu}_i k_w}{D} = \frac{76.4(0.584)}{0.01} = 4462 \text{ W/m}^2 \text{ K} \tag{28.24}$$

28.8 Heat Exchangers

Overall heat-transfer rate

The overall heat-transfer coefficient is now calculated using Eq. (28.1):

$$U = [(0.0534)(0.0793) + 1/9037 + 1/4462]^{-1} = 219 \text{ W/m}^2\text{ K} \qquad (28.25)$$

Using an LMTD analysis, the total heat-transfer rate q the heat exchanger can be expressed as

$$q = FUA\Delta T_{\text{lm}} = FUA \left[\frac{(T_{1,\text{dp}} - T_{w,i}) - (T_{1,\text{dp}} - T_{w,o})}{\ln\left(\dfrac{T_{1,\text{dp}} - T_{w,i}}{T_{1,\text{dp}} - T_{w,o}}\right)} \right]$$

$$= \dot{m}_w c_{p,w}(T_{w,o} - T_{w,o}) \qquad (28.26)$$

where $T_{w,i}$ and $T_{w,o}$ are the inlet and outlet temperatures of the domestic cold water. Note that this situation can be treated as a pure counterflow, since the temperature on the outside of the condensate film is essentially constant. So, the correction factor is unity ($F \approx 1$).

Solving Eq. (28.26) for the domestic water outlet temperature $T_{w,o}$, we have

$$T_{w,o} = T_{1,\text{dp}} - (T_{1,\text{dp}} - T_{w,i}) \exp\left(-\frac{UA}{\dot{m}_w c_{p,w}}\right) \qquad (28.27)$$

Evaluating Eq. (28.27) yields

$$T_{w,o} = 43.0 - (43.0 - 8.0)\exp\left(-\frac{(219)(0.0534)}{(0.1)(4196)}\right) = 8.96°\text{C} \qquad (28.28)$$

Using this result, we can now get an improved estimate of the total heat-transfer rate to the condenser coils:

$$q = \dot{m}_w c_{p,w}(T_{w,o} - T_{w,i}) = (0.1)4196(8.96 - 8.0) = 403 \text{ W} \qquad (28.29)$$

Now a better estimate of the interface temperature $\overline{T}_i$ can be made. Referring to the thermal resistance diagram in Fig. 28.2, the improved interface temperature is

$$\overline{T}_i = T_{\text{dp},1} - q\,R_{\text{mass}} = 43.0 - (403)(0.0793) = 11.0°\text{C} \qquad (28.30)$$

This is significantly different from the initial guess of $\overline{T}_i = 15°\text{C}$. So, at least one more iteration will be required.

To proceed with the next iteration, an improved estimate of the temperature drop across the condensate film is also needed. This quantity can be obtained using the improved heat-transfer rate estimate and the

TABLE 28.1 Various Quantities after First Three Iterations and for Converged Solution

Quantity	Iterations			Converged solution
	First	Second	Third	
Heat-transfer rate q, W	404	380	378	378
Vapor-condensate interface temperature $\overline{T}_i$, °C*	11.0	10.7	10.6	10.6
Temperature difference across condensate $\overline{T}_i - \overline{T}_S$, °C†	0.84	0.65	0.61	0.59
Condensation heat-transfer coefficient h_{cond}, W/m² K	9037	10,978	11,675	11,932
Internal convection coefficient h_i, W/m² K	4462	4195	4195	4195
Condensation rate $\dot{m}_v$, kg/s	1.43×10^{-4}	1.52×10^{-4}	1.53×10^{-4}	1.53×10^{-4}

*Initial assumed value $\overline{T}_i = 15°\text{C}$.
†Initial assumed value $\overline{T}_i - \overline{T}_S = 2°\text{C}$.

condensation heat-transfer coefficient (h_{cond}), as follows:

$$\overline{T}_i - \overline{T}_s = \frac{q}{h_{\text{cond}} A} = \frac{403}{(9037)(0.0534)} = 0.835°\text{C} \quad (28.31)$$

Improved values have now been calculated for all the parameters that were "guessed" in the first iteration. So, the procedure can be repeated until convergence. Of course, an alternate method for solving this set of equations is to use commercial equation solving software which has built in functions for vapor properties (e.g., Engineering Equation Solver [5]).

Table 28.1 shows some of the key parameters after the first three iterations and the final converged solution. The converged value for the total heat-transfer rate is $q = 378$ W. It can be seen from the table that the solution converges quickly, in about three iterations.

Conditions after the coil

The flow rate and inlet conditions for this problem are based on an industrial electrical clothes dryer with a rated power of 20 kW. So, for this application, 378 W is a relatively low quantity of heat to recover. To be practical, more coils would likely have to be added. In this case, the analysis would follow along the same lines, with the outlet conditions of the first coil becoming the inlet conditions for the second coil, and so on. The conditions at state 2 (after the coil) can be calculated using the condensation rate $\dot{m}_v$ on the first coil, as follows:

$$w_2 = w_1 - \frac{\dot{m}_v}{\dot{m}_A} \quad (28.32)$$

where w_1 and w_2 are the values of specific humidity before and after the coil. *Specific humidity w is the mass fraction of vapor per kilogram of dry air.* The specific humidity before the coil can be calculated, assuming ideal-gas behavior as

$$w_1 = \frac{P_{v,1} V M_v / R T_1}{P_{A,1} V M_A / R T_1} = \frac{M_v P_{v,1}}{M_A P_{A,1}} = 0.622 \frac{P_{v,1}}{P_{A,1}} \tag{28.33}$$

Evaluating Eq. (28.33), the specific humidity at the inlet is

$$w_1 = 0.622 \frac{8.634}{100 - 8.634} = 0.05878 \tag{28.34}$$

The mass flow rate of dry air is

$$\dot{m}_A = \rho_A Q = \frac{P_{A,1} M_A}{RT} Q = \frac{(100 - 8.634)29}{(8.314)(318)} 0.13 = 0.13 \text{ kg/s} \tag{28.35}$$

Evaluating Eq. (28.32), using the condensation rate for the converged solution, we obtain

$$w_2 = w_1 - \frac{\dot{m}_v}{\dot{m}_A} = 0.05878 - \frac{1.53 \times 10^{-4}}{0.13} = 0.0576 \tag{28.36}$$

Equation (28.33) can be written at state 2 (after the coil) as follows:

$$w_2 = 0.622 \frac{P_{v,2}}{P_{\text{atm}} - P_{v,2}} \tag{28.37}$$

Solving Eq. (28.37) for $P_{v,2}$ gives

$$P_{v,2} = \frac{P_{\text{atm}} w_2}{0.622 + w_2} = \frac{(100)(0.0576)}{0.622 + 0.0576} = 8.47 \text{ kPa} \tag{28.38}$$

Since the heat transfer to the pipe is assumed to be all sensible heat, the temperature of the flow can be assumed to remain at 45°C. So, relative humidity after the first coil is

$$\phi_2 = \frac{P_{v,2}}{P_{\text{sat},2}} = \frac{8.47}{9.593} = 88.3\% \tag{28.39}$$

This outlet condition could be used as the inlet condition for the next condenser coil, if one existed.

Concluding Remarks

It should be noted that the current analysis probably underestimates the total heat-transfer rate. The main reason is the conservative assumption that the heat transfer to the coil occurs by condensation only.

In reality, there will also be a component of sensible-heat transfer, due to convection and radiation. These effects could be added quite easily to the current method.

In addition, the effects of vapor shear have been neglected. The flow in the duct can be expected to thin the condensate film on the upstream side of the tube, thereby enhancing the heat-transfer rate. However, in the present case, mass-transfer effects are the main "resistance" to condensation. So, improvements to the condensation model would not be expected to greatly improve the estimated heat-transfer rate.

As a final comment, one should take note of the need to include the effects of noncondensable vapors in this type of calculation. As shown in Table 28.1, the high concentration of air suppresses the temperature of the vapor-condensate interface far below the freestream dewpoint temperature. In fact, if this analysis is repeated with the interface temperature at the saturation temperature corresponding to the freestream vapor pressure (i.e., a standard Nusselt-type analysis), the heat-transfer rate is overpredicted by a factor of more than 10. This general result has been shown before. For example, the reader is referred to the finite-volume numerical study of the forced convection on a flat plate with condensation by Chin et al. [6].

References

1. Incropera, F. P., and D. P. DeWitt, *Fundamentals of Heat and Mass Transfer*, 4th ed., Wiley, New York, 1996.
2. Oosthuizen, P. H., and D. Naylor, *Introduction to Convective Heat Transfer Analysis*, WCB/McGraw-Hill, New York, 1999.
3. Ebadian, M. A., and Z. F. Dong, "Forced Convection, Internal Flow in Ducts," in *Handbook of Heat Transfer*, 3d ed., McGraw-Hill, New York, 1999, chap. 5.
4. Gnielinksi, V., "New Equations for Heat and Mass Transfer in Turbulent Pipe and Channel Flow," *Int. Chem. Eng.*, **16**, 359–368, 1976.
5. Engineering Equation Solver, F-Chart Software, Madison, Wis., 2004.
6. Chin, Y. S, Ormiston, S. J., and Soliman, H. M., "Numerical Solution of the Complete Two-phase Model for Laminar Film Condensation with a Noncondensable Gas," *Proceedings of 10th International Heat Transfer Conference*, Brighton, U.K., 1994, vol. 3, pp. 287–292.

Chapter 29

Sizing of a Crossflow Compact Heat Exchanger

Dusan P. Sekulic
Department of Mechanical Engineering
University of Kentucky
Lexington, Kentucky

Summary

This text offers a rigorous, step-by-step methodology for calculating core dimensions of a compact heat exchanger. Considering the analytical complexity of implemented calculations, the most intricate basic flow arrangement situation in a single-pass configuration would be a crossflow in which fluids do not mix orthogonal to the respective flow directions. Such a flow arrangement is selected to be considered in this calculation. The set of known input data is provided in the problem formulation. Subsequently, calculations are executed using an explicit step-by-step routine. The procedure follows a somewhat modified thermal design (sizing) procedure derived from the routine advocated in Shah and Sekulic (2003). The modification is related to the fact that iterative steps are listed and executed for both intermediate and overall refinements of all assumed and/or estimated entities. The overall pressure drop constraints are ultimately satisfied without further optimization of the design. The main purpose of the calculation sequence is to illustrate the procedure, usually hidden behind a user-friendly, but content-non-revealing, platform of any existing commercial software package. Such a black-box approach executed by a computer, although very convenient for handling by a not necessarily sophisticated user, is utterly nontransparent for a deeper insight. Consequently, this example

29.2 Heat Exchangers

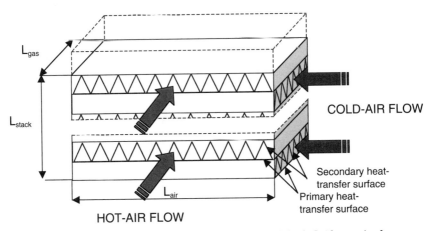

Figure 29.1 Schematic of a cross-flow heat exchanger with both fluids unmixed.

calculation is not intended to discuss a particular design; rather, it illustrates the procedure. Therefore, the numerical values calculated do not have importance for any particular design.

Problem Formulation

A task at hand is to design ("to size") a heat exchanger, specifically, to determine principal heat-exchanger core dimensions (width, length, and height of the core having specified heat-transfer surfaces) (see Fig. 29.1). The heat exchanger has to cool a hot-air gas stream, available at an elevated temperature, with a cold-air stream, available at a significantly lower temperature. Terminal states of both fluid streams are known, except for the outlet of the hot stream as follows:

Fluid → Property ↓	Cold fluid			Hot fluid		
	Symbol	Unit	Value	Symbol	Unit	Value
Inlet temperature	$T_{c,i}$	K	500	$T_{h,i}$	K	700
Outlet temperature	$T_{c,o}$	K	620	$T_{h,o}$	K	—
Inlet pressure	$p_{c,i}$	kPa	500	$p_{h,i}$	kPa	100
Mass flow rate	$\dot{m}_c$	kg/s	20	$\dot{m}_h$	kg/s	20
Pressure drop	Δp_c	kPa	5	Δp_c	kPa	4.2
Fluid type	Air	—	—	Air	—	—

Given data

The following information is provided: cold fluid—air; hot fluid—air, $T_{c,i}$, $T_{h,i}$, $T_{c,o}$, $p_{c,i}$, $p_{h,i}$, $\dot{m}_c$, $\dot{m}_h$, Δp_c, Δp_h; heat exchanger flow arrangement type—single-pass, crossflow unmixed-unmixed flow arrangement.

Find

The principal dimensions of the core must be determined: (1) fluid flow lengths (core dimensions) in directions of hot and cold fluid flows and (2) the dimension of the stack of alternate layers of flow passages in the direction orthogonal to crossflow planes.

Assumptions

Determination of the core dimensions assumes an a priori decision regarding selection of heat-transfer surface types on both sides of a heat exchanger. This selection is, as a rule, within the realm of an engineer's decisions for any sizing problem; thus, it is not necessarily given in the problem formulation. In the present calculation, a decision regarding the surface selection will be made at a point when geometric and heat-transfer and/or hydraulic characteristics of the core need to be assessed for the first calculation iteration. That decision may always be modified and calculation repeated. The types of heat-transfer surfaces will be selected, and data involving geometric, heat-transfer, and hydraulic properties will be obtained from a database given in Kays and London (1998). The assumptions on which the calculation procedure is based are listed and discussed in detail in Shah and Sekulic (2003, chap. 3, p. 100) and will not be repeated here (standard assumptions for designing a compact heat exchanger).

Calculation Procedure

Design procedure for a sizing problem features two distinct segments of calculation. The first one delivers the magnitude of the thermal size of the core, expressed as an overall heat-transfer area A, and/or formulated as a product of the overall heat-transfer coefficient and the heat-transfer area UA. Determination of this quantity should be based on an application of thermal energy balance [i.e., the heat-transfer rate delivered by one fluid is received by the other; no losses (gains) to (from) the surroundings are present]. Formulation of this balance involves a fundamental analysis of heat-transfer phenomena within the heat exchanger core, which can be summarized through a concept of heat exchanger effectiveness, Shah and Sekulic (2003). The resulting design procedure is the "effectiveness number of heat-transfer units" method. The effectiveness is expressed in terms of known (or to be determined) inlet/outlet temperatures, and mass flow rates (for known fluids). The unknown temperatures (for some problem formulations) must be determined (i.e., they may not be known a priori —such as the outlet temperature of the hot fluid in this problem), and any assumed thermo-physical properties should be re-calculated multiple times (i.e., an iterative procedure is inherent). This feature of the calculation is only one aspect of the design methodology that ultimately leads to an iterative calculation sequence. The second reason for an iterative nature of this procedure is, as a rule, inherently transcendent structure of the effectiveness-number-of-heat-transfer correlation (for the crossflow unmixed-unmixed arrangement, as in the case that will be revealed in step TDG-8 in the tabular list below). Finally, the third and main reason for an iterative procedure is a constraint imposed on pressure drops. The magnitudes of pressure drops must be obtained from the hydraulic part of the design procedure. The hydraulic design part of the procedure cannot be decoupled from the thermal part, which leads to the calculation of pressure drops *after* thermal calculations are completed, and hence is followed by a comparison of calculated pressure drops with the imposed limits. As a rule, these limits are not necessarily satisfied after the first iteration, and subsequent iterations are needed.

In this routine calculation presentation, determination of the thermal size of the heat exchanger will be termed the "targeting the design goal" (TDG) procedure. Each step will be separately marked for the purpose of cross-referencing. The second segment of the calculation is devoted to the determination of actual overall dimensions of the core, in a manner to satisfy the required overall heat-transfer area and to achieve the overall heat-transfer coefficient, that is, to satisfy the required thermal size. This segment is inherently iterative because it requires a satisfaction of pressure drop constraints as emphasized above. This segment of calculation will be termed "matching (the design goal and) geometric characteristics" (MGC) procedure.

Both procedures will be organized as a continuous sequence of calculations and presented in a tabular format for the sake of compactness and easy access to various steps. The most important comments will be given as the notes to the respective calculation steps immediately after the equation(s) defining the step. A detailed discussion of numerous aspects of these calculations, and the issues involving relaxation of the assumptions, are provided in Shah and Sekulic (2003). The reader is advised to consult that source while following the step-by-step calculation procedure presented here. Some numerical values of the derivative variables presented may differ from the calculated values because of rounding for use elsewhere within the routinely determined data.

Targeting the design goal (TDG) procedure

Step	Calculation	Value	Units
TDG-1a	$T_{c,\text{ref}} = \dfrac{T_{c,i} + T_{c,o}}{2} = \dfrac{500 + 620}{2}$	560	K
TDG-1b	$T_{h,\text{ref}} = T_{h,i} = 700$	700	K
	To initiate the iterative procedure for a sizing problem like the one given in this problem formulation, a determination of referent temperatures of both fluids is needed. As a first guess, either an arithmetic mean of temperature terminal values or a given temperature value (if single) for each fluid may be selected.		
TDG-2a	$c_{p,c} = c_{p,\text{air}}(T_{c,\text{ref}}) = c_{p,\text{air}}(560)$	1.041	kJ/kg K
TDG-2b	$c_{p,h} = c_{p,\text{air}}(T_{h,\text{ref}}) = c_{p,\text{air}}(700)$	1.075	kJ/kg K
	The specific heat of either of the two fluids (for single-phase flows) is determined at the calculated referent temperatures. Since both fluids are gases in this case, and since both are considered as air, an ideal-gas thermodynamic properties data source) properties will be assumed.		
TDG-3a	$C_c = (\dot{m}c_p)_c = 20 \times 1.041$	20.82	kW/K
TDG-3b	$C_h = (\dot{m}c_p)_h = 20 \times 1.075$	21.50	kW/K
	Heat capacity rates of the fluid streams represent the products of respective mass flow rates and corresponding specific heats, calculated at the estimated referent temperatures.		
TDG-4	Designation of fluid streams $C_h = C_2 > C_c = C_1$	—	—
	At this point, it is convenient to determine which of the two fluid streams has a larger heat capacity (for a nonbalanced case). Note that the designator 1 denotes the "weaker" fluid and the designator 2, the "stronger" fluid.		
TDG-5	$C^* = \dfrac{C_1}{C_2} = \dfrac{20.82}{21.50}$	0.9685	—

Step	Calculation	Value	Units
	The heat capacity rate ratio is not equal to 1; therefore, the heat exchanger operates with nonbalanced fluid streams.		
TDG-6	$$\varepsilon = \frac{T_{1,o} - T_{1,i}}{T_{2,i} - T_{1,i}} = \frac{620 - 500}{700 - 500}$$	0.600	—
	Heat-exchanger effectiveness represents the dimensionless temperature of the weaker fluid (the one with $C_1 = C_c$) (Sekulic, 1990, 2000). Note that the current decision on which fluid is weaker was based on the rough estimate, namely, a first iteration of referent temperatures. These are not necessarily the best assumptions, in particular for the hot fluid in this case. So, the outlet temperature of the hot fluid must be determined with more precision (see TDG-7), and steps TDG-1 through TDG-7 should be repeated.		
TDG-7	$T_{2,o} = T_{h,o}$ $$T_{h,o} = T_{c,i} + (1 - C^*\varepsilon)(T_{h,i} - T_{c,i})$$ $$= 500 + (1 - 0.9685 \times 0.6)(700 - 500)$$	583.8	K
	Note that the relationship used for determining the outlet temperature of the hot fluid is a straightforward consequence of adopted definitions of the heat-exchanger effectiveness and heat capacity rate ratio, both expressed as functions of terminal temperatures. Now, having this originally unknown temperature estimated, a new value of the referent temperature for the hot fluid can be determined; that is, steps TDG-1 through TDG-7 should be repeated.		
TDG-1a/2	$T_{1,\text{ref}} = T_{c,\text{ref}}$ $$T_{c,\text{ref}} = \frac{T_{c,\text{in}} + T_{c,\text{out}}}{2} = \frac{500 + 620}{2}$$	560	K
	The second iteration for $T_{c,\text{ref}}$ is identical to the first because both terminal temperatures were already provided in the problem formulation and an arithmetic average of the two was used (i.e., not, say, an integral mean value for which the distribution throughout the core would be needed).		
TDG-1b/2	$$T_{h,\text{ref}} = \frac{T_{h,\text{in}} + T_{h,\text{out}}}{2} = \frac{700 + 583.8}{2}$$	641.9	K

Note that the second iteration of $T_{h,\text{ref}}$ differs significantly from the first iteration: 641.9 K instead of 700 K.

TDG-2a/2	$c_{p,c} = c_{p,\text{air}}(T_{c,\text{ref}}) = c_{p,\text{air}}(560)$	1.041	kJ/kg K
	There is no change in this iteration for the cold-fluid specific heat.		
TDG-2b/2	$c_{p,h} = c_{p,\text{air}}(T_{h,\text{ref}}) = c_{p,\text{air}}(641.9)$	1.061	kJ/kg K
	The result obtained represents a new value for the specific heat of the hot fluid.		
TDG-3a/2	$C_c = (\dot{m}c_p)_c = 20 \times 1.041$	20.82	kW/K
	There is no change in this iteration for the cold fluid.		
TDG-3b/2	$C_h = (\dot{m}c_p)_h = 20 \times 1.061$	21.22	kW/K
	This is a new value for the heat capacity rate of the hot fluid.		
TDG-4/2	Designation of fluid streams $C_h = C_2 > C_c = C_1$	—	—
	Note that the implemented refinement of the value of the heat capacity rate on the hot-fluid side did not change the designations of the fluids.		
TDG-5/2	$C^* = \dfrac{C_1}{C_2} = \dfrac{20.82}{21.22}$	0.9814	—
	This is a new, refined value of the heat capacity rate. The new value is slightly larger than the value in the first iteration.		
TDG-6/2	$\varepsilon = \dfrac{T_{1,o} - T_{1,i}}{T_{2,i} - T_{1,i}} = \dfrac{620 - 500}{700 - 500}$	0.600	—
	This represents a refined value of the heat-exchanger effectiveness; there is no change in this value because all the values involved were already known (i.e., all were given in the problem formulation), and are not affected by intermediate calculations.		
TDG-7/2	$T_{h,o} = T_{c,i} + (1 - C^*\varepsilon)(T_{h,i} - T_{c,i})$	582.2	K
	$= 500 + (1 - 0.9814 \times 0.6)(700 - 500)$		

Step	Calculation	Value	Units
	We obtained a new value of originally unknown outlet temperature of the hot fluid. The criterion for a termination of the iterative procedure may involve either a sufficiently small change of two successive values of this temperature, or a change of the successive values for heat-exchanger effectiveness. In this case, these comparisons indicate that either no change or a very small change takes place, and the iterative procedure is terminated at this point.		
TDG-8	$NTU = NTU(\varepsilon = 0.600; C^* = 0.9814;$ unmixed-unmixed)	1.811	—
	For the crossflow unmixed-unmixed arrangement the relationship between effectiveness and the number of units (NTU) (explicit in terms of effectiveness but not explicit in terms of NTU) is as follows (Baclic and Heggs, 1985):		
	$$\varepsilon = 1 - \frac{e^{-NTU(1+C^*)} \sum_{n=1}^{\infty} \binom{n}{1}(C^*)^{n/2} I_n(2NTU\sqrt{C^*})}{C^*NTU}$$		
	Therefore, the exact expression for the heat-exchanger effectiveness of an unmixed-unmixed crossflow arrangement is algebraically very complex, regardless of the fact that it may be represented in different forms (Baclic and Heggs, 1985). See Shah and Sekulic (2003) for an approximate equation and Sekulic et al. (1999) for methods of determining the effectiveness relationships for other (not necessarily crossflow) flow arrangements. Graphical representation, as well as tabular data for the crossflow unmixed-unmixed flow arrangement can be found in Baclic and Heggs (1985). It is important to realize that regardless of which correlation is used, the determination of NTU versus effectiveness is algebraically an intractable task without some kind of iterative procedure. In the correlation presented, $I_n, n = 1, 2, \ldots, \infty$, represent modified Bessel functions of the integer n order.		
TDG-9	$UA = NTU \cdot C_1 = 1.811 \times 20.82$	37.71	kW/K
	The product UA, also termed the "thermal size," is a compounded thermal and physical size of the unit. This size involves the physical size (area of the heat-transfer surface A) and heat-transfer size (U is the overall heat-transfer coefficient). The subsequent procedure (i.e., MGC) untangles intricate relations between these entities and physical size of the heat exchanger, leading to explicit values of all core dimensions for selected types of both (hot- and cold-side) heat-transfer surfaces.		

Matching geometric characteristics (MGC) procedure

Step	Calculation	Value		Units
		Fluids 1 and 2		
MGC-1	Determination of fluids' thermophysical properties			
	■ Referent temperature $T_{c,ref} = T_{1,ref}$, $T_{h,ref} = T_{2,ref}$	560	642	K
	■ Inlet pressures $p_{1,i}$, $p_{2,i}$	500	100	kPa
	■ Outlet pressures $p_{1,o}$, $p_{2,o}$	>495	>96	kPa
	■ Specific heats $c_{p,c} = c_{p,1}$, $c_{p,h} = c_{p,2}$	1.041	1.061	kJ/kg K
	■ Viscosities μ_1, μ_2	28.95	31.67	10^6 Pa s
	■ Thermal conductivities k_1, k_2	4.32	4.80	10^2 W/m K
	■ Prandtl numbers Pr_1, Pr_2	0.698	0.699	—
	■ Densities (inlet) $\rho_{1,i}$, $\rho_{2,i}$	3.484	0.498	kg/m^3
	■ Densities (outlet) $\rho_{1,o}$, $\rho_{2,o}$	2.782	0.574	kg/m^3
	■ Densities (bulk mean) $\rho_{1,m}$, $\rho_{2,m}$	3.093	0.533	kg/m^3
	The fluid properties are usually determined at arithmetic (or integral) mean values of fluid temperatures. In this calculation, we will adopt the arithmetic mean values from the second iteration. Certain data (temperatures, pressures) are provided in the problem formulation (see above), and/or devised from the inlet data and known pressure drops. Specific heats and viscosities are based on the mean temperatures. Densities are calculated assuming the ideal-gas assumption. The mean density is based on the following relationship: $$\rho_{i,m} = \left[\frac{1}{2} \left(\frac{1}{\rho_i} + \frac{1}{\rho_o} \right) \right]^{-1}$$			
MGC-2	$NTU_1 = NTU_c = 2NTU = 2 \times 1.811$	3.622		—
	$NTU_2 = NTU_h = C^* NTU_1 = 0.9814 \times 3.622$	3.555		—
	Distribution of the total dimensionless thermal size between two fluid sides is determined in this step. Since both fluids are gases, both thermal resistances are to be assumed as equal in the first iteration. That leads to the given distribution of NTU_1 and NTU_2 versus NTU (Shah and Sekulic, 2003).			

Step	Calculation	Value	Units
MGC-3	Selection of heat surface types: plane plate fin surface 19.86 (Kays and London, 1998)		
	■ Plate spacing b	6.350×10^{-3}	m
	■ Number of fins n_f	782	m^{-1}
	■ Hydraulic diameter D_h	1.875×10^{-3}	m
	■ Fin thickness δ	0.152×10^{-3}	m
	■ Uninterrupted flow length l_f	63.75×10^{-3}	m
	■ Heat-transfer area per volume between passes β	1841	m^{-1}
	■ Fin area per total area A_f/A	0.849	—
	■ Plate thickness a (under designer's control)	2×10^{-3}	m
	The sizes and shapes of heat-transfer surfaces are correlated with the heat-transfer and hydraulic characteristics. However, these characteristics in turn are needed to determine the sizes and shapes of the heat-transfer surfaces! This interrelation renders the calculation procedure iterative. A selection of the surface geometry (i.e., selection of both fluid flow area geometries) should be done first a priori. Subsequently, calculation of heat-transfer and fluid flow characteristics may be conducted to establish whether the surfaces selection fits the thermal size distribution and the overall thermal size (but in a manner to satisfy the pressure drop constraints). We will select the surface designated as 19.86 (Kays and London, 1998) for both fluid sides. See also Webb (1994) for further discussion of issues involving enhanced heat-transfer surfaces.		
MGC-4	Surface characteristics involve the ratio of Colburn and Fanning factors j/f for plane plate fin surface 19.86 (Kays and London, 1998). For this surface, the ratio of j and f over the wide range of Reynolds numbers is approximately constant: $$\left(\frac{j}{f}\right)_{c,h} \approx \text{const} = 0.29$$	0.29	—

	Although selection of surface types leads to the known heat-transfer geometries on both fluid sides, the calculation of $j(\text{Re})$ and $f(\text{Re})$ parameters (i.e., heat-transfer and friction factors in dimensionless form) cannot be performed straightforwardly at this point. This is because the Reynolds numbers for fluid flows are still unknown. This hurdle can be resolved by first guessing at the magnitude of the ratio of j to f (i.e., j/f). Since this ratio is nearly constant in a wide range of Reynolds numbers, a unique value can be suggested as indicated above. In the first iteration (which will follow) only the value of j/f (rather than separate j and f values) would be needed for calculation of both fluid core mass velocities. Subsequently, these core mass velocities (see step MGC-6) will be used to determine the first iteration of Reynolds numbers, leading to the corresponding values of j and f. Subsequently, the second iteration for j/f can be calculated from known j and f values. In this case, $j/f \in (0.25, 0.37)$ for $\text{Re} \in (500, 4000)$, so an educated guess would lead to $j/f \approx 0.29$.	
MGC-5	An initial guess of the first estimate for the total surface temperature effectiveness for both sides should be made at this point (i.e., $\eta_{0,c} = \eta_{0,h}$). It is generally assumed that the total surface temperature effectiveness for a compact heat-transfer surface (for a good design) must be within the range of 0.7 to 0.9. Let us assume the value to be at the high end of this range for both fluid sides (note that the same geometry was suggested for both surfaces).	0.9 —
MGC-6	$$G_c \approx \left[\left(\frac{j}{f}\right)\left(\frac{\Delta p}{p_{\text{in}}}\right)\frac{2p_{\text{in}}\eta_0 p_m}{\text{NTU Pr}^{2/3}}\right]_c^{1/2} = \left[0.29\,\frac{5\times 10^3}{500\times 10^3}\,\frac{2\times 500\times 10^3 \times 0.9 \times 3.093}{3.622 \times 0.698^{2/3}}\right]^{1/2}$$ $$G_h \approx \left[\left(\frac{j}{f}\right)\left(\frac{\Delta p}{p_{\text{in}}}\right)\frac{2p_{\text{in}}\eta_0 p_m}{\text{NTU Pr}^{2/3}}\right]_h^{1/2} = \left[0.29\,\frac{4\times 10^3}{100\times 10^3}\,\frac{2\times 100\times 10^3 \times 0.9 \times 0.533}{3.555 \times 0.699^{2/3}}\right]^{1/2}$$ In this step, the first estimates of core mass velocities are based on the estimates of j/f and η parameters as discussed in MGC-4 and MGC-5. This estimate, as given above, is based on a simplified expression for G that takes into account the assumptions as follows: (1) only friction contributes to the pressure drop, (2) fouling resistances are neglected, (3) thermal resistance of the heat-transfer wall is neglected, and (4) thermal resistances caused by formations of convective boundary layers on both fluid sides are equal.	53.22 kg/m² s 19.94 kg/m² s

Step	Calculation	Value	Units
MGC-7	$\text{Re}_c = \dfrac{G_c D_{h,c}}{\mu_c} = \dfrac{53.22 \times 1.875 \times 10^{-3}}{28.95 \times 10^{-6}}$	3446	—
	$\text{Re}_h = \dfrac{G_h D_{h,h}}{\mu_h} = \dfrac{19.94 \times 1.875 \times 10^{-3}}{31.67 \times 10^{-6}}$	1181	—
	Note that uncertainties involved with an experimental determination of the Reynolds values, and subsequently j and f, are ± 2, ± 14, and ± 3 percent, respectively. So, the first estimates for Reynolds numbers must be refined later (in subsequent iterations) up to the margin of ± 2 percent. One iteration would likely suffice.		
MGC-8	$j_c = \exp(a_0 + r_c(a_1 + r_c(a_2 + r_c(a_3 + r_c(a_4 + a_5 r_c)))))$	0.003718	—
	$j_h = \exp(a_0 + r_h(a_1 + r_h(a_2 + r_h(a_3 + r_h(a_4 + a_5 r_h)))))$	0.00506	—
	$f_c = \exp(b_0 + r_c(b_1 + r_c(b_2 + r_c(b_3 + r_c(b_4 + b_5 r_c)))))$	0.01006	—
	$f_h = \exp(b_0 + r_h(b_1 + r_h(b_2 + r_h(b_3 + r_h(b_4 + b_5 r_h)))))$	0.01707	—
	$r_c = \ln(\text{Re}_c) = \ln(3446)$	8.145	—
	$r_h = \ln(\text{Re}_h) = \ln(1181)$	7.074	—
	This explicit calculation of the refined values for j and f is conducted by using $j(\text{Re})$ and $f(\text{Re})$ correlations for the selected geometry as listed in MGC-4 (note that both fluid sides have the same geometry, based on a decision elaborated in MGC-3). The values are calculated for Reynolds numbers as determined in step MGC-7. The form of the correlation is dictated by the curve fitting of the experimentally obtained data listed in Kays and London (1998).		

MGC-9

$$\left(\frac{j}{f}\right)_c = \frac{0.003718}{0.01006}$$ 0.3696 —

$$\left(\frac{j}{f}\right)_h = \frac{0.00506}{0.01707}$$ 0.2965 —

Refined values of j/f ratios are now based on newly calculated values of j and f factors for the estimated Reynolds numbers.

MGC-6/2

$$G_c \approx \left[\left(\frac{j}{f}\right)\left(\frac{\Delta p}{p_{\text{in}}}\right)\frac{2 p_{\text{in}} \eta_0 \rho_m}{\text{NTU}\,\text{Pr}^{2/3}}\right]_c^{1/2} = \left[0.3696 \cdot \frac{5 \times 10^3}{500 \times 10^3} \cdot \frac{2 \times 500 \times 10^3 \times 0.9 \times 3.093}{3.622 \times 0.698^{2/3}}\right]^{1/2}$$ 60.08 kg/m² s

$$G_h \approx \left[\left(\frac{j}{f}\right)\left(\frac{\Delta p}{p_{\text{in}}}\right)\frac{2 p_{\text{in}} \eta_0 \rho_m}{\text{NTU}\,\text{Pr}^{2/3}}\right]_h^{1/2} = \left[0.2965 \cdot \frac{4 \times 10^3}{100 \times 10^3} \cdot \frac{2 \times 100 \times 10^3 \times 0.9 \times 0.533}{3.555 \times 0.699^{2/3}}\right]^{1/2}$$ 20.16 kg/m² s

These are the refined values for core mass velocities, but still based on an approximate G–(j/f) relationship.

MGC-7/2

$$\text{Re}_c = \frac{G_c D_{h,c}}{\mu_c} = \frac{60.08 \times 1.875 \times 10^{-3}}{28.95 \times 10^{-6}}$$ 3890 —

$$\text{Re}_h = \frac{G_h D_{h,h}}{\mu_h} = \frac{20.16 \times 1.875 \times 10^{-3}}{31.67 \times 10^{-6}}$$ 1194 —

Refined values of Reynolds numbers can now be calculated again. Note that Re_c has increased in this iteration; that is, it differs from the previous iteration by 13 percent—this is obviously far more than a standard uncertainty margin of 2 percent allowed for determining the Reynolds number. Although Re_h differs from the previous iteration for less than 2 percent, we will continue iterations until both Reynolds number changes reduce to below that margin.

Step	Calculation	Value	Units
MGC-8/2	$j_c = \exp(a_0 + r_c(a_1 + r_c(a_2 + r_c(a_3 + r_c(a_4 + a_5 r_c)))))$	0.003649	—
	$j_h = \exp(a_0 + r_h(a_1 + r_h(a_2 + r_h(a_3 + r_h(a_4 + a_5 r_h)))))$	0.005029	—
	$f_c = \exp(b_0 + r_c(b_1 + r_c(b_2 + r_c(b_3 + r_c(b_4 + b_5 r_c)))))$	0.009779	—
	$f_h = \exp(b_0 + r_h(b_1 + r_h(b_2 + r_h(b_3 + r_h(b_4 + b_5 r_h)))))$	0.01692	—
	$r_c = \ln(\text{Re}_c) = \ln(3890)$	8.266	—
	$r_h = \ln(\text{Re}_h) = \ln(1194)$	7.085	—
	In this step, the next estimation of the j and f factors is executed. Note that j factors differ from the ones determined in the previous iteration for less than 2 percent, respectively, for both fluids the f factors differ a bit more. As indicated in the MGC-7 step, the actual values of j factors (determined experimentally) usually have a margin of error for an order of magnitude larger than calculated here in two successive calculations. The f factors usually have the experimental margin of error at the calculated level. So, an additional iteration would probably be sufficient.		
MGC-9/2	$\left(\dfrac{j}{f}\right)_c = \dfrac{0.003649}{0.009779}$	0.3731	—
	$\left(\dfrac{j}{f}\right)_h = \dfrac{0.005029}{0.01692}$	0.2972	—
	This is the next iteration for j/f ratios.		
MGC-6/3	$G_c \approx \left[\left(\dfrac{j}{f}\right)\left(\dfrac{\Delta p}{p_{\text{in}}}\right)\dfrac{2p_{\text{in}}\eta_0\rho_m}{\text{NTU Pr}^{2/3}}\right]_c^{1/2} = \left[0.3731\,\dfrac{5\times 10^3}{500\times 10^3}\,\dfrac{2\times 500\times 10^3\times 0.9\times 3.093}{3.622\times 0.698^{2/3}}\right]^{1/2}$	60.36	kg/m² s
	$G_h \approx \left[\left(\dfrac{j}{f}\right)\left(\dfrac{\Delta p}{p_{\text{in}}}\right)\dfrac{2p_{\text{in}}\eta_0\rho_m}{\text{NTU Pr}^{2/3}}\right]_h^{1/2} = \left[0.2972\,\dfrac{4\times 10^3}{100\times 10^3}\,\dfrac{2\times 100\times 10^3\times 0.9\times 0.533}{3.555\times 0.699^{2/3}}\right]^{1/2}$	20.19	kg/m² s
	These are refined values for the core mass velocities.		

MGC-7/3

$$\mathrm{Re}_c = \frac{G_c D_{h,c}}{\mu_c} = \frac{60.36 \times 1.875 \times 10^{-3}}{28.95 \times 10^{-6}}$$ 3909 —

$$\mathrm{Re}_h = \frac{G_h D_{h,h}}{\mu_h} = \frac{20.19 \times 1.875 \times 10^{-3}}{31.67 \times 10^{-6}}$$ 1195 —

These are the new values of Reynolds numbers. They differ well below the margin of 2 percent from the previously calculated values (adopted here as a criterion for termination of the iterative procedure). Therefore, this should be considered as the last iteration.

MGC-8/3

$j_c = \exp(a_0 + r_c(a_1 + r_c(a_2 + r_c(a_3 + r_c(a_4 + a_5 r_c)))))$ 0.003646 —
$j_h = \exp(a_0 + r_h(a_1 + r_h(a_2 + r_h(a_3 + r_h(a_4 + a_5 r_h)))))$ 0.005026 —
$f_c = \exp(b_0 + r_c(b_1 + r_c(b_2 + r_c(b_3 + r_c(b_4 + b_5 r_c)))))$ 0.009769 —
$f_h = \exp(b_0 + r_h(b_1 + r_h(b_2 + r_h(b_3 + r_h(b_4 + b_5 r_h)))))$ 0.01691 —
$r_c = \ln(\mathrm{Re}_c) = \ln(3909)$ 8.271 —
$r_h = \ln(\mathrm{Re}_h) = \ln(1195)$ 7.086 —

This is the last estimation of j and f factors. Calculated values are practically identical to the ones calculated in the previous iteration.

MGC-9/3

$$\left(\frac{j}{f}\right)_c = \frac{0.003646}{0.009769}$$ 0.3732 —

$$\left(\frac{j}{f}\right)_h = \frac{0.005026}{0.01691}$$ 0.2972 —

These are the last estimations for j/f ratios.

MGC-10

$$T_w = \frac{T_{h,\mathrm{ref}} + \dfrac{\mathrm{NTU}_c C^*}{\mathrm{NTU}_h} T_{c,\mathrm{ref}}}{1 + \dfrac{\mathrm{NTU}_c C^*}{\mathrm{NTU}_h}} = \frac{T_{c,\mathrm{ref}} + T_{h,\mathrm{ref}}}{2} = \frac{560 + 642}{2}$$ 601 K

Step	Calculation	Value	Units
	The temperature of the heat-transfer surface wall between the fluids is calculated from a balance equation that relates heat-transfer rates delivered from one fluid to those received by the other. These heat-transfer rates are expressed in terms of fluid-to-wall and wall-to-fluid temperature differences and the respective thermal resistances on both fluid sides. The wall temperature is needed to perform a correction of thermophysical properties. This correction is due to temperature gradients between the fluids and the heat-transfer surface wall across the respective boundary layers on both fluid sides.		
MGC-11	$$j_{c,\text{corr}} = j_c \left(\frac{T_w}{T_{c,\text{ref}}}\right)^n = 0.003646 \left(\frac{601}{560}\right)^{-0.1185}$$	0.003616	—
	$$n = 0.3 - \left[\log_{10}\left(\frac{T_w}{T_{c,\text{ref}}}\right)\right]^{1/4} = 0.3 - \left[\log_{10}\frac{601}{560}\right]^{1/4}$$	-0.1185	—
	Note that cold air (i.e., gas) is exposed to heating, and that its flow regime is turbulent. For details of the alternate exponent determination, see Shah and Sekulic (2003, table 7.13, p. 531). The conditions to be satisfied are $1 < T_{w,\text{ref}}/T_{c,\text{ref}} < 5$; $0.6 < \Pr < 0.9$.		
MGC-12	$$j_{h,\text{corr}} = j_h \left(\frac{T_w}{T_{h,\text{ref}}}\right)^n = j_h = 0.005026$$	0.005026	—
	The hot fluid experiences cooling conditions. The flow regime is in the laminar region. Therefore, the exponent $n = 0$ in Shah and Sekulic (2003, table 7.12, p. 531).		
MGC-13	$$f_{c,\text{corr}} = f_c \left(\frac{T_w}{T_{c,\text{ref}}}\right)^m = 0.009769 \left(\frac{601}{560}\right)^{-0.1}$$	0.00970	—
	$m = -0.1$		

Cold fluid is heated, and the flow regime is turbulent. The suggested calculation of the exponent in the correction term is valid for the range of temperature ratios as follows:

$$1 < \frac{T_{w,\text{ref}}}{T_{c,\text{ref}}} < 5$$

In our case, this ratio is 1.07; therefore the calculation of the exponent m is performed as indicated. Fluid at the cold side is air; therefore, it is treated as an ideal gas.

MGC-14	$f_{h,\text{corr}} = f_h \left(\dfrac{T_w}{T_{h,\text{ref}}} \right)^m = 0.01691 \left(\dfrac{601}{642} \right)^{0.81}$	0.01603	—
	$m = 0.81$		

For a fluid cooling case and laminar flow, the exponent is equal to 0.81. The conditions to be satisfied are: $0.5 < T_{w,\text{ref}}/T_{h,\text{ref}} = 0.94 < 1$; $0.6 < \Pr = 0.699 < 0.9$.

MGC-15	$h_c = j_{c,\text{corr}} \dfrac{G_c c_{p,c}}{\Pr_c^{2/3}} = 0.003616 \dfrac{60.36 \times 1041}{0.698^{2/3}}$	288.7	W/m² K

The heat-transfer coefficient for the cold fluid is determined from the definition of the Colburn factor.

MGC-16	$h_h = j_{h,\text{corr}} \dfrac{G_h c_{p,h}}{\Pr_h^{2/3}} = 0.005026 \dfrac{20.19 \times 1061}{0.699^{2/3}}$	136.7	W/m² K

The heat-transfer coefficient for the hot fluid is determined from definition of the Colburn factor.

MGC-17	$\eta_{f,c} = \dfrac{\tanh(ml)_c}{(ml)_c} = \dfrac{\tanh(137.8 \times 0.00302)}{137.8 \times 0.00302}$	0.9460	—
	$m_c = \sqrt{2h_c/k\delta} = \sqrt{2 \times 288.7/(200 \times 0.152 \times 10^{-3})}$	137.8	m⁻¹
	$l_c = \dfrac{b}{2} - \delta = \dfrac{6.35 \times 10^{-3}}{2} - 0.152 \times 10^{-3}$	0.00302	m

Step	Calculation	Value	Units
	Note that δ and l_c differ from each other more than for an order of magnitude, so the exposed strip edge of the fin is not taken into account and the m parameter is calculated in the first approximation as indicated [see Shah and Sekulic (2003), pp. 280 and 627) for an alternative]. Thermal conductivity of the fin is assumed to be 200 W/m K for an alloy at the given temperature. Note that the resulting fin efficiency becomes very high. Actual fin efficiency in a brazed heat exchanger throughout the core may be significantly smaller (Zhao et al., 2003).		
MGC-18	$\eta_{f,h} = \dfrac{\tanh(ml)_h}{(ml)_h} = \dfrac{\tanh(94.8 \times 0.00302)}{94.8 \times 0.00302}$	0.9735	—
	$m_h = \sqrt{2h_h/k\delta} = \sqrt{2 \times 136.7/(200 \times 0.152 \times 10^{-3})}$	94.8	m^{-1}
	$l_h = \dfrac{b}{2} - \delta = \dfrac{6.35 \times 10^{-3}}{2} - 0.152 \times 10^{-3}$	0.00302	m
	Note that the fin geometry is the same on both fluid sides; therefore the lengths are the same.		
MGC-19	$\eta_{0,c} = 1 - (1 - \eta_{f,c})\dfrac{A_f}{A} = 1 - (1 - 0.9460) \times 0.849$	0.9542	—
	A detailed discussion of the meaning of the total extended surface efficiency can be found in Shah and Sekulic (2003, p. 289).		
MGC-20	$\eta_{0,h} = 1 - (1 - \eta_{f,h})\dfrac{A_f}{A} = 1 - (1 - 0.9735) \times 0.849$	0.9775	—
	Note that the extended surface efficiencies must differ for both fluid sides regardless of the same geometry of the fins used. This is due to the difference in the heat-transfer coefficients.		
MGC-21	$U = \left[\dfrac{1}{(\eta_0 h)_c} + \dfrac{A_c/A_h}{(\eta_0 h)_h}\right]^{-1} = \left[\dfrac{1}{(\eta_0 h)_c} + \dfrac{1}{(\eta_0 h)_h}\right]^{-1} = \left[\dfrac{1}{0.9542 \times 288.7} + \dfrac{1}{0.9775 \times 136.7}\right]^{-1}$	89.93	W/m^2 K

Because of a high thermal conductivity of wall material, thermal resistance of the wall is neglected in this calculation. Also, it is assumed that no significant fouling is present on either side of the heat-transfer surface. Therefore, the overall heat-transfer coefficient is defined by heat-transfer conductance due to convection on both fluid sides only. Note, again, that the heat-transfer surface areas are the same on both fluid sides because the same fin geometry is used.

MGC-22
$$A_c = A_h = \frac{(\dot{m}c_p)_c \, \text{NTU}}{U} = \frac{20.82 \times 10^3 \times 1.811}{89.93}$$
419.3 m^2

The heat-transfer surface areas are the same on both fluid sides.

MGC-23
$$A_{0,c} = \frac{\dot{m}_c}{G_c} = \frac{20}{60.36}$$
0.3313 m^2

The free-flow area on the cold fluid side is determined from the definition of the mass velocity, $G = \dot{m}/A_0$.

MGC-24
$$A_{0,h} = \frac{\dot{m}_h}{G_h} = \frac{20}{20.19}$$
0.9908 m^2

The free-flow area on the hot fluid side is determined analogously to the same entity on the cold fluid side; see step MGC-23.

MGC-25
$$\sigma_c = \sigma_h = \frac{b\beta D_h}{8(b+a)} = \frac{6.35 \times 10^{-3} \times 1841 \times 1.875 \times 10^{-3}}{8(6.35 + 2) \times 10^{-3}}$$
0.3281 —

The minimum free-flow area : frontal area ratio is the same on both fluid sides (because of the same heat-transfer surface geometry). Note that the equation used represents a reduced form of a general expression for that ratio, assuming that N_p passages exist on the hot side and N_{p+1} passages exist on the cold side.

MGC-26
$$A_{\text{fr},c} = \frac{A_{0,c}}{\sigma_c} = \frac{0.3313}{0.3281}$$
1.0097 m^2

The free-flow area on the cold fluid side is determined from the definition of the free-flow area : frontal area ratio for the cold fluid side.

Step	Calculation	Value	Units
MGC-27	$A_{\text{fr},h} = \dfrac{A_{0,h}}{\sigma_h} = \dfrac{0.9908}{0.3281}$	3.0198	m^2
	The free-flow area on the hot fluid side is determined from the definition of the free-flow area : frontal area ratio for the hot fluid side.		
MGC-28	$L_c = \dfrac{D_h A_c}{4 A_{0,c}} = \dfrac{1.875 \times 10^{-3} \times 419.3}{4 \times 0.3313}$	0.593	m
	The fluid flow length on the cold fluid side represents the principal core dimension in this direction.		
MGC-29	$L_h = \dfrac{D_h A_h}{4 A_{0,h}} = \dfrac{1.875 \times 10^{-3} \times 419.3}{4 \times 0.9908}$	0.198	m
	The fluid flow length on the hot fluid side represents the principal core dimension in this direction.		
MGC-30	$L_{\text{stack}} = \dfrac{A_{\text{fr},c}}{L_h} = \dfrac{1.0097}{0.198}$	5.099	m
	$L_{\text{stack}} = \dfrac{A_{\text{fr},h}}{L_c} = \dfrac{3.0198}{0.593}$	5.092	m
	The core dimension in the third direction can be calculated by using the frontal area of either the cold fluid or the hot fluid. If the calculation were conducted correctly, both values would have to be within the margin of error only as a result of rounding of the numbers. Note that no constraint regarding the aspect ratio of any pair of the core side dimensions is imposed. Such a nonconstrained calculation may lead, as is the case here, to a heat-exchanger core with relatively large aspect ratios. An optimization procedure would be needed to execute this calculation with this constraint imposed. In that case, an additional iterative procedure would be needed. Such a procedure would require a reconsideration of the heat-transfer surface geometry on both fluid sides (in this case, for simplicity, it is assumed that these geometries are the same).		

MGC-31 $\left(\dfrac{\Delta p}{p_i}\right)_c = \dfrac{G_c^2}{2(p_{\text{in}}\rho_{\text{in}})_c}\left[(1-\sigma^2+K_c)+f\dfrac{L\rho_i}{r_h\rho_m}+2\left(\dfrac{\rho_i}{\rho_o}-1\right)-(1-\sigma^2-K_e)\dfrac{\rho_i}{\rho_o}\right]_c$ — —

$K_{c,c} = f_1(\sigma_c, \text{Re}_c, \text{surface geometry})$
$K_{e,c} = f_2(\sigma_c, \text{Re}_c, \text{surface geometry})$

$K_{c,c} = f_1(\sigma_c = 0.3281, \text{Re}_c = 3909, \text{surface geometry} = 19.86)$ 0.51
$K_{e,c} = f_2(\sigma_c = 0.3281, \text{Re}_c = 3909, \text{surface geometry} = 19.86)$ 0.45

The relative pressure drop calculations require determination of both entrance and exit pressure loss coefficients K_c and K_e. These coefficients can be determined from Kays and London (1998) data for a given set of information concerning (1) the ratio of the minimum free-flow area to the frontal area $\sigma_{c(h)}$ for both fluid sides (in this case, these values are identical; see calculation step MGC-25), (2) Reynolds numbers, and (3) surface geometry [i.e., $(L/D_h)_{c(h)}$]. Fanning friction factors should, as a rule, be determined by accounting for corrections for the referent wall and fluid temperatures. Note that the referent wall temperature may be calculated by taking into account thermal resistances on both fluid sides. In step MGC-10, the wall temperature is determined in the first approximation without accounting for this factor (thermal resistances on both sides were assumed as equal).

MGC-32 $\left(\dfrac{\Delta p}{p_i}\right)_h = \dfrac{G_h^2}{2(p_{\text{in}}\rho_{\text{in}})_h}\left[(1-\sigma^2+K_c)+f\dfrac{L\rho_i}{r_h\rho_m}+2\left(\dfrac{\rho_i}{\rho_o}-1\right)-(1-\sigma^2-K_e)\dfrac{\rho_i}{\rho_o}\right]_h$ — —

$K_{c,h} = f_1(\sigma_h, \text{Re}_h, \text{surface geometry})$
$K_{e,h} = f_2(\sigma_h, \text{Re}_h, \text{surface geometry})$

$K_{c,h} = f(\sigma_h = 0.3281, \text{Re}_h = 1195, \text{surface geometry} = 19.86)$ 1.22
$K_{e,h} = f(\sigma_h = 0.3281, \text{Re}_h = 1195, \text{surface geometry} = 19.86)$ 0.20

Step	Calculation	Value	Units
	Correction of the magnitude of a Fanning friction factor is conducted as suggested in steps MGC-13 and MGC-14. A refinement of the Fanning friction factor may not be justified in cases when the actual change of its value is below the margin of uncertainty typical for an experimental estimation of its value. We will assume that the margin is at the 3 percent level. Note that the geometry of the surface is a plane plate fin with a designation of 19.86 (Kays and London, 1998). An additional comment regarding the correction of the Fanning friction factors is appropriate at this point. This correction was performed in steps MGC-13 and MGC-14, assuming that thermal resistances on both sides of the wall are considered as being the same. This may not be the case, and in such a situation the correction would require determination of the wall temperature (see step MGC-10), assuming that resistances are not the same, specifically $$T_w = \frac{T_{h,\text{ref}} + \frac{R_h}{R_c}T_{c,\text{ref}}}{1 + \frac{R_h}{R_c}}$$ where $R_h = 1/(\eta_o h A)_h$ and $R_c = 1/(\eta_o h A)_c$. If this correction is not performed, the Fanning friction factors will be the same as determined in steps MGC-13 and MGC-14.		
MGC-33	$$\left(\frac{\Delta p}{p_i}\right)_c = \frac{60.36^2}{2(500 \times 10^3 \times 3.484)}$$ $$\times \left[(1 - 0.3281^2 + 0.51) + 0.00970\frac{0.593 \times 3.484}{\frac{1.875 \times 10^{-3}}{4}}3.093\right.$$ $$\left. + 2\left(\frac{3.484}{2.782} - 1\right) - (1 - 0.3281^2 - 0.45)\frac{3.484}{2.782}\right]$$ The relative pressure drops can now be calculated by using all the previously determined variables and parameters. Note that the hydraulic radius is represented as one-quarter of the hydraulic diameter. The analytical expression for this calculation is given in the step MGC-31 (above).	0.0159	—

MGC-34

$$\Delta p_c = p_{i,c}\left(\frac{\Delta p}{p_i}\right)_c = 500 \times 10^3 \times 0.0159$$

7.9   kPa

From the input data, an allowed pressure drop is 5 kPa < 7.9 kPa. This indicates that the imposed condition is not satisfied! This prompts a need to reiterate the calculation with a new value of the mass velocity (in the first iteration, the mass velocity was calculated in step MGC-6 by using the first approximation based on a weak dependence of j/f on the Reynolds number).

MGC-35

$$\left(\frac{\Delta p}{p_i}\right)_h = \frac{20.19^2}{2(100 \times 10^3 \times 0.498)}\left[(1 - 0.3281^2 + 1.22) + 0.01603\frac{0.198 \times 0.498}{\frac{1.875 \times 10^{-3}}{4}}0.5\right.$$

$$\left. + 2\left(\frac{0.498}{0.574} - 1\right) - (1 - 0.3281^2 - 0.2)\frac{0.498}{0.574}\right]$$

0.031   —

The hot fluid side pressure drop divided by the inlet fluid pressure at the hot fluid side can now be calculated in the same manner as for the cold fluid. Of course, one may decide to calculate the pressure drop right away (step MGC-36). The analytical expression for this calculation is given in the step MGC-32.

MGC-36

$$\Delta p_h = p_{i,h}\left(\frac{\Delta p}{p_i}\right)_h = 100 \times 10^3 \times 0.031$$

3.1   kPa

The imposed maximum allowable pressure drop is 4.2 kPa > 3.1 kPa. Therefore, this requirement is satisfied. However, since the pressure drop for the cold fluid was not satisfied, the new iteration is needed, as emphasized in step MGC-34.

MGC-37

$$G_c = \left[\frac{2(p_{in}\rho_{in})\left(\frac{\Delta p}{p_i}\right)_c}{\left[(1-\sigma^2+K_c)+f\frac{L\rho_i}{r_h\rho_m}+2\left(\frac{\rho_i}{\rho_o}-1\right)-(1-\sigma^2-K_e)\frac{\rho_i}{\rho_o}\right]_c}\right]^{1/2}$$

$$= \left[\frac{2 \times 500 \times 10^3 \times 3.484 \times \frac{5 \times 10^3}{500 \times 10^3}}{(1-0.3281^2+0.51)+0.00970\frac{0.593 \times 3.484}{\frac{1.875 \times 10^{-3}}{4}3.093}+2\left(\frac{3.484}{2.782}-1\right)-(1-0.3281^2-0.45)\frac{3.484}{2.782}}\right]^{1/2}$$

47.91   kg/m² s

29.23

Step	Calculation	Value	Units
	The new iteration loop starts with the determination of the set of new mass velocities. These values will be used to calculate the refined values of Reynolds numbers in step MGC-7. Subsequently, steps MGC-8 through MGC-36 should be revisited. The new mass velocities G should be calculated from the exact expression for the pressure drop, as given in steps MGC-33 and MGC-35, assuming G values as unknown and the other numerical values in these equations as given. The convergence would be very fast. The new value of the mass velocity of 53.61 kg/m² s is significantly smaller than the assessed value in the first iteration (60.37 kg/m² s); thus, a reduction of almost 10 percent is achieved.		
MGC-38	$$G_h = \frac{\left[2(p_{\text{in}}\rho_{\text{in}})\left(\frac{\Delta p}{p_i}\right)\right]_h^{1/2}}{\left[(1-\sigma^2+K_c)+f\frac{L\rho_i}{r_h\rho_m}+2\left(\frac{\rho_i}{\rho_o}-1\right)-(1-\sigma^2-K_e)\frac{\rho_i}{\rho_o}\right]_h^{1/2}}$$ $$= \frac{\left[2\times100\times10^3\times0.498\times\dfrac{4\times10^3}{100\times10^3}\right]^{1/2}}{\left[(1-0.3281^2+1.22)+0.01603\,\dfrac{0.198\times0.498}{\dfrac{1.875\times10^{-3}}{4}\cdot 0.5}+2\left(\dfrac{0.498}{0.574}-1\right)-(1-0.3281^2-0.2)\dfrac{0.498}{0.574}\right]^{1/2}}$$ The new value of the mass velocity is increased from 20.19 to 27.24 kg/m² s, that is, for roughly 7 percent.	22.93	kg/m² s
MGC-7/4	$$\text{Re}_c = \frac{G_c D_{h,c}}{\mu_c} = \frac{47.91\times 1.875\times 10^{-3}}{28.95\times 10^{-6}}$$	3103	—
	$$\text{Re}_h = \frac{G_h D_{h,h}}{\mu_h} = \frac{22.93\times 1.875\times 10^{-3}}{31.67\times 10^{-6}}$$	1357	—
	The new iteration cycle requires determination of new Reynolds numbers (with changed G values, including all the earlier local iterations; this is the fourth iteration of Reynolds numbers). This iteration repeats step MGC-7.		

MGC-8/4

$$j_c = \exp(a_0 + r_c(a_1 + r_c(a_2 + r_c(a_3 + r_c(a_4 + a_5 r_c)))))\quad 0.003777$$
$$j_h = \exp(a_0 + r_h(a_1 + r_h(a_2 + r_h(a_3 + r_h(a_4 + a_5 r_h)))))\quad 0.004709$$
$$f_c = \exp(b_0 + r_c(b_1 + r_c(b_2 + r_c(b_3 + r_c(b_4 + b_5 r_c)))))\quad 0.01034$$
$$f_h = \exp(b_0 + r_h(b_1 + r_h(b_2 + r_h(b_3 + r_h(b_4 + b_5 r_h)))))\quad 0.01539$$

$$r_c = \ln(\mathrm{Re}_c) = \ln(3103) \quad 8.040$$
$$r_h = \ln(\mathrm{Re}_h) = \ln(1357) \quad 7.213$$

The new iteration for the MGC-8 estimation of j and f factors takes the values for Reynolds numbers from MGC-7/4.

MGC-9/4

$$\left(\frac{j}{f}\right)_c = \frac{0.003777}{0.01034} \quad 0.3651$$

$$\left(\frac{j}{f}\right)_h = \frac{0.004709}{0.01539} \quad 0.306$$

The ratios of Colburn and Fanning factors are calculated exactly for new Reynolds numbers. Note that these values did change, but as in any other iteration for determining this ratio, these changes are not significant beyond the first two decimal places.

MGC-10/2

$$T_w = \frac{T_{h,\mathrm{ref}} + \dfrac{R_h}{R_c} T_{c,\mathrm{ref}}}{1 + \dfrac{R_h}{R_c}} = \frac{T_{h,\mathrm{ref}} + \dfrac{1/(\eta_0 h A)_h}{1/(\eta_0 h A)_c} T_{c,\mathrm{ref}}}{1 + \dfrac{1/(\eta_0 h A)_h}{1/(\eta_0 h A)_c}}$$

$$= \frac{642 + \dfrac{1/(0.9775 \times 136.7 \times 419.3)}{1/(0.9542 \times 288.7 \times 419.3)} 560}{1 + \dfrac{1/(0.9775 \times 136.7 \times 419.3)}{1/(0.9542 \times 288.7 \times 419.3)}} = \frac{642 + \dfrac{1.785 \times 10^{-5}}{8.657 \times 10^{-6}} 560}{1 + \dfrac{1.785 \times 10^{-5}}{8.657 \times 10^{-6}}} \quad 586.7 \quad \mathrm{K}$$

The wall temperature is calculated this time by taking into account the difference in thermal resistances on both fluid sides (compare this with step MGC-10). With a difference of slightly more than 10 K (vs. the last iteration) for air, the correction of Colburn and Fanning friction factors will still remain relatively small.

Step	Calculation	Value	Units
MGC-11/2	$j_{c,\text{corr}} = j_c \left(\dfrac{T_w}{T_{c,\text{ref}}}\right)^n = 0.003777 \left(\dfrac{586.7}{560}\right)^{-0.077}$	0.003764	—
	$n = 0.3 - \left(\log_{10}\dfrac{T_w}{T_{c,\text{ref}}}\right)^{1/4} = 0.3 - \left[\log_{10}\dfrac{586.7}{560}\right]^{1/4}$	-0.077	—
	Note that cold gas is being exposed to heating and that the flow regime is turbulent. For details of the alternate exponent determination, see Shah and Sekulic (2003, table 7.13, p. 531). The conditions to be satisfied are $1 < T_{w,\text{ref}}/T_{c,\text{ref}} < 5$; $0.6 < \text{Pr} < 0.9$.		
MGC-12/2	$j_{h,\text{corr}} = j_h \left(\dfrac{T_w}{T_{h,\text{ref}}}\right)^n = j_h$	0.004709	—
	For the hot fluid, we have the fluid cooling conditions. The flow regime is in the laminar region. Therefore, the exponent $n = 0$ (Shah and Sekulic, 2003, table 7.12, p. 531).		
MGC-13/2	$f_{c,\text{corr}} = f_c \left(\dfrac{T_w}{T_{c,\text{ref}}}\right)^m = 0.01034 \left(\dfrac{586.7}{560}\right)^{-0.1}$	0.01029	—
	$m = -0.1$		
	Cold fluid is heated, and the flow regime is turbulent. The suggested calculation of the exponent in the correction term is valid for the range of temperature ratios as follows:		
	$1 < \dfrac{T_{w,\text{ref}}}{T_{c,\text{ref}}} < 5$		
	In this case, this ratio is 1.04; therefore, calculation of the exponent m is performed as indicated. Fluid at the cold side is air; therefore, it is treated as an ideal gas.		
MGC-14/2	$f_{h,\text{corr}} = f_h \left(\dfrac{T_w}{T_{h,\text{ref}}}\right)^m = 0.01539 \left(\dfrac{586.7}{642}\right)^{0.81}$	0.01431	—
	$m = 0.81$		
	For the fluid cooling case and laminar flow, the exponent is equal to 0.81. The conditions to be satisfied are $0.5 < T_{w,\text{ref}}/T_{h,\text{ref}} = 0.91 < 1$; $0.6 < \text{Pr} = 0.699 < 0.9$.		

MGC-15/2	$h_c = j_{c,\text{corr}} \dfrac{G_c c_{p,c}}{\text{Pr}_c^{2/3}} = 0.003764 \dfrac{47.91 \times 1041}{0.698^{2/3}}$	238.5	W/m² K

The heat-transfer coefficient on the cold fluid side is determined from definition of the Colburn factor.

MGC-16/2	$h_h = j_{h,\text{corr}} \dfrac{G_h c_{p,h}}{\text{Pr}_h^{2/3}} = 0.004709 \dfrac{22.93 \times 1061}{0.699^{2/3}}$	145.4	W/m² K

The heat-transfer coefficient on the hot fluid side is determined from definition of the Colburn factor.

MGC-17/2	$\eta_{f,c} = \dfrac{\tanh(ml)_c}{(ml)_c} = \dfrac{\tanh(125.3 \times 0.00302)}{125.3 \times 0.00302}$	0.9549	—
	$m_c = \sqrt{2h_c/k\delta} = \sqrt{2 \times 238.5/(200 \times 0.152 \times 10^{-3})}$	125.3	m⁻¹
	$l_c = \dfrac{b}{2} - \delta = \dfrac{6.35 \times 10^{-3}}{2} - 0.152 \times 10^{-3}$	0.00302	m

Note that δ and l_c differ more than for an order of magnitude, so the exposed strip edge of the fin is not taken into account and the m parameter is calculated using this approximation [see Shah and Sekulic (2003, pp. 280 and 627) for an alternative]. Thermal conductivity of the fin is assumed to be 200 W/m K for an alloy at the given temperature.
Note that the resulting fin efficiency is a bit larger than in the previous iteration (MGC-17). Actual fin efficiency in a brazed aluminum heat exchanger may be significantly smaller within the core (Zhao et al., 2003).

MGC-18/2	$\eta_{f,h} = \dfrac{\tanh(ml)_h}{(ml)_h} = \dfrac{\tanh(97.8 \times 0.00302)}{97.8 \times 0.00302}$	0.9719	—
	$m_h = \sqrt{2h_h/k\delta} = \sqrt{2 \times 145.4/(200 \times 0.152 \times 10^{-3})}$	97.8	m⁻¹
	$l_h = \dfrac{b}{2} - \delta = \dfrac{6.35 \times 10^{-3}}{2} - 0.152 \times 10^{-3}$	0.00302	m

Note that the fin geometry is the same on both fluid sides; therefore the lengths are the same. The fin efficiency is quite large.

Step	Calculation	Value	Units
MGC-19/2	$\eta_{o,c} = 1 - (1 - \eta_{f,c})\dfrac{A_f}{A} = 1 - (1 - 0.9549) \times 0.849$	0.9617	—
	A discussion of the meaning of the total extended surface efficiency can be found in Shah and Sekulic (2003, p. 289).		
MGC-20/2	$\eta_{o,h} = 1 - (1 - \eta_{f,h})\dfrac{A_f}{A} = 1 - (1 - 0.9719) \times 0.849$	0.9761	—
	Note that the extended surface efficiencies differ for both sides regardless of the same geometry of the fins. This is due to the difference in heat-transfer coefficients.		
MGC-21/2	$U = \left[\dfrac{1}{(\eta_o h)_c} + \dfrac{A_c/A_h}{(\eta_o h)_h}\right]^{-1} = \left[\dfrac{1}{(\eta_o h)_c} + \dfrac{1}{(\eta_o h)_h}\right]^{-1} = \left[\dfrac{1}{0.9617 \times 238.5} + \dfrac{1}{0.9761 \times 145.4}\right]^{-1}$	87.67	W/m² K
	Because of the high thermal conductivity of wall material, the thermal resistance of the wall is neglected in this calculation. Also, it is again assumed that no significant fouling is present on either side of the heat-transfer surface. Therefore, the overall heat-transfer coefficient is defined by heat-transfer conductance due only to convection on both fluid sides. Note, again, that the heat-transfer surface areas are the same on both fluid sides because of the use of the same fin geometry.		
MGC-22/2	$A_c = A_h = \dfrac{(\dot{m}c_p)_c \, \text{NTU}}{U} = \dfrac{20.82 \times 10^3 \times 1.811}{87.67}$	430.1	m²
	The heat-transfer surface areas are the same on both fluid sides.		
MGC-23/2	$A_{0,c} = \dfrac{\dot{m}_c}{G_c} = \dfrac{20}{47.91}$	0.4175	m²
	The free-flow area on the cold side is determined from the definition of the mass velocity, $G = \dot{m}/A_0$.		
MGC-24/2	$A_{0,h} = \dfrac{\dot{m}_h}{G_h} = \dfrac{20}{22.93}$	0.8724	m²
	The free-flow area on the hot fluid side is determined analogously to determination of the same entity on the cold fluid side; see step MGC-23.		

MGC-25/2	$\sigma_c = \sigma_h = \dfrac{b\beta D_h}{8(b+a)} = \dfrac{6.35 \times 10^{-3} \times 1841 \times 1.875 \times 10^{-3}}{8(6.35+2) \times 10^{-3}}$	0.3281	—
	This step is identical to the one performed in the first iteration (the heat-transfer surface type was not changed).		
MGC-26/2	$A_{\mathrm{fr},c} = \dfrac{A_{0,c}}{\sigma_c} = \dfrac{0.4175}{0.3281}$	1.272	m²
	The free-flow area on the cold fluid side is determined from definition of the free-flow area : frontal area ratio for the cold fluid side.		
MGC-27/2	$A_{\mathrm{fr},h} = \dfrac{A_{0,h}}{\sigma_h} = \dfrac{0.8724}{0.3281}$	2.659	m²
	The free-flow area on the hot fluid side is determined from definition of the free-flow area : frontal area ratio for the hot fluid side.		
MGC-28/2	$L_c = \dfrac{D_h A_c}{4 A_{0,c}} = \dfrac{1.875 \times 10^{-3} \times 430.1}{4 \times 0.4175}$	0.4829	m
	The fluid flow length on the cold fluid side represents the principal core dimension in this direction.		
MGC-29/2	$L_h = \dfrac{D_h A_h}{4 A_{0,h}} = \dfrac{1.875 \times 10^{-3} \times 430.1}{4 \times 0.8724}$	0.2311	m
	The fluid flow length on the hot fluid side represents the principal core dimension in this direction.		
MGC-30/2	$L_{\mathrm{stack}} = \dfrac{A_{\mathrm{fr},c}}{L_h} = \dfrac{1.272}{0.2311}$	5.504	m
	$L_{\mathrm{stack}} = \dfrac{A_{\mathrm{fr},h}}{L_c} = \dfrac{2.659}{0.4829}$	5.506	m
	The core size in the direction orthogonal to the flow directions is apparently reduced when compared to the previous iteration.		

Step	Calculation	Value	Units
MGC-31/2	$\left(\dfrac{\Delta p}{p_i}\right)_c = \dfrac{G_c^2}{2(p_{in}\rho_{in})_c} \left[(1-\sigma^2+K_c) + f\dfrac{L\rho_i}{r_h\rho_m} + 2\left(\dfrac{\rho_i}{\rho_o}-1\right) - (1-\sigma^2-K_e)\dfrac{\rho_i}{\rho_o}\right]_c$		
	$K_{c,c} = f_1(\sigma_c, \text{Re}_c, \text{surface geometry})$		
	$K_{e,c} = f_2(\sigma_c, \text{Re}_c, \text{surface geometry})$		
	$K_{c,c} = f_1(\sigma_c = 0.3281, \text{Re}_c = 3103, \text{surface geometry} = 19.86)$	0.52	
	$K_{e,c} = f_2(\sigma_c = 0.3281, \text{Re}_c = 3103, \text{surface geometry} = 19.86)$	0.42	
	The change of both K parameters takes place because of the change in Reynolds number values in this iteration.		
MGC-32/2	$\left(\dfrac{\Delta p}{p_i}\right)_h = \dfrac{G_h^2}{2(p_{in}\rho_{in})_h} \left[(1-\sigma^2+K_c) + f\dfrac{L\rho_i}{r_h\rho_m} + 2\left(\dfrac{\rho_i}{\rho_o}-1\right) - (1-\sigma^2-K_e)\dfrac{\rho_i}{\rho_o}\right]_h$		
	$K_{c,h} = f_1(\sigma_h, \text{Re}_h, \text{surface geometry})$		
	$K_{e,h} = f_2(\sigma_h, \text{Re}_h, \text{surface geometry})$		
	$K_{c,h} = f(\sigma_h = 0.3281, \text{Re}_h = 1357, \text{surface geometry} = 19.86)$	1.22	—
	$K_{e,h} = f(\sigma_h = 0.3281, \text{Re}_h = 1357, \text{surface geometry} = 19.86)$	0.20	—
	Note that the hot fluid stays in the laminar flow region; therefore, the K's coefficients do not change.		
MGC-33/2	$\left(\dfrac{\Delta p}{p_i}\right)_c = \dfrac{47.91^2}{2(500\times 10^3 \times 3.484)} \times \left[(1-0.3281^2+0.52) + 0.01029\,\dfrac{\dfrac{0.4829\times 3.484}{1.875\times 10^{-3}}}{\dfrac{4}{3.093}} + 2\left(\dfrac{3.484}{2.782}-1\right) - (1-0.3281^2-0.42)\dfrac{3.484}{2.782}\right]$	0.0087	—
	The relative pressure drops can now be calculated by using all the previously determined variables and parameters. Note that hydraulic radius is again represented as one-quarter of the hydraulic diameter.		

MGC-34/2

$$\Delta p_c = p_{i,c}\left(\frac{\Delta p}{p_i}\right)_c = 500 \times 10^3 \times 0.0087 \qquad 4.3 \quad \text{kPa}$$

From the input data, the allowed pressure drop is 5 kPa > 4.3 kPa. This result indicates that this condition is now safely satisfied. Consequently, because of this pressure drop, there is no need for further iterations.

MGC-35/2

$$\left(\frac{\Delta p}{p_i}\right)_h = \frac{22.93^2}{2(100 \times 10^3 \times 0.498)} \qquad 0.04 \quad —$$

$$\times \left[(1 - 0.3281^2 + 1.22) + 0.01431 \frac{0.2311 \times 0.498}{\frac{1.875 \times 10^{-3}}{4} 0.533} \right.$$

$$\left. + 2\left(\frac{0.498}{0.574} - 1\right) - (1 - 0.3281^2 - 0.2)\frac{0.498}{0.574}\right]$$

The hot-side fluid pressure drop divided by the inlet fluid pressure at the hot fluid side can now be calculated in the same manner as for the cold fluid. Of course, one may decide to calculate the pressure drop right away (see step MGC-36).

MGC-36/2

$$\Delta p_h = p_{i,h}\left(\frac{\Delta p}{p_i}\right)_h = 100 \times 10^3 \times 0.04 \qquad 4.0 \quad \text{kPa}$$

The imposed maximum allowable pressure drop is 4.2 kPa, which is virtually equal to the value determined in this step. Therefore, this requirement is also satisfied. The calculated pressure drop is exactly at the level of the allowed pressure drop limit. Therefore, the calculation procedure is completed.

Conclusion

The step-by-step heat-exchanger design procedure clearly demonstrates how intricate the sizing of a compact heat exchanger may be. However, an inevitably iterative routine converges very rapidly. The main dimensions of this heat-exchanger core are determined to be $L_c = 48$ cm, $L_h = 23$ cm, and $L_{\text{stack}} = 550$ cm. The core is made of plane triangular plate fin surfaces [surface designation 19.86 (Kays and London, 1998)]. Imposed limitations on the pressure drops are both satisfied. If, because of say, space considerations, the core dimensions must satisfy certain a priori imposed aspect ratios (fluid flow vs. stack length), further iterations would be needed. Calculation is presented in a most explicit manner, by listing each step (regardless of whether it may consist merely of a repeated calculation, already exercised in a previous step). All algebraic operations are, as a rule, included. This is done keeping in mind a need for full transparency of the calculation algorithm. Such design is conducted, as a rule, in practice by using a computer routine (what would never be fully transparent but eliminates any calculation errors that may often be present in a calculation as given here). Still, following the procedure as presented here, one can easily devise such a routine and execute the calculation.

References

Baclic, B. S., and P. J. Heggs, 1985, "On the Search for New Solutions of the Single-Pass Crossflow Heat Exchanger Problem," *Int. J. Heat Mass Transfer*, vol. 28, no. 10, pp. 1965–1976.

Kays, W. M., and A. L. London, 1998, *Compact Heat Exchangers*, reprint 3d ed., Krieger, Malabar, Fla.

Sekulic, D. P., 1990, "A Reconsideration of the Definition of a Heat Exchanger," *Int. J. Heat Mass Transfer*, vol. 33, pp. 2748–2750.

Sekulic, D. P., 2000, "A Unified Approach to the Analysis of Unidirectional and Bidirectional Parallel Flow Heat Exchangers," *Int. J. Mech. Eng. Educa.*, vol. 28, pp. 307–320.

Sekulic, D. P., R. K. Shah, and A. Pignotti, 1999, "A Review of Solution Methods for Determining Effectiveness-NTU Relationships for Heat Exchangers with Complex Flow Arrangements," *Appl. Mech. Reviews*, vol. 52, no. 3, pp. 97–117.

Shah, R. K., and D. P. Sekulic, 2003, *Fundamentals of Heat Exchanger Design*, Wiley, Hoboken, N.J.

Webb, R. L., 1994, *Principles of Enhanced Heat Transfer*, Wiley, New York.

Zhao, H., A. J. Salazar, and D. P. Sekulic, 2003, "Influence of Topological Characteristics of a Brazed Joint Formation on Joint Thermal Integrity," *Int. Mech. Eng. Congress*, vol. 1, ASME paper IMECE2003-43885, Washington, D.C.

Chapter 30

Single-Phase Natural Circulation Loops

An Analysis Methodology with an Example (Numerical) Calculation for an Air-Cooled Heat Exchanger

Daniel M. Speyer
Mechanical Engineering Department
The Cooper Union for the Advancement of Science
  and Art
New York, New York

Introduction

When heat transfer in a fluid (flow) circuit results in a vertical temperature variation (a nonsymmetric variation), the resulting density differences and effects of friction will establish a circulation flow. In effect, the density differences will "pump" the fluid.

This is called *natural circulation* (not to be confused with *natural convection*), and it is used by design, or is present, in many systems or situations involving heat transfer and a single-phase gas, single-phase liquid, or two-phase liquid and gas; in these situations the usual forced-convection heat-transfer equations are applicable.

Engineering examples of these are air-cooled heat exchangers used to cool the dielectric oil in electric transformers, hot-water circulation heating systems in residential buildings and steam-liquid water

30.2 Heat Exchangers

circulation (recirculation) in steam boilers in electricity-generating power stations.

In the case of single-phase gas or liquid systems, typically the analysis is not difficult, conceptually or in practice; and the calculations are not time-consuming. However, the possibility of, applications of, and analysis of such natural circulation systems are not discussed in many general heat-transfer and fluid flow texts, and are seldom realized by engineers.

The present analysis develops the equations and applies them to an example problem. In addition, the following text also discusses methods to obtain optimum solutions—and as such represents a methodology for single-phase natural circulation calculations.

Description of Single-Phase Natural Circulation

A single-phase natural circulation fluid flow circuit is shown in Fig. 30.1. The system is distributed vertically, with fluid heated in the riser and cooled in the downcomer. As a consequence of density differences and gravity, the fluid is subject to (initially) an unbalanced buoyant force.

As the fluid velocity increases, the temperature and resulting density differences will decrease, and friction effects in the form of minor losses and viscous shear will increase. These effects, with regard to minor losses and density effects, reduce the buoyant force, while the viscous

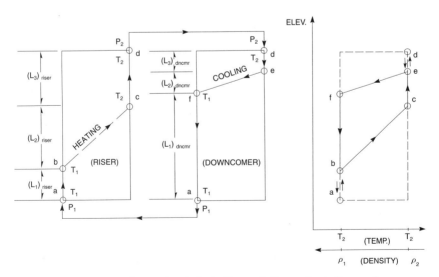

Figure 30.1 Natural circulation circuit, including modeling and notation used.

drag force is increased, and a steady state is reached when these forces are numerically equal.

By itself the heating and cooling of the fluid does not establish a net buoyant force. A symmetric (temperature) distribution would create no unbalanced force. Basically it is the existence of isothermal (adiabatic) sections in the upper portion of the riser (the chimney) and lower portion of the downcomer that give rise to the net force—as their respective fluid densities are at a minimum and maximum.[1]

Major Analysis Assumptions

The following restrictions and assumptions are made in order to develop relatively simple equations, which are nevertheless broadly applicable.

The fluid is single-phase (liquid or gas), with a steady one-dimensional flow; and the (spatial) differences in pressure, temperature, and velocity (and the velocity itself) are small. As a result, the differences in density are small and the fluid flow can be considered to be incompressible.

The fluid density is assumed to be a linear function of temperature, and other fluid properties, including specific heat c_p and viscosity μ, are constant, or the use of an average value is satisfactory.

The rate of heat transfer per length is assumed to be a symmetric function (in elevation), symmetric with respect to the vertical midpoint of the heated and cooled regions. The resulting equations (for flow rate and inlet or outlet temperature) are identical to those obtained for the simpler case of a constant heat transfer (per unit length).

If the heat transfer is constant, the temperature and density will be linear functions of elevation. Figure 30.1 exhibits this trend; however, this behavior is shown for simplicity—it is not a requirement.

In two-phase liquid vapor systems the basic approach is similar, although the details can be considerably more complex. Effects of compressibility (changes in momentum, critical flow, etc.), flow regime, differing gas and liquid (phase) velocity in a mixture, flooding, and other phenomena may be significant. These may even preclude a steady state.

---

[1] In Fig. 30.1 the area $\overline{bcef}$ in the density-versus-elevation plot, multiplied by acceleration due to gravity g, is the total buoyant force—without the local pressure losses due to friction. This can be compared to the larger rectangular area which represents the maximum buoyant force, if the heating could be accomplished over negligible height at the bottom of the riser, and cooling over negligible height at the top of the downcomer. The figures and description are consistent with a fluid that expands when heated (i.e., $\partial \rho / \partial T < 0$).

30.4 Heat Exchangers

Development of Natural Circulation Equations for Flow Rate and Temperature

The change in pressure between two locations is the same for any path; thus (from Fig. 30.1) the pressure difference $(P_2 - P_1)$ will be identical for fluid flowing up the riser from P_1 to P_2 and fluid flowing down the downcomer from P_2 to P_1.

The pressure varies in the riser and downcomer as a result of hydrostatic head, viscous shear, and minor losses:

Hydrostatic head
$$\frac{dP}{dz} = -\rho g \rightarrow \int dP = -g \int \rho \, dz \qquad (30.1)$$

For the viscous shear and minor losses, it is convenient to use the mass flux, calculated at a single convenient area, as the flow variable: $G_0 = \dot{m}/A_0$. Since loss coefficients (K_{local}) and friction factors (f) are in general based on different areas, a ratio of the flow area A_0 to the local flow area is included:

Viscous shear
$$\int dP_i = -\int_{L_i} \frac{4 f_i}{D_i} \left(\frac{A_0}{A_i}\right)^2 \frac{G_0}{2\rho_i} dz \rightarrow \Delta P_i$$

$$= -\frac{4 f_i L_i}{D_i} \left(\frac{A_0}{A_i}\right)^2 \frac{G_0^2}{2\rho_i} \qquad (30.2)$$

Local minor losses
$$\Delta P_j = -(K_{\text{local}})_j \left(\frac{A_0}{A_j}\right)^2 \frac{G_0^2}{2\rho_j} \qquad (30.3)$$

Integrating along the height of the riser and downcomer, we obtain

Riser
$$P_2 = P_1 - \int_{\text{riser}} \rho g \, dz - G_0^2 \sum_{\text{riser}} \left[\frac{4 f_i L_i}{D_i} \left(\frac{1}{2\rho_i}\right)\left(\frac{A_0}{A_i}\right)\right.$$

$$\left. + (K_{\text{local}})_j \left(\frac{1}{2\rho_j}\right)\left(\frac{A_0}{A_j}\right)^2 \right] \qquad (30.4)$$

Downcomer
$$P_2 = P_1 - \int_{\text{dncmr}} \rho g \, dz + G_0^2 \sum_{\text{dncmr}} \left[\frac{4 f_i L_i}{D_i} \left(\frac{1}{2\rho_i}\right)\left(\frac{A_0}{A_i}\right)\right.$$

$$\left. + (K_{\text{local}})_j \left(\frac{1}{2\rho_j}\right)\left(\frac{A_0}{A_j}\right)^2 \right] \qquad (30.5)$$

The energy equation (first law of thermodynamics) provides the fluid temperature, and thus the density, versus height:

$$\frac{\dot{q}}{L}dz = \dot{m}c_p\,dT \rightarrow \int \frac{\dot{q}}{L}dz = \dot{m}c_p \int dT = G_0 A_0 c_p \int dT \qquad (30.6)$$

If the heat transfer per length is a function of height z and is a symmetric function with respect to the midpoint of a heated or cooled section, then the average temperature T_{avg} is the mean of the inlet and outlet temperatures (T_1 and T_2 for the riser) and the average density is the mean of the inlet and outlet densities, ρ_1 and ρ_2. For $T(z)$ symmetric about $z = 0$, we obtain

$$\frac{\int_{-L/2}^{L/2} T(z)dz}{L} \equiv T_{\text{avg}} = \frac{T_1 + T_2}{2} \rightarrow \rho_{\text{avg}} = \frac{\rho_1 + \rho_2}{2} \qquad (30.7)$$

We will divide the riser and downcomer (each) into three vertically stacked sections (regions) as shown in Fig. 30.1: two isothermal (adiabatic) sections, one above and one below the heated (and cooled) sections. These are isothermal sections of height $(L_1)_{\text{riser}}$ and $(L_1)_{\text{dncmr}}$ with fluid at density ρ_1; using Eq. (30.7), heated and cooled sections $(L_2)_{\text{riser}}$ and $(L_2)_{\text{dncmr}}$ at ρ_{avg}; and isothermal sections $(L_3)_{\text{riser}}$ and $(L_3)_{\text{dncmr}}$ at ρ_2.

Substituting these heights and densities in Eqs. (30.4) and (30.5) results in expression (30.8) for the mass flow per unit area G_0, and integration of the energy equation [Eq. (30.6)] provides a second equation [Eq. (30.9)] in the common variables T_1, T_2, and G_0:

$$G_0^2 = \frac{\rho_1 g[(L_1)_{\text{dncmr}} - (L_1)_{\text{riser}}] + \rho_{\text{avg}}\,g[(L_2)_{\text{dncmr}} - (L_2)_{\text{riser}}] + \rho_2 g[(L_3)_{\text{dncmr}} - (L_3)_{\text{riser}}]}{\sum\limits_{\text{all}}\left[(4f_i L_i/D_i)(1/2\rho_i)(A_0/A_i)^2 + (K_{\text{local}})_j (1/2\rho_j)(A_0/A_j)^2\right]}$$

(30.8)

$$\dot{q}_{\text{riser}} = -\dot{q}_{\text{dncmr}} = \dot{m}c_p(T_2 - T_1) = G_0 A_0 c_p(T_2 - T_1) \qquad (30.9)$$

These two equations contain the heat transfer $\dot{q}$, the riser or downcomer outlet temperature (T_1 or T_2), and the mass flux G_0; and can be (numerically) solved since the fluid properties versus temperature, friction factors, and loss coefficients are readily available—often with correlations in the form of equations, tables, or graphs.

However, the solution to Eqs. (30.8) and (30.9) presents some difficulties, particularly for a "casual analysis." It is typically iterative, numerical accuracy issues are present (in particular, obtaining accurate density differences), and tables and graphs are time-consuming and difficult to implement in computer programs.

30.6 Heat Exchangers

These problems can be resolved, or significantly lessened, if the density, friction factors, and loss coefficients are expressed by equations of a particular functional form. The remainder of this analysis is directed largely toward the particular equations chosen to address these problems and the equations obtained from Eqs. (30.8) and (30.9).

Particular Equations (Correlations) for Density, Loss Coefficients, and Friction Factors

It should be noted that the buoyant (force) terms, the numerator in Eq. (30.8), are equivalent to differences in density. Consider a typical case: the configuration used to analyze an air-cooled heat exchanger in the following example calculation. Assume a three-section riser with unheated height $(L_1)_{riser}$ at (fluid density) ρ_1, heated height $(L_2)_{riser}$ at ρ_{avg}, and unheated height $(L_3)_{riser}$ at ρ_2 and a downcomer with its entire length unheated at density ρ_1,[2] resulting in Eq. (30.8) with the numerator replaced by $(\rho_1 - \rho_{avg})gL_2 + (\rho_1 - \rho_2)gL_3$.

This expression is typical; in general, the expression for the buoyant force will include density differences. In a typical problem (e.g., the following example calculation) the density values (ρ_1, ρ_{avg}, and ρ_2) are the same to two significant digits; thus to obtain an accuracy of only 10 percent, it would be necessary to use density values accurate to four digits. Most property tables only provide three or four significant digits! However, if the density differences are replaced by temperature differences, the large errors in evaluating density differences are avoided, and the resulting equation will be more easily solved.

Such a linear dependence (of density) on temperature is satisfactory for gases, as well as liquids, since the pressure and temperature differences are generally small.

This linear dependence is conveniently expressed by the coefficient of thermal expansion (α)—the fractional change in volume per unit change in temperature:

$$\left. \begin{array}{c} \alpha = -\dfrac{1}{\rho}\left(\dfrac{\partial \rho}{\partial T}\right)_P \cong -\dfrac{1}{\rho_1}\dfrac{\rho - \rho_1}{T - T_1} \to \rho - \rho_1 = -\alpha\rho_1(T - T_1) \\ \rho_2 - \rho_1 = -\alpha\rho_1(T_2 - T_1) \qquad \rho_{avg} - \rho_1 = -1/2\,\alpha\rho_1(T_2 - T_1) \end{array} \right\} \qquad (30.10)$$

Values of α for common liquids are often tabulated; however, α is easily calculated from the values of density at two temperatures. For

---

[2] This corresponds to air heated, rising, and mixing at the top of the riser or downcomer with cooler ambient air, which then flows down the downcomer. Mixing is equivalent to heat removal over a negligible height: $(L_2)_{dncmr} = (L_3)_{dncmr} = 0$, and $(L_1)_{dncmr} = (L_1 + L_2 + L_3)_{riser}$.

the ubiquitous liquid water and ideal gas (at ambient temperature), we obtain

Liquid water at 25°C $\alpha = 2.1 \times 10^{-4}$ K^{-1}

Ideal gas at 25°C $\alpha = \dfrac{1}{T} = \dfrac{1}{298 \text{ K}} = 0.00336$ K^{-1}

With regard to the loss coefficients (and friction factors) in prior analysis—with equations such as Eqs. (30.4) and (30.5), or Eq. (30.8)—Speyer [9] found it useful to describe loss coefficients as the sum of the laminar equation (varies as Re^{-1}) plus a constant turbulent value.[3] For several of the loss coefficients, such an equation was realistic,[4] and the expectation was it would be applicable to friction factors as well. In the present application this would lead to polynomials which can easily be solved[5]:

$$K_{\text{local}} = \frac{K\text{Re}}{\text{Re}} + CCK \quad \text{where} \quad K\text{Re} = \text{const}, CCK = \text{const}' \quad (30.11)$$

$$f = \frac{f\text{Re}}{\text{Re}} + CCf \quad \text{where} \quad f\text{Re} = \text{const}, CCf = \text{const}' \quad (30.12)$$

With regard to loss coefficients (K_{local}), although the minor losses frequently represent the majority of the pressure drop, laminar coefficients (KRe) are seldom required—the fully turbulent value, which is equivalent to the CCK value in Eq. (30.11),[6] is sufficient. Also, a large number of geometries are involved, and laminar data are scarce. In any

---

[3] As this calculation was being completed, three references in Idelchik [5] were found that apparently describe the use of an equation [e.g., Eq. (30.11)] for loss coefficients. Two references are for A. D. Altshul, and one for Idelchik. These were not available and are not included in the references list.

[4] Orifices (particularly tangent orifices) and tube banks with a staggered square or triangular pitch are some examples where the form of Eq. (30.11) works well, as are screens and flow normal to a circular disk; however, flow normal to a sphere, a single tube, and multiple tubes, all in a line, are some examples where it seems to be a poor model (due to boundary-layer separation, etc.).

[5] Laminar flow in pipes, and similar uniform interior flow geometries, are common in natural circulation calculations, and for these conditions, a constant friction factor is a particularly poor representation. For transitional or fully turbulent Reynolds numbers, if Eq. (30.11) or (30.12) and the associated constants are not sufficiently accurate for a particular flow regime, the constants can be revised (once an approximate solution for flow rate and Reynolds number is obtained). Thus it is an improvement over simply using a constant value for friction factor and loss coefficient—or at least no worse, as a constant value is obtained if the laminar-flow-based constants (KRe and fRe) are set to zero: $K_{\text{local}} = CCK$, $f = CCf$.

[6] The local Reynolds number is based on a velocity and characteristic dimension appropriate to the flow geometry. It is unrelated to that used for the friction factor.

30.8 Heat Exchangers

event the laminar component to the loss coefficients ($K\mathrm{Re}$) were not required for the example problem (based on a calculation of the Reynolds numbers—developed as part of that calculation), and are not reviewed herein.

In laminar flow the friction factor is inversely proportional to the Reynolds number ($f = f\mathrm{Re}/\mathrm{Re}$). For two common cases, flow inside a circular duct (pipe flow), and flow in the narrow passageway between parallel surfaces (slit flow), the constants ($f\mathrm{Re}$) are

$$\text{Laminar flow, } f = \frac{f\mathrm{Re}}{\mathrm{Re}} : \quad \begin{array}{l} f\mathrm{Re} = 16 \quad \text{(pipe flow)} \\ f\mathrm{Re} = 24 \quad \text{(slit flow)} \end{array} \tag{30.13}$$

For other geometries the values are readily available; for instance, Holman [4] summarizes the results of Shah and London [8], for values of $f\mathrm{Re}$ (his $f\mathrm{Re}_{DH}/4$), as well as Nusselt number Nu, for various geometries.[7]

For turbulent flow in smooth and commercial piping, a number of authors have suggested a constant value of (about) 0.005

$$\text{Turbulent flow} \quad f \cong 0.005 \tag{30.14}$$

which is used for smooth tubing, and for commercial piping, $f \cong 0.0075$ is used.[8]

The laminar and turbulent equations [Eqs. (30.13) and (30.14)] are now combined, and to assess how realistic this combined equation is, the laminar flow [Eq. (30.13)] is used, and for turbulent flow in smooth and rough pipes the equation [Eq. (30.15)] due to Colebrook [2] is used:

$$\text{Turbulent flow} \quad \frac{1}{\sqrt{4f}} = -0.869 \ln\left(\frac{\varepsilon/D}{3.7} + \frac{2.523}{\mathrm{Re}\sqrt{4f}}\right) \tag{30.15}$$

This equation, with a surface roughness-to-diameter of zero ($\varepsilon/D = 0$), reproduces *Prandtl's universal law of friction for smooth pipes*; and (for $\varepsilon/D = 0.01$), the equation of Drew and Genereaux [3], which McAdams [6] recommends as being applicable for "commercial pipes, steel, cast iron, etc., ±10 percent."

---

[7] If using an earlier edition of Holman [4], the values should be checked. There are errors in the fifth edition (1981).

[8] Although the difference is rarely large, prior experience with a particular application, known or estimated surface roughness, and known diameter(s), may suggest another turbulent friction factor constant (CCf). For large surface roughness and small diameter an order of magnitude larger value is possible, in which case Eq. (30.12) would be particularly inaccurate for laminar and intermediate Reynolds numbers ($2300 > \mathrm{Re} \gtrsim 100$).

Equation (30.13) is found to be realistic, with an error of 25 percent or less, for most Reynolds numbers of interest. In particular, as a general guide (for pipe and slit flow), $f = (f\text{Re}/\text{Re}) + CCf$: $f\text{Re} \approx 16$ to 24 and $CCf \approx 0.005$ to 0.0075, for which

1. Generally error in $f < 25$ percent
2. f value generally conservative
3. Large overestimates likely for $\approx 1000 \leq \text{Re} \leq 2300$

Incorporating Density, Friction Factor, and Loss Coefficient Correlations in Natural Circulation Equations

Equation (30.8) is now modified to include temperature differences rather than density differences, and the friction factor and loss coefficient equations. Equations (30.8) and (30.9) are iterative if solving for flow (G_0) and outlet temperature (T_1 or T_2); however, the revised equations are usually not iterative,[9] unless there are large changes (with temperature) in fluid properties (principally viscosity)—which is not typical of natural circulation in air or liquid water.

Combining Eqs. (30.8) and (30.10) to (30.12) results in a second-order polynomial:

$$G_0^2 = \frac{[(L_2)_{\text{riser}} + 2(L_3)_{\text{riser}} - (L_2)_{\text{dncmr}} - 2(L_3)_{\text{dncmr}}]^{1/2} \alpha \rho_1 g(T_2 - T_1)}{\sum_{\text{all}} \left[\left(\frac{A_0}{A_i}\right)^2 \left(\frac{1}{2\rho_i}\right)\left(\frac{4L_i}{D_i}\right)\left(\frac{f\text{Re}_i \mu_i A_i}{G_0 D_i A_0} + CCf_i\right) + \left(\frac{A_0}{A_j}\right)^2 \left(\frac{1}{2\rho_j}\right)\left(\frac{K\text{Re}_j \mu_j A_j}{G_0 D_j A_0} + CCK_j\right)\right]} \times \frac{2\rho_1}{2\rho_1}$$

(30.16)

The solution for G_0 is Eq. (30.17), and the sign of the coefficients are $a \geq 0$, $b \geq 0$, and $c \leq 0$.

$$\left. \begin{aligned} aG_0^2 + bG_0 + c = 0 &\rightarrow G_0 = \frac{-b + \sqrt{b^2 - 4ac}}{2a} \\ c = \big[&-(L_2)_{\text{riser}} - 2(L_3)_{\text{riser}} + (L_2)_{\text{dncmr}} + 2(L_3)_{\text{dncmr}}\big] \alpha \rho_1^2 g(T_2 - T_1) \\ b = \sum_{\text{all}} &\left[\frac{4L_i f \text{Re}_i \mu_i}{D_i^2}\left(\frac{\rho_1}{\rho_i}\right)\left(\frac{A_0}{A_i}\right) + \frac{K\text{Re}_j \mu_j}{D_j}\left(\frac{\rho_1}{\rho_j}\right)\left(\frac{A_0}{A_j}\right)\right] \\ a = \sum_{\text{all}} &\left[\frac{4L_i CCf_i}{D_i}\left(\frac{\rho_1}{\rho_i}\right)\left(\frac{A_0}{A_i}\right)^2 + CCK_j\left(\frac{\rho_1}{\rho_j}\right)\left(\frac{A_0}{A_j}\right)^2\right] \end{aligned} \right\} \quad (30.17)$$

---

[9] When the natural circulation equations are applied to a heat exchanger, as in the following example, the solution is often iterative. But, this is due to the additional heat-transfer rate equation.

30.10 Heat Exchangers

If the Reynolds number is large, coefficient b is negligible, resulting in the following equation, where a and c are as defined in (30.17):

$$aG_0^2 + c = 0 \rightarrow G_0 = \sqrt{-\frac{c}{a}} \qquad (30.18)$$

Although unlikely, if the flow is laminar (for all the important friction factors and loss coefficients), the coefficient a is negligible, resulting in the following equation, where b and c are as defined in Eq. (30.17):

$$bG_0 + c = 0 \rightarrow G_0 = -\frac{c}{b} \qquad (30.19)$$

If the riser outlet temperature T_2 and heat transfer are both unknown, then $T_2 - T_1$ can be replaced by the heat transferred, by combining Eqs. (30.9) and (30.16). This results in a third-order polynomial in G_0, Eq. (30.20).[10]

As in Eqs. (30.18) and (30.19), if applicable, the solutions may be simplified by setting $a = 0$ or $b = 0$ [and b and a are as defined in Eq. (30.17)]:

$$\left.\begin{array}{l} aG_0^3 + bG_0^2 + c' = 0 \\[4pt] c' = c\dfrac{\dot{q}_{\text{riser}}}{A_0 c_p (T_2 - T_1)} \\[4pt] = [-(L_2)_{\text{riser}} - 2(L_3)_{\text{riser}} + (L_2)_{\text{dncmr}} + 2(L_3)_{\text{dncmr}}]\dfrac{\alpha\rho_1^2 g \dot{q}_{\text{riser}}}{A_0 c_p} \end{array}\right\} \qquad (30.20)$$

The equations (to this point) are written assuming that the riser inlet temperature T_1 is known. If it is the downcomer inlet temperature T_2 that is known (and not the riser inlet), a more convenient equation, equivalent to Eq. (30.16), is as follows:

$$G_0^2 = \frac{[-(L_2)_{\text{riser}} - 2(L_1)_{\text{riser}} + (L_2)_{\text{dncmr}} + 2(L_1)_{\text{dncmr}}]^{1/2} \alpha \rho_2 g (T_2 - T_1)}{\sum_{\text{all}} \left[\left(\dfrac{A_0}{A_i}\right)^2 \left(\dfrac{1}{2\rho_i}\right)\left(\dfrac{4L_i}{D_i}\right)\left(\dfrac{f \text{Re}_i \mu_i A_i}{G_0 D_i A_0} + CC f_i\right) + \left(\dfrac{A_0}{A_j}\right)^2 \left(\dfrac{1}{2\rho_j}\right)\left(\dfrac{K \text{Re}_j \mu_j A_j}{G_0 D_j A_0} + CC K_j\right) \right]} \times \frac{2\rho_2}{2\rho_2} \qquad (30.21)$$

---

[10] A number of the commercial computer programs will easily find the roots of a cubic polynomial.

The solutions to Eq. (30.21) are similar to those for Eq. (30.16).[11] Equation (30.22) is analogous to Eq. (30.17):

$$a'G_0^2 + b'G_0 + c'' = 0 \rightarrow G_0 = \frac{-b' + \sqrt{b'^2 - 4a'c''}}{2a'}$$

$$c'' = [(L_2)_{\text{riser}} + 2(L_1)_{\text{riser}} - (L_2)_{\text{dncmr}} - 2(L_1)_{\text{dncmr}}]\alpha\rho_2^2 g(T_2 - T_1)$$

$$b' = \sum_{\text{all}} \left[\frac{4L_i f \text{Re}_i \mu_i}{D_i^2} \left(\frac{\rho_2}{\rho_i}\right)\left(\frac{A_0}{A_i}\right) + \frac{K \text{Re}_j \mu_j}{D_j}\left(\frac{\rho_2}{\rho_j}\right)\left(\frac{A_0}{A_j}\right) \right] \qquad (30.22)$$

$$a' = \sum_{\text{all}} \left[\frac{4L_i CC f_i}{D_i} \left(\frac{\rho_2}{\rho_i}\right)\left(\frac{A_0}{A_i}\right)^2 + CCK_j\left(\frac{\rho_2}{\rho_j}\right)\left(\frac{A_0}{A_j}\right)^2 \right]$$

Similarly, Eq. (30.23) is analogous to Eq. (30.19):

$$a'G_0^3 + b'G_0^2 + c''' = 0$$

$$c''' = c'' \frac{\dot{q}_{\text{riser}}}{A_0 c_p (T_2 - T_1)}$$

$$= [(L_2)_{\text{riser}} + 2(L_1)_{\text{riser}} - (L_2)_{\text{dncmr}} - 2(L_1)_{\text{dncmr}}]\frac{\alpha\rho_2^2 g \dot{q}_{\text{riser}}}{A_0 c_p} \qquad (30.23)$$

and b' and a' are as given in Eq. (30.22).

Example Calculation for an Air-Cooled Heat Exchanger

Air-cooled heat exchangers are used to cool the oil in large electric transformers, as shown in Fig. 30.2. This calculation develops the air natural circulation flow, its outlet temperature, and the heat transfer.[12] The source of the heat is resistance heating due principally to current in the transformer's metal windings. This is generally referred to as *load loss*.

To calculate the heat transfer, the air natural circulation flow and air outlet temperature must be calculated. It is assumed that the oil pumps are operating, the air-side heat transfer is controlling, and the oil and ambient air temperatures are known.

The natural circulation process shown in Fig. 30.2 differs from the earlier configuration in Fig. 30.1, in which the entire natural circulation

---

[11] In both Eqs. (30.20) and (30.23), $\dot{q}_{\text{riser}}$ is used. If the standard thermodynamic convention is observed (with reference to a system, heat flow in is numerically positive, and heat flow out is negative), the equation would be correct with the heat transfer $\dot{q}_{\text{riser}}$ replaced by $-\dot{q}_{\text{dncmr}}$.

[12] The calculation uses approximate (made up, but reasonable) numerical values to describe the geometry.

30.12 Heat Exchangers

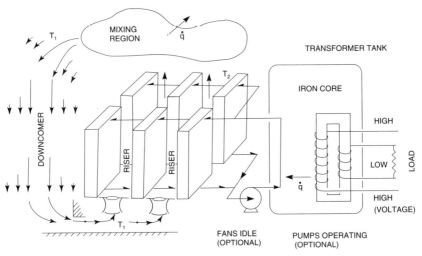

Figure 30.2 Schematic view of electric transformer air-cooled heat exchanger for example calculation.

flow was contained by a physical boundary. In this case, except for the inlet at the bottom of the heat exchanger, the airflow is (often) not enclosed by actual physical boundaries. Because of this difference it may not be immediately evident that a natural circulation analysis applies, or it may not be evident how it is to be applied. The following description should clarify matters.

A mass (per unit time) of air enters the bottom of the heat exchanger, and as it rises, it is heated. It is assumed that the air does not mix with ambient air at the periphery of the riser section. At the top of the heat exchanger the heated air mixes with the ambient cooler air. (If the imagined enclosure were made real, this would be the location of a second heat exchanger.) To satisfy steady-state conservation of mass, the identical mass of air flows down external to the heat exchanger, and it enters the bottom, thus completing the flow circuit.

As the air flows, it encounters several locations where pressure loss will occur. These include contractions and expansions in flow area and viscous drag.

With a solid vertical boundary, and assuming radial variation in heat transfer is small, the riser flow can be described by a one-dimensional (1D) model, as assumed herein. Without the vertical boundary, an infinitely wide riser (compared to its height) remains in 1D flow; however, reducing the width increases lateral flow. This warmer fluid (air) can mix with cooler ambient fluid at the periphery—resulting in increased heat transfer.

In the following numerical example (and typical of electric transformers) the riser width and height are similar, and lateral flow resistance is not large (the geometry is rather "open"). The 1D assumption is not realistic and an underprediction (in heat transfer) by a factor of 2 would not be unexpected; however, if a solid vertical boundary were present the 1D calculation would be expected to be realistic.

The heat exchanger comprises parallel flattened tubes, roughly rectangular in shape: 10 cm long, 1.1 cm wide, and 2 m high. Above and below the 2-m tubes are structural members and a manifold. These entry and exit regions are each an additional 0.25 m high.

Hot oil from the transformer enters a manifold at the top of the heat exchanger, splitting the oil flow between the (flattened) tubes (3400 tubes). The oil flows down the interior of the tubes, and a manifold at the bottom collects the cooled oil, which is returned to the transformer.

The spacing between these tubes provides a partially enclosed flow area for the cooler air—in counterflow to the hotter oil. The tubes are arranged in rows; the distance between the ends of successive tubes is 1.5 cm, and the distance between the sides of adjacent tubes is 3.5 cm. In the horizontal plane a repeating unit (unit cell) can be assumed to be the rectangular region connecting the centers of four neighboring tubes.

The heat exchanger has a 3- × 6-m footprint and a 2.5-m height. The long sides of the flattened tubes are parallel to the 6-m dimension.

For riser:

Flow area in unit cell	$(10 + 1.5) \times 3.5 + (1.1 \times 1.5) \, \text{cm}^2 = 41.9 \, \text{cm}^2$
Equivalent diameter of unit cell (see below)	$\dfrac{4 \times 52.9 \, \text{cm}^2 \times 10^{-2} \, \text{m/cm}}{2 \times (10 + 1.1) \, \text{cm}} = 0.0755 \, \text{m}$
Height of isothermal sections	$(L_1)_{\text{riser}} = (L_3)_{\text{riser}} = 0.25 \, \text{m}$
Height of heated section	$(L_2)_{\text{riser}} = 2.00 \, \text{m}$
Flow area	$A_0 = 3 \times 6 \, \text{m}^2 \times \dfrac{49.1 \, \text{cm}^2}{52.9 \, \text{cm}^2} = 14.2 \, \text{m}^2$
Heat-transfer (ht) area	$A_{\text{ht}} = 2 \times (10 + 1.1 \, \text{cm}) \times 2 \, \text{m} \times \dfrac{3 \times 6 \, \text{m}^2}{52.9 \, \text{cm}^2}$ $\times \dfrac{100 \, \text{cm}}{\text{m}} = 1510 \, \text{m}^2$

For heat exchanger overall:

Projected area of unit cell $(10 + 1.5) \times (1.1 + 3.5) \, \text{cm}^2 = 52.9 \, \text{cm}^2$

The mixing of air with ambient (as was discussed earlier) is modeled by removing all the heat at the top of the downcomer.

For downcomer:

Height of upper isothermal section	$(L_3)_{\text{dncmr}} = 0$
Height of cooling section	$(L_2)_{\text{dncmr}} = 0$
Height of bottom isothermal section	$(L_1)_{\text{dncmr}} = 2.5 \, \text{m}$

30.14 Heat Exchangers

The bottom entry is modeled as an orifice with an area that is 20 percent of the projected horizontal area.

There is a sudden (area) contraction at a 0.6-m clearance between the bottom of the riser and the concrete pad. In addition, the presence of the manifold results in a reduced area, representing 34 percent of the total projected area, at the top and bottom of the riser.

For heat exchanger overall:

Total projected (horizontal) area (top area) $A_\tau = 3 \text{ m} \times 6 \text{ m} = 18 \text{ m}^2$

Minimum flow area (orifice) at inlet $A_{\text{orifice}} = 0.2 \times 18 \text{ m}^2 = 3.6 \text{ m}^3$

For the calculation of loss coefficients, the permanent pressure loss for an orifice and sudden contraction and sudden expansion (in flow area) are required.

The following (for turbulent flow) are adapted from a summary in Bird, Stewart, and Lightfoot [1]. For a sudden contraction, the coefficient is increased from 0.45 to 0.5. Defining an area ratio β, the ratio of smaller cross sectional area to larger cross sectional area, we obtain

$$\Delta P_{\text{orifice}} = -\frac{K_{\text{orifice}}}{2\rho_{\text{local}}} \left(\frac{\dot{m}}{A_{\text{downstream}}}\right)^2 \tag{30.24}$$

$$\beta = \frac{A_{\text{orifice}}}{A_{\text{downstream}}} \qquad K_{\text{orifice}} = \frac{2.7(1-\beta^2)(1-\beta)}{\beta}$$

$$\Delta P_{\text{SE}} = -\frac{K_{\text{SE}}}{2\rho_{\text{local}}} \left(\frac{\dot{m}}{A_{\text{upstream}}}\right)^2 \qquad K_{\text{SE}} = (1-\beta)^2 \tag{30.25}$$

$$\Delta P_{\text{SC}} = -\frac{K_{\text{SC}}}{2\rho_{\text{local}}} \left(\frac{\dot{m}}{A_{\text{downstream}}}\right)^2 \qquad K_{\text{SC}} = 0.5(1-\beta) \tag{30.26}$$

(where subscripts SC and SE denote sudden compression and contraction, respectively).

For the natural circulation flow G_0, the riser flow area is the most convenient area to use (for A_0) in Eq. (30.17).

The loss coefficients are often the major part of a fluid flow calculation. Typically, drawings and other sources provide the detail, and then a large number of individual resistances must be developed and documented. Thus the calculations should be developed in a clear, compact manner—in this case a tabular format, such as Table 30.1.

The effect of the density ratio ($\rho_1/\rho_{\text{local}}$) will be small, and thus it will be simplest if the losses are combined into a single sum (this simplifies the bookkeeping). About two-thirds of the losses occur at the inlet conditions, and $\rho_1/\rho_{\text{local}} = \rho_1/\rho_{\text{avg}}$ is used (this is slightly conservative).

TABLE 30.1 Calculation of Loss Coefficients for Sample Calculation of Air-Cooled Heat Exchanger

Description	A_i, m²	K_i for	β_i	K_i	$(A_o/A_i)^2$	$K(A_o/A_i)^2$
Clearance at entry	$0.6 \times 2(3+6) = 10.8$	SC*	$10.8/18 = 0.6$	$0.5(1-\beta) = 0.20$	$(14.2/10.8)^2 = 1.73$	$0.2 \times 1.73 = 0.35$
Minimum entry area	$0.25 \times 18 = 4.5$	Orifice	0.25	$(1-\beta)^2(1-\beta)/\beta = 4.56$	$(14.2/18)^2 = 0.622$	$4.56 \times 0.789 = 2.84$
Bottom manifold	$0.34 \times 18 = 6.12$	SC	0.34	$0.5(1-\beta) = 0.33$	$(14.2/6.12)^2 = 5.38$	$0.33 \times 5.38 = 1.78$
		SE†	$6.12/14.2 = 0.43$	$(1-\beta)^2 = 0.32$		$0.32 \times 5.38 = 1.74$
Top manifold		SC		$0.5(1-\beta) = 0.29$		$0.29 \times 5.38 = 1.53$
		SE	0.34	$(1-\beta)^2 = 0.44$		$0.44 \times 5.38 = 2.35$

*SC = sudden compression.
†SE = sudden expansion.

All loss coefficients are adjusted by the area ratio (A_0/A_{local}) and then summed. The resulting sum of the loss coefficients (10.6) is rounded to 11:

$$\sum CCK_i \left(\frac{\rho_1}{\rho_i}\right)\left(\frac{A_0}{A_i}\right)^2 \cong \sum K_i \left(\frac{\rho_1}{\rho_{\text{avg}}}\right)\left(\frac{A_0}{A_i}\right)^2 \cong 11\left(\frac{\rho_1}{\rho_{\text{avg}}}\right)$$

Since the laminar loss coefficient contribution was not included, it is a good idea to develop a justification particularly when a laminar result (for friction is obtained and/or in new situations), although a viscous oil, or the equivalent, is typically required to obtain laminar flow for the minor losses. The following provides the justification.

The Reynolds number for flow in the downcomer (Re_0) can be related to the local Reynolds number (Re_{local}) for the evaluation of the K_{local} values:

$$\text{Re}_{\text{local}} = \frac{\dot{m}D_{\text{local}}}{A_{\text{local}}\mu} \times \frac{A_0}{A_0} \times \frac{D_0}{D_0} = \frac{\dot{m}D_0}{A_0\mu}\frac{D_{\text{local}}}{D_0}\frac{A_0}{A_{\text{local}}} = \text{Re}_0 \frac{D_{\text{local}}}{D_0}\frac{A_0}{A_{\text{local}}} \quad (30.27)$$

From Table 30.1, the principal losses are at the manifold and the inlet orifice. The equivalent diameter of the manifold will be similar to the tubes ($D_{\text{manifold}} \approx D_0$), and the equivalent diameter of the orifice is assumed to be 1.5 m.

Looking ahead, the result of this example calculation is $\text{Re}_0 = 1430$, and we obtain

$$\text{Re}_{\text{fan}} = \text{Re}_0 \frac{D_{\text{orifice}}}{D_0}\frac{A_0}{A_{\text{orifice}}} = 1430 \times \frac{0.15 \text{ m}}{0.07 \text{ m}}\sqrt{0.622} \approx 2400$$

$$\text{Re}_{\text{manifold}} = \text{Re}_0 \frac{D_{\text{manifold}}}{D_0}\frac{A_0}{A_{\text{manifold}}} = 1430 \times \frac{0.07 \text{ m}}{0.07 \text{ m}}\sqrt{5.38} \approx 3340$$

The manifold regions are essentially orifices,[13] and the orifice K values (for $\beta < 0.5$) vary by about ±15 percent for $\text{Re}_{\text{orifice}} > 40$.[14] Thus the use of constant turbulent K values is justified.

---

[13] The values of β are 0.25, 0.33, and 0.43, from Table 30.1. The average is $\beta \sim 0.35$, and the orifice K is 4.4; however, to compare this to the sudden expansion and sudden compression, the areas need to be put on an equal basis [see Eqs. (30.24) to (30.26)]. If all are based on the smaller area, the orifice K value is $K = 4.4 \times 0.35^2 = 0.54$. This compares reasonably well to the sum of the sudden compression and sudden expansion values, $0.33 + 0.42 = 0.75$.

[14] Perry and Chilton [7, figure 5-18] has discharge coefficient varying from 0.62 to 0.72, for $\beta < 0.5$ and $\text{Re}_{\text{orifice}} > 40$ (Reynolds number calculated with D = orifice diameter). The equivalent K values are 2.60 (0.62^{-2}) and 1.93 (0.72^{-2}), or $K \sim 2.3 \pm 15$ percent.

Now returning to our (main) calculation for the natural circulation, in order to calculate the laminar flow friction and heat transfer, it is necessary to adjust the flow geometry to a case where a laminar flow solution is readily available. The unit cell is similar to slit flow ($D = 2 \times$ gap width), and this is used. Note that the equivalent diameter calculated for the unit cell, and the value for a slit, are quite similar (0.0755 and 0.07 m).

From Holman [4], $f \mathrm{Re}_{DH}/4 = f\mathrm{Re} = 24$; and for constant temperature $\mathrm{Nu}_T = 7.541$ and constant heat flux $\mathrm{Nu}_H = 8.235$. Constant heat flux is consistent with expected poor heat transfer and a large, and thus relatively constant, tube wall-to-air temperature difference:

Equivalent diameter (for slit flow) $\quad D_{\text{riser}} = D_0 = 2 \times 3.5\,\text{cm} \times 10^{-2}\,\text{m/cm}$
$\qquad\qquad\qquad\qquad\qquad\qquad\qquad = 0.07\,\text{m}$

Laminar friction factor $\qquad\qquad\qquad f\mathrm{Re} = 24$

Laminar Nusselt number (constant heat flux) $\qquad (\mathrm{Nu}_{\text{riser}})_{\text{laminar}} = \mathrm{Nu}_H = 8.235$

For turbulent flow the friction factor is independent of geometry, but it depends on surface roughness. Heat exchangers such as this are placed in the outdoors and the exterior surfaces are painted, and (thus) not hydraulically smooth.

Friction factor (turbulent and not smooth) $\qquad (\varepsilon/D \approx 0.01) \rightarrow CCf = 0.0075$

The unknowns are the natural circulation flow and the outlet air temperature—and assuming a large oil flow rate the inlet and outlet oil temperatures will be essentially the same. For illustrative purposes the oil was assumed to change by 2°C; and air inlet temperature = $T_1 = 25°\text{C}$, oil inlet temperature = $(T_{\text{oil}})_{\text{in}} = 61°\text{C}$, and oil outlet temperature = $(T_{\text{oil}})_{\text{out}} = 59°\text{C}$.

In order to calculate the airflow and outlet temperature, Eq. (30.17) is used. As the friction (flow area) is chosen to be A_0 and the loss coefficients are calculated, including the adjustment to the area A_0 [i.e., $(A_0/A_{\text{local}})^2 \rightarrow (A_0/A_i)^2$], the calculation will proceed with $(A_0/A_i)^2 = 1$. To start the calculation, a value of the air outlet temperature (T_2) is assumed.

The first estimate (guess) is $T_2 = 40°\text{C}$:

$$T_{\text{avg}} = \frac{25 + 40°\text{C}}{2} = 32.5°\text{C}$$

Air properties are $\rho_1 = 1.168\,\text{kg/m}^3$, $\mu_{\text{avg}} = 1.872 \times 10^{-5}\,\text{kg/(m s)}$, $\rho_{\text{avg}} = 1.139\,\text{kg/m}^3$, and $k_{\text{avg}} = 2.672 \times 10^{-5}\,\text{kW/m/°C}$, with

$$\alpha = (32.5 + 273.15\,\text{K})^{-1} = 3.272 \times 10^{-3}\,\text{K}^{-1}$$

30.18 Heat Exchangers

Then

$$c = [-2.0 - 2 \times 0.25\,\text{m} + 0 + 0]3.272 \times 10^{-3}\,\text{K}^{-1}\,(1.139\,\text{kg/m}^3)^2$$
$$\times 9.8\,\text{m/s}^2(40 - 25°\text{C}) = -1.640[\text{kg/(m}^2\,\text{s})]^2$$

$$\frac{\rho_1}{\rho_{avg}} = \frac{1.168\,\text{kg/m}^3}{1.139\,\text{kg/m}^3} = 1.025 \qquad \frac{A_0}{A_i} = 1 \quad \text{(see discussion in text)}$$

$$b = \sum_{\text{all}} \frac{4 \times 2.5\,\text{m} \times 24 \times 1.872 \times 10^{-5}\,\text{kg/(m}^2\,\text{s})}{(0.07\,\text{m})^2} \times 1.025 \times 1$$

$$= 0.9402\,\text{kg/(m}^2\,\text{s})$$

$$a = \sum_{\text{all}} \left[\frac{4 \times 2.5\,\text{m} \times 0.0075}{0.07\,\text{m}} \times 1.025 + 11 \times 1.025 \right] \times 1^2 = 12.37$$

$$G_0 = \frac{-0.9402\,\text{kg/(m}^2\,\text{s}) + \sqrt{(0.9402^2[\text{kg/(m}^2\,\text{s})]^2 - 4 \times 12.37 \times (-1.640)}}{2 \times 12.37}$$

$$= \frac{0.3281\,\text{kg}}{\text{m}^2\,\text{s}}$$

A review of this calculation reveals that an incorrect tube length of 2.5 m, not 2 m, was used in the calculation of the coefficients a and b, immediately above. However, the effect of this minor error is not significant (and it is conservative), and the calculation was not revised:

$$(\text{Velocity}_0)_{avg} = \frac{G_0}{\rho_{avg}} = \frac{0.3281\,\text{kg/(m}^2\,\text{s})}{1.168\,\text{kg/m}^3} = 0.2809\,\frac{\text{m}}{\text{s}}$$

$$(\text{Re}_0)_{avg} = \frac{G_0 D_0}{\mu} = \frac{0.3281\,\text{kg/(m}^2\,\text{s})\,0.07\,\text{m}}{1.872 \times 10^{-5}\,\text{kg/(m s)}} = 1227$$

One version of the heat transfer is obtained from conservation of energy [Eq. (30.9)]:

$$\dot{q}_{riser} = G_0 A_0 c_p (T_2 - T_1) = \frac{0.3316\,\text{kg}}{\text{m}^2\,\text{s}} \times 14.2\,\text{m}^2 \times \frac{1.008\,\text{kJ}}{\text{kg °C}}$$

$$\times \frac{\text{kW}}{1\,\text{kJ/s}} \times (40 - 25°\text{C}) = 71.2\,\text{kW}$$

The other version of the heat transfer is obtained from the (heat exchanger) rate equation. As the Reynolds number is less than 2300, the laminar flow convective heat-transfer coefficient is used:

$$(\text{Nu}_{\text{riser}})_{\text{laminar}} = \text{Nu} = \frac{hD}{k} = 8.235 \rightarrow h = \frac{\text{Nu}\, k}{D}$$

$$= \frac{8.235 \times 2.672 \times 10^{-5}\ \text{kW/(m\ °C)}}{0.07\ \text{m}} = 0.003143\ \frac{\text{kW}}{\text{m}^2\ °\text{C}}$$

The oil and air are in counterflow:

$$\dot{q} = U \times A \times \text{LMTD} \cong \frac{(A_{\text{ht}})_{\text{air}}}{\dfrac{1}{h_{\text{air}}} + \dfrac{1}{h_{\text{oil}}}\dfrac{(A_{\text{ht}})_{\text{air}}}{(A_{\text{ht}})_{\text{oil}}} + \dfrac{1}{h_{\text{fouling}}} + \dfrac{\Delta x_{\text{wall}}}{k}}$$

$$\times \frac{[(T_{\text{oil}})_{\text{in}} - T_2] - [(T_{\text{oil}})_{\text{out}} - T_1]}{\ln\dfrac{(T_{\text{oil}})_{\text{in}} - T_2}{(T_{\text{oil}})_{\text{out}} - T_1}} \qquad (30.28)$$

For a very small air-side heat-transfer coefficient, the fouling factor, wall conductance, and oil-side heat-transfer coefficient will be negligible:

$$\dot{q} = \frac{0.003143\ \text{kW}}{\text{m}^2\ °\text{C}} \times 1510\ \text{m}^2 \frac{(61 - 40°\text{C}) - (59 - 25°\text{C})}{\ln\left(\dfrac{61 - 40°\text{C}}{59 - 25°\text{C}}\right)} = 128.0\ \text{kW}$$

The two values of the heat transfer (71.2 and 128.0 kW) are in poor agreement, and a new value of the air outlet temperature is estimated (guessed). Once the second calculation is completed, the difference in heat transfer for each calculation can be used to extrapolate to a better estimate of heat transfer.

The results for the first three values of air outlet temperature are summarized in Table 30.2.

TABLE 30.2 Summary of Outlet Temperature T_2 and Heat Transfer $\dot{q}$ for Three Iterations for Sample Calculation of Air-Cooled Heat Exchanger

Air outlet temperature T_2, °C	Basis of T_2	G_0, kg/(m² s)	q, kW Eq. (30.9)	q, kW Eq. (30.27)	$\Delta \dot{q}$
40	Guess	0.3281	70.44	128.0	−45.0
45	Guess	0.3805	108.9	114.1	−4.5
45.56	$\Delta \dot{q} \rightarrow 0$	0.3859	113.6	112.4	1.0

30.20 Heat Exchangers

The final (third iteration) results are as follows:

Riser outlet temperature $T_2' = 45.56°C$
Mass flux of air in riser $G_0 = 0.3846$ kg/(m² s)
Total forced-convection heat transfer $\dot{q} = 1/2(113.0 + 112.9 \text{ kW}) = 113.0$ kW

Summarizing some of the other numerical results, we obtain riser air velocity (Velocity$_{avg}$) of 0.3418 m/s; outside heat-transfer coefficient (h_{conv}) of 0.003167 kW/(m² °C); LMTD of 23.51°C; and skin friction (ΔP_{fric}) and minor loss ($\Delta P_{εκ}$) pressure drop of 0.2285 and 0.7253 Pa, respectively.

Finally, the radiation heat transfer, and natural convection heat transfer from the peripheral tubes, should be calculated. As the manifold is also at the oil temperature, the total height (2.5 m) is used.

The radiation heat transfer is from the four sides and top. The graybody radiation is given by

$$\dot{q}_{radiation} = \sigma A_{radiation} \varepsilon (T_{wall}^4 - T_{ambient}^4) \qquad (30.29)$$

The tube wall temperature is the average oil temperature, and paints typically have values of emissivity of 0.8 to about 0.95, and a value of 0.90 is used. The outside (envelope) area is applicable—four sides and the top:

$$A_{radiation} = 2 \times (6 \text{ m} \times 2.5 \text{ m} + 3 \text{ m} \times 2.5 \text{ m}) + 6 \text{ m} \times 3 \text{ m} = 63 \text{ m}^2$$

$$\dot{q} = \frac{5.67 \times 10^{-11} \text{ kW}}{\text{m}^2 \text{ K}^4} 63 \text{ m}^2 \, 0.90[(60 + 273.15 \text{ K})^4$$
$$- (25 + 273.15 \text{ K})^4] = 14.20 \text{ kW}$$

Assuming turbulent free convection at the outer surface of the peripheral tubes (i.e., vertical plates), McAdams [6] recommends

$$(\text{Nu})_{film} = \frac{hL}{k_{film}} = 0.13(\text{Gr Pr}_{film})^{1/3} \quad 10^9 < \text{Gr Pr}_{film} < 10^{12} \qquad (30.30)$$

and the Grashof and Prandtl numbers are evaluated at the film temperature (average of wall and ambient temperatures):

$$\text{Gr Pr}_{film} = (\text{Gr} \times \text{Pr})_{film} = gL^3(T_{wall} - T_{ambient})(\alpha \rho^2/\mu^2)_{film} \text{Pr}_{film}$$

$$T_{film} = \frac{1}{2}\left[\frac{25 + 45.42}{2} + 60°C\right] = 47.61°C$$

Air properties are $\rho_{film} = 1.086$ kg/m³, $\mu_{film} = 1.944 \times 10^{-5}$ kg/(m s), $\text{Pr}_{film} = 0.0703$, $k_{film} = 2.783 \times 10^{-5}$ kW/(m °C), and $\alpha = (47.61 + 273.15 \text{ K})^{-1} = 0.003118$ K⁻¹.

Then

$$A_{\text{nat conv}} = 2\left(6\text{ m} \times 2.5\text{ m} \times \frac{10\text{ cm}}{10+1.5\text{ cm}} + 3\text{ m} \times 2.5\text{ m} \times \frac{1.1\text{ cm}}{3.5+1.1\text{ cm}}\right)$$

$$= 29.67\text{ m}^2$$

$$\text{Gr Pr}_{\text{film}} = \frac{9.8\text{ m/s}^2\, 2.5^3\text{ m}^3\, 3.118 \times 10^{-3}\text{ K}^{-1}(60-25°\text{C})(1.086\text{ kg/m}^3)^2}{3.118 \times 10^{-3}\text{ K}[1.944 \times 10^{-5}\text{ kg/(m s)}]^2}$$

$$\times\, 0.703 = 3.665 \times 10^{10}$$

$$h = 0.13 \times (3.665 \times 10^{10})^{1/3} \times \frac{2.783 \times 10^{-5}\text{ kW/(m °C)}}{2.5\text{ m}}$$

$$= 0.004807 \frac{\text{kW}}{\text{m}^2\text{ s}}$$

This heat-transfer coefficient is similar in magnitude to the air-side natural convection value of 0.003167 kW/(m² °C)—thus the contribution of the peripheral tube heat transfer will be small. The additional heat transfer and the total are

$$\Delta \dot{q} = [0.004807 - 0.003167\text{ kW/(m}^2\text{ °C)}]\, 23.7\text{ m}^2(60-25°\text{C}) = 1.361\text{ kW}$$

The total heat transfer is $\dot{q}_{\text{total}} = 113.0 + 14.20 + 1.361\text{ kW} = 128.6\text{ kW}$.

References

1. Bird, R. B., W. E. Steward, and E. N. Lightfoot, *Transport Phenomena*, Wiley, p. 217 (1960).
2. Colebrook, C. F., "Turbulent Flow in Pipes, with Particular Reference to the Transition Region between Smooth Rough Pipe Laws," *J. Inst. Civil Eng.* (London), **11**, 133–156 (1938–1939).
3. Drew, T. B., and R. P. Genereaux, *Trans. Am. Inst. Chem. Eng.*, **32**, 17–19 (1936).
4. Holman, J. P., *Heat Transfer*, 9th ed., McGraw-Hill, p. 273 (2002).
5. Idelchik, I. E., *Handbook of Hydraulic Resistance*, 2d ed., Hemisphere Publishing, p. 3 (1986).
6. McAdams, W. H., *Heat Transmission*, 3d ed., McGraw-Hill, pp. 56–157 (1954).
7. Perry, R. H., and C. H. Chilton, eds., *Chemical Engineers' Handbook*, 5th ed., pp. 5–13 (1973).
8. Shah, R. K., and A. L. London, *Laminar Flow: Forced Convection in Ducts*, Academic Press (1978).
9. Speyer, D. M., *Shell-Side Pressure Drop in Baffled Shell-and-Tube Heat Exchangers*, Ch.E. Ph.D. thesis, New York University (1973).
10. Streeter, V. L., and E. B. Wylie, *Fluid Mechanics*, 8th ed., McGraw-Hill, p. 215 (1985).

References Colebrook [2] and Drew [3] were not available. The information was taken from secondary sources: Streeter [10] and McAdams [6], respectively.

Chapter 31

Evaluation of Condensation Heat Transfer in a Vertical Tube Heat Exchanger

Karen Vierow
School of Nuclear Engineering
Purdue University
West Lafayette, Indiana

Problem

Some nuclear reactors are equipped with "passive" condensers to remove heat when the reactor is isolated from the turbine and other normal operation systems. In this context "passive" means that the heat exchangers are driven by natural forces and do not require any power supplies, moving parts, or operator actions. The condensers are vertical tube bundles that sit in a pool of water and condense steam on the tube side. The condensers are designed to take steam generated in the reactor and return condensate back to the reactor vessel. Noncondensable gases such as air may enter the system and seriously degrade heat-transfer performance.

To design one of these heat exchangers, evaluate the local condensation heat-transfer coefficient in a test loop from experimental data provided in Table 31.1.

Estimate how many tubes would be needed if the total heat removal rate were 1 MW and the tube length were reduced by a factor of 2.

31.2 Heat Exchangers

TABLE 31.1 Temperature Measurements

Distance from tube inlet z, m	Condenser tube wall outer surface temperature, °C	Coolant bulk temperature, °C
0.03	57.0	21.0
0.25	52.9	—
0.50	41.9	12.6
0.75	22.9	—
1.00	12.3	10.7
1.25	11.8	—
1.50	10.9	10.1
1.75	11.5	—
2.00	11.3	10.0

Given information

Geometry
- Double-pipe, concentric tube vertical heat exchanger
- Steam-noncondensable gas mixture flowing into center tube from top
- Condensate draining out the bottom of center tube
- Coolant water flowing upward in the coolant annulus around condenser tube
- Stainless-steel condenser tube 2 m in length, 1 in. o.d., 1.5-mm tube wall thickness
- Well-insulated outer pipe

Conditions
- Steam inlet mass flux = 6.75 kg/(m² s)
- Noncondensable gas inlet mass flux = 0.75 kg/(m² s)
- Primary side inlet pressure = 220 kPa
- Coolant flow rate = 0.13 kg/s
- Temperature measurements as in Table 31.1

Assumptions

Saturated conditions at condenser tube inlet

Noncondensable gas and steam in thermal equilibrium

Coolant temperature well represented by bulk temperature due to turbulent mixing

Steady-state operation

Each tube in prototype tube bundle carrying an equal heat load

Nomenclature

A	Heat-transfer area
C_1, C_2	Constants in coolant axial temperature profile fit

c_p	Specific heat
D_{in}	Condenser tube inner diameter
D_{out}	Condenser tube outer diameter
h_c	Condensation heat-transfer coefficient
h_{fg}	Heat of vaporization
h'_{fg}	Modified heat of vaporization
k_{wall}	Condenser tube wall thermal conductivity
$\dot{m}$	Mass flow rate
P	Pressure
P_{inlet}	Total pressure at condenser tube inlet
P_{total}	Total pressure
P_v	Vapor partial pressure
q	Heat rate
q''	Heat flux
R_c	Condensation thermal resistance
$R_{coolant}$	Coolant thermal resistance
R_{wall}	Condenser tube wall thermal resistance
T	Temperature
T_{CL}	Condenser tube centerline temperature
$T_{coolant}$	Coolant temperature
T_{sat}	Saturation temperature
$T_{wall,in}$	Condenser tube wall inner surface temperature
$T_{wall,out}$	Condenser tube wall outer surface temperature
X	Mole fraction
z	Axial distance downstream from condenser tube inlet

Calculation Method

The heat-transfer coefficient is obtained from experimental data because there are no widely applicable mechanistic analysis models for the condensation heat-transfer coefficient in tubes under forced convection with noncondensable gases. Empirical correlations are available for given ranges of steam flow rates, noncondensable gas flow rates, pressures, and tube diameters against which the calculation results could be compared; for example, see the paper by Vierow and Schrock (1991).

Definition of the heat-transfer coefficient

The condensation heat-transfer coefficient must first be defined by identifying the thermal resistances. The steam–noncondensable gas

31.4 Heat Exchangers

mixture enters the condenser tube from the top. Steam directly contacts the cold tube wall and immediately begins to condense, forming a thin condensate film along the inner surface of the condenser tube. This film thickens with distance from the inlet and represents an increasing thermal resistance. As the steam–noncondensable gas mixture is drawn to the condensate surface, a noncondensable gas boundary layer develops and provides additional thermal resistance. There is also a thermal resistance in the tube wall and the coolant. Noting that the coolant is assumed to be well mixed, the radial profile for the coolant temperature at any axial location may be taken as a constant temperature over the cross section.

The *condensation heat-transfer coefficient* is theoretically defined as the heat-transfer coefficient at the vapor-liquid interface. The heat-transfer coefficient between the condenser tube centerline and the condenser tube inner surface is more practical to calculate because information on the condensate thickness is seldom available and the heat-transfer resistance radially within the steam-gas mixture is small enough to be neglected. The heat-transfer resistances considered in the calculation are as in Fig. 31.1.

If the local heat flux and the temperatures are known, the thermal resistances and therefore the heat-transfer coefficients may be calculated. The condensation heat-transfer coefficient can then be obtained as below, where z is the axial distance from the condenser tube inlet and $T_{sat}(z)$ is the local saturation temperature in the condenser

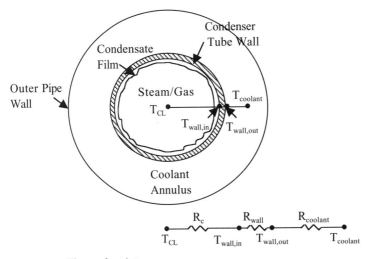

Figure 31.1 Thermal resistances.

tube at z:

$$h_c(z) = \frac{q''(z)\frac{D_{out}}{D_{in}}}{T_{sat}(z) - T_{wall,\,in}(z)} \tag{31.1}$$

The diameter ratio factor is necessary because the heat flux will be calculated at the condenser tube outer surface and must be corrected to correspond to the tube inner surface.

Estimation of local heat flux from axial coolant temperature profile

An energy balance on the coolant provides the local heat flux:

$$q''(z)dA = -\dot{m}_{coolant}\,c_p dT(z) \tag{31.2}$$

Note that the right-hand side is negative since the coolant is in countercurrent flow with the primary side, from bottom to top:

$$q''(z) = -\dot{m}_{coolant}\,c_p \frac{dT(z)}{\pi D_{out} dz} = -\frac{\dot{m}_{coolant}\,c_p}{\pi D_{out}}\frac{dT(z)}{dz} \tag{31.3}$$

The axial temperature gradient can be obtained by fitting the coolant temperature as a function of z. The top four points of the coolant data provided in Table 31.1 can be fit by an exponential. Because the temperature change is minimal between 2.0 and 1.5 m, the heat flux is smaller than can be expected from condensation heat transfer and the heat-transfer modes are most likely forced convection and conduction.

Since the axial temperature rise rather than the actual temperature is of interest, an equation for the temperature rise is calculated. For $0.03 \leq z \leq 1.5$ m, we obtain

$$\Delta T(z) = T(z) - T(1.5\text{ m}) = C_1 \times e^{-z/C_2} \tag{31.4}$$

For the current data, $C_1 = 12.37$, $C_2 = 0.32$, and $\Delta T(z) = 12.37 e^{-z/0.32}$:

$$q''(z) = -\frac{\dot{m}_{coolant}\,c_p}{\pi D_{out}}\frac{dT(z)}{dz} = -\frac{\dot{m}_{coolant}\,c_p}{\pi D_{out}}\frac{d\Delta T(z)}{dz} = \frac{\dot{m}_{coolant}\,c_p}{\pi D_{out}}\frac{C_1 e^{-z/C_2}}{C_2} \tag{31.5}$$

Determination of local saturation temperature

The local saturation temperature is obtained from knowledge of the local steam partial pressure. To determine local steam partial pressure, the local steam flow rate and noncondensable gas flow rate must be known.

31.6 Heat Exchangers

The local steam flow rate is the difference between the inlet steam flow rate and the amount of steam that has been condensed. The latter is found from the heat removal rate between the tube inlet and the desired distance from the tube inlet z:

$$q_{i \to i+1} = \int_{z_i}^{z_{i+1}} q''(z) dA = -\dot{m}_{\text{coolant}} c_p 12.37 \left[e^{-z_{i+1}/0.32} - e^{-z_i/0.32} \right] \quad (31.6)$$

$$\dot{m}_{\text{condensate}}(z) = \frac{q_{0 \to z}}{h'_{fg}(z)} \quad (31.7)$$

To account for condensate subcooling, the modified heat of vaporization suggested by Rosenhow is employed:

$$h'_{fg}(z) = h_{fg}(z) + 0.68 c_{p,l}(T_{\text{sat}}(z) - T_{\text{wall, in}}(z)) \quad (31.8)$$

The local saturation temperature is obtained in later steps and must be iterated on:

$$\dot{m}_{\text{steam}}(z) = \dot{m}_{\text{steam, inlet}} - \dot{m}_{\text{condensate}}(z) \quad (31.9)$$

The local noncondensable gas flow rate is not a function of z. It is vented out the bottom of the condenser tube under steady-state conditions, and the local mass flow rate may be considered constant throughout the tube.

The local steam partial pressure is obtained as

$$P_v(z) = P_{\text{total}}(z) \times X_{\text{steam}}(z) = P_{\text{inlet}} \times X_{\text{steam}}(z) \quad (31.10)$$

where $X_{\text{steam}}(z)$ is the local steam mole fraction. The total pressure at any location in the tube can be assumed the same as the inlet pressure. This assumption arises from the requirement that the pressure drop through the condenser tube be very small so that steam can easily enter the condenser in a passive mode.

Let the noncondensable gas be air. The local steam fraction is approximately

$$X_{\text{steam}}(z) = \frac{\dfrac{\dot{m}_{\text{steam}}(z)}{18.02}}{\dfrac{\dot{m}_{\text{air}}}{28.97} + \dfrac{\dot{m}_{\text{steam}}(z)}{18.02}} \quad (31.11)$$

The local saturation temperature is obtained from Incropera and DeWitt (2002):

$$T_{\text{sat}}(z) = T_{\text{sat}}(P_v(z)) \quad (31.12)$$

This saturation temperature and h_{fg} at $P_v(z)$ are substituted back into Eq. (31.8).

Calculation of tube wall inner surface temperature

The axial profile of the tube wall outer surface temperature has been measured and provided in the given information. The inner surface temperature is much more difficult to measure because the instrumentation can easily affect the condensation phenomena. Hence, the inner surface temperature is deduced from the outer surface temperature by a simple conduction heat-transfer calculation:

$$q(z) = \frac{T_{\text{wall, in}}(z) - T_{\text{wall, out}}(z)}{R_{\text{wall}}} \qquad R_{\text{wall}} = \frac{\ln\left(\dfrac{D_{\text{out}}}{D_{\text{in}}}\right)}{2\pi k_{\text{wall}} dz} \qquad (31.13)$$

Rearranging, we obtain

$$T_{\text{wall, in}}(z) = q''(z) \times R_{\text{wall}} \, dA + T_{\text{wall, out}} \qquad (31.14)$$

where

$$R_{\text{wall}} dA = \frac{\ln\left(\dfrac{D_{\text{out}}}{D_{\text{in}}}\right) D_{\text{out}}}{2k_{\text{wall}}} \qquad (31.15)$$

Note that the wall temperature data do not monotonically decrease. This is due to the uncertainty in the thermocouple measurement of approximately ±0.5°C.

The parameters necessary to calculate the local condensation heat-transfer coefficient are now available.

Example calculation at z = 0.03 m

From Eq. (31.5), the local heat flux on the condenser tube outer surface is

$$q''(z) = \frac{(0.13 \text{ kg/s}) \left(4.18 \text{ kJ/(kg K)}\right)}{\pi(1 \text{ in.}) \left(\dfrac{1 \text{ m}}{39.3 \text{ in.}}\right)} \frac{12.37 e^{-(0.03 \text{ m})/0.32}}{0.32}$$

$$= 239 \text{ kW/m}^2$$

From Eq. (31.8), the modified heat of vaporization is

$$h'_{fg}(0.03 \text{ m}) = 2202 \text{ kJ/kg} + 0.68 \left(4.18 \text{ kJ/(kg K)}\right) (121°C - 57°C)$$

$$= 2384 \text{ kJ/kg K}$$

31.8 Heat Exchangers

Note that h_{fg} and T_{sat} must be taken at the vapor partial pressure calculated below. This must be iterated on once the partial pressure is known. The procedure is simple to do in software such as an Excel spreadsheet.

The local condensate flow rate is obtained from Eqs. (31.6) and (31.7):

$$\dot{m}_{cond}(0.03 \text{ m}) = -\frac{(0.13 \text{ kg/s})(4.18 \text{ kJ/kg K})12.37 \text{ K}[e^{-(0.03 \text{ m})/0.32} - e^{-(0 \text{ m})/0.32}]}{2384 \text{ kJ/kg}}$$

$$= 2.5 \times 10^{-4} \text{ kg/s}$$

Equation (31.9) provides the local steam flow rate:

$$\dot{m}_{steam}(0.03 \text{ m}) = (6.75 \text{ kg/m}^2\text{s})\frac{\pi}{4}\left[\frac{1 \text{ in.}}{39.3 \text{ in./m}} - 2(0.0015 \text{ m})\right]^2$$

$$- 2.5 \times 10^{-4} \text{ kg/s}$$

$$= 0.0027 \text{ kg/s} - 0.00025 \text{ kg/s} = 0.0024 \text{ kg/s}$$

To obtain the steam mole fraction, the noncondensable gas mass flow rate is first calculated.

$$\dot{m}_{air} = (0.75 \text{ kg/m}^2\text{s})\frac{\pi}{4}\left[\frac{1 \text{ in.}}{39.3 \text{ in./m}} - 2(0.0015 \text{ m})\right]^2 = 0.00030 \text{ kg/s}$$

From Eq. (31.11), we obtain

$$X_{steam}(0.03 \text{ m}) = \frac{\dfrac{0.0024 \text{ kg/s}}{18.02 \text{ kg/kg mol}}}{\dfrac{0.00030 \text{ kg/s}}{28.97 \text{ kg/kg mol}} + \dfrac{0.0024 \text{ kg/s}}{18.02 \text{ kg/kg mol}}}$$

$$= 0.93$$

The local steam partial pressure can now be estimated from Eq. (31.10):

$$P_v(0.03 \text{ m}) = (220 \text{ kPa}) \times (0.9242) = 204 \text{ kPa}$$

From ASME steam tables, we have

$$T_{sat}(0.03 \text{ m}) = 121°C$$

The solution is iterated on between h'_{fg} and T_{sat}.

The condenser tube inner wall temperature is as follows, from Eqs. (31.14) and (31.15):

$$R_{\text{wall}} \, dA = \frac{\ln\left(\dfrac{1 \text{ in.}}{1 \text{ in.} - 2(0.0015 \text{ m})\left(\dfrac{39.3 \text{ in.}}{1 \text{ m}}\right)}\right)\left(\dfrac{1 \text{ in.}}{\dfrac{39.3 \text{ in.}}{1 \text{ m}}}\right)}{2(15.5 \text{ W/m K})}$$

$$= 1.0 \times 10^{-4} \frac{\text{K m}^2}{\text{W}}$$

$$T_{\text{wall, in}}(0.03 \text{ m}) = (239 \text{ W/m}^2 \text{ K}) \times (1.0 \times 10^{-4} \text{ K m}^2/\text{W}) + 57°\text{C}$$

$$= 57.03°\text{C}$$

From Eq. (31.1), the local condensation heat-transfer coefficient becomes

$$h_c(0.03 \text{ m}) = \frac{239 \text{ kW/m}^2 \dfrac{0.0254 \text{ m}}{0.0224 \text{ m}}}{121°\text{C} - 57°\text{C}}$$

$$= 4.23 \text{ kW/m}^2 \text{ K}$$

The local heat-transfer coefficient and related parameters are tabulated in Table 31.2 at each tube wall temperature measurement location. At 1.5 m, the condensate mass flow rate was calculated to be greater than the inlet mass flow rate. From this point on, the calculation procedure becomes invalid because the condensation rate is very small.

To estimate the number of tubes needed for a heat removal rate of 1 MW at half the tube length, note that essentially all the condensation

TABLE 31.2 Heat-Transfer Calculation Results

Parameter z, m	0.03	0.25	0.5	0.75	1.0	1.25	1.5
$q''(z)$ on tube outer surface, kW/m^2	239	120	55	25	12	5	2
$T_{\text{sat}}(z)$, °C	121	119	112	104	94	74	—
$h'_{fg}(z)$, kJ/kg	2384	2394	2422	2476	2505	2500	—
Steam flow rate (z), kg/s	0.0024	0.0011	0.00048	0.00022	0.00011	3.61E-05	—
Steam mole fraction (z)	0.93	0.86	0.72	0.54	0.36	0.16	—
$P_{\text{steam}}(z)$, kPa	204	189	159	119	80	36	—
$h_c(z)$, kW/m^2 K	4.23	2.06	0.89	0.35	0.16	0.098	—

31.10 Heat Exchangers

takes place within the first meter from the tube inlet. The local heat-transfer coefficients and heat fluxes from about 1.0 m on are too low to be due primarily to condensation; the tubes in this example appear to be oversized.

The number of tubes required is

$$\text{Number of tubes} = \frac{1 \text{ MW}}{\text{heat removal/tube}}$$

Each tube condenses the total steam flow coming in within 1 m from the tube inlet. Heat removal per tube is found from Eq. (31.6) to be 6.42 kW per tube. Thus, 162 tubes are required.

References

Incropera, F. P., and D. P. DeWitt, *Fundamentals of Heat and Mass Transfer*, 5th ed., Wiley, New York, 2002.

Vierow, K., and V. E. Schrock, "Condensation in a Natural Circulation Loop with Noncondensable Gases, Part I—Heat Transfer," *Proceedings of International Conference on Multiphase Flow '91* (Japanese Society of Multiphase Flow, ANS), Tsukuba, Japan, Sept. 1991, pp. 183–186.

Chapter

32

Heat-Exchanger Design Using an Evolutionary Algorithm

Keith A. Woodbury
Mechanical Engineering Department
University of Alabama
Tuscaloosa, Alabama

Abstract

Heat-exchanger design is achieved by satisfying a required heat transfer while not exceeding an allowable pressure drop. Iterative procedures for hand or calculator operation have been used for many years, but the common availability of powerful desktop computers admits a wider class of design approaches. In this chapter, a design methodology based on optimization via an evolutionary algorithm is detailed. The evolutionary algorithm manipulates populations of solutions by mating the best members of the current population (natural selection) and randomly introducing perturbations into the newly created population (mutation). Heat-exchanger design by optimization allows inclusion of additional constraints such as minimum volume or pressure drop.

Introduction

Heat-exchanger design

Hodge and Taylor [1] state that "heat exchanger design requirements can present multifaceted *heat transfer* and *fluid-dynamic* requirements on both the hot and cold fluid sides" and that these "Heat exchanger specifications are normally centered about the [heat-transfer] *rating* and the allowable *pressure drop*." Heat-exchanger design requires satisfaction of the heat-transfer (rating) specification while not exceeding the allowable pressure drop. Note that allowable pressure drop specifications for both hot and cold streams could be specified.

Shah [2] outlines a procedure suitable for hand calculations to design heat exchangers. In Shah's procedure applied to a shell-and-tube exchanger, a number of tubes of a chosen diameter are assumed implicitly through the mass velocity G, and the length of these tubes is chosen to satisfy the heat-transfer requirement. If this geometry combination exceeds the allowable pressure drop, then a new number of tubes is chosen in accordance with the allowable pressure drop and the process is repeated until convergence is attained. Hodge and Taylor [1] provide a synopsis of Shah's method.

Design procedures such as Shah's are simple and effective, but cannot be extended to include other design constraints, such as total volume. Procedures based on optimization methods are easily extensible and can obtain results quickly with a desktop computer.

Genetic and evolutionary algorithms

Goldberg [3] offers the following definition: "Genetic algorithms are search algorithms based on the mechanics of natural selection and natural genetics." Most genetic algorithm (GA) purists hold that a GA must contain all the elements of binary encoding, natural selection, crossover breeding, and mutation. However, binary encoding imposes a priori precision limitations and an extra layer of coding/decoding when the unknown parameters are continuous real numbers.

A slightly different approach is to use strings of real numbers (arrays) to directly represent the parameters in an optimization problem. An *evolutionary algorithm* (EA) manipulates a population of solutions through the essential elements of natural selection and mutation, but does not utilize binary encoding to represent the real-valued parameters of the process. This is the approach used in the present application.

Overview

In the remainder of this chapter the EA is first described in detail. Next the spiral plate heat exchanger (SPHE) is briefly described, along with the design constraints and properties for a particular application. Then the EA is applied to the design of a spiral plate heat exchanger. An *ensemble design* is performed, and some useful statistical information about the design parameters is extracted from these results. Finally, the fitness function is modified to allow minimization of the exchanger volume and the SPHE is redesigned. Some conclusions close out the chapter.

Evolutionary Optimization Algorithm

The EA used here is the same as the real-number encoding algorithm described by Woodbury [4]. However, a substantial change has been made in the selection of parents for the creation of a new generation [5]. A summary of the basic algorithm is presented below.

Outline of the algorithm

The evolutionary algorithm incorporates the following basic steps:

1. A population of possible solutions is initialized randomly.
2. The fitness of each solution is determined according to an appropriate performance index (also known as the *fitness function*).
3. Pairs of the population are selected as parents of the new generation. This selection of parents is biased toward the most-fit members of the current population.
4. The selected parents mate to form children, which constitute the new population.
5. Variation is introduced into the new population through mutation, crossover, and creep.
6. Steps 2 to 5 are repeated until a specified number of generations are processed.

These basic steps are amplified in the following sections.

Initial population

The number of unknown parameters (n_{parm}) and the allowable range for each must first be specified, as well as a number of solutions to be contained in the population (n_{pop}). The "population" is a collection of n_{pop} vectors, and each vector contains n_{parm} values of the parameters. A fitness value that indicates the "goodness" of each member of the population is calculated from the performance index for this initial generation.

Selection of parents

The process of selection is implemented to allow better members of the population to contribute the most to the new generation. This

32.4 Heat Exchangers

helps ensure that the characteristics of the better solutions persist through future generations, and that poor solutions do not exert influence over subsequent solutions. During selection, parents are chosen from the current population to reproduce children for the next generation.

Tournament selection is utilized to create a mating pool for each generation, where two or more members of the population are randomly selected to compete. The winner of the tournament is declared as the population member with the best fitness value, and this population member is copied into the mating pool for subsequent reproduction [6].

In the present algorithm, three members of the current population are randomly selected from the entire population, and the member with the best fitness is added to the mating pool. This process is repeated until the mating pool is the same size as the initial population (i.e., contains n_{pop} members). Note that it is possible, and actually very likely, that the mating pool will have repeat members. After the pool has been created, two parents will be randomly chosen from the pool for reproduction to create a new offspring for the next generation.

Reproduction

Reproduction in a traditional GA is accomplished by crossover recombination of binary strings in the population. However, crossover of the real-valued arrays to produce children may not be a good approach if the elements of the vector represent drastically different quantities. An alternative method of mating the parents is through a weighted average [4]. A random weight w between 0 and 1 is chosen for each child. The child is then produced as follows:

$$\text{Offspring} = w * \text{Parent}_1 + (1 - w) * \text{Parent}_2 \quad (32.1)$$

where Parent is the vector array of parameters constituting a single individual. Enough children are created to fill the new population with as many members as the original population.

Crossover

It is also possible to use randomly triggered crossover in real-number encoding to introduce variation into a new generation [4]. This is done by generating a random number between zero and one for each member of the new population. If that number is below a user-specified crossover threshold (p_{cross}), the population members are recombined analogous to crossover of binary strings. Thus, a crossover site is selected within the vector array, and another member of the new population is selected as a crossover partner, and the two segments of the corresponding arrays are switched at the crossover site. Crossover is meant to introduce large-scale variation in the each generation.

Mutation

After a new generation of children is produced, mutations may occur. Mutations are important because they widen the search space of the genetic algorithm by introducing new values not contained in the current population. A random number between zero and one is generated for each population member, and if that number is below a user-specified mutation threshold (p_{mutation}), a mutation will occur to that member. In the real-number encoding of the present algorithm, mutations occur by changing the value of the population member to a random number that is within the allowable (specified) range of that parameter.

Creep

The notion of creep [4] can be applied only to real-number encoding. Unlike mutation, which generally introduces large changes into the population, creep introduces small changes into the population. The parameters that control the creep process are a specified maximum creep value ($\text{Creep}_{\text{max}}$) and the probability that creep will occur (p_{creep}). As in mutation, a random number between zero and one is selected for each population member, and if the random number is below the specified threshold p_{creep}, the member will creep according to the following formula

$$\text{New\_member} = (1 + c) * \text{Original\_member} \tag{32.2}$$

where c is a random number generated in the range of [$-\text{Creep}_{\text{max}}$, $\text{Creep}_{\text{max}}$].

Elitism

After creating a new generation and applying the techniques of crossover, mutation, and creep, one may find that the best member in the new population is not as good as the best member in the previous generation. The technique of *elitism* preserves the best member of the population and lets it persist into the new generation, ensuring that the best fitness for the new generation is never worse than that of the previous generation. Elitism enhances the performance of the evolutionary algorithm by guaranteeing that the solutions will not diverge.

The present EA utilizes elitism by preserving the best member of the population. At each generation the population is sorted, best-to-worst by fitness value, and during reproduction the first member of the population is not replaced. Therefore, this elite population member will persist unchanged throughout generations until a better member is created.

Termination

The EA is terminated after a specified number of generations have been created. However, an alternate stopping criterion is to continue until a specific fitness level is achieved.

Design of a Spiral Plate Heat Exchanger

Reference 8 depicts a *spiral plate heat exchanger* (SPHE), which is a counterflow heat exchanger with rectangular flow passages created by wrapping separation plates into a spiral. The assembly process is described by Martin [7, p. 74]: "Each spiral channel is sealed by welding on one edge, the other side being accessible for inspection and cleaning." Martin reports that this type of exchanger has advantage in applications of energy recovery and is especially advantageous in handling solid suspensions. SPHEs are also claimed to be more resistant to fouling [8].

As mentioned previously, any heat-exchanger design requires satisfaction of both a heat-transfer requirement and a pressure drop requirement. Thus, suitable phenomenological models relating these quantities to the conditions of the flow are needed. Martin [7] provides the following heat-transfer and pressure drop relations for a SPHE.

Heat transfer

The heat-transfer coefficient for flow in the rectangular channel in the SPHE can be found from the correlation

$$\mathrm{Nu} = 0.04\, \mathrm{Re}^{0.74}\, \mathrm{Pr}^{0.4} \tag{32.3}$$

which is valid for $400 < \mathrm{Re} < 30{,}000$. Here the Reynolds and Nusselt numbers are based on the hydraulic diameter of the rectangular cross-sectional area channel.

The performance of a heat exchanger is typically characterized either by its effectiveness or using the log mean temperature difference (LMTD). Martin gives the correction factor for use with the LMTD for

SPHEs as

$$F_c = \frac{n}{\text{NTU}} \tanh\left(\frac{\text{NTU}}{n}\right) \qquad (32.4)$$

where n is the number of turns (wraps) in the exchanger and NTU is the number of transfer units:

$$\text{NTU} = \frac{UA}{C_{\min}} \qquad (32.5)$$

Here U is the overall heat-transfer coefficient, A is the heat-exchanger surface area, and $C_{\min}$ is the lesser of $\dot{m}c_p$ for the hot and cold fluids.

Pressure drop

The pressure drop in the SPHE is computed as

$$\Delta p = \frac{fL}{D_h}\frac{V^2}{2} \qquad (32.6)$$

where L is the length of the flow passage, D_h is the hydraulic diameter, V is the velocity of the flow, and f is the friction factor, which is given by Martin as follows:

$$f = 1.5\frac{64}{\text{Re}} + 0.2\,\text{Re}^{-0.1} \qquad (32.7)$$

Properties and design data

Property and design data are summarized in Table 32.1. The target application requires heating of $\dot{m}_c = 16$ kg/h of a high-heat-capacity fluid using $\dot{m}_h = 50$ kg/h of hot waste exhaust gas. The cold fluid enters

TABLE 32.1 Properties and Data for SPHE Design

	Cold fluid	Hot fluid
T_i, °C	20	500
T_o, °C	225	156.7
$\dot{m}$, kg/h	16	50
c_p, J/kg-K	6960	1330
ρ, kg/m³	0.3774	0.4526
Pr	0.96	0.70
μ, N-s/m²	1.3×10^{-5}	2.5×10^{-5}
k, W/m-°C	0.0270	0.0346

32.8 Heat Exchangers

at 20°C and is to be heated to 225°C, resulting in a heating load of 6342 W. The hot fluid enters at 500°C and, accounting for the heating load, must exit at 157.6°F in this counterflow exchanger.

Design parameters

The spiral heat exchanger is characterized by the following geometric parameters:

Width W	Width of rectangular channel
Hot-space thickness t_h	Height of hot-flow passage
Cold-space thickness t_c	Height of cold-flow passage
Core diameter D_c	Diameter of inner core of spiral
Number of turns n	Number of wraps of spiral
Thickness t	Thickness of metal sheet separating passages

All other geometric parameters (such as flow path length and total heat exchanger volume) can be computed from this set of input parameters.

Design fitness function

There are already two constraints that must be met for a successful design: the heat rating and to-be-specified pressure drop. Additional constraints on the design might be desirable, for example, a specified volume for the exchanger. A general fitness function for the EA optimization is utilized in the form

$$F = \sum_{i=1}^{N} W_i \left(\frac{\text{desired}_i - \text{actual}_i}{\text{desired}_i} \right)^2 \qquad (32.8)$$

where "desired" is the required value of the constraint and "actual" is the value resulting from the present set of design parameters. Here W_i is an optional relative weight attached to a particular constraint. For example, it might be very important that the heat-transfer requirement be met, so a larger weighing could be applied to that term. For the design performed here, a specific form of Eq. (32.8) is employed:

$$F = 100 \left(\frac{Q_{\text{desired}} - Q_{\text{act}}}{Q_{\text{desired}}} \right)^2 + \left(\frac{\Delta P_{\text{desired}} - \Delta P_{\text{hot}}}{\Delta P_{\text{desired}}} \right)^2$$

$$+ \left(\frac{\Delta P_{\text{desired}} - \Delta P_{\text{cold}}}{\Delta P_{\text{desired}}} \right)^2 + \left(\frac{\text{Vol}_{\text{desired}} - \text{Vol}_{\text{act}}}{\text{Vol}_{\text{desired}}} \right)^2 \qquad (32.9)$$

The larger weight is applied to the heat-transfer term to ensure that this requirement is met.

A single design

Because the EA relies heavily on random processes, different searches of the same parameter space will result in different, but similar, designs. A design was obtained with the Matlab routine EAez.m [10] with the following inputs and search ranges:

$Q_{desired} = 6341.5$ W

$\Delta P_{desired} = 25$ Pa

$Vol_{desired} = 35,000$ cm^3

1.0 cm $\leq W \leq 10$ cm

$1 \leq n \leq 50$

0.1 cm $\leq t_c \leq 5.0$ cm

0.1 cm $\leq t_h \leq 5.0$ cm

1.0 cm $\leq D_c \leq 10.0$ cm

0.1 cm $\leq t \leq 1.0$ cm

The convergence history from this design search is shown in Fig. 32.1 as the best fitness value from the population [computed from Eq. (32.9)] at each generation of the EA. After 500 generations, the best fitness value has reached almost 10^{-5}, indicating good satisfaction of the constraints in Eq. (32.9).

The results from the EA for this single design are $W = 4.16$ cm, $n = 12$, $t_c = 0.857$ cm, $t_h = 1.791$ cm, $D_c = 5.911$ cm, and $t = 0.6947$ cm.

Ensemble design

Because the EA relies heavily on random procedures during the search, different designs are achieved for each execution of the search. This fact may seem unsettling, but it can be exploited to gain some insight about the design.

Table 32.2 lists the results from a series of 10 designs obtained from the EA starting with the parameters from the previous section. The first column gives only the design (sequence) number, while the second column gives the best fitness value from the population [from Eq. (32.9)] for each design. The next four columns report the actual values of the constraints Q, Δp_c, Δp_h, and Vol corresponding to the best member of each population. The last six columns give the values of the design parameters obtained for each design.

32.10 Heat Exchangers

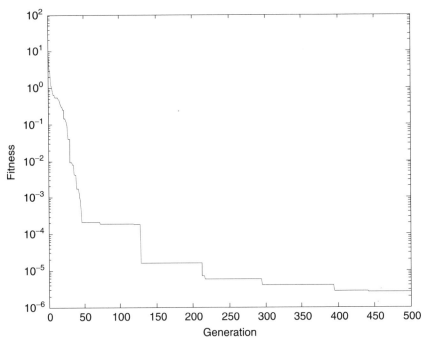

Figure 32.1 Typical convergence history for evolutionary algorithm SPHE design.

The last four lines of Table 32.2 are statistics computed from the 10 values in each column. The first is the mean or expected value of the 10 values; the second is the sample standard deviation. The third line is the 95 percent confidence interval for the mean value reported on the first of the last four lines. This confidence interval is computed

TABLE 32.2 Results from 10 EA Designs

Design	F	Q	ΔP_c	ΔP_h	Vol	W	n	t_c	t_h	D_c	t
1	4.27×10^{-5}	6342.8	24.9	25.0	34,841	3.85	13	0.943	1.993	4.703	0.495
2	1.13×10^{-5}	6340.5	25.0	25.1	35,032	4.15	12	0.860	1.799	6.136	0.689
3	2.52×10^{-5}	6339	25.0	25.0	35,074	4.05	12	0.885	1.860	8.318	0.629
4	1.18×10^{-5}	6340.8	25.1	25.0	34,956	3.84	13	0.945	2.001	4.615	0.500
5	2.38×10^{-5}	6340.7	25.1	25.1	34,899	3.84	13	0.945	2.001	4.705	0.497
6	1.79×10^{-6}	6342.1	25.0	25.0	34,990	3.57	14	1.034	2.212	3.084	0.311
7	3.20×10^{-5}	6343.2	25.0	25.0	35,158	3.74	13	0.979	2.079	7.202	0.429
8	8.70×10^{-6}	6341.8	25.0	24.9	35,003	3.83	13	0.947	2.009	4.859	0.493
9	2.92×10^{-6}	6342.0	25.0	25.0	34,961	3.83	13	0.950	2.011	5.045	0.488
10	2.52×10^{-6}	6341.9	25.0	25.0	35,030	3.81	13	0.955	2.022	5.325	0.481
Mean	1.63×10^{-5}	6341.5	25.0	25.0	34,994	3.85	12.9	0.944	1.999	5.399	0.501
Standard deviation	1.40×10^{-5}	1.24	0.05	0.04	89	0.16	0.57	0.047	0.112	1.48	0.102
95% confidence ±		1.2	0.9	0.04	63.6	0.11	0.41	0.034	0.080	1.06	
± percent, %		0.01	0.1	0.03 0.2	2.9	3.1	3.6	4.0	19.6	14.6	

for small samples (< 30) with unknown standard deviation [9] using Student's t distribution. Note that this confidence interval is the "±" uncertainty associated with the mean value reported, in the units of the parameter. Thus, the average heat-transfer rating of the 10 designs is 6341.5 W ± 1.2 W. Finally, the last line of the table shows the percentage uncertainty in the corresponding column, which is simply the plus/minus variation amount divided by the mean.

Looking at the last line of Table 32.2, it can be seen that the design constraints are met to a high degree, as the 10 different designs vary by 0.2 percent or less for all the specified constraints. One can also observe that some of the design parameters have less variation than do others. Specifically, W, t_c, t_h, and n vary by less than 5 percent over the 10 designs, while D_c and t vary by 15 percent or more. These results demonstrate that the design of the SPHE is relatively *insensitive* to the values of D_c and t; that is, similar values of the fitness function F result for relatively wide variations in these two parameters. This means that the design is very forgiving with regard to these parameters, and that some convenient values of these two parameters could be selected and the design process repeated excluding them.

"Asking for too much..."

It's possible for the designer to ask too much of the EA, and it is important to be able to recognize this when it happens. Figure 32.2 shows a convergence history from a single execution of the EAez.m [10] routine with all the same input parameters as before *except* that $\text{Vol}_{\text{desired}} = 3500 \text{ cm}^3$ (an order of magnitude less than before). Note in Fig. 32.2 that the fitness function decreases exponentially (linearly on the semilog arithmetic plot) until about generation 200; then the F value has "hit the wall" and cannot be reduced further. The values of the design constraints after 500 generations for this case are $Q = 6114.6$ W, $\Delta p_c = 31.59$ Pa, $\Delta p_h = 31.12$ Pa, and Vol = 5165.3 cm^3, which do not agree well with the corresponding constraint values. As can be seen, the algorithm is now trying to minimize the volume to satisfy the unrealistic constraint, but is unable to reduce the fitness F any further without increasing the penalty added to F by Q, Δp_c, and Δp_h. This is because the Vol term of the fitness function now dominates the value of F, and no combination of design parameters can lower F. Physically, this is because an exchanger with the desired volume cannot also satisfy the heat-transfer and pressure drop requirements for the process.

Inequality constraints

A reasonable request would be to design the heat exchanger to satisfy the heat-transfer requirement and the pressure drop constraint, but to

32.12 Heat Exchangers

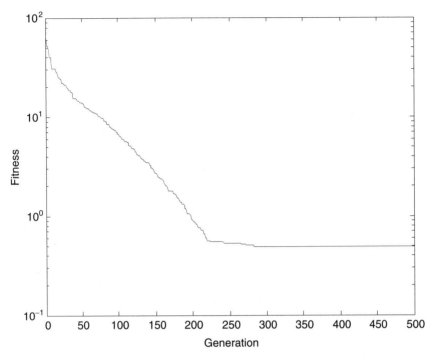

Figure 32.2 Convergence history when "asking for too much."

make the total volume as small as possible. This can be accomplished by modifying the fitness function [Eq. (32.9)] as

$$F = 100\left(\frac{Q_{\text{desired}} - Q_{\text{act}}}{Q_{\text{desired}}}\right)^2 + \left(\frac{\Delta P_{\text{desired}} - \Delta P_{\text{hot}}}{\Delta P_{\text{desired}}}\right)^2$$
$$+ \left(\frac{\Delta P_{\text{desired}} - \Delta P_{\text{cold}}}{\Delta P_{\text{desired}}}\right)^2 + \left(\frac{\text{Vol}_{\text{act}}}{\text{Vol}_{\text{nominal}}}\right)^2 \quad (32.10)$$

where $\text{Vol}_{\text{nominal}}$ is chosen so that the last term will be on the same order as the others when converged. This may require a try-it-and-see approach.

Figure 32.3 shows the convergence history for 500 generations of a design search using Eq. (32.10) with $\text{Vol}_{\text{nominal}} = 1 \times 10^5$ cm^3. It is not obvious from the convergence history in Fig. 32.3 that the best solution has been obtained. Table 32.3 shows results from this search after 500, 1000, and 1500 generations. Note that only modest improvement in the fitness function F [Eq. (32.10)] is made after the original 500 generations,

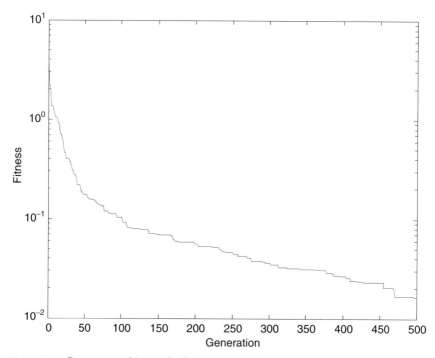

Figure 32.3 Convergence history for first 500 generations using Eq. (32.9).

but that further refinement does result in a smaller-volume exchanger (9825 cm^3 vs. $12{,}300$ cm^3).

Execution times

The EA search for the best design can be very quick. However, in the present design, a nonlinear subsearch [involving the ln() function] must be performed to find the outlet temperatures for each flow stream for each member of the population at each generation. This subsearch requirement is due entirely to the LMTD approach used by Martin [7] to characterize heat-exchanger performance. If an effectiveness-NTU

TABLE 32.3 Results for Design Search Using Inequality Constraint [Eq. (32.10)]

N_{gen}	F	Q	ΔP_c	ΔP_h	Vol	W	n	t_c	t_h	D_c	t
500	1.56×10^{-2}	6332.9	25.1	24.6	12,297	6.6	13	0.498	1.006	3.675	0.112
1000	1.02×10^{-2}	6339	25.0	25.0	10,095	7.56	13	0.430	0.855	2.465	0.101
1500	9.72×10^{-3}	6344.5	24.9	24.9	9,824	7.77	13	0.419	0.832	2.247	0.101

correlation were available for the SPHE, this subsearch could be eliminated and the design time significantly shortened. As it stands, a single design using EAez.m [10] requires about 55 s of CPU time on a 2.0-GHz Pentium 4 computer with 1 Gbyte of RAM, so that the ensemble design requires about 10 min to achieve.

Conclusions

A spiral plate heat exchanger has been designed for a particular application. By performing an ensemble design, statistical information about the design parameters can be recovered. Inequality constraints on total exchanger volume can be incorporated into the fitness function to minimize the size of the designed exchanger. Designs can be obtained in about 1 min on a 2.0-GHz Pentium 4 computer.

Nomenclature

A	Area, m^2
Creep$_{max}$	Maximum amount of creep for a parameter
C_{min}	Smaller of values of $\dot{m}c_p$ for the two streams of an exchanger
c_p	Constant-pressure specific heat
D_h	Hydraulic diameter
D_c	SPHE core diameter
F	Fitness value
F_c	Correction factor for LMTD method
f	Friction factor
G	Mass velocity $G = \dot{m}/A$, kg/m^2-s
k	Thermal conductivity
L	Length of flow passage
LMTD	log mean temperature difference
$\dot{m}$	Mass flow rate, kg/s
n_{parm}	Number of parameters in solution
n_{pop}	Number of solutions in population
n	Number of turns in SPHE
N	Number of design constraints
NTU	Number of transfer units, UA/C_{min}
p_{cross}	Probability of crossover
p_{creep}	Probability of creep mutation

$p_{mutation}$ Probability of mutation
Δp Pressure drop
Pr   Prandtl number $= \nu/\alpha$
Re   Reynold's number VD/ν
t Thickness of separating plate
t_c Thickness of cold space
t_h Thickness of hot space
U Overall heat-transfer coefficient for exchanger
V Velocity
w Weight for reproduction
W Width of SPHE flow passage

Greek

α Thermal diffusivity $k/\rho c_p$
μ Dynamic viscosity
ν Kinematic viscosity μ/ρ
ρ Density

Subscripts and superscripts

c Cold fluid
h Hot fluid

References

1. Hodge, B. K., and Taylor, R. P., *Analysis and Design of Energy Systems*, 3d ed., Upper Saddle River, N.J.: Prentice-Hall, 1999.
2. Shah, R. K., "Compact Heat Exchanger Design Procedures," in *Heat Exchangers: Thermal-Hydraulic Fundamentals and Design*, Kakac, S., Bergles, A. E., and Mayinger, F. (eds.), New York: Hemisphere, 1981, pp. 495–536.
3. Goldberg, D. E., *Genetic Algorithms in Search, Optimization, and Machine Learning*, Boston: Addison-Wesley, 1989.
4. Woodbury, K. A., "Application of Genetic Algorithms and Neural Networks to the Solution of Inverse Heat Conduction Problems: A Tutorial," in *Inverse Problems in Engineering*, Proceedings of the 4th International Conference on Inverse Problems in Engineering, Angra dos Reis, Brazil, May 2002, pp. 73–88.
5. Woodbury, K. A., Graham, C., Baker, J., and Karr, C., "An Inverse Method Using a Genetic Algorithm to Determine Spatial Temperature Distribution from Infrared Tranmissivity Measurements in a Gas," 2004 ASME Heat Transfer/Fluids Engineering Summer Conference, paper HT-FED04-56779, Charlotte, N.C., July 2004.
6. Bäck, T., Fogel, D. B., and Michalewicz, Z., *Evolutionary Computation 1: Basic Algorithms and Operators*, Philadelphia, Pa.: Institute of Physics Publishing, 2000.
7. Martin, H., *Heat Exchangers*, Washington, D.C.: Hemisphere, 1992.

8. Crutcher, M., "No More Fouling: The Spiral Heat Exchanger," *Process Heating*, 1999 (http://www.process-heating.com/CDA/ArticleInformation/coverstory/BNPCoverStoryItem/0,3154,18383,00.html).
9. Bluman, A. G., *Elementary Statistics*, 5th ed., Boston: McGraw-Hill, 2004.
10. Woodbury, K. A., SPHE Matlab Functions, www.me.ua.edu/inverse/SPHEdesign, 2005.

Part 7

Fluidized Beds

Gas-fluidized beds are addressed in many heat-transfer books because they are a common reactor scheme for a wide range of applications where intimate gas-solid contact is required. Gas-fluidized beds are used for such diverse applications as coating, pharmaceutical production, coal combustion, and petroleum cracking. The extensive surface area of particles, coupled with the steady particle mixing induced by bubbles, ensures nearly isothermal operation and impressive convective heat-transfer coefficients.

This part is composed of a single chapter, by two U.S. academics, that presents two typical design problems, one in the bubbling regime and the other for a more dilute circulating fluidized bed. Both illustrate very common gas-solid contacting methods with quite different hydrodynamics, so each is described briefly.

Chapter 33

Fluidized-Bed Heat Transfer

Charles J. Coronella and Scott A. Cooper
Chemical and Metallurgical Engineering Dept.
University of Nevada
Reno, Nevada

Background

Gas-fluidized beds are a common reactor scheme for a wide range of applications where intimate gas-solid contact is required, and are used for such diverse applications as coating, pharmaceutical production, coal combustion, and petroleum cracking. The extensive surface area of particles, coupled with the steady particle mixing induced by bubbles, ensures nearly isothermal operation and impressive convective heat-transfer coefficients.

In order to predict heat-transfer rates in fluidized beds, it is necessary to have a basic understanding of fluidization hydrodynamics, so a brief summary is given here. In the design of an actual reactor, the engineer is advised to consult additional publications, starting with the excellent text by Kunii and Levenspiel (1991) or the text edited by Geldart (1986). Fluidized beds are composed of a large collection (a bed) of solid particles, as small as 40 μm, up to a maximum of about 5 mm, with a gas rising through them. You can imagine that at a low gas throughput, the particles are still, in the mode of a packed bed, but as the gas flow is increased, eventually a point is reached such that the drag force on the particles is sufficient to overcome the gravitational force. This gas flow corresponds to the minimum fluidization velocity, and the particles are lifted and become mobile. The fluidized particles are not carried out of the reactor, but move around, resembling a boiling liquid. As the gas flow is increased further, the particles mix quite rapidly, with gas bubbles rising through the bed. If the gas flow is increased still further,

33.4 Fluidized Beds

particles are slowly ejected from the bed, at first slowly, until a gas flow is reached where all the particles in the bed are carried up (elutriated) and removed from the bed. In this mode of operation, the chamber has a low solids volume fraction, perhaps only 1 to 3 percent, and solids are continuously injected to make up a bed.

Two typical design problems are given below, one in the bubbling regime and the other for a more dilute circulating fluidized bed (CFB). Both illustrate very common gas-solid contacting methods with quite different hydrodynamics, so each is described briefly.

In the bubbling regime, all important aspects are dominated by the flow of gas bubbles, including heat transfer. The phase map given by Geldart (1973) is useful in characterizing this phenomenon. Small and light particles, typically catalysts, are classified as group A, in the range of 40 μm up to approximately 100 μm. These particles fluidize smoothly at low gas flow rates, and bubbles appear only in beds fluidized at a gas flow rate somewhat higher than the minimum required for fluidization. Group B particles are those larger than 100 μm and smaller than 800 μm; bubbles are always present when these particles are fluidized. Fluidization of the largest particles, group D, is characterized by slow-moving bubbles, and much of the gas flow bypasses the bubbles, unlike in groups A and B, and flows through the interstices between particles. The smallest particles, group C, are not fluidized easily, due to strong interparticle forces.

A bubbling fluidized bed is often characterized as consisting of two phases: a bubble phase, consisting primarily of gas, and a dense "emulsion" phase, consisting of both gas and solids. The dense phase is often said to have a void fraction ε_{mf} identical to that of a bed fluidized at the minimum flow necessary for fluidization U_{mf} and is independent of gas flow. The volume fraction of the bed occupied by bubbles ε_b is a function of the size of bubbles and the speed at which they travel through the bed.

It is quite common to control the temperature and remove heat from a fluidized bed by contact between the bed and either the reactor walls themselves or heat-transfer tubes immersed within the bed horizontally or vertically. Note that tubes in a fluidized bed are subject to substantial abrasion, and if the environment is also corrosive, tube failure is likely without special care.

At any moment, a tube may be surrounded by a rising gas bubble, therefore undergoing very slow heat transfer, since the gas is characterized by a relatively small thermal conductivity k_g. When a tube is in contact with the dense phase, heat is transferred into an emulsion consisting of both gas and solid. The relatively high thermal conductivity and heat capacity of the solid substantially enhances heat transfer. A "packet" of particles may be in contact with a surface briefly, and heats (or cools) relatively quickly from the bed temperature to the tube surface temperature. The passage of the next bubble replaces that packet with another packet of particles, this one, like the previous packet, initially at the reactor temperature, reinvigorating the heat-transfer rate. After a packet of solids is removed from a surface, it mixes into the dense emulsion phase, and rapidly changes temperature to that of the bed itself, since each particle is surrounded by many others at the reactor temperature.

Thus, the overall rate of heat transfer between a fluidized bed and a surface is a function of the relative size of each phase, and also a function of the rate of replacement of packets. Starting at a relatively low gas flow, increasing gas flow causes greater bubble flow, which in turn causes an increase in ε_b and in the frequency of packet movement. As packets of solids are replaced more frequently, the rate of heat transfer should increase. Simultaneously, however, the fraction of heat-transfer surface surrounded by gas increases with ε_b, with a slower rate of heat transfer relative to the rate to particle packets. Typically, a plot of heat-transfer rate versus gas flow shows a rapid increase from the minimum bubbling gas flow and then a broad maximum, so that the rate of heat transfer is near the maximum over a fairly broad range of gas flow, and finally decreases with further increased gas flow.

Circulating fluidized beds (CFBs) are characterized by very high gas flow rates, causing the particles within the bed to be elutriated out of the bed. The gas-solid flow leaving the bed goes through a cyclone separator, so that the solids may be returned to the base of the bed, as illustrated in Fig. 33.1. There is significant variation in the distribution of solids between the bottom gas distributor and the top exit of the CFB riser. Typically, fresh and recycled solids are introduced into the riser near the bottom. Here, the solids concentration is the greatest, resembling a turbulently bubbling bed. The rapidly moving gas carries particles out of this dense region and up into the riser, lifting the solids up through the riser to the exit at the top. Werther and Hirschberg (1997) report that the region immediately above this is dilute, having an average cross-sectional concentration of solids typically on the order of 1 to 2 percent. Along the length of the dilute zone, particles tend to loosely accumulate into clusters. These clusters tend to regularly settle out of the gas suspension, falling downward along the riser walls. After descending along the walls for some distance, the clusters break apart and are reentrained into the upward-flowing gas stream. The rapid mixing of solids and gas helps maintain a relatively even temperature profile through the CFB riser, even for highly exothermic reactions.

The hydrodynamics within the upper region of a CFB is complex, and is well described by Werther and Hirschberg (1997). To summarize, the gas rapidly moves upward through the center of the riser, carrying with it a dilute concentration of solids. Along the wall there exists a higher concentration of solids streaming downward in clusters. This downward flow is downward relative to the upward gas flow, and may be upward

33.6 Fluidized Beds

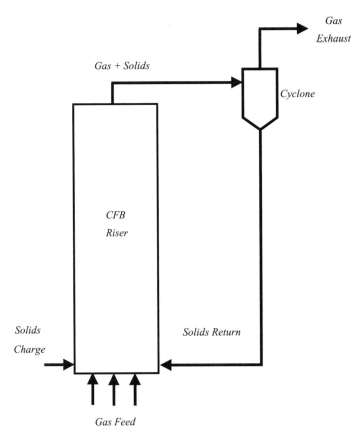

Figure 33.1 This schematic depicts the arrangement of a typical circulating fluidized bed. Solids are fed into the bottom of the CFB riser, where they are carried upward by the rapidly moving gas stream. On exiting the riser, the solids are separated from the gas by a cyclone and are eventually returned to the bottom of the riser.

or downward relative to the riser wall. A time-averaged cross section of the riser would show a relatively solid-laden annulus that surrounds a relatively solid-free core. Accordingly, this description is known as the core-annulus model.

Heat is transported from the bed to the wall by three mechanisms following this core-annulus flow description:

1. Particles move from the hot core of the combustor to the downward-moving annulus, exchanging heat with the nearby wall before returning to the hot core. Some correlations describe this particle convection as a single-particle mechanism, while others model the movement of particle clusters.

2. Heat transfer occurs by gas convection along the exposed walls.
3. Radiation is an important mechanism of heat transfer when the CFB is operated at high temperatures, as are all combustors.

Glicksman (1997) describes how the contributions of these three heat-transfer mechanisms are summed according to the amount of contact that each mechanism has with the riser wall.

Problem Statement I: Bubbling Fluidized Bed

A fluidized bed is being designed to heat and volatilize bitumen from tar sands. The bed will operate in the bubbling regime, the fluidizing gas consists primarily of nitrogen at 450°C, and the particles are essentially round sand of 200 µm diameter. Heat is supplied to the reactor through 3/4-in. tubes immersed vertically in the fluidized bed. The gas flow is 200 standard cubic feet per minute (SCFM). The reactor is 1 m in diameter, is fluidized by gas coming through eight distinct nozzles, and is filled with sand fluidized to a height 0.8 m. There are 20 heat-transfer tubes pushing up through the gas distributor at the base of the reactor to a height of 0.3 m. Calculate the convective heat-transfer coefficient between the tubes and the bed.

Solution

Sand in this size range fluidized by a gas is typical of Geldart group B fluidization, and therefore we can be sure that the bed is bubbling (provided the gas flow is below the particle terminal velocity). Thus, we should expect a fairly large heat-transfer coefficient. First, we should calculate some preliminary variables that will be needed in the calculation of the convective coefficient. We'll need the minimum fluidization velocity, which is calculated from the (dimensionless) Ergun equation, set equal to the weight of solids

$$\mathrm{Ar} = 150 \frac{1 - \varepsilon_{\mathrm{mf}}}{\varepsilon_{\mathrm{mf}}^2} \cdot \mathrm{Re}_{\mathrm{mf}} + \frac{1.75}{\varepsilon_{\mathrm{mf}}^3} \cdot \mathrm{Re}_{\mathrm{mf}}^2 \qquad (33.1)$$

where Ar is the Archimedes number and $\mathrm{Re}_{\mathrm{mf}}$ is the particle Reynolds number at minimum fluidization conditions:

$$\mathrm{Ar} = \frac{\rho_g (\rho_s - \rho_g) \cdot g \cdot d_p^3}{\mu^2} \qquad \mathrm{Re}_{\mathrm{mf}} = \frac{d_p \cdot U_{\mathrm{mf}} \cdot \rho_g}{\mu}$$

The viscosity of N_2 at 450°C is $\mu = 33.5 \times 10^{-6}$ Pa · s, and the gas density is calculated from the ideal-gas law $\rho_g = 0.472$ kg/m³. Sand density is $\rho_s = 2600$ kg/m³, the particle diameter is $d_p = 200$ µm, and

33.8 Fluidized Beds

so the Archimedes number is calculated as $Ar = 85.8$. In practice, the void fraction at minimum fluidization ε_{mf} should always be measured, but here it is estimated as $\varepsilon_{mf} = 0.45$, a typical value for relatively round, small particles. Thus Re_{mf} is calculated from Eq. (33.1), $Re_{mf} = 0.0946$, from which the minimum fluidization velocity can be found, $U_{mf} = 0.0336$ m/s. The actual superficial gas velocity U is also required:

$$U = \frac{F}{A} = \frac{200 \frac{\text{ft}^3}{\text{min}}}{\frac{\pi}{4} \cdot (1 \text{ m})^2} \cdot \left(\frac{450 + 273}{293}\right) = 0.297 \frac{\text{m}}{\text{s}}$$

The second term accounts for the raised temperature, and unit conversions aren't shown. The gas flow is nearly 9 times that required for fluidization, which means that the bed is indeed fluidized, and bubbling will be vigorous.

An empirical correlation presented by Zabrodsky (1966) is quite good at predicting h_{max}, the maximum heat-transfer coefficient to be expected at some gas velocity:

$$h_{max} = 35.8 \rho_s^{0.2} \cdot k_g^{0.6} \cdot d_p^{-0.36} \tag{33.2}$$

All variables are given in SI (International System; metric) units in this correlation. For this particular application, it predicts $h_{max} = 627$ W/(m$^2 \cdot$ K).

It is tempting to stop here, since we know that the overall heat transfer is near the maximum rate over a broad range of U. However, this gas flow is fairly small, so we might suspect that the actual h is somewhat below the maximum value. There is no uniformly accepted correlation for predicting h in bubbling beds. We will use the correlation given by Molerus et al. (1995). However, the interested reader should consult one or more of the excellent published reviews (Kunii and Levenspiel, 1991, chapter 13; Botterill, 1986; Xavier and Davidson, 1985).

The correlation presented by Molerus et al. (1995) is given below. Despite its complex appearance, it is not difficult to use, and is reasonably accurate for bubbling fluidization systems over a wide range of operating conditions:

$$Nu_\lambda \equiv \frac{h \cdot \lambda}{k_g} = Nu_{pc} + Nu_{gc} \tag{33.3}$$

The dimensionless convective coefficient is given as the sum of a component from particle convection and a component from gas convection. The former accounts for heat transfer to moving packets, and the latter accounts for heat transfer directly to the gas phase, which becomes more important at high gas velocities and in beds composed of larger particles:

$$\mathrm{Nu}_{gc} = \frac{0.165\sqrt[3]{\mathrm{Pr}\cdot R \cdot V}}{V + 0.05} \qquad (33.4)$$

$$\mathrm{Nu}_{pc} = \frac{0.125(1-\varepsilon_{mf})\cdot\left(1+\dfrac{33.3}{\sqrt[3]{V\cdot U_e}}\right)^{-1}}{1+\left(\dfrac{k_g}{2\cdot C_s\cdot \mu}\right)\cdot\left[1+\dfrac{0.28}{V}\cdot(1-\varepsilon_{mf})^2\cdot\sqrt{R}\cdot U_e^{\,2}\right]} \qquad (33.5)$$

Intermediate parameters are defined as follows,

$$V \equiv \frac{U-U_{mf}}{U_{mf}} \qquad R \equiv \frac{\rho_g}{\rho_s - \rho_g} \qquad U_e \equiv \sqrt[3]{\frac{\rho_s\cdot C_s}{k_g\cdot g}}\cdot(U-U_{mf}) \qquad (33.6)$$

and the length constant λ is defined as

$$\lambda \equiv \left[\frac{\mu}{\sqrt{g}\cdot(\rho_s-\rho_g)}\right]^{2/3} \qquad (33.7)$$

Both V and U_e are two ways to represent dimensionless excess gas velocity and are evaluated as follows:

$$V = \frac{U-U_{mf}}{U_{mf}} = \frac{0.297\,\dfrac{m}{s} - 0.034\,\dfrac{m}{s}}{0.034\,\dfrac{m}{s}} = 7.74$$

$$U_e = \sqrt[3]{\frac{\rho_s\cdot C_s}{k_g\cdot g}}\cdot(U-U_{mf})$$

$$= \sqrt[3]{\frac{\left(2600\,\dfrac{kg}{m^3}\right)\cdot\left(740\,\dfrac{J}{kg\cdot K}\right)}{\left(0.0518\,\dfrac{W}{m\cdot K}\right)\cdot\left(9.81\,\dfrac{m}{s^2}\right)}}\cdot\left(0.297\,\frac{m}{s} - 0.034\,\frac{m}{s}\right) = 41.0$$

The term R is simply a density ratio, and is evaluated easily:

$$R = \frac{\rho_g}{\rho_s-\rho_g} = \frac{0.472\,\dfrac{kg}{m^3}}{2600\,\dfrac{kg}{m^3} - 0.472\,\dfrac{kg}{m^3}} = 1.816\times 10^{-4}$$

33.10 Fluidized Beds

We also need the gas-phase Prandtl number:

$$\Pr \equiv \frac{C_g \cdot \mu}{k_g} = \frac{\left(1104 \, \frac{J}{kg \cdot K}\right) \cdot \left(3.35 \times 10^{-5} \, \frac{kg}{m \cdot s}\right)}{0.0518 \, \frac{W}{m \cdot K}} = 0.714$$

Now we can evaluate the contribution of heat transfer directly to the gas phase from Eq. (33.4):

$$\mathrm{Nu}_{gc} = \frac{0.165 \sqrt[3]{\Pr \cdot R} \cdot V}{V + 0.05}$$

$$= \frac{0.165 \sqrt[3]{(0.714) \cdot (1.816 \times 10^{-4})} \cdot (7.74)}{7.74 + 0.05} = 8.298 \times 10^{-3}$$

We can also evaluate the contribution of particle packets from Eq. (33.5):

$$\mathrm{Nu}_{pc} = \frac{0.125(1 - \varepsilon_{mf}) \cdot \left(1 + \frac{33.3}{(\sqrt[3]{7.74})(41.0)}\right)^{-1}}{1 + \left(\frac{k_g}{2 \cdot C_s \cdot \mu}\right) \cdot \left[1 + \frac{0.28}{V} \cdot (1 - \varepsilon_{mf})^2 \cdot \sqrt{R} \cdot U_e^2\right]}$$

$$\mathrm{Nu}_{pc} = \frac{0.125(1 - 0.45) \cdot \left(1 + \frac{33.3}{(\sqrt[3]{7.74})(41.0)}\right)^{-1}}{1 + \left[\frac{\left(0.0518 \, \frac{W}{m \cdot K}\right)}{2 \cdot \left(740 \, \frac{J}{kg \cdot K}\right) \cdot \left(3.35 \times 10^{-5} \, \frac{kg}{m \cdot s}\right)}\right] \cdot \left[1 + \left(\frac{0.28}{7.74}\right) \cdot (1 - 0.45)^2 \cdot \sqrt{1.816 \times 10^{-4}} \cdot (41.0)^2\right]}$$

$$\mathrm{Nu}_{pc} = 0.02116$$

The Nusselt number is evaluated from Eq. (33.3):

$$\mathrm{Nu}_\lambda = \mathrm{Nu}_{gc} + \mathrm{Nu}_{pc} = 8.298 \times 10^{-3} + 0.02116 = 0.0295$$

To find the convective coefficient, we need to evaluate λ from Eq. (33.7):

$$\lambda = \left[\frac{\mu}{\sqrt{g} \cdot (\rho_s - \rho_g)}\right]^{2/3} = \left[\frac{3.35 \times 10^{-5} \, \frac{kg}{m \cdot s}}{\sqrt{9.81 \, \frac{m}{s^2}} \cdot (2600 - 0.472) \, \frac{kg}{m^3}}\right]^{2/3} = 2.57 \, \mu m$$

Finally, we obtain

$$h = \frac{\mathrm{Nu}_\lambda \cdot k_g}{\lambda} = \frac{(0.0295) \cdot \left(0.0518 \, \frac{W}{m \cdot K}\right)}{2.57 \, \mu m} = 595 \, \frac{W}{m^2 \cdot K}$$

This coefficient, then, is the one to use to predict the rate of heat transfer between the fluidized bed and the tubes in the bed. The prediction indicates that the rate is near the maximum rate predicted by Eq. (33.2). In fact, the maximum coefficient predicted by this correlation is 635 W/(m² K) at a gas flow rate 85 percent greater than that specified in this design, remarkably similar to the maximum coefficient predicted by Eq. (33.2). Economics will determine the preferred gas flow rate.

Note that, according to this correlation, the predicted convective coefficient is independent of the size, location, and number of tubes. Of course, the total rate of heat transfer to the bed is proportional to the tube area, using $q = hA_t\Delta T$. The effect of tubes on bubble flow in a fluidized bed is well documented, causing bubbles to be smaller, rise more slowly, and therefore to reduce the flow of particle "packets." Like all correlations for convective coefficients, this prediction should be used cautiously.

In this calculation, we've neglected any radiation effects. For these conditions, that is reasonable, but for higher temperatures (>800°C) radiation might make a substantial contribution to overall heat transfer.

Problem Statement II: Circulating Fluidized Bed

Circulating fluidized beds are often used as coal combustion furnaces. One such unit consists of a 24-m-tall riser with a 4.0 × 4.0 m² cross section. The bottom 10 m of the riser is covered in a refractory material. The walls of the upper 14 m of the riser are made up of membrane tubes through which steam is heated. On average, the fluidized particles in the riser consist of round coal particles with a diameter of 200 μm. From pressure drop measurements, it is estimated that solids represent approximately $c_b = 1.8$ percent of the volume in the dilute zone of the riser where the membrane tubes are located. The fluidizing medium is air with a superficial velocity of 6.5 m/s. If the bed temperature is uniform at 850°C, what is the heat-transfer coefficient from the bed to the membrane wall? The coal properties include the density specified as $\rho_s = 1350$ kg/m³ and the heat capacity of $C_s = 1260$ J/(kg·K). At the given temperature, the air has a density of $\rho_g = 0.31$ kg/m³, a viscosity of $\mu = 4.545 \times 10^{-5}$ kg/(m·s), and a heat capacity of $C_g = 1163$ J/(kg·K).

Solution

We will give two solution techniques, noting that predicting heat transfer in CFB risers is still not well developed. The first method is a very simple empirical method:

$$h = a\rho_b^n \tag{33.8}$$

33.12   Fluidized Beds

On the basis of data taken from commercial boilers, Basu and Nag (1996) suggest $a = 40$ and $n = 0.5$ when ρ_b is given in kg/m³ and h is predicted in W/(m² K). These parameter values are restricted to $d_p \approx 200$ μm, 750°C $< T_b <$ 850°C, and 5 kg/m³ $< \rho_b <$ 25 kg/m³. First, we calculate the average bed density, to ensure that the correlation applies:

$$\rho_b = c_b \rho_s + (1 - c_b)\rho_g = 24.6 \text{ kg/m}^3$$

Thus, the correlation is valid, and we predict an average heat-transfer coefficient:

$$h = a\rho_b^n = 40\,(24.6)^{0.5} = 198\,\frac{\text{W}}{\text{m}^2\text{K}}$$

Another far more detailed, mechanistic method is also given here. Most such CFB heat-transfer correlations are derived from experimental studies of relatively small, laboratory-scale units. Care must be taken when applying these correlations to large CFB combustors as differences in hydrodynamics between small and large units are considerable and not yet well predicted. The heat-transfer coefficient from the bed suspension to the wall is calculated by summing the contributions from particle convective heat transfer, gas convective heat transfer, and radiation. The heat-transfer contributions from the particle and gas convection are dependent on the fraction of the wall covered by the cluster phase f

$$h = f h_c + (1 - f) h_g + h_r \tag{33.9}$$

where h_c is the contribution of particle clusters, h_g is the effect of the gas phase, and h_r is the radiation contribution, assumed to apply to the entire surface.

Particle cluster convective term.   Lints and Glicksman (1994) present a model for calculation of the particle convective heat-transfer coefficient. According to this model, particles coalesce into clusters as they exit the upward-moving core into the dense annular region within the riser. As the cluster approaches the wall and travels downward, heat is transferred from the cluster to the wall. After descending along the wall for some distance, the cluster is reentrained into the hot riser core:

$$\frac{1}{h_c} = \frac{\delta \cdot d_p}{k_g} + \sqrt{\frac{\pi \cdot t_c}{k_c \cdot C_s \cdot \rho_s \cdot c_c}} \tag{33.10}$$

where δ is a dimensionless cluster thickness, predicted empirically:

$$\delta = 0.0287(c_b)^{-0.581} = 0.0287(0.018)^{-0.581} = 0.296 \tag{33.11}$$

The fraction of solids in the cluster is also predicted empirically from the average solids concentration:

$$c_c = 1.23(c_b)^{0.54} = 1.23(0.018)^{0.54} = 0.141 \tag{33.12}$$

The thermal conductivity of a cluster is predicted as a linear average of its components:

$$k_c = c_c k_s + (1 - c_c) k_g \tag{33.13}$$

$$= (0.141)\left(0.26\,\frac{W}{m\,K}\right) + (1 - 0.141)\left(0.073\,\frac{W}{m\,K}\right) = 0.099\,\frac{W}{m\,K}$$

The final parameter required for evaluating h_c from Eq. (33.10) is t_c, the time of contact between the cluster and the wall. It is evaluated from the ratio of the distance traveled to the speed at which a cluster travels

$$t_c = \frac{L_c}{U_c} \tag{33.14}$$

where L_c is the distance that a cluster travels along a riser wall before being entrained back into the upward bulk suspension, and is correlated empirically:

$$L_c = 0.0178(\rho_b)^{0.596} = 0.0178(24.6)^{0.596} = 0.120\,\text{m} \tag{33.15}$$

The cluster speed is assumed to always equal 1 m/s. Thus

$$t_c = \frac{0.120\,\text{m}}{1\,\text{m/s}} = 0.120\,\text{s}$$

and h_c can now be evaluated from Eq. (33.10):

$$\frac{1}{h_c} = \frac{\delta \cdot d_p}{k_g} + \sqrt{\frac{\pi \cdot t_c}{k_c \cdot C_s \cdot \rho_s \cdot c_c}} = \frac{(0.296) \cdot (0.0002\,\text{m})}{0.073\,\dfrac{W}{m \cdot K}}$$

$$+ \sqrt{\frac{\pi \cdot (0.120\,\text{s})}{\left(0.099\,\dfrac{W}{m \cdot K}\right) \cdot \left(1260\,\dfrac{J}{kg \cdot K}\right) \cdot \left(1350\,\dfrac{kg}{m^3}\right) \cdot (0.141)}}$$

$$h_c = 209\,\frac{W}{m^2 K}$$

The resistance to conduction through the cluster ($\delta d_p / k_g$) is somewhat less than the resistance in the transient term, and has only a slight effect on the cluster convective coefficient.

33.14 Fluidized Beds

Gas convection term. One simple way to determine the gas convective heat-transfer coefficient is to use the correlation presented by Martin (1984), applicable when gas velocities are greater than the minimum fluidization velocity:

$$\text{Nu} = 0.009 \, \text{Pr}^{1/3} \text{Ar}^{1/2} \tag{33.16}$$

The Prandtl number is evaluated from its definition:

$$\text{Pr} = \frac{\mu \cdot C_g}{k_g} = \left[\frac{\left(4.545 \times 10^{-5} \frac{\text{kg}}{\text{m} \cdot \text{s}}\right) \cdot \left(1163 \frac{\text{J}}{\text{kg} \cdot \text{K}}\right)}{\left(0.073 \frac{\text{W}}{\text{m} \cdot \text{K}}\right)} \right] = 0.724 \tag{33.17}$$

Additionally, the Archimedes number must be calculated:

$$\text{Ar} = \left[\frac{\rho_g \cdot (\rho_s - \rho_g) \cdot g \cdot (d_p)^3}{\mu^2} \right]$$

$$= \left[\frac{\left(0.3105 \frac{\text{kg}}{\text{m}^3}\right) \cdot \left(1350 \frac{\text{kg}}{\text{m}^3} - 0.3105 \frac{\text{kg}}{\text{m}^3}\right) \cdot \left(9.81 \frac{\text{m}}{\text{s}^2}\right) \cdot (0.0002 \, \text{m})^3}{\left(4.545 \times 10^{-5} \frac{\text{kg}}{\text{m} \cdot \text{s}}\right)^2} \right] = 15.92$$

We calculate the Nusselt number from Eq. (33.16):

$$\text{Nu} = 0.009 \, \text{Pr}^{1/3} \text{Ar}^{1/2} = (0.009)(0.724)^{1/3}(15.92)^{1/2} = 0.032$$

The gas convective heat-transfer coefficient is calculated from the Nusselt number:

$$h_g = \frac{\text{Nu} \cdot k_g}{d_p} = \left[\frac{(0.032) \cdot \left(0.073 \frac{\text{W}}{\text{m} \cdot \text{K}}\right)}{0.0002 \, \text{m}} \right] = 11.68 \, \frac{\text{W}}{\text{m}^2 \cdot \text{K}}$$

Radiation term. To calculate the radiation heat-transfer coefficient, one must first determine the effective emissivity of the bed. This is related to the bed volume where heat transfer occurs and to the surface area of the suspended particles, the mean beam length, and the mean extinction coefficient as described by Glicksman (1997). Looking at the region of the CFB riser whose walls are composed of membrane tubes, the volume of this region of the bed is

$$V_{\text{bed}} = (24 \, \text{m} - 10 \, \text{m}) \times (4 \, \text{m})(4 \, \text{m}) = 224 \, \text{m}^3 \tag{33.18}$$

Recalling the volume fraction of solids in the riser $c_b = 0.018$, the volume of particles in this region of the riser can be found:

$$V_{TS} = V_{bed} \times c_b = (224 \, \text{m}^3) \times (0.018) = 4.032 \, \text{m}^3 \tag{33.19}$$

If the particles are assumed to be smooth spheres, then the volume of a single particle is

$$V_p = \frac{\pi}{6} d_p^3 = 4.19 \times 10^{-12} \, \text{m}^3 \tag{33.20}$$

and the surface area of a single particle is

$$S_p = \pi d_p^2 = 1.26 \times 10^{-7} \, \text{m}^2 \tag{33.21}$$

The total number of particles in this region of the riser is found from the total volume occupied by the particles and the volume of a single particle:

$$N_p = \frac{V_{TS}}{V_p} = \frac{4.032 \, \text{m}^3}{4.189 \times 10^{-12} \, \text{m}^3} = 9.63 \times 10^{11} \, \text{particles} \tag{33.22}$$

The surface area of all of the particles can then be calculated:

$$S_{bed} = N_p \cdot S_p = (9.63 \times 10^{11}) \cdot (1.257 \times 10^{-7} \, \text{m}^2) = 1.21 \times 10^5 \, \text{m}^2 \tag{33.23}$$

From this information, an estimation of the mean beam length L_m, can be made.

$$L_m = 3.5 \cdot \left(\frac{V_{bed}}{S_{bed}}\right) = (3.5) \cdot \frac{224 \, \text{m}^3}{1.21 \times 10^5 \, \text{m}^2} = 0.00648 \, \text{m} \tag{33.24}$$

If the particle emissivity is estimated to be $\varepsilon_{rp} = 0.5$, as suggested by Glicksman (1997), then the mean extinction coefficient K can be found:

$$K = \left(\frac{3}{2}\right) \cdot \varepsilon_{rp} \cdot \left(\frac{c_b}{d_p}\right) = \left(\frac{3}{2}\right) \cdot (0.5) \cdot \left(\frac{0.018}{0.0002 \, \text{m}}\right) = 67.5 \, \text{m}^{-1} \tag{33.25}$$

If the bed is sufficiently dilute and the effective bed emissivity ε_{rb} is less than approximately 0.5 to 0.8, then the following correlation, suggested by Glicksman (1997), can be used:

$$\varepsilon_{rb} = 1 - \exp(-K \cdot L_m) = 1 - \exp[(-67.5 \, \text{m}^{-1}) \cdot (0.00648 \, \text{m})] = 0.354 \tag{33.26}$$

If the wall emissivity ε_{rw} is estimated to be 0.3, and the wall temperature is estimated at 600°C, then the radiation heat-transfer coefficient can be calculated as reported by Basu and Nag (1996):

$$h_r = \frac{(T_b^4 - T_w^4) \cdot \sigma}{\left(\dfrac{1}{\varepsilon_{rb}} + \dfrac{1}{\varepsilon_w} - 1\right) \cdot (T_b - T_w)}$$

$$= \frac{[(1123\ \text{K})^4 - (873\ \text{K})^4] \cdot \left(5.67 \times 10^{-8}\ \dfrac{\text{W}}{\text{m}^2 \cdot \text{K}^4}\right)}{\left(\dfrac{1}{0.354} + \dfrac{1}{0.3} - 1\right) \cdot (1123\ \text{K} - 873\ \text{K})}$$

$$= 44.4\ \frac{\text{W}}{\text{m}^2 \text{K}} \tag{33.27}$$

Total bed-to-wall heat-transfer coefficient. At this point, the heat-transfer coefficient for each of the three mechanisms has been calculated. All that is remaining is to determine the fractional wall coverage of particle clusters f, as required in Eq. (33.9). Lints and Glicksman (1994) present the following correlation:

$$f = 3.5 \cdot (c_b)^{0.37} = 3.5 \cdot (0.018)^{0.37} = 0.79 \tag{33.28}$$

Along this fraction of surface area f, particle convective heat transfer is occurring. Along the remainder of the surface area $(1 - f)$, gas convective heat transfer occurs. Radiation acts over the entire surface.

Recalling the results calculated previously, we have

$$h_c = 209\ \frac{\text{W}}{\text{m}^2 \cdot \text{K}} \qquad h_g = 11.68\ \frac{\text{W}}{\text{m}^2 \cdot \text{K}} \qquad h_r = 44.4\ \frac{\text{W}}{\text{m}^2 \cdot \text{K}}$$

The total bed-to-wall heat-transfer coefficient is found from Eq. (33.9):

$$h = f \cdot h_c + (1 - f) \cdot h_g + h_r$$

$$= 0.79(209) + (1 - 0.79) \cdot (11.68) + 44.4 = 212\ \frac{\text{W}}{\text{m}^2 \cdot \text{K}}$$

The simple empirical solution for this case is similar to the more rigorous solution, differing by less than 7 percent. Such close agreement is unusual. The total rate of heat transfer from the CFB to the membrane walls is found from $q = hA_t \Delta T$, with A_t consisting of the area of the membrane wall $= (4 \times 4\ \text{m}) \times (14\ \text{m}) = 224\ \text{m}^2$.

Nomenclature

a	Empirical constant in Eq. (33.8)
A	Fluidized-bed cross-sectional area
A_t	Area of all heat-transfer surfaces

Ar	Dimensionless Archimedes number
c_b	Volume fraction of solids in the riser
c_c	Volume fraction of solids in clusters
C_g	Heat capacity of gas
C_s	Heat capacity of solids
d_p	Harmonic mean particle diameter
f	Fraction of riser wall covered by particle clusters
F	Gas volumetric flow rate
g	Acceleration due to gravity
h	Mean convective coefficient between a fluidized bed and a surface
h_c	Convective heat-transfer coefficient through particle clusters
h_g	Convective heat-transfer coefficient through fluidizing gas
h_r	Radiation heat-transfer coefficient
k_c	Thermal conductivity of a particle cluster
k_g	Thermal conductivity of fluidizing gas
k_s	Thermal conductivity of solid particles
K	Mean extinction coefficient
L_c	Distance that a cluster travels along the wall
L_m	Mean beam length
n	Empirical constant in Eq. (33.8)
N_p	Number of particles
Nu	Dimensionless Nusselt number
Pr	Dimensionless Prandtl number of the gas
q	Total heat-transfer rate between surface and a fluidized bed
R	Dimensionless density defined in Eq. (33.6)
Re_{mf}	Particle Reynolds number at minimum fluidization gas velocity
S_{bed}	Surface area of all particles in riser section
S_p	Surface area of a single particle
t_c	Cluster contact time with wall
T_b	Bed temperature
T_w	Wall temperature
U	Superficial gas velocity
U_c	Speed at which a cluster travels along a wall
U_e	Dimensionless excess gas velocity defined in Eq. (33.6)
U_{mf}	Minimum superficial gas velocity necessary for fluidization
V	Dimensionless excess gas velocity defined in Eq. (33.6)
V_{bed}	Volume of circulating fluidized-bed riser subject to heat transfer
V_p	Volume of a single particle
V_{TS}	Volume of total number of solid particles in riser

Greek

δ	Dimensionless gas-layer thickness at riser wall
ΔT	Temperature difference between fluidized bed and heat-transfer surface
ε_{mf}	Void fraction of a bed at minimum fluidization conditions
ε_b	Volume fraction of a fluidized bed occupied by gas bubbles
ε_{pb}	Radiation emissivity of single particle
ε_{rb}	Radiation emissivity of particle bed
ε_{wb}	Radiation emissivity of riser wall
λ	Characteristic length in a bubbling fluidized bed
μ	Gas viscosity
ρ_b	Bulk suspension density in a fluidized bed
ρ_g	Gas density
ρ_s	Particle density
σ	Stefan-Boltzmann constant

References

P. Basu and P. K. Nag, "Heat Transfer to Walls of a Circulating Fluidized-Bed Furnace," *Chem. Eng. Sci.*, **51**(1), 1–26 (1996).

J. S. M. Botterill, "Fluid Bed Heat Transfer," in *Gas Fluidization Technology*, D. Geldart (ed.), Wiley, New York, 1986.

D. Geldart (ed.), *Gas Fluidization Technology*, Wiley, New York, 1986.

D. Geldart, "Types of Gas Fluidisation," *Powder Technol.*, **7**, 285–292 (1973).

L. R. Glicksman, "Heat Transfer in Circulating Fluidized Beds," in *Circulating Fluidized Beds*, J. R. Grace, A. A. Avidan, and T. M. Knowlton (eds.), Blackie Academic and Professional, New York, 1997, pp. 261–311.

D. Kunii and O. Levenspiel, *Fluidization Engineering*, 2d ed., Butterworth-Heinemann, Boston, 1991.

M. C. Lints, and L. R. Glicksman, "Parameters Governing Particle-to-Wall Heat Transfer in a Circulating Fluidized Bed," in *Circulating Fluidized Bed Technology IV*, A. Avidan, (ed.), American Institute of Chemical Engineers (AIChE), New York, 1994, pp. 297–304.

H. Martin, "Heat Transfer between Gas Fluidized Beds of Solid Particles and the Surfaces of Immersed Heat Exchanger Elements, Part I," *Chem. Eng. Process.*, **18**, 157–169 (1984).

O. Molerus, A. Burschka, and S. Dietz, "Particle Migration at Solid Surfaces and Heat Transfer in Bubbling Fluidized Beds—II. Prediction of Heat Transfer in Bubbling Fluidized Beds," *Chem. Eng. Sci.*, **50**, 879–885 (1995).

J. Werther and B. Hirschberg, "Solids Motion and Mixing," in *Circulating Fluidized Beds*, J. R. Grace, A. A. Avidan, and T. M. Knowlton (eds.), Blackie Academic and Professional, New York, 1997, pp. 119–148.

A. M. Xavier and J. F. Davidson, "Heat Transfer in Fluidized Beds: Convective Heat Transfer in Fluidized Beds," in *Fluidization*, 2d ed., J. F. Davidson, R. Clift, and D. Harrison (eds.), Academic Press, New York, 1985.

S. S. Zabrodsky, *Hydrodynamics and Heat Transfer in Fluidized Beds*, MIT Press, Cambridge, Mass., 1966.

Part 8

Parameter and Boundary Estimation

The three chapters in this part, two by U.S. academics and one by an engineer employed in industry in Japan, present calculations that vary from being conjectural to representing an industrial situation. The first chapter, which covers parameter estimation, is arguably the most theoretical in the entire handbook. The second chapter deals with the thermal design of boxes whose dimensions do not exceed 1 m and whose function is to house heat-generating components that must be prevented from overheating. Examples include desktop computers, laptop computers, and cell phones. Relevant calculations involve the upper bounds of heat transfer from sealed boxes, from boxes having vent ports for a natural draft, and from boxes cooled by fans. The third chapter in this part presents calculations that can be used to predict the freezing time of strawberries being frozen in a fluidized-bed freezer.

Chapter 34

Estimation of Parameters in Models

Ashley F. Emery
Mechanical Engineering Department
University of Washington
Seattle, Washington

Consider the one-dimensional heat transfer through a plane layer whose temperature satisfies

$$\rho c \frac{\partial T}{\partial t} = k \frac{\partial^2 T}{\partial x^2} \qquad (34.1)$$

with boundary conditions

$$x = 0: \quad h(T_0 - T(0, t)) = -k \frac{\partial T}{\partial x} \qquad (34.2\text{a})$$

$$x = L: \quad -k \frac{\partial T}{\partial x} = h(T(L, t) - T_L) \qquad (34.2\text{b})$$

and initial conditions

$$t = 0: \quad T(x, 0) = 0 \qquad (34.2\text{c})$$

We wish to estimate the thermal diffusivity $\kappa = k/\rho c$. We will do this by measuring the temperature at a selected location $x = x^*$ as a function of time. The idea is to choose a value of the conductivity for which the predictions of the model [Eq. (34.1)] match the measurements as closely as possible.

34.4 Parameter and Boundary Estimation

We will do this by minimizing the least-squares error E:

$$E = \sum_{j}^{M}\sum_{i}^{N}(T_e(x_j, t_i) - T_p(x_j, t_i, \kappa))^2 = \sum_{j}^{M}\sum_{i}^{N} e_{i,j}^2(\kappa) \qquad (34.3)$$

where T_e and T_p represent the experimental and predicted temperatures, respectively; M is the number of measurement locations; N is the number of measurements at each location; and $e_{i,j}(\kappa)$ are called the *residuals*, which are functions of κ. We could simply choose different values of the diffusivity κ until we find a value that minimizes E by plotting $E(\kappa)$ as a function of κ and choosing that value of κ for which $E(\kappa)$ is a minimum. Note that in Eq. (34.3) all the residuals are given equal weight. We could associate a different weight with each residual, but for simplicity in this problem we will treat each measurement as being of equal importance. Readers interested in a more detailed treatment should consult one of the references listed.

Estimation and Sensitivity

Unfortunately, although plotting $E(\kappa)$ works, it doesn't tell us much about where and when to make the measurements. A better way is to investigate the sensitivity of our estimate of κ to the measurements. We do this as follows. Since $T_p(x, t, \kappa)$ is a nonlinear function of κ, we expand $T(x, t, \kappa)$ about some initial estimate κ_0, giving a residual of

$$e_{i,j}(\kappa) = T_e(x_i, t_j) - T_p(x_i, t_j, \kappa_0) - \frac{\partial T}{\partial \kappa}\bigg|_{\kappa_0}(\kappa - \kappa_0) \qquad (34.4a)$$

Substituting Eq. (34.4a) into Eq. (34.3) and differentiating with respect to κ and setting the derivative equal to 0 gives

$$\kappa - \kappa_0 = \frac{\sum_i^N \sum_j^M (T_e(x_i, t_j) - T_p(x_i, t_j, \kappa_0))A_{i,j}(\kappa_0)}{\sum_i^N \sum_j^M A_{i,j}^2(\kappa)} \qquad (34.4b)$$

where $A_{i,j} = \partial T(\kappa)/\partial \kappa$ evaluated at x_i, t_j, κ_0 and is called the *sensitivity* of the model. Now this will give us an improved value of κ, but since T_p is a function of κ, we will have to iterate. Our procedure is then

$$\kappa_{p+1} - \kappa_p = \frac{\sum_i^N \sum_j^M (T_e(x_i, t_j) - T_p(x_i, t_j, \kappa_p))A_{i,j}(\kappa_p)}{\sum_i^N \sum_j^M A_{i,j}^2(\kappa_p)} \qquad (34.5a)$$

where

$$A_{i,j}(\kappa_p) = \frac{\partial T(\kappa)}{\partial \kappa} \qquad (34.5b)$$

evaluated at x_i, t_j, κ_p, and we iterate until we have achieved convergence. From Eq. (34.5a) it is apparent that the rate of convergence will depend strongly on the denominator that is a function of the sensitivities. The higher the sensitivity (i.e., the larger the absolute value of A), the faster our method will converge. In fact, it is known that $\sigma(\hat{\kappa})$, the standard deviation of our final estimate, is given by

$$\sigma(\hat{\kappa}) = \frac{\sigma(T_e)}{\sqrt{\sum_i^N \sum_j^M A_{i,j}^2(\hat{\kappa})}} \quad (34.6)$$

where $\sigma(T_e)$ is the noise associated with our experimental measurements. If there is no noise, then $\hat{\kappa}$ will be the exact value.

Thus the first thing that we want to do before making the measurements is to compute the sensitivities $A_{i,j}$ and to choose measurement locations and times that give us large absolute values. Another way to look at it is to avoid measurements for which the sensitivity is low.

The Matlab program Slab_Sensitivity computes and plots the sensitivities for a slab subjected to convective boundary conditions. The call is

[T,S] = Slab_Sensitivity(Fo,kappa,bc,Nt)

where Fo is the time duration in terms of the Fourier number, kappa is the true thermal diffusivity, bc defines the boundary conditions, and Nt is the number of time steps; bc is in the form of a vector with components in terms of the Nusselt number and nondimensional temperature, bc = [$h_0 L/k$; T_0; $h_L L/k$; T_L]. Figure 34.1 depicts the temperatures and the sensitivities for Fo = 1, kappa = 1, and bc = [10;0;100;1] at $x/L = 0, 0.25, 0.50, 0.75$, and 1.00.

In Fig. 34.1a and b, note how the sensitivities change with time and position. The sensitivity at $x/L = 1$ is essentially zero because the high value of the convective heat-transfer coefficient h_L causes the temperature to reach a constant value very quickly and to be insensitive to the diffusivity. On the other hand, the temperature at $x/L = 0$, where h_0 is low, continuously changes and the sensitivity is significantly larger than that at $x/L = 1$, and eventually becomes the greatest of all positions. For moderate times, the best location to measure the temperature is near the center, someplace between 0.25 and 0.75. We define information about our estimate of κ to be

$$\text{Information about } \hat{\kappa} = \frac{\sigma^2(\hat{\kappa})}{\sigma^2(\text{noise})} \quad (34.7)$$

34.6 Parameter and Boundary Estimation

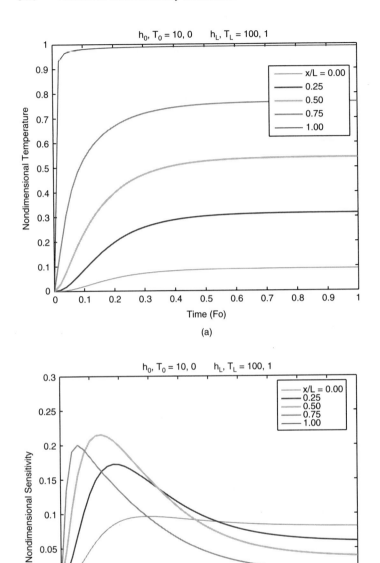

Figure 34.1 (a) Temperature histories; (b) sensitivity histories.

Figure 34.2a illustrates the instantaneous and cumulative information versus time. While it might appear from Fig. 34.1b that a measurement at $x/L = 0.5$ is the most sensitive, the contributions of the different measurements to the information are shown in Fig. 34.2b and it is easy to see how each contributes differently to the total information. At early times, the measurement made near the boundary where the heat is added is the most informative. As time goes on, the measurements made toward the boundary at $x/L = 0$ become more important and finally at long times, the sensor at $x/L = 0$ is the most valuable, but contributes little to overall accuracy of the estimation.

Estimation

We simulate an actual experiment by adding noise to the temperature computed using the exact value of κ. The diffusivity is estimated by comparing the measured (simulated) temperatures to that predicted from the model using the program Estimate_kappa. The call is

[kappahat, sigma_khat]
            = Estimate_kappa(Fo, kappa, bc, Nt, Ixm, noise)

where Ixm is the node number of the measured temperature. The noise variable can be entered in one of two ways: (1) noise > 0, where the noise is in terms of the nondimensional value of temperature (T/T_L); or (2) noise < 0, where the noise is a percentage of the nondimensional temperature. Using the same values for bc as in the sensitivity calculation, measuring the temperature at the midpoint Ixm = 11, and using a noise of 0.02 gives the simulated measured temperature shown in Fig. 34.3a.

The estimated thermal diffusivity is 1.0215, and its standard deviation is 0.0410.

Figure 34.3b shows the convergence as a function of the iterations.

Noise

The noise is generated using a random-number generator. Each time the program is run, a different set of random numbers is generated, simulating different experiments. If you want the measurement noise to be constant, you should modify the program as shown in the code.

Programs

Pseudocode and Matlab programs for these calculations are available at www.me.washington.edu/people/faculty/emery/Handbook_Programs.

34.8 Parameter and Boundary Estimation

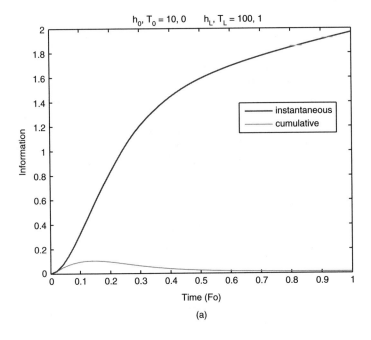

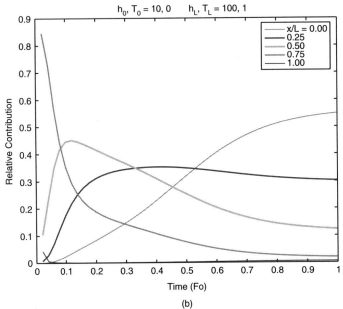

Figure 34.2 (a) History of the information; (b) contributions to the information.

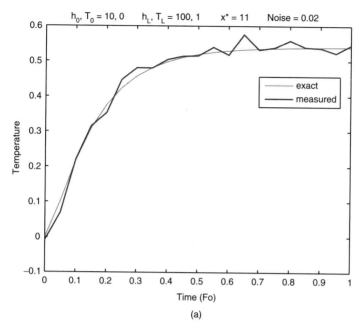

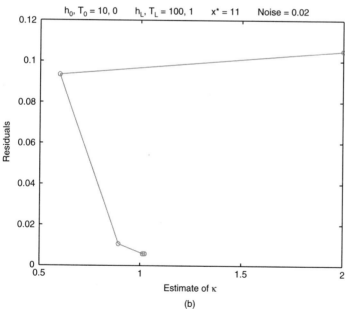

Figure 34.3 (a) Temperature histories; (b) convergence of estimated value of κ.

34.10 Parameter and Boundary Estimation

References

Beck, J. V., and Arnold, K. J., 1977, *Parameter Estimation in Engineering and Science,* Wiley, New York.

Norton, J. P., 1986, *An Introduction to Identification,* Academic Press, New York.

Ross, Gavin J. S., 1990, *Nonlinear Estimation,* Springer-Verlag, New York.

Sorenson, H. W., 1980, *Parameter Estimation: Principles and Problems,* Marcel Dekker, New York.

Strang, G., 1993, *Introduction to Linear Algebra,* Wellesley-Cambridge Press, Wellesley, Mass.

Chapter

35

Upper Bounds of Heat Transfer from Boxes

Wataru Nakayama
ThermTech International
Kanagawa, Japan

Introduction

In this chapter we consider boxes that house heat-generating components. We confine our attention to those boxes that we can readily handle. So, their dimensions are of the order of centimeters and do not exceed approximately one meter. The items of our primary interest are electronic boxes and gadgets. The most familiar examples are desktop computers, laptop computers, and cell phones. Many other electronic boxes are integrated in various machines and systems. Common to all those electronic boxes is the need for thermal design to prevent the components in the box from overheating.

The first step of thermal design is the determination of cooling mode. The designer faces a question as to whether a particular future product can be encased in a sealed box, or a box having vent ports to allow cooling air through it, or to use one or more fans to drive air mechanically out of the box. In this first phase of the design analysis the rule-of-thumb calculation suffices, and this chapter explains such calculations. The upper bounds of heat transfer from sealed boxes, boxes having vent ports for natural draft, and those cooled by the fan(s) are studied here. Knowledge about the upper-bound heat transfer serves as a useful guide in assessment of different cooling modes.

35.2 Parameter and Boundary Estimation

The calculation of upper-bound heat transfer is performed using simplified models. Internal organization of the box, often complex in actual equipment, is omitted from consideration here. We focus our attention on the exterior envelope of the box. The "envelope" is the exterior surface in the case of a sealed box, and includes the vent ports in the case of a ventilated box. The upper-bound heat transfer is determined by several factors. In the case of sealed boxes, the laws of natural convection heat transfer and radiation heat transfer, and the surface area of the box are the determinants. To calculate the upper-bound heat transfer from ventilated boxes, we need only the enthalpy transport equation applied to the airflow from the vent port. To calculate upper-bound values we also need to specify the allowable temperature potential to drive heat flow from the box. In the thermal design of consumer electronic equipment, 30°C is commonly assumed as a maximum environmental temperature. Meanwhile, the surface temperature of those commercial products that have accidental contact with or are worn by human users needs to be held below 45°C. This criterion also applies to the temperature of exhaust air from the vent port, which may also come in contact with human users. So, the temperature potential available for heat dissipation from the box to the environment is assumed to range from 10 to 25 K.

Passive Cooling of Sealed Box

It is assumed that the materials filling the box have infinite thermal conductivity and hence that heat generated from any parts of the interior instantaneously diffuse to the surface of the box. The model is an isothermal lump (Fig. 35.1). The heat flow from the box surface Q [in watts (W)] is the sum of heat flow by natural convection Q_{NC} (W), and

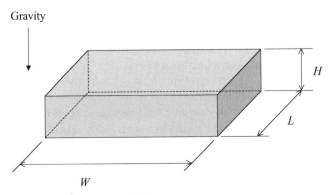

Figure 35.1 Isothermal solid lump.

radiation heat flow Q_R (W):

$$Q = Q_{NC} + Q_R \qquad (35.1)$$

The convective heat flow Q_{NC} is the sum of heat flows from different surfaces

$$Q_{NC} = Q_V + Q_U + Q_L \qquad (35.2)$$

where Q_V, Q_U, and Q_L are the heat flow from the vertical, the upper, and the lower surfaces, respectively.

Each term on the right-hand side of Eq. (35.2) is written in the form

$$Q_i = h_i A_i \Delta T \qquad (35.3)$$

where h_i is the heat-transfer coefficient (W/m² K), A_i is the surface area (m²), ΔT is the difference between the surface temperature and the environment temperature (K), and the subscript i is V, U, or L. The sum of vertical surface areas is $A_V = 2(W + L)H$. The horizontal surface area is $A_U = A_L = WL (\equiv A_H)$.

The heat-transfer coefficient is found from the Nusselt number–Rayleigh number correlations. These numbers are defined as

$$\overline{Nu}_i \equiv \frac{h_i L_B}{k} \qquad (35.4)$$

$$Ra \equiv \frac{g\beta L_B^3 \Delta T}{\nu^2} \cdot Pr \qquad (35.5)$$

where the overbar on the Nusselt number means the average value over the surface, L_B is the characteristic length (m) [$L_B = H$ for vertical surface; $L_B = A_H/2(W + L)$ for horizontal surface], and g is the natural gravity, 9.807 m/s². The physical properties of air in these equations are as follows: k is the thermal conductivity (W/m K), β is the volumetric expansion coefficient (K⁻¹), ν is the kinematic viscosity (m²/s), and Pr is the Prandtl number.

The Nu-Ra formulas used to estimate the natural convection heat transfer from the isothermal surface are as follows. For the sake of simplicity, the subscript i (V, U, or L) will be omitted in the following:

- Vertical surface (Refs. 1 and 2)

$$\overline{Nu} = 0.68 + \frac{0.670\, Ra^{1/4}}{\left[1 + (0.492/Pr)^{9/16}\right]^{4/9}} \qquad Ra \leq 10^9 \qquad (35.6)$$

35.4 Parameter and Boundary Estimation

- Upper horizontal surface (Ref. 1)

$$\overline{\mathrm{Nu}} = 0.54\,\mathrm{Ra}^{1/4} \qquad 10^4 \leq \mathrm{Ra} \leq 10^7 \qquad (35.7)$$

$$\overline{\mathrm{Nu}} = 0.15\,\mathrm{Ra}^{1/3} \qquad 10^7 \leq \mathrm{Ra} \leq 10^{11} \qquad (35.8)$$

- Lower horizontal surface (Ref. 1)

$$\overline{\mathrm{Nu}} = 0.27\,\mathrm{Ra}^{1/4} \qquad 10^5 \leq \mathrm{Ra} \leq 10^{10} \qquad (35.9)$$

The radiation heat transfer is computed from

$$Q_R = A_s h_R \Delta T \qquad (35.10)$$

where A_s is the total surface area exposed to the environment (m^2), h_R is the radiation heat-transfer coefficient (W/m^2 K) given as (Ref. 1)

$$h_R = 4\varepsilon\sigma\,T_0^3 \qquad (35.11)$$

where ε is the emissivity of the box surface, σ is the Stefan-Boltzmann constant (5.670 × 10^{-8} W/m^2 K^4), and T_0 is the temperature in absolute scale (K). The T_0 is, by definition, the mean temperature of the surface and the environment; however, for most practical purposes and ease of calculation, it may be equated to the environment temperature. Using h_R and h_i (restoring the subscript i), we may write Eq. (35.1) as

$$Q = \Delta T \left(\sum_i (h_i + h_R) A_i \right) \qquad (35.12)$$

where the summation extends over the heat-transfer areas.

Calculations of heat flow were carried out for a wide range of box dimensions. Table 35.1 shows the dimensions of model boxes. The model designation consists of two numbers; one before a hyphen signifies the model, and that after the hyphen, the orientation of the box. In this dimension range are included those from cell phones, handheld calculators, laptop computers, desktop computers, and medium-sized server computers. It is assumed that the box is placed on a large adiabatic floor or attached to an adiabatic ceiling; hence, five surfaces are available for heat dissipation. This means that, for convection heat-transfer estimation, either Q_U or Q_L is set to zero in Eq. (35.2).

For convenience of explanation, we first calculate the radiation heat-transfer coefficient using Eq. (35.11). Except for shiny metal surfaces, the emissivity of surfaces of commercial and industrial products is in the range 0.8 to 0.9. For upper-bound estimation we set $\varepsilon = 1$ and $T_0 = 300$ K. Then

$$h_R = 6.12\,\mathrm{W/m^2\,K}$$

TABLE 35.1 Dimensions of Model Boxes (m)

Model no.	W	L	H
1-1	0.6	0.2	1.0
1-2	0.6	1.0	0.2
1-3	1.0	0.2	0.6
2-1	0.5	0.2	0.5
2-2	0.5	0.5	0.2
3-1	0.5	0.2	0.1
3-2	0.5	0.1	0.2
3-3	0.2	0.1	0.5
4-1	0.3	0.2	0.01
4-2	0.3	0.01	0.2
4-3	0.2	0.01	0.3
5-1	0.1	0.05	0.02
5-2	0.1	0.02	0.05
5-3	0.05	0.02	0.1
6-1	0.1	0.05	0.01
6-2	0.1	0.01	0.05
6-3	0.05	0.01	0.1

In calculation of convective heat-transfer coefficient, the properties of air listed in Table 35.2 are used. Substituting these values into Eq. (35.5), we have

$$\text{Ra} = 9.063 \times 10^7 L_B^3 \, \Delta T \tag{35.13}$$

The range of ΔT is practically 5 to 25 K. For the 17 dimensions listed in Table 35.1, the Rayleigh number takes a wide range of values. On the vertical surfaces, Ra = 453 ($H = 0.01$ m, $\Delta T = 25$ K) $\sim 2.27 \times 10^9$ ($H = 1.0$ m and $\Delta T = 25$ K). On the horizontal surfaces, Ra = 33 ($L_B = 0.0042$ m, $\Delta T = 5$ K) $\sim 1.49 \times 10^7$ ($L_B = 0.188$ m, $\Delta T = 25$ K). These extreme values are outside the applicability range of some of the correlations [Eqs. (35.6) to (35.9)]. However, those cases of Rayleigh numbers

TABLE 35.2 Physical Properties of Air at 300 K and Other Constants Used in Calculations

Property	Symbol	Value
Density	ρ	1.161 kg/m^3
Specific heat at constant pressure	c_p	1007 J/kg K
Kinematic viscosity	ν	1.589×10^{-5} m^2/s
Thermal conductivity	k	0.0263 W/m K
Volumetric thermal expansion coefficient	β	0.00333 K^{-1}
Prandtl number	Pr	0.707
Surface emissivity	ε	1
Stefan-Boltzmann constant	σ	5.670×10^{-8} W/m^2 K^4
Gravity acceleration	g	9.807 m/s^2

35.6 Parameter and Boundary Estimation

outside the applicability range occupy only small corners in the matrix of dimensions and temperature difference. From Eq. (35.6) we compute the heat-transfer coefficient on the vertical surface h_V, which is reduced to

$$h_V = 0.0179 + 1.317 H^{-1/4} \Delta T^{1/4} \tag{35.14}$$

On the small horizontal surfaces the Rayleigh number is much smaller than the lower bound of Eqs. (35.7) and (35.9). Fortunately, the contribution from small areas to the overall heat transfer is small; the following formulas from Eqs. (35.7) and (35.9), respectively, serve the purpose of estimating the upper-bound heat transfer:

$$h_U = 1.39 L_B^{-1/4} \Delta T^{1/4} \tag{35.15}$$

$$h_L = 0.693 L_B^{-1/4} \Delta T^{1/4} \tag{35.16}$$

Equation (35.12) is now rewritten to calculate the lower-bound thermal resistance $\Theta_{\min}$ (K/W):

$$\Theta_{\min} \equiv \frac{\Delta T}{Q} = \frac{1}{\sum_i (h_i + h_R) A_i} \tag{35.17}$$

Substituting the dimensional data of Table 35.1 into Eqs. (35.14) to (35.16) and setting $h_R = 6.12$ W/m² K, we find the overall thermal resistance as shown in Fig. 35.2. The horizontal axis (abscissa) is the total heat-transfer area A_s. The heat-transfer area depends on the

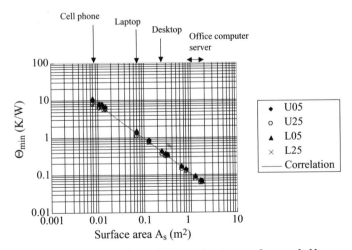

Figure 35.2 Lower-bound overall thermal resistance from sealed boxes.

orientation of the box; it is large when the box is erected with its narrow area on the adiabatic floor. Hence, the results for the same model box spread over a certain range of the area in Fig. 35.2. The areas of typical electronic equipment are indicated assuming their usual orientations. In the data designations, U and L indicate the surface exposed to the environment in addition to the vertical surfaces that are exposed in all cases; the subscripts U and L denote the upper and lower surfaces, respectively. The numerals indicate ΔT; 05 is 5 K, and 25 is 25 K. The calculated thermal resistances fall in a narrow band. The line in Fig. 35.2 represents the following correlation

$$\Theta_{min} = \frac{1}{A_s h_{eq}} = \frac{1}{A_s(9.18 - 0.6 \ln A_s)} \text{ K/W} \qquad (35.18)$$

where h_{eq} is the equivalent heat-transfer coefficient (W/m^2 K) obtained from line fitting of the calculated results: $h_{eq} = 9.18 - 0.6 \ln A_s$. Equation (35.18) correlates the calculated results in a range -23 to $+30$ percent, and provides a means to perform the rule-of-thumb calculation of lower-bound thermal resistance. Some examples are as follows.

1. For a portable gadget having a 0.01 m^2 heat-transfer area, the lower-bound thermal resistance is calculated from Eq. (35.18) as 8.4 K/W. If we allow a 15 K ($= \Delta T$) temperature rise on the surface, the upper-bound heat flow is 1.8 W.
2. For a medium-sized computer having a 1 m^2 heat-transfer area, the lower-bound thermal resistance is 0.11 K/W. For $\Delta T = 15$ K, the upper-bound heat flow is 138 W.

Enthalpy Transport from Vent Ports

Natural draft ventilation

Figure 35.3 shows a box having top and bottom sides available as vent ports. The following derivation of equations is commonly employed in the analysis of natural draft systems (e.g., see Ref. 3). The static pressure immediately below the bottom of the box is

$$p_i = p_0 - \tfrac{1}{2}\rho_0 U^2 \qquad (35.19)$$

where p_0 is the pressure of the environment at the level of the box bottom and U is the velocity that is assumed to be uniform over the area $W \times L$. The static pressure immediately above the top of the box is

$$p_e = p_i - \Delta p_H \qquad (35.20)$$

35.8 Parameter and Boundary Estimation

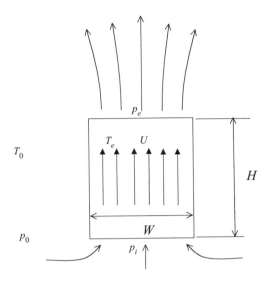

Figure 35.3 Natural draft through a box.

where Δp_H is the pressure difference between the bottom and the top of the box.

From Eqs. (35.19) and (35.20), we have

$$p_e = p_0 - \Delta p_H - \tfrac{1}{2}\rho_0 U^2 \qquad (35.21)$$

In the atmosphere above the box the dynamic pressure given to the plume is lost by viscous diffusion. Hence, considering the force balance outside the box, we have

$$p_e = p_0 - \rho_0 g H \qquad (35.22)$$

This boundary condition at the exit, together with the inlet boundary condition (35.19), is commonly employed in the numerical analysis of ventilated systems. Canceling out p_0 and p_e from Eqs. (35.21) and (35.22), we have

$$\Delta p_H = -\tfrac{1}{2}\rho_0 U^2 + \rho_0 g H \qquad (35.23)$$

Meanwhile, Δp_H is the sum of the frictional pressure drop Δp_f and the static-pressure head $\rho g H$, where ρ is the air density that varies along the box height:

$$\Delta p_H = \Delta p_f + \rho g H \qquad (35.24)$$

The static-pressure head depends on the temperature distribution in the block. However, for the sake of simple formulation, we suppose a fictitious situation; the air in the box is uniformly heated up to the temperature T_e. Then, the density ρ in Eq. (35.24) is written as $\rho_e = \rho_0\{1 - \beta(T_e - T_0)\}$. Eliminating Δp_H from Eqs. (35.23) and (35.24), we have

$$\Delta p_f = -\tfrac{1}{2}\rho_0 U^2 + \rho_0 g \beta H (T_e - T_0) \qquad (35.25)$$

This is a nonlinear equation, since the terms on the left-hand side are functions of the velocity U. Suppose that we solved Eq. (35.25) for U; then, the enthalpy equation would give the heat flow Q from the box:

$$Q = \rho_0 c_p \, U A_e \, (T_e - T_0) \quad \text{W} \qquad (35.26)$$

where A_e is the cross-sectional area of the exit vent ($= W \times L$) (m²). Denoting the volumetric flow rate as $V = 60 U A_e$ (m³/min), we write the thermal resistance as

$$\Theta = \frac{T_e - T_0}{Q} = \frac{60}{\rho_0 c_p V} \quad \text{K/W} \qquad (35.27)$$

Note that the variation of air density is taken into account only in the buoyancy term (the Boussinesq approximation); hence, the density in Eqs. (35.26) and (35.27) is ρ_0. Also, note that Eqs. (35.26) and (35.27) are valid for mechanical draft systems as well; thus, they will be used in the subsequent sections.

The upper bound of heat transfer is now derived assuming the flow resistance along the path of airflow to be negligibly small. Then, from Eq. (35.25) we have

$$U_{\max} = \sqrt{2 g \beta H (T_e - T_0)} \qquad (35.28)$$

Using $U_{\max}$ to compute V in Eq. (35.27), we obtain the lower bound of thermal resistance as

$$\Theta_{\min} = \frac{1}{\rho_0 c_p A_e \sqrt{2 g \beta H (T_e - T_0)}} \qquad (35.29)$$

Substituting the numerical values of Table 35.2 and setting $T_e - T_0 = 20$ K in Eq. (35.29), the rule-of-thumb estimation of the lower-bound thermal resistance is obtained as

$$\Theta_{\min} = \frac{1}{1337 A_e \sqrt{H}} \quad \text{K/W} \qquad (35.30)$$

The dimensions of the models of Table 35.1 are now substituted in Eq. (35.30). Some examples are plotted in Fig. 35.4 by solid triangles,

35.10 Parameter and Boundary Estimation

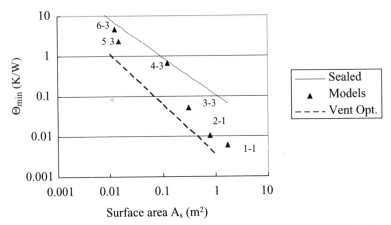

Figure 35.4 Comparison of lower-bound thermal resistances in natural draft cooling: Sealed box (upper line); some models of Table 35.1 with vents (triangles); boxes of optimum dimensions for natural draft cooling (lower line). Note that A_s is the surface area of five sides of the sealed box.

where the horizontal axis is the external heat-transfer area of the sealed box A_s, defined in the previous section. These results belong to the boxes having their longest dimensions as the heights. Also shown in Fig. 35.4 is a line representing the correlation in Eq. (35.18) for sealed boxes. The Θ_{min} for case 4-3 is very close to that for the sealed box, indicating that, for a thin, flat box, the ventilation effect is comparable to the natural cooling from the larger vertical sides.

The Θ_{min} for a given A_s is actually lower than the results calculated for the model boxes. In order to find the lower bound for a given A_s, we allow the variation of the dimensions, H, W, and L, to minimize Θ_{min}. Geometric consideration defines the maximum height as given by Eq. (35.31), so Θ_{min} of Eq. (35.30) can be written in terms of A_e or H alone. The optimization condition is as follows:

$$H = \frac{A_s - A_e}{4\sqrt{A_e}} \qquad (35.31)$$

$$W = L = \sqrt{A_e} \qquad (35.32)$$

$$A_e = \tfrac{3}{5} A_s \qquad (35.33)$$

Substitution of optimum A_e and H into Eq. (35.30) yields the following equation:

$$\Theta_{min} = 0.00347 A_s^{-5/4} \qquad (35.34)$$

The lower line in Fig. 35.4 represents Eq. (35.34). The vertical separation between the sealed box correlation (the upper line) and the ventilated box correlation indicates the order of magnitude of maximum benefits obtained by natural ventilation. For a box having a 0.01-m² heat-transfer area, ventilation reduces the lower-bound thermal resistance from 8.4 to 1.1 K/W. For a 1-m² box, the corresponding figures are 0.11 and 0.0035 K/W. However, as we will see in the next section, the presence of flow resistance inside the actual box seriously compromises the benefit of ventilation.

In terms of the volume of the box having the optimum dimensional proportion [V_B (m³)], Eq. (35.34) is rewritten as

$$\Theta_{\min} = 4.12 \times 10^{-4} V_B^{-5/6} \quad (35.35)$$

Mechanical draft ventilation

We suppose appropriate fans for the models of Table 35.1. First, consider the largest box, model 1. The fans used in most electronic boxes of interest here are propeller fans driven by brushless direct-current (dc) motors. Among the commercially readily available fans, the largest one has the dimensions 12×12 cm. In order to maximize the mechanical draft, four fans are accommodated in the top plane of the model 1-1 box, as shown in Fig. 35.5. Figure 35.6 shows a typical characteristic curve of a fan of this size (curve pFAN); the vertical axis (ordinate) is the static-pressure rise Δp (Pa) produced by the fan, and the horizontal axis (abscissa) is the collective volumetric flow rate of four fans. Following the definition set for natural draft cooling, the upper-bound heat transfer is again determined assuming negligible flow resistance in the box. Then, the upper-bound heat flow is determined by the flow rate at

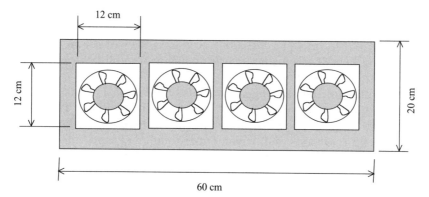

Figure 35.5 Top view of model 1-1 with four propeller fans.

35.12 Parameter and Boundary Estimation

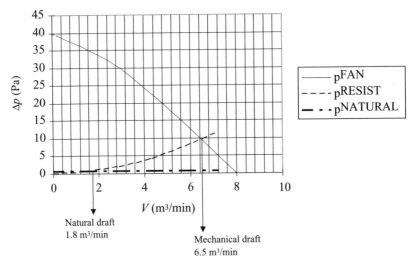

Figure 35.6 Fan curve (pFAN; V = collective flow rate of four fans), resistance curve (pRESIST), and natural draft potential (pNATURAL) for model 1-1.

the free-delivery point of the fan curve ($\Delta p = 0$), V_F (m³/min):

$$Q_F = \rho c_p V_F (T_e - T_0)/60 \quad \text{W} \tag{35.36}$$

where T_e is the temperature of exhaust air and T_0 is the environment temperature. From Fig. 35.6 we find $V_F = 8$ m³/min. The corresponding lower-bound thermal resistance [Eq. (35.27)] is

$$\Theta_{\min} = \frac{60}{1.161 \times 1007 \times 8} = 0.0064 \quad \text{K/W} \tag{35.37}$$

This value almost coincides with the lower bound of natural draft thermal resistance for model 1-1 (0.0062 K/W). As we will see in the next section, the benefit of mechanical draft lies in its ability to overcome flow resistance in the box.

For the smallest box among the models of Table 35.1, that is, model 6, a fan having a 4-cm side length is commercially available. The fan is designed to be accommodated in a thin space; the air is sucked in the direction of a rotor axis and driven out through a slot provided on the side of the fan frame. Figure 35.7 shows such a fan accommodated in the model 6 box, with a sketch of a stack of cards to be cooled. A typical characteristic curve is shown in Fig. 35.8 (pFAN). The free-delivery

Upper Bounds of Heat Transfer from Boxes 35.13

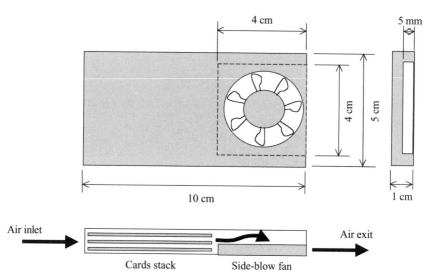

Figure 35.7 Model 6 box equipped with a side-blow fan.

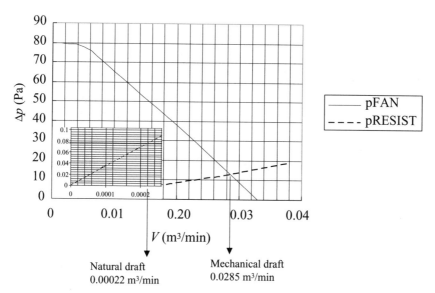

Figure 35.8 Fan curve (pFAN), resistance curve (pRESIST), and natural draft potential (horizontal line in the inset) for model 6-3.

35.14 Parameter and Boundary Estimation

flow rate is $V_F = 0.033$ m³/min. The corresponding lower-bound thermal resistance is

$$\Theta_{min} = \frac{60}{1.161 \times 1007 \times 0.033} = 1.56 \text{ K/W} \quad (35.38)$$

This value is almost one-third the lower-bound thermal resistance of natural draft cooling for model 6-3 (4.73 K/W).

The Effects of Flow Resistance on Thermal Resistance

In this section we consider situations that are a little closer to reality. Again, we take the largest model (model 1) and the smallest (model 6) as our illustrative examples. The resistance to airflow is defined in terms of the static-pressure drop from the inlet to the exit of the box and the volumetric flow rate. Large flow resistance means a large pressure drop for a given flow rate, or a low flow rate under a given pressure difference. In Figs. 35.6 and 35.8, typical resistance curves are shown (pRESIST). The actual flow resistance is a function of the dimensions and configurations of internal airflow paths. To determine the flow resistance curve of actual electronic equipment, we need to conduct an experiment. In the experiment, the test object is placed in a wind tunnel. The wind tunnel provides a means to accurately measure the airflow rate and the pressures upstream and downstream of the test object. The intersection of the resistance curve and the fan curve gives an actual operating point of the flow system; that is, the fan provides a driving potential in the form of pressure rise that matches the pressure drop caused in the box.

From Fig. 35.6, we read the actual volumetric flow rate in the model 1 box as $V = 6.5$ m³/min. The corresponding thermal resistance is $\Theta = 0.0079$ K/W, representing an approximately 20 percent increase from the lower bound calculated earlier in the "Mechanical Draft Ventilation" section. In Fig. 35.6 a line showing the maximum driving potential of natural draft is also included (for the posture of model 1-1, see Table 35.1); that is, for $H = 1$ m and $(T_e - T_0) = 20$ K, we have

$$\rho_0 g \beta H(T_e - T_0) = 1.161 \times 9.807 \times 0.00333 \times 20 = 0.759 \text{ Pa} \quad (35.39)$$

The intersection of this line with the resistance curve gives the actual natural draft flow rate as $V = 1.8$ m³/min. The corresponding thermal resistance is $\Theta = 0.029$ K/W, almost fivefold the lower bound calculated earlier in the "Natural Draft Ventilation" section.

As illustrated above, the natural draft has very low driving potential. Low potential of natural draft is amplified as the box size reduces. Figure 35.8 shows a typical resistance curve for equipment of model 6

TABLE 35.3 Comparison of Thermal Resistance Values (K/W)

Model no.	Completely sealed	Natural draft		Mechanical draft	
		Lower bound	With flow resistance	Lower bound	With flow resistance
1-1	0.066	0.0062	0.0286	0.0064	0.0079
6-3	6.8	4.73	233	1.56	1.80

size. The intersection of the resistance curve with the fan curve gives $V = 0.0285$ m³/min; hence, $\Theta = 1.8$ K/W. The natural draft potential for $H = 0.1$ m and $(T_e - T_0) = 20$ K is 0.0759 Pa. From the inset of Fig. 35.8 we read the volumetric flow rate as $V = 0.00022$ m³/min. The thermal resistance of model 6-3 in natural draft cooling with flow resistance is calculated as $\Theta = 233$ K/W.

Table 35.3 compares the thermal resistance values of models 1 and 6. The values for sealed boxes are calculated using Eq. (35.18). Table 35.3 provides some insight into the cooling technique for small electronic gadgets. For small boxes of the size of model 6 or smaller, if natural convection cooling is favored, heat dissipation from the surfaces is the realistic mode of cooling. Heat generated in the box has to be spread in the box and conducted to the surface. Vent holes bring only marginal enhancement of cooling. To benefit from mechanical draft, the fan must be made smaller than currently available models; occupation of almost 40 percent of the box space as illustrated in Fig. 35.7 is unacceptable. With the reduction of fan size, the rotational speed of the fan needs to be raised to provide a sufficiently high fan performance. This has to be accompanied by the efforts to mitigate acoustic noise from the fan.

References

1. F. P. Incropera and D. P. DeWitt, *Fundamentals of Heat and Mass Transfer*, 4th ed., Wiley, New York, 1996, pp. 493–498.
2. S. W. Churchill and H. H. S. Chu, "Correlating Equations for Laminar and Turbulent Free Convection from a Vertical Plate," *International Journal of Heat and Mass Transfer*, **18**, 1323–1329, 1975.
3. T. S. Fisher and K. E. Torrance, "Free Convection Cooling Limits for Pin-Fin Cooling," *ASME Journal of Heat Transfer*, **120**, 633–640, 1998.

Chapter 36

Estimating Freezing Time of Foods

R. Paul Singh
Department of Biological and Agricultural Engineering
University of California
Davis, California

Freezing of foods is a common unit operation employed in the food industry. Many fruits and vegetables, such as peas, strawberries, diced carrots, and green beans, are frozen in fluidized-bed freezers where the product comes into direct contact with air at subfreezing temperatures. As the food undergoes freezing, there is a change in phase of water into ice; this complicates the heat-transfer computations required to estimate freezing times. In this example, we will predict freezing time of strawberries being frozen in a fluidized-bed freezer. The initial temperature of a strawberry is 15°C, and it has a moisture content of 75 percent (wet basis). The shape of the strawberry is assumed to be a sphere with a diameter of 2.5 cm. The final desired center temperature of the strawberry is −18°C. The air temperature in the fluidized-bed freezer is −40°C. The convective heat-transfer coefficient is measured to be 80 W/(m² K). The properties of the strawberry are assumed as follows. The density of an unfrozen strawberry is 1130 kg/m³, the density of a frozen strawberry is 950 kg/m³, the specific heat of an unfrozen strawberry is 3.55 kJ/(kg °C), and the specific heat of a frozen strawberry is 1.5 kJ/(kg °C). The thermal conductivity of the frozen

36.2 Parameter and Boundary Estimation

strawberry is 1.5 W/(m K). Calculate how long it will take to reduce the center temperature of the strawberry from 15 to −18°C.

Approach

To solve this problem, we will use a method proposed by Pham (1986). Pham's method is useful in calculating freezing time of foods with high moisture content (>55 percent). The following assumptions were made by Pham: (1) the temperature of the air used in the freezer is constant, (2) the initial temperature of the product is constant, (3) the value of the final temperature is fixed, and (4) the convective heat transfer on the surface of the food is described by Newton's law of cooling. Pham divides the total heat removed during freezing from some initial temperature to a desired final temperature into two components. As seen in Fig. 36.1, a plot of center temperature versus heat removal is divided into two parts using a mean freezing temperature T_m. The first part of this curve represents heat removal in precooling and some initial change of phase as the product begins to freeze; the second part represents the remaining heat removal during phase change and additional cooling to reach the final desired temperature.

The mean freezing temperature T_m is obtained using the following equation, which has been empirically obtained for foods with moisture content exceeding 55 percent

$$T_m = 1.8 + 0.263 T_c + 0.105 T_a \qquad (36.1)$$

where T_c is the final center temperature of the product (°C) and T_a is the air temperature used in freezing (°C).

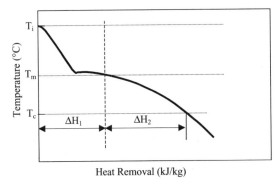

Figure 36.1 A plot of temperature at the product center versus heat removal during freezing of foods.

The time to freeze the product from an initial temperature T_i (°C) to a final temperature of T_c is obtained using the following equation:

$$t = \frac{d}{Eh}\left(1 + \frac{N_{Bi}}{2}\right)\left[\frac{\Delta H_1}{\Delta T_1} + \frac{\Delta H_2}{\Delta T_2}\right] \quad (36.2)$$

where d is the characteristic dimension, which is the shortest dimension from the surface to the center [for a sphere it is the radius (m)]; h is the convective heat transfer [W/(m² K)]; and E is the dimensional shape factor (it is 1 for an infinite slab, 2 for an infinite cylinder, and 3 for a sphere).

The term ΔH_1 represents the change in the volumetric enthalpy (J/m³) for the precooling and initial periods of phase change:

$$\Delta H_1 = \rho_u c_u (T_i - T_m) \quad (36.3)$$

where ρ_u is the density of unfrozen material, c_u is the specific heat of the unfrozen material [J/(kg K)], and T_i is the initial temperature of the product (°C).

The term ΔH_2 is the change in the volumetric enthalpy (J/m³) for the period involving the remaining phase change and postcooling of the product to the final center temperature:

$$\Delta H_2 = \rho_f [L_f + c_f(T_m - T_c)] \quad (36.4)$$

where ρ_f is the density of the frozen product, c_f is the specific heat of the frozen product [kJ/(kg K)], and L_f is the latent heat of fusion of the food undergoing freezing [J/(kg K)].

The Biot number N_{Bi} is defined as

$$N_{Bi} = \frac{hd}{k_f} \quad (36.5)$$

where h is the convective heat-transfer coefficient [W/(m² K)], d is the characteristic dimension (m), and k_f is the thermal conductivity of the frozen material [W/(m K)].

The temperature gradient ΔT_1 is obtained from

$$\Delta T_1 = \left(\frac{T_i + T_m}{2}\right) - T_a \quad (36.6)$$

where T_i is the initial temperature (°C), T_a is the temperature of the air (°C), and T_m is the mean temperature as defined in Eq. (36.1).

36.4 Parameter and Boundary Estimation

The temperature gradient ΔT_2 is obtained from

$$\Delta T_2 = T_m - T_a \tag{36.7}$$

For more information on estimating thermal properties of foods required in Eqs. (36.3) to (36.5), refer to Singh and Heldman (2001) and Singh (2004). Next, we will use the preceding equations to calculate freezing time for the given problem.

Solution

Given

Initial temperature of strawberry = 15°C
Moisture content of strawberry = 75 percent (wet basis)
Diameter of strawberry = 2.5 cm
Final desired center temperature of strawberry = −18°C
Air temperature in fluidized-bed freezer = −40°C
Convective heat-transfer coefficient = 80 W/(m² K)
Density of unfrozen strawberry = 1130 kg/m³
Density of frozen strawberry = 950 kg/m³
Specific heat of unfrozen strawberry = 3.55 kJ/(kg °C)
Specific heat of frozen strawberry = 1.5 kJ/(kg °C)
Thermal conductivity of frozen strawberry = 1.5 W/(m K)

Procedure

1. Using Eq. (36.1), calculate T_m:

$$T_m = 1.8 + 0.263 \times (-18) + 0.105 \times (-40)$$
$$= -7.134°C$$

2. Using Eq. (36.3), calculate ΔH_1:

$$\Delta H_1 = 1130 \text{ kg/m}^3 \times 3.55 \text{ kJ/kg K} \times 1000 \text{ J/kJ} \times (15 - (-7.134))\ °C$$
$$= 88{,}790{,}541 \text{ J/m}^3$$

3. The latent heat of fusion of strawberries is obtained as a product of moisture content and latent heat of fusion of water (333.2 kJ/kg):

$$L_f = 0.75 \times 333.2 \text{ kJ/kg} \times 1000 \text{ J/kJ}$$
$$= 249{,}900 \text{ J/kg}$$

4. Using Eq. (36.4), calculate ΔH_2:

$$\Delta H_2 = 950 \text{ kg/m}^3 \times \begin{pmatrix} 249{,}900 \text{ J/kg} + 1.5 \text{ kJ/kg K} \times 1000 \text{ J/kJ} \\ \times (-7.134 - (-18))°\text{C} \end{pmatrix}$$

$$= 252{,}889{,}050 \text{ J/m}^3$$

5. Using Eq. (36.6), calculate ΔT_1:

$$\Delta T_1 = \left(\frac{15 + (-7.134)}{2}\right) - (-40)$$

$$= 43.93°\text{C}$$

6. Using Eq. (36.7), calculate ΔT_2:

$$\Delta T_2 = [-7.134 - (-40)]$$

$$= 32.87°\text{C}$$

7. The Biot number is calculated using Eq. (36.5) as follows:

$$N_{\text{Bi}} = \frac{80 \text{ W/m}^2 \text{ K} \times 0.0125 \text{ m}}{1.5 \text{ W/m}^2 \text{ K}}$$

$$= 0.667$$

8. Substituting results of steps 1 through 7 in Eq. (36.2), noting that for a sphere $E_f = 3$, we have

$$t = \frac{0.0125 \text{ m}}{3 \times 80 \text{ W/m}^2 \text{ K}} \left(\frac{88790541 \text{ J/m}^3}{43.93°\text{C}} + \frac{252889050 \text{ J/m}^3}{32.87°\text{C}}\right)$$

$$\times \left(1 + \frac{0.667}{2}\right)$$

Time = 674.7 s = 11.2 min

Result

The estimated time to freeze strawberries in a fluidized-bed freezer is 11.2 min.

References

Pham, Q. T., 1986. "Simplified Equation for Predicting the Freezing Time of Foodstuffs," *J. Food Technol.* **21**:209–219.

Singh, R. P., and Heldman, D. R., 2001. *Introduction to Food Engineering*, Elsevier, Amsterdam, The Netherlands.

Singh R. P., 2004. *Food Properties Database*, RAR Press, Davis, Calif.

Part

9

Temperature Control

In the single chapter in this part, the author, who works in industry in the United States, describes how precise temperature control of an electronic device can be accomplished by mounting it on a thermoelectric module (TEM). Depending on the direction of the current flowing through a TEM, heat is released or absorbed at the interface between it and the electronic device by virtue of the Peltier effect. Because the rate of heat released or absorbed by the Peltier effect is proportional to the magnitude of the current, TEMs are effective temperature controllers of devices such as lasers and optical routers.

Chapter 37

Precision Temperature Control Using a Thermoelectric Module

Marc Hodes
Bell Laboratories
Lucent Technologies
Murray Hill, New Jersey

Introduction

Depending on the direction of the current flowing through a thermoelectric module (TEM), heat is absorbed or released at the interface between it and an electronic device mounted on it by virtue of the Peltier effect. Moreover, the rate of heat absorbed or released by the Peltier effect is directly proportional to the magnitude of the current. Hence, TEMs are well suited to precision temperature control of electronic devices. Here the equations governing the operation of a TEM are presented and used to calculate parameters relevant to precision temperature control of an optical router.

Background

Thermal energy may be reversibly converted to electrical energy and vice versa in electrically conducting materials (including semiconductors) by thermoelectric effects (see, e.g., Foiles [1]). Thermoelectric effects are small in metals but moderate in certain types of semiconductors. The moving charge carriers constituting an electric current I carry different amounts of electrical energy in different conductors.

37.4 Temperature Control

Thus heat is evolved or absorbed when current flows through the interface between two conductors. This is known as the *Peltier effect*. The rate of (reversible) heat absorption at the interface between two conductors q due to the Peltier effect is

$$q = I(\alpha_B - \alpha_A)T \tag{37.1}$$

where I is current, α is the Seebeck coefficient (a material property), and T is absolute temperature. Current is positive when positive charge carriers flow from conductor A to conductor B. The irreversible (bulk) effects of Ohmic (I^2R) heating and heat conduction must also be considered in the analysis of thermoelectricity.

Thermoelectric modules are solid-state devices that exploit the Peltier effect in order to cool, heat, or generate electrical power. They are commonly used to cool electronics or to precisely control their temperature through cooling and heating. A single-stage TEM is shown in Fig. 37.1. It consists of an array of positively doped (p-type) and negatively doped (n-type) semiconductor pellets connected electrically in series and thermally in parallel between ceramic substrates. Each adjacent pair of p-type and n-type pellets is referred to as a *thermocouple*, and there are N thermocouples in a TEM. Here it is assumed that the controlled side of the TEM is that which does not connect to the external leads according to Fig. 37.1. The other side is defined as the *uncontrolled*

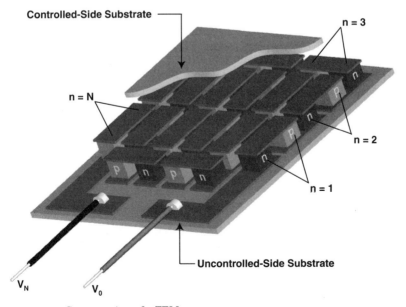

Figure 37.1 Cutaway view of a TEM.

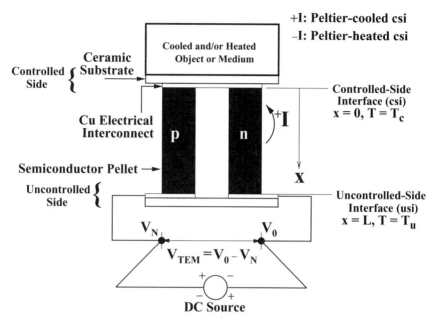

Figure 37.2 Schematic representation of a single-thermocouple TEM.

side. The *controlled-side interface* (csi) of a TEM is defined as that between the thermocouples inside it and the controlled side substrate and the *uncontrolled-side interface* (usi) is defined analogously according to Fig. 37.2, which shows a single thermocouple ($N = 1$) TEM.

Currents are taken to be positive when positive charge flows from n-type pellets to p-type pellets at the controlled side of a TEM, that is, counterclockwise according to Fig. 37.2. Since α_p and α_n are positive and negative quantities, respectively, heat is absorbed by the Peltier effect at the csi and released by the Peltier effect at the usi when current is positive as per Eq. (37.1). Analogously, when current is negative, heat is absorbed and released by the Peltier effect at the usi and the csi, respectively. When a TEM operates in cooling or heating modes, dc power must be supplied to it. Moreover, for precision temperature control applications, the polarity of the power source must be reversible in order to switch between cooling and heating modes.

Thermoelectric modules posses a number of highly advantageous attributes. For example, they have no moving parts and are of very simple construction, acoustically silent, compact, rugged, lightweight, moderate-cost, and modular. The major disadvantage of TEMs is that their thermodynamic efficiencies are smaller than those of competing technologies, such as vapor compression refrigerators.

37.6 Temperature Control

Governing Equations

The equations resulting from the one-dimensional analysis of a single-stage TEM by Hodes [2] are presented in this section. The assumptions made by Hodes [2] are commonplace in the one-dimensional analysis of TEMs. Seifert et al. [3] and Yao et al. [4] attained good agreement between equivalent one-dimensional models and numerical solutions to more rigorous formulations. Hence, solutions to the present formulation provide sufficiently accurate results for many TEM applications such as the one described below.

The net rate of heat conducted out of a differential control volume normal to the x direction defined in Fig. 37.2 equals the rate of Ohmic heat generation inside it. Hence the form of the heat equation governing conduction inside the pellets is

$$\frac{d^2T}{dx^2} + \frac{I^2\rho}{kA_P^2} = 0 \qquad (37.2)$$

where x is distance from the csi, k is thermal conductivity of a pellet, and A_P is the cross-sectional area of a pellet. The boundary conditions on the heat equation at the csi and the usi are $T = T_c$ at $x = 0$ and $T = T_u$ at $x = L$, respectively, where L is the height of a pellet, as per Fig. 37.2. It follows that the (parabolic) temperature distribution in the pellets is:

$$T = \frac{-I^2\rho}{2kA_P^2}x^2 + \left(\frac{I^2\rho L}{2kA_P^2} + \frac{T_u - T_c}{L}\right)x + T_c \qquad (37.3)$$

The rates of heat transfer from the controlled-side substrate into the csi q_c and from the usi into the uncontrolled-side substrate q_u may be determined from surface energy balances. The sign convention adopted in the surface energy balances is that q_c and q_u are positive in the positive x-direction defined in Fig. 37.2. Physically, the rate of heat transfer from the controlled-side substrate into the csi q_c equals the rate of energy absorbed by the Peltier effect at the csi of $[I(\alpha_p - \alpha_n)T_c]$ per thermocouple plus the rate of heat transfer by conduction into the pellets at the csi of $(-2kA_P dT/dx|_{x=0})$ per thermocouple. The quantity $(\alpha_p - \alpha_n)$ is subsequently denoted by $\alpha_{p,n}$. To accommodate appreciable amounts of heat at realistic supply currents and voltages, a TEM must contain an array of N thermocouples as shown in Fig. 37.1. For an N-thermocouple TEM, the surface energy balances at the csi and the usi are

$$q_c = N\left[I\alpha_{p,n}T_c - K(T_u - T_c) - \frac{I^2R}{2}\right] \qquad (37.4)$$

$$q_u = N\left[I\alpha_{p,n}T_u - K(T_u - T_c) + \frac{I^2R}{2}\right] \qquad (37.5)$$

where the Ohmic resistance R and the thermal conductance K of a thermocouple are defined conventionally as

$$R = \frac{2\rho L}{A_P} \tag{37.6}$$

$$K = \frac{2kA_P}{L} \tag{37.7}$$

An energy balance on a control volume around the interconnected pellets forming a TEM shows that the rate of electrical work done on it ($\dot{W}_{\text{TEM}}$) is equal to ($q_u - q_c$). Hence

$$\dot{W}_{\text{TEM}} = N[I\alpha_{p,n}(T_u - T_c) + I^2 R] \tag{37.8}$$

$\dot{W}_{\text{TEM}}$ is positive when a dc power supply does work on a TEM in cooling and heating modes. An energy balance on a control volume around the power supply in Fig. 37.2 shows that $(IV)_N + \dot{W}_{\text{TEM}} = (IV)_0$, where the voltages V_0 and V_N in a TEM are defined according to Fig. 37.2. Hence, the voltage across a TEM is

$$V_{\text{TEM}} = V_0 - V_N = N[\alpha_{p,n}(T_u - T_c) + IR] \quad \text{for} \quad I \neq 0 \tag{37.9}$$

Despite the utility of the results developed thus far, the boundary conditions at the csi and the usi in a TEM are usually more complicated in practice; that is, T_c and T_u are not specified but must be computed (see, e.g., the analysis by Yamanashi [5]). This necessitates accounting for the three thermal resistances shown in Fig. 37.3. First, the thermal resistance between the location on the object or in the medium where constant temperature is being maintained, subsequently referred to as the *control point* (cp), and its local ambient ($R_{\text{cp}-\infty,c}$) must be accounted for. This necessitates the addition of a surface energy balance at the control point to the balance equations. Also, the thermal resistances between the control point and the csi ($R_{\text{cp}-c}$) and between the usi and its local ambient ($R_{u-\infty,u}$) must be accounted for. Constriction/spreading resistances in the controlled and uncontrolled substrates of a TEM are embedded in $R_{\text{cp}-c}$ and $R_{u-\infty,u}$, respectively. The temperatures of the local ambients around the control point ($T_{\infty,c}$) and the usi ($T_{\infty,u}$) may or may not be the same. Specifying all the aforementioned thermal resistances by their corresponding overall conductances (K), the surface energy balances at the control point, csi and usi are given by Eqs. (37.10) through (37.12). The overall energy balance given by Eq. (37.8)

37.8 Temperature Control

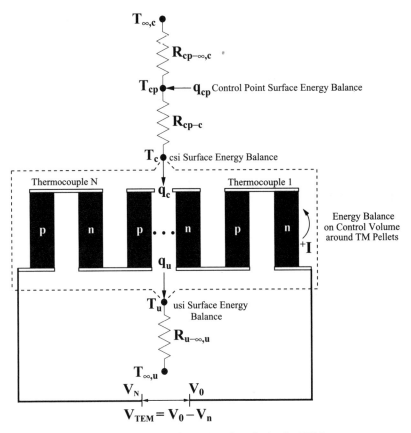

Figure 37.3 Representation of one-dimensional analysis of a TEM.

remains valid:

$$q_{cp} = K_{cp-c}(T_{cp} - T_c) + K_{cp-\infty,c}(T_{cp} - T_{\infty,c}) \quad (37.10)$$

$$K_{cp-c}(T_{cp} - T_c) = N\left[I\alpha_{p,n}T_c - K(T_u - T_c) - \frac{I^2R}{2}\right] \quad (37.11)$$

$$K_{u-\infty,u}(T_u - T_{\infty,u}) = N\left[I\alpha_{p,n}T_u - K(T_u - T_c) + \frac{I^2R}{2}\right] \quad (37.12)$$

Calculational Example

Problem

Reconfigurable optical add/drop modules (ROADMs) are analog devices that route light containing information (voice, video, and/or data) among optical fibers in telecommunications networks. Light from incoming optical fibers travels along waveguides (often silica) fabricated on a large die (often silicon) and is routed to the appropriate outgoing fiber by various optical components. ROADMs use thermooptic phase shifters to locally heat the waveguides in order to change their temperature. Since the index of refraction of the waveguides is temperature-dependent, the phase of the light traveling through them may be controlled. This phase control in combination with interferometers enables light to be switched between different paths. Integrated on the same die as the heat-dissipating thermooptic phase shifters, however, are temperature-sensitive optical components, such as optical filters, which must be maintained within narrow operating temperature ranges (e.g., $\pm 1°C$) to function properly. In order to maintain the die at constant temperature except in the vicinity of the thermooptic phase shifters as, for example, the ambient temperature changes, it may be mounted on a heat spreader which, in turn, is mounted on a TEM.

A TEM has been selected to maintain the temperature of the die in a ROADM at 75°C for ambient temperatures ranging from −5 to 65°C. The TEM consists of a uniform array of 142 Bi_2Te_3 pellets (71 thermocouples), and the pellet properties are $\alpha_p = 2.0 \times 10^{-4}$ V/K, $\alpha_n = -2.0 \times 10^{-4}$ V/K, $\rho = 1.25 \times 10^{-5}$ Ω m, and $k = 1.5$ W/(m K). Pellet height L equals 1.14 mm and the pellet footprint is 1.40 × 1.40 mm. The electrical interconnects for the pellets and the external leads of the TEM are copper, which may be assumed to contribute no electrical or thermal contact resistances and have a Seebeck coefficient of 0 V/K. The pellets are located between 30-mm × 30-mm × 0.75-mm-thick alumina [$k = 36.0$ W/(m K)] substrates. The die has the same footprint as the TEM on which it is mounted, and its power dissipation may be approximated as uniform. Assume that the die is thermally insulated from the environment except through a layer of thermally conductive epoxy of interfacial resistance $R''_{ti} = 1.70 \times 10^{-5}$ m^2 W/K coupling it to the control-side substrate of the TEM and that the uncontrolled-side substrate of the TEM transfers heat to the ambient by conduction through a layer of thermal grease ($R''_{ti} = 2.58 \times 10^{-5}$ m^2 W/K) in series with a 1.0°C/W thermal resistance representing a heat sink.

(A). Assuming that the die dissipates 10 W of power, calculate the ambient temperatures at which it is unnecessary to supply power to the TEM when its external leads terminate in an open circuit, when they are electrically short-circuited together and when +4 V is applied across them. In each case quantitatively describe the temperature distribution in the pellets inside the TEM.

(B). What are the minimum and maximum powers that the die can dissipate when the power available to the TEM is limited to 10 W and when it is unlimited?

37.10 Temperature Control

(C). Assume that the size of the heat sink and, hence, its thermal resistance, may be varied. When the die dissipates 10 W, what is the optimum thermal resistance of the heat sink insofar as minimizing the power consumed by the TEM?

(D). Assume, for the sake of this example, that the objective is to bring the die to 200°C. Is it more efficient (i.e., is less electrical power required) when the die is used as a resistive heater or when the TEM is used to heat the die? Assume that all electrical power is used to drive either the die or the TEM. Perform the calculations assuming that $T_{\infty,u} = 65.0°C$ and $R_{HS} = 1.0°C/W$.

Solution

Known

Thermocouple properties, geometry, and number: $\alpha_p = 2.0 \times 10^{-4}$ V/K, $\alpha_n = -2.0 \times 10^{-4}$ V/K, $\alpha_{p,n} = 4.0 \times 10^{-4}$ V/K, $\rho = 1.25 \times 10^{-5}$ Ω m, $k = 1.5$ W/(m K), $L = 1.14$ mm, $A_P = 30 \times 30$ mm, $N = 71$.

30- × 30-mm footprint by 0.75-mm-thick TEM substrates with $k = 36.0$ W/(m K).

Control-point temperature and uncontrolled-side ambient temperature range: $T_{cp} = 75°C$, $-5°C \leq T_{\infty,u} \leq 65°C$.

Data required to compute thermal resistances between control point and csi of the TEM and the usi of the TEM and its local ambient.

Schematic. A schematic representation of the problem is shown in Fig. 37.4.

Assumptions

Only steady-state conditions are relevant.

All of the assumptions in the one-dimensional analysis of a TEM by Hodes [2] are valid.

No heat transfer between the control point and its local ambient ($K_{cp-\infty,c} = 0$).

Analysis. Surface energy balances at the control point (the silicon die), the csi of the TEM, and the usi of the TEM [Eqs. (37.10) through (37.12), respectively] and the overall energy balance on a control volume around the interconnected pellets within the TEM [Eq. (37.8)] are necessary to solve this problem. [Note that the surface energy balance at the control point reduces to $q_{cp} = K_{cp-c}(T_{cp} - T_c)$ because it has been assumed that heat transfer from the control point to its local ambient is negligible,

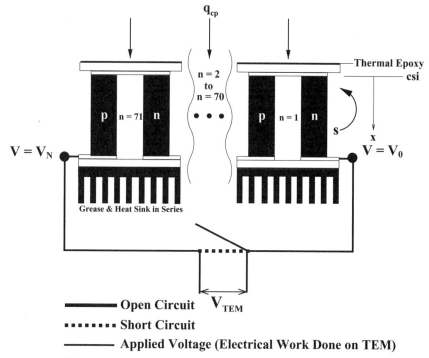

Figure 37.4 Schematic representation of the problem.

i.e., $K_{cp-\infty,c} = 0$.] It is convenient to compute the parameters required to solve the foregoing equations before proceeding with the analysis. The electrical resistance R and thermal conductance K of each thermocouple in the TEM are given by Eqs. (37.6) and (37.7), respectively. The result is that

$$R = 1.45 \times 10^{-2} \ \Omega \quad \text{and} \quad K = 5.16 \times 10^{-3} \ \text{W/K}$$

The overall conductance between the die (control point) and the csi (K_{cp-c}) is the reciprocal of the sum of the thermal resistance across the thermal epoxy between the die and the alumina substrate ($R_{epoxy} = R''_{ti}/A_{substrate} = 1.89 \times 10^{-2}$ K/W) and that due to conduction through the alumina substrate. The thermal resistance to conduction through the substrate equals the one-dimensional resistance through the substrate [$R_{1d} = t/(kA)$] plus the constriction resistance associated with conduction into the pellets. Each pellet is surrounded by four lines of symmetry forming a corresponding unit cell from which the constriction resistance can be calculated. Approximating the square unit cell

37.12 Temperature Control

and square pellets as concentric circles with equivalent cross-sectional areas, it follows that the pellet radius a and the unit cell radius b are 0.79 and 1.42 mm, respectively. Then the constriction resistance associated with conduction through the alumina in each unit cell may be calculated from the results of Yovanovich et al. [6]. The result is that the constriction resistance equals 2.30°C/W.[1] The corresponding one-dimensional resistance is 3.29°C/W. Since the 142 pellets are thermally in parallel, it follows that the overall conductance between the control point and control side of the TEM ($K_{\text{cp}-c}$) is

$$K_{\text{cp}-c} = \frac{1}{R_{\text{ep}} + (R_{1d} + R_c)/142} = 17.17 \text{ W/K}$$

The overall conductance between the usi and its local ambient ($K_{u-\infty,u}$) may be calculated analogously. The thermal resistance to conduction through the uncontrolled-side substrate (including spreading) is identical to that through the controlled-side substrate (including constriction). This conduction resistance is in series with the thermal resistances of the grease layer between the uncontrolled-side substrate and the heat sink and the heat sink itself. The result is that

$$K_{u-\infty,u} = 0.936 \text{ W/K}$$

(A). When the external leads of the TEM are open-circuited, the current I through the TEM equals 0. Setting $I = 0$ and simultaneously solving the surface energy balances [Eqs. (37.10) through (37.12)] results in T_c, T_u, and $T_{\infty,u}$ values of 74.42, 47.11, and 36.43°C, respectively. When the external leads of the TEM are short-circuited together but no external power is supplied to the TEM, the voltage across the TEM as given by Eq. (37.9) must equal 0. The values of I, T_c, T_u, and $T_{\infty,u}$ which result from the simultaneous solution of the surface energy-balance equations subject to this constraint are 0.436 A, 74.42°C, 58.58°C, and 47.90°C, respectively. It is noted that when the external leads of the TEM are short-circuited together, both the voltage across them and the rate of external electrical work done on the TEM ($\dot{W}_{\text{TEM}}$) equal 0. However, the voltage constraint must be used to determine the state of the TEM rather than the fact that $\dot{W}_{\text{TEM}} = 0$. Otherwise, the open-circuit

---

[1]The boundary condition at the thermal epoxy-alumina interface was taken to be the heat flux distribution corresponding to an almost isothermal boundary condition ($\mu = -1/2$ in the Yovanovich et al. [6] nomenclature), and the alumina-pellet interface was assumed to be isothermal. The constriction resistance calculation was done by a spreading-resistance calculator available on the University of Waterloo's Microelectronics Heat Transfer Laboratory (MHTL) Website [7]. It is noted that Yovanovich [8] recommends the use of the results of Yovanovich et al. [6] provided that $(a/b < 0.6)$ and $(t/b < 0.72$, where t is the thickness of the alumina), which is the case here.

solution is mathematically valid but physically incorrect, because current must flow through the short circuit. When the leads are short-circuited together, the induced current through the TEM is positive. Hence, Peltier cooling occurs at the csi and the ambient temperature on the uncontrolled side of the TEM ($T_{\infty,u}$) for which the die temperature equals 75°C while dissipating 10 W is 11.47°C higher than in the open-circuit case. Finally, when +4 V is applied across the TEM, the surface energy-balance equations may be solved subject to this constraint according to Eq. (37.9). The result is that I, T_c, T_u, and $T_{\infty,u}$ equal 2.84 A, 74.42°C, 112.20°C, and 89.41°C, respectively. Active Peltier cooling at this condition maintains the control point at 75°C at an ambient temperature of 89.41°C. In summary, the ambient temperatures for which the die operates at 75°C when the TEM leads are open-circuited, are short-circuited and have a 4 V potential across them are 36.43, 47.90, and 89.41°C, respectively.

The temperature distribution in all the pellets is given by Eq. (37.3), where $x = 0$ corresponds to the csi and $x = L$ corresponds to the usi [Eq. (37.3) applies for any external lead configuration]. From the foregoing results, the temperature distributions in the pellets corresponding to open-circuit conditions, short-circuit conditions, and an applied voltage of 4 V are, respectively:

$$T = 74.42 - 23.95x \qquad \text{Open circuit}$$

$$T = 74.42 - 13.65x - 0.21x^2 \qquad \text{Short circuit}$$

$$T = 74.42 + 43.08x - 8.72x^2 \qquad \text{+4 V across TEM}$$

where x is in millimeters and T is in degrees Celsius. The temperature distribution for all three conditions is plotted in Fig. 37.5. The temperature distribution is linear when the current is zero (open-circuit conditions), as must be the case for one-dimensional heat conduction in the absence of any bulk cooling or heating effects. Moreover, the temperature distribution is pragmatically linear when the leads are short-circuited together because the rate of Ohmic heating (NI^2R) is very small: 0.20 mW. Modest curvature is observed when 4 V is applied across the TEM as the rate of Ohmic heating equals 8.30 W.

(B). When the TEM operates in heating mode, the lower the ambient temperature, the more power it requires to maintain the die at 75°C. Hence, the minimum power that the die must dissipate when the TEM power is limited to 10 W may be computed by solving the balance equations for q_{cp} subject to the constraint that $\dot{W}_{\text{TEM}} = 10$ W when the uncontrolled-side ambient temperature ($T_{\infty,u}$) equals −5°C, its lowest possible value. The result is that q_{cp} is equal to −0.07 W. The negative

37.14 Temperature Control

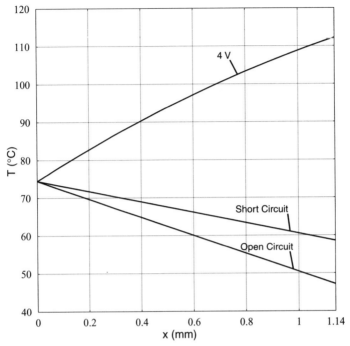

Figure 37.5 Temperature distribution in the TEM's pellet when its external leads are open-circuited ($I = 0$), short-circuited ($V_{TEM} = 0$) and have a 4 V potential applied across them ($V_{TEM} = 4$ V) such that $T_{\infty,u}$ equals 36.43, 47.90, and 89.41°C, respectively ($T_{cp} = 75.0$°C, $q_{cp} = 10$ W).

value of q_{cp} indicates that the die could be maintained at its operating temperature (75°C) even if it were to absorb up to 0.07 W of heat rather than dissipate heat. Hence, the die must not dissipate any power when the TEM power is limited to 10 W. Of course, when the power to the TEM is unlimited, the minimum die power remains 0 W.

Cooling of the die by the TEM requires the most power at the maximum ambient temperature of 65°C. The rate of heat absorbed by the TEM (q_{cp}) versus the current supplied to it when $T_{\infty,u} = 65$°C is plotted in Fig. 37.6, based on the solution of the surface energy-balance equations. Also plotted in Fig. 37.6 is the TEM power corresponding to a given current supplied to it calculated from the overall energy balance on the interconnected pellets of the TEM [Eq. (37.8)]. This plot shows that the maximum powers which may be dissipated by the die equal 16.8 and 19.4 W when the TEM power is limited to 10 W and unlimited, respectively. It is noted that the TEM power corresponding to 19.4 W dissipation on the die equals 28.4 W. The small amount of additional die power (2.6 W) that may be accommodated as the TEM

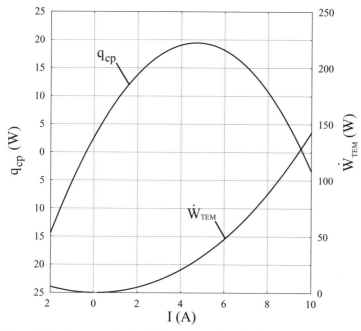

Figure 37.6 Die power that could be accommodated and corresponding TEM power as a function of current supplied to the TEM when $T_{cp} = 75.0°C$ and $T_{\infty,u} = 65.0°C$.

power is increased from 10 to 28.4 W results because Ohmic heating increases as the square of the TEM current, whereas the Peltier cooling rate at the csi varies linearly with current (for fixed T_c) as per Eq. (37.4). A *cooling efficiency* may be defined as the rate of heat transfer from the die into the TEM (q_c) divided by the power supplied to the TEM ($\dot{W}_{TEM}$). It equals 1.7 when 16.8 W is dissipated by the die, but only 0.7 when the die dissipates 19.4 W.

(C). The lower the value of the thermal resistance of the heat sink attached to the uncontrolled side of the TEM, the less power it consumes in cooling mode. However, as the resistance of the heat sink is decreased, the TEM consumes more power in heating mode. If the resistance of the heat sink were sufficiently high, heating mode would, in fact, be unnecessary. Both cooling and heating of the die by the TEM are required for the specified conditions. Moreover, the maximum powers required to cool or heat the die occur when the ambient temperature is highest ($T_{\infty,u} = 65°C$) and lowest ($T_{\infty,u} = -5°C$), respectively. Hence, the maximum amount of power required for the TEM to maintain the die temperature at 75°C over the entire ambient temperature operating range ($-5°C \leq T_{\infty,u} \leq 65°C$) is minimized when the heat sink

37.16 Temperature Control

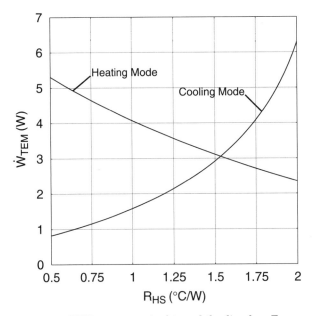

Figure 37.7 TEM power required to cool the die when $T_{\infty,u} = 65°C$ and heat the die when $T_{\infty,u} = -5°C$ as a function of the thermal resistance of the heat sink attached to the uncontrolled-side substrate of the TEM ($T_{cp} = 75.0°C$, $q_{cp} = 10$ W).

is sized such that the TEM requires the same amount of power to cool the die in a 65°C ambient as to heat it in a −5°C ambient. This optimal thermal resistance is determined by plotting the maximum powers required by the TEM for cooling ($T_{\infty,u} = 65°C$) and heating ($T_{\infty,u} = -5°C$) as a function of the heat-sink resistance based on a solution of the balance equations as shown in Fig. 37.7. The result is that the optimal heat-sink resistance equals 1.55°C/W and the corresponding maximum TEM power equals 3.1 W. It is important to minimize the power required for the thermal management of optical devices by properly sizing heat sinks. Since such devices reside in circuit packs that have a limited total electrical power budget, the less power which is required for thermal management, the more optical functionality that may be provided.

(D). For $T_{\infty,u} = 65.0°C$ and $R_{HS} = 1.0°C/W$, a plot of T_{cp} when all external power is consumed by the TEM in cooling mode, the TEM in heating mode, and the die is shown in Fig. 37.8. The uncontrolled-side temperature T_u is also shown on this plot and is identical for all three conditions because all power is ultimately dissipated through the heat sink. Increasing the temperature of the die to 200°C requires 35.0 W of external power when the die is used as a heater, but only 17.8 W is

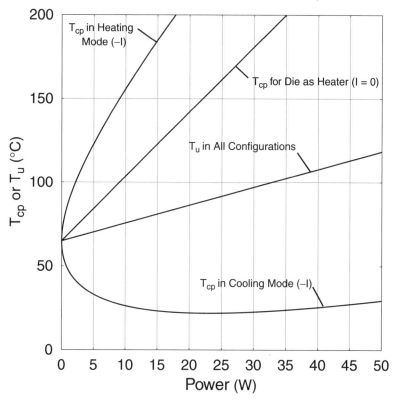

Figure 37.8 Control-point temperature (T_{cp}) as a function of TEM power when TEM operates in cooling mode ($+I$) and heating mode ($-I$) with the die dissipating no power ($q_{cp} = 0$). Also shown is T_{cp} when zero power is supplied to the TEM ($I = 0$) as a function of die power (q_{cp}) and the temperature of the usi (T_u) for all conditions. $T_{\infty,u} = 65.0°C$ and $R_{HS} = 1.0°C/W$.

necessary when the TEM is used in heating mode. It is noted that Ohmic heating does not dominate Peltier heating of the csi when the TEM operates in heating mode because T_c increases as more power is supplied to the TEM; thus, Peltier heating is not linearly proportional to current. When the die is used as a heater, all the heat must conduct through the entire TEM, but when the TEM is used in heating mode, Ohmic heating occurs within the thermocouples and the conduction length scale is smaller. Despite this, the TEM is still more efficient because of the heat released by the Peltier effect at the usi in heating mode.

Comments

1. Temperatures must be in kelvins when evaluating terms accounting for the Peltier effect, for example, $I\alpha_{p,n}T_c$.

37.18 Temperature Control

2. When the die operates at 75°C in a −5°C ambient, the temperature difference (heat-transfer driving force) between it and the ambient is very large (i.e., 80°C). Thus the assumption of zero heat transfer between the control point and the ambient ($K_{cp-\infty,c} = 0$) is valid only if the die is very well insulated, except where it is attached to the TEM.
3. Many commercially available TEMs cannot operate at the target temperature of 200°C in part D). However, "high temperature" TEMs that operate at temperatures up to about 225°C may be obtained.

Nomenclature

A_P	Cross-sectional area of pellet, m²
I	Current through TEM, A
k	Thermal conductivity, W/(m K)
K	Thermal conductance of a thermocouple, W/°C
K_{i-j}	Thermal conductance from node i to node j, W/°C
L	Pellet height, m
n	Number of a particular thermocouple in a TEM
N	Total number of thermocouples in a TEM
q	Heat rate, W
q_c	Heat-transfer rate from controlled-side substrate into csi, W
q_{cp}	Rate of heat generation at control point, W
q_u	Heat-transfer rate from usi into uncontrolled-side substrate, W
R	Ohmic resistance of a thermocouple, Ω
R_{i-j}	Thermal resistance from node i to node j, °C/W
R''_{ti}	Thermal interface resistance, m² K/W
t	Thickness of TEM substrate, mm or m
T	Temperature, °C or K
V	Voltage, V
V_0	Voltage in external lead connected to n-type thermocouple, V
V_N	Voltage in external lead connected to p-type thermocouple, V
V_{TEM}	Voltage difference across TEM ($V_0 - V_N$), V
$\dot{W}_{TEM}$	Rate of electrical work done on a TEM, W
x	Distance from controlled-side interface, m

Greek

α	Seebeck coefficient, V/K
$\alpha_{p,n}$	$\alpha_p - \alpha_n$, V/K
ρ	Electrical resistivity, Ω m

Subscripts

∞, c   Controlled-side ambient temperature, °C
c   Controlled-side interface, csi
cp   Control point
u   Uncontrolled-side interface, usi
∞, u   Uncontrolled-side ambient temperature, °C

References

1. C. L. Foiles, "Thermoelectric Effects," in *Encyclopedia of Physics*, 2d ed., R. G. Lerner and G. L. Trigg (eds.). New York: VCH Publishers, 1991, pp. 1263, 1264.
2. M. Hodes, "On One-Dimensional Analysis of Thermoelectric Modules (TEMs)," *IEEE Trans. Components Packaging*, vol. 28, no. 2, pp. 218–229, 2005.
3. W. Seifert, M. Ueltzen, C. Strümpel, W. Heiliger, and E. Müller, "One-Dimensional Modeling of a Peltier Element," in *Proceedings of 20th International Conference on Thermoelectrics* (ICT), Beijing, China, June 2001.
4. D. Yao, C. J. Kim, G. Chen, J. P. Fleurial, and H. B. Lyon, "Spot Cooling Using Thermoelectric Microcoolers," in *Proceedings of 18th International Conference on Thermoelectrics* (ICT), Baltimore, Aug.–Sept. 1999.
5. M. Yamanashi, "A New Approach to Optimum Design in Thermoelectric Cooling Systems," *J. Appl. Phys.*, vol. 80, no. 9, pp. 5494, 5502, 1996.
6. M. M. Yovanovich, C. H. Tien, and G. E. Schneider, "General Solution of Constriction Resistance within a Compound Disk," *AIAA Progress in Astronautics and Aeronautics*, vol. 70, pp. 47–62, 1980.
7. The University of Waterloo Microelectronics Heat Transfer Laboratory (MHTL), "Spreading Resistance of Circular Siyrces in Compound and Isotropic Disks," 2000 (the URL for this spreading-resistance calculator is http://www.mhtl.uwaterloo.ca/tools.html#).
8. M. M. Yovanovich, "Conduction and Thermal Contact Resistances (Conductances)," in *Handbook of Heat Transfer*, 3d ed., W. M. Rohsenow, J. P. Hartnett, and Y. I. Cho (eds.). New York: McGraw-Hill, 1998, chap. 3.

Part 10

Thermal Analysis and Design

Heat-transfer calculations are often used for thermal analysis and design. The contributors of the five chapters in this part are engineers working at aerospace installations operated by the U.S. government and a university, as well as academics at universities in the United States and Israel. All of the chapters deal with the real world:

- *Thermal analysis of a large telescope mirror*
- *Cooling a real-size room with a latent-heat-storage unit designed as a shell-and-tube heat exchanger, the tubes of which are filled with a phase-change material*
- *Assuring the operational functionality of a commercial, air-cooled computer used aboard the gondola of a helium balloon at an altitude of 100,000 ft above sea level*
- *Thermal analysis of convectively cooled heat-dissipating components on printed-circuit boards*
- *Design of a fusion-bonding process for fabricating thermoplastic-matrix composites*

Chapter 38

Thermal Analysis of a Large Telescope Mirror

Nathan E. Dalrymple
Air Force Research Laboratory
Space Vehicles Directorate
National Solar Observatory
Sunspot, New Mexico

Introduction

Telescope mirrors have rather stringent temperature requirements, since a mirror surface that is either hotter or colder than the surrounding air temperature will produce buoyant flows, or "seeing," that disturb image quality. In addition, lateral and through-thickness thermal gradients cause mirror substrates to warp by differential expansion. Thermal warpage has been solved to a large extent by the advent of ultra-low-expansion (ULE) glasses (Corning's ULE and Schott's Zerodur) and actuator-controlled surface figure. However, the problem of mirror seeing remains especially important for the primary mirror (M1) of large telescopes.

Many large modern telescopes use a primary mirror configuration known as the *thin-meniscus mirror*. A thin meniscus mirror looks a lot like a scaled-up contact lens—it is approximately uniform in thickness, with a large ratio of diameter to thickness. The 8-m Gemini telescopes use 200-mm-thick menisci, and the Advanced Technology Solar Telescope (ATST) is planned for use with a 4.3-m-diameter, 80-mm-thick mirror. (The ATST is a project led by the National Solar Observatory and largely funded by the National Science Foundation.) Such a mirror is supported on its rear face and edges by an array of actively controlled

38.4 Thermal Analysis and Design

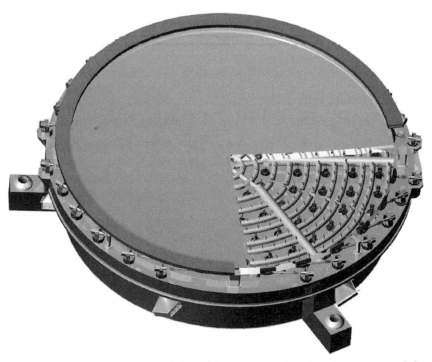

Figure 38.1 The ATST M1 assembly. The thin-meniscus mirror is cut away to reveal the support array and air-cooling distribution network. (*Courtesy National Solar Observatory.*)

supports. The ATST has 144 supports, 120 of them axial (on the substrate backside) and 24 of them radial (on the meniscus edges). The mirror supports react against a stiff structure called the *cell* that surrounds the backside of the M1 substrate. An artist's concept of the ATST M1 cell is shown in Fig. 38.1. The front (shiny) side of M1 is generally open to the surrounding air, which may in fact be the ambient wind at the telescope site, if the telescope is placed in a collapsible or passively ventilated dome.

As the ambient air on the frontside of the mirror changes temperature in its diurnal temperature cycle, the M1 front surface is required to change temperature also, so that the M1 surface temperature always remains in some small band of tolerance about the ambient air temperature. Solar telescopes, pointed at the sun during the daytime, have an additional thermal load since mirror coatings typically absorb about 10 percent of the incoming irradiance. Like the ambient temperature, the solar irradiance varies with the diurnal and annual cycles, from 500 to 1000 W/m^2 or so, depending on the site latitude, weather conditions, time of day, and time of year.

Frontside convective cooling is not very effective, since convection is proportional to the surface-air temperature difference, which we want to be zero. Our thermal control mechanism must then be located behind the M1 substrate. This is not a very efficient location, since the substrate material has both low thermal conductivity k and large thermal mass $\rho c V$. As one might intuit, the M1 frontside responds very slowly to changes in the thermal control system at the backside.

In the past, mirror seeing has been avoided by enclosing the mirror in a vacuum, or by enclosing the mirror in a sealed chamber filled with helium (the index of refraction of helium changes very little with temperature, so buoyant flows do not disturb image quality as they do in air). Both of these solutions alleviate mirror seeing without tackling the thermal problem. However, both solutions require sealed entrance windows, which are impractical above approximately 1 m aperture. In the future, highly conductive mirror materials such as silicon carbide or aluminum may become available in large sizes. For the present, however, we are stuck with a difficult and interesting thermal problem.

Problem Distillation

Figure 38.2 shows M1 in its thermal environment. On the frontside, a uniform heat flux $\dot{q}''_{abs}(t)$ is imposed over the whole surface as a result of solar irradiance absorption (or radiation to the cold night sky for nighttime telescopes). Heat is also transferred convectively to the surrounding air [at temperature $T_1(t)$] with some coefficient h_1. Within the M1 substrate, heat is transferred by conduction. On the rear side of M1, convection to a coolant at temperature $T_2(t)$ occurs with coefficient h_2. In addition, radiation plays a role as the unpolished mirror backside usually has a high emissivity ϵ and the surrounding cell structure can be quite cold [temperature $T_r(t)$].

The goal of this analysis is to understand how the mirror surface temperature responds to changes in $\dot{q}''_{abs}(t)$, $T_1(t)$, $T_2(t)$, and $T_r(t)$. Ultimately,

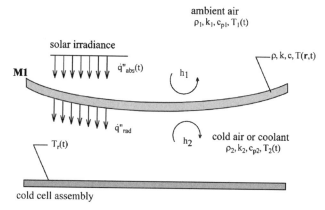

Figure 38.2 The M1 thermal environment.

38.6 Thermal Analysis and Design

we would like to determine how to vary $T_2(t)$ so as to keep the surface temperature within some small tolerance of $T_1(t)$. Since this optimization process is likely to be iterative, we need a mirror response analysis that runs quickly on small desktop computers.

We expect that heat conduction through the thickness of the mirror substrate will play a significant role in the mirror surface temperature behavior, while radial conduction from the center to the edges will play less of a role. Thus, most of the behavior of a real mirror should be captured in a one-dimensional, time-dependent (1D,t) model. We will begin, however, with a lumped-capacity, or zero-dimensional model based on integrated energy considerations. As we shall see, this is very easy to implement and gives quick qualitative guidance and even quantitative results in special cases. After developing some physical intuition with this model, we will proceed to develop the (1D,t) model.

For detailed final design, including edge effects, three-dimensional models are quite useful. A number of commercial numerical solvers perform (3D,t) calculations nicely (e.g., NASTRAN, FEMLab, COSMOS-Works). We will describe only the first two types of models in the following sections.

A Lumped-Capacity Model

When the internal thermal resistance is much smaller than the external heat-transfer resistance, the temperature of M1 will change as a unit, with little variation through the solid. For convectively cooled solids, the Biot number Bi quantifies the resistance ratio:

$$\text{Bi} = \frac{h}{k/L} = \frac{hL}{k} \tag{38.1}$$

where L is a characteristic length (thickness) of the solid, h is the heat-transfer coefficient, and k is the solid conductivity. When Bi $\ll$ 1, the internal conduction resistance is much smaller than the external convection resistance and the mirror can be considered as a unit, or a lumped capacity. If, on the other hand, Bi $\gg$ 1, the conductive heat transfer is located near the surfaces of the solid and solutions of the "semi-infinite solid" variety may be employed [see, e.g., Mills (1992, p. 155)]. Evaluating for the case of a 100-mm-thick ULE meniscus with $h = 5$ W/m²-K, $L = 0.05$ m (using the half-thickness), and $k = 1.3$ W/m-K results in Bi = 0.2. This model should work pretty well for cases with small values of h (calm air). For solar loading, however, the effective h is very large (since we regard $\dot{q}''_{\text{abs}}$ as being imposed, with very little external resistance), reducing the validity of a lumped-capacity model. We expect, then, that the mirrors of nighttime telescopes in calm air with no solar heat load will be described fairly well by lumped-capacity models, while those of solar telescopes will be described less accurately.

Proceeding nevertheless, and regarding M1 as a single unit with area A, thickness d, and temperature T, the change in internal energy U of M1 is equal to the sum of the heat inputs:

$$\frac{dU}{dt} = \dot{Q}_{abs} - \dot{Q}_{conv} - \dot{Q}_{rad} \tag{38.2}$$

where

$$\frac{dU}{dt} = A\rho dc \frac{dT}{dt} \tag{38.3}$$

$$\dot{Q}_{abs} = A\dot{q}''_{abs} \tag{38.4}$$

$$\dot{Q}_{conv} = Ah_1[T(t) - T_1(t)] + Ah_2[T(t) - T_2(t)] \tag{38.5}$$

and the backside radiation transfer is

$$\dot{Q}_{rad} = A\epsilon\sigma[T^4(t) - T_r^4(t)] \tag{38.6}$$

Here, σ is the Stefan-Boltzmann constant and we have assumed the mirror cell is large and nearly black. We can cast the radiation in terms of a third convective coefficient by linearizing:

$$A\epsilon\sigma[T^4(t) - T_r^4(t)] = A\epsilon\sigma[T^2(t) + T_r^2(t)][T(t) + T_r(t)][T(t) - T_r(t)]$$

$$= Ah_r[T(t) - T_r(t)] \tag{38.7}$$

with

$$h_r \equiv \epsilon\sigma[T^2(t) + T_r^2(t)][T(t) + T_r(t)] \tag{38.8}$$

Rewriting Eq. (38.2), we have

$$\rho dc \frac{dT}{dt} = \dot{q}''_{abs} - h_1[T - T_1] - h_2[T - T_2] - h_r[T - T_r] \tag{38.9}$$

Equation (38.9) is mildly nonlinear in T because of the T dependence of the convective h values. For this calculation, however, we presume that h_1, h_2, and h_r do not vary with temperature over the range of interest.

Letting $T_0 \equiv T(t = 0)$, $\theta \equiv T - T_0$, and $\theta_x \equiv T_x - T_0$, where $x = 1, 2$, or r, we obtain

$$\frac{d\theta}{dt} + M(t)\theta = N(t) \tag{38.10}$$

$$\theta(0) = 0 \tag{38.11}$$

38.8 Thermal Analysis and Design

where

$$M(t) \equiv \frac{h_1(t) + h_2(t) + h_r(t)}{\rho dc} \tag{38.12}$$

$$N(t) \equiv \frac{1}{\rho dc}[\dot{q}''_{\text{abs}}(t) + h_1(t)\theta_1(t) + h_2(t)\theta_2(t) + h_r(t)\theta_r(t)] \tag{38.13}$$

For generality, we allow h_1, h_2, and h_r to vary in time. Note that $M(t)$ is an inverse time constant, so that increasing h shortens the time response of the mirror. The constant $N(t)$ is the driving function that includes the time-changing boundary conditions. Equation (38.10) is an ordinary first-order differential equation, and can be solved along with Eq. (38.11) using an integrating factor (see, e.g., Greenberg (1978, p. 395)].

Writing Eq. (38.10) as

$$d\theta + (M\theta - N)dt = 0 \tag{38.14}$$

we seek $\mu(t)$ such that the product of Eq. (38.14) and $\mu(t)$ will be an exact differential. This will be so if

$$\frac{d\mu}{dt} = \frac{d}{d\theta}[\mu(M\theta - N)] \tag{38.15}$$

Solving for $\mu(t)$, we obtain

$$\frac{d\mu}{dt} = M(t)\mu(t) \tag{38.16}$$

$$\frac{d\mu}{\mu} = M(t)dt \tag{38.17}$$

$$\ln \mu = \int M(t)dt \tag{38.18}$$

$$\mu(t) = e^{\int M(t)dt} \tag{38.19}$$

Multiplying Eq. (38.14) through by μ, we now have the exact differential

$$\mu(t)d\theta + \mu(t)[M(t)\theta - N(t)]dt \equiv d\phi = 0 \tag{38.20}$$

$$\frac{\partial \phi}{\partial \theta} = \mu(t) \tag{38.21}$$

$$\frac{\partial \phi}{\partial t} = \mu(t)[M(t)\theta - N(t)] \tag{38.22}$$

Now, since $d\phi = 0$, ϕ must integrate to a constant, C. Integrating Eq. (38.21), we have

$$\phi = \theta e^{\int M dt} + A(t) = C \qquad (38.23)$$

Substituting this result into Eq. (38.22), and solving for A gives

$$A(t) = -\int N(t) e^{\int M(t) dt} dt \qquad (38.24)$$

Plugging this result back into Eq. (38.23) and applying Eq. (38.11) gives the desired solution:

$$\theta(t) = e^{-\int_0^t M(\xi) d\xi} \int_0^t N(\tau) e^{\int_0^\tau M(\eta) d\eta} d\tau \qquad (38.25)$$

Here, ξ, η, and τ are dummy variables of integration.

For the special case $M = $ constant, the solution reduces to

$$\theta(t) = \int_0^t N(\tau) e^{M(\tau - t)} d\tau \qquad (38.26)$$

This solution [in Eq. (38.26)] is just a weighted integral of $N(t)$, which includes all the boundary conditions and can be specified arbitrarily. This is a very nice result, since we can quickly calculate the integral (e.g., on a spreadsheet program) for a given set of boundary conditions.

Lumped-Capacity Model Test Case

Consider a meniscus mirror 100 mm thick, made of ULE with $\rho = 2300$ kg/m^3, $c_p = 766$ J/kg-K, and $k = 1.3$ W/m-K. The mirror is initially at 0°C. Both sides of the mirror are exposed to calm, ambient air ($h = 5$ W/m^2-K). The ambient air temperature varies sinusoidally from 0 to 15°C at 6 h, and back down to 0°C at 12 h, after which the temperature remains constant at 0°C. There is no solar heat load. Solving Eq. (38.26) using a simple numerical integration program, we obtain the solution shown in Fig. 38.3. The upper panel shows the input curves, with temperature on the left axis and heat flux on the right axis. The solar load $\dot{q}''_{abs}$ is zero throughout this simulation. Temperatures T_1 and T_2 overlay. In this test case, the change in ambient temperature drives the mirror to change temperature. The lower panel shows the result of the calculation (by a dot-dashed curve; the other curves are described below). Temperature T rises and falls qualitatively in the same manner as does the ambient air temperature. However, there is a significant time lag between T and T_1, T_2. The peak T lags the maximum T_1 by over

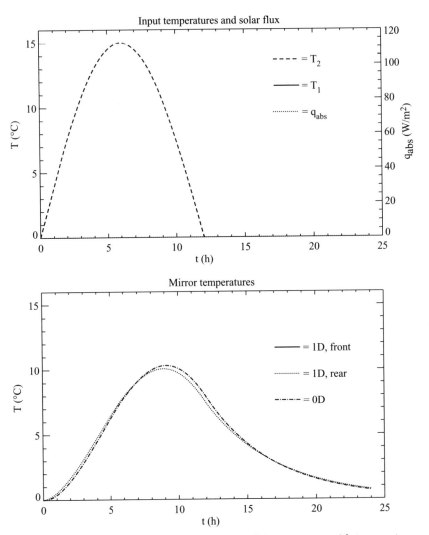

Figure 38.3 Lumped-capacity test case. Upper panel: input curves, with temperatures (dashed curve and solid curve) on the left axis and heat flux (dotted curve) on the right axis. Temperatures T_1 and T_2 overlay. The solar load $\dot{q}''_{abs}$ is zero throughout this simulation. In this test case, the change in ambient temperature drives the mirror to change temperature. Lower panel: result of the calculations. Lumped-capacity results are shown by a dot-dashed curve. Results from the 1D,t model are shown by a solid curve and a dotted curve.

3 h. In addition, the sharp slope changes in T_1 at 0 h and 12 h have been smeared out. Temperature T does not return to its initial value until more than 24 h have elapsed.

Other results of the lumped-capacity model are presented below in "Some Examples Using 0D and 1D Models."

A 1D,t Model

When the Biot number Bi is not small, we must consider conductive heat transfer through the thickness of the mirror. As mentioned above, Bi will be of order unity or larger when h is large (windy days) or when the mirror is exposed to solar radiation. Since both of these conditions are encountered in solar telescopes, we need a model that accurately describes the conduction problem. As mentioned in "Problem Distillation" above, we expect most of the physics to be captured by a one-dimensional, time-dependent (1D,t) treatment.

Fourier's equation is satisfied within M1:

$$\frac{\partial T}{\partial t} = \alpha \nabla^2 T = \alpha \frac{\partial^2 T}{\partial x^2} \tag{38.27}$$

(in one dimension). Here, α is the thermal diffusivity of the M1 material, T is the M1 temperature, and t is time (refer to Fig. 38.2). At time $t = 0$ the substrate is assumed to be uniform at a temperature T_i. On the front surface, we obtain a boundary condition using the continuity of heat flux:

$$\dot{q}''_{\text{cond,in}}(t) = \dot{q}''_{\text{abs}}(t) + \dot{q}''_{\text{conv,in}}(t) \tag{38.28}$$

where subscripts "abs," "conv," and "cond" denote absorbed, convective, and conductive heat flux, respectively. Using Newton's law of cooling and Fourier's law of heat conduction, we have

$$k \left.\frac{\partial T}{\partial x}\right|_{x=L} = \dot{q}''_{\text{abs}}(t) + h_1[T_1(t) - T(x = L, t)] \tag{38.29}$$

Here, we assume that h_1 is constant for simplicity and L is the substrate thickness.

The rear boundary condition is similar:

$$\dot{q}''_{\text{cond,out}}(t) = \dot{q}''_{\text{rad,out}}(t) + \dot{q}''_{\text{conv,out}}(t) \tag{38.30}$$

or

$$k \left.\frac{\partial T}{\partial x}\right|_{x=0} = h_r[T(x = 0, t) - T_r(t)] + h_2[T(x = 0, t) - T_2(t)] \tag{38.31}$$

Here we have linearized the radiation flux as above, and h_r and h_2 are assumed to be constant. At this point we simplify the problem slightly by recognizing that T_r, the cold surface temperature at the rear of the substrate, is likely to be very close to T_2, the rear coolant temperature. If $T_r \approx T_2$, then we can combine h_r and h_2 into a single effective h_2.

Rewriting the rear boundary condition, we obtain

$$k \left.\frac{\partial T}{\partial x}\right|_{x=0} = h_2[T(x=0,t) - T_2(t)] \tag{38.32}$$

All the boundary conditions will vary temporally, both with the earth's rotation and with sporadic events such as passing clouds or changing weather conditions. The challenge of thermal modeling is to solve the set of equations Eqs. (38.27), (38.29), and (38.32) for the internal temperature distribution $T(x,t)$, given the boundary curves $\dot{q}''_{abs}(t)$, $T_1(t)$, and $T_2(t)$ and the initial value T_i.

Since the boundary curves may vary arbitrarily, it is difficult to arrive at a simple analytical solution. One practical approach is to set up a numerical solution using either a spreadsheet program like Microsoft Excel or a high-level programming and plotting environment like IDL or Matlab.

The first step in our numerical solution is to divide the substrate volume into M equal subvolumes of width Δx, represented by $M+1$ node points. We also divide time into intervals of length Δt. We approximate the time derivative in Eq. (38.27) using a first-order forward difference:

$$\frac{\partial T}{\partial t} \simeq \frac{T_m^{i+1} - T_m^i}{\Delta t} \tag{38.33}$$

Here the superscript is the time index and the subscript is the node (spatial) index. The spatial derivative is approximated using a second-order central difference:

$$\frac{\partial^2 T}{\partial x^2} \simeq \frac{T_{m+1}^i - 2T_m^i + T_{m-1}^i}{(\Delta x)^2} \tag{38.34}$$

Plugging into Eq. (38.27) and solving, we obtain

$$T_m^{i+1} = (1 - 2\text{Fo})T_m^i + \text{Fo}\left(T_{m+1}^i + T_{m-1}^i\right) \tag{38.35}$$

Here, Fo is the grid Fourier number, defined as

$$\text{Fo} \equiv \frac{\alpha \Delta t}{(\Delta x)^2} \tag{38.36}$$

The result in Eq. (38.35) is called an "explicit scheme," since the nodal temperature at a new time step is expressible in terms of nodal temperatures at previous time steps. This means that we can "march out" a solution by successive calculations. For this scheme to be numerically stable, however, we must keep the coefficient of T_m^i positive. This places a restriction on the grid Fourier number, and hence on the size of the time step:

$$1 - 2\text{Fo} \geq 0 \qquad (38.37)$$

$$\text{Fo} \leq \frac{1}{2} \qquad (38.38)$$

$$\Delta t \leq \frac{(\Delta x)^2}{2\alpha} \qquad (38.39)$$

For example, if we simulate a 100-mm-thick mirror substrate using 21 nodes (20 thickness intervals of 5 mm) in a material with $\alpha \simeq 7.4 \times 10^{-7} \text{m}^2/\text{s}$, then Δt must be less than approximately 16.9 s.

To discretize the boundary conditions, we must add nodal capacity terms to the energy balances in Eqs. (38.29) and (38.31), since the nodes representing the surfaces of the solid are not infinitely thin but have thickness $\Delta x/2$ and, hence, a finite heat capacity. For the front surface (node M), we write

$$\rho c \frac{\Delta x}{2} \frac{T_M^{i+1} - T_M^i}{\Delta t} = q_{\text{abs}}^i - h_1\left(T_M^i - T_1^i\right) - \frac{k}{\Delta x}\left(T_M^i - T_{M-1}^i\right) \qquad (38.40)$$

where we have approximated the partial time derivative as before and have dropped all extraneous notation on q_{abs}. Rearranging terms and solving for the boundary node temperature at time $i+1$, we have

$$T_M^{i+1} = (1 - 2\text{Fo} - 2\text{Fo}\text{Bi}_1)T_M^i + 2\text{Fo}\left(T_{M-1}^i + \text{Bi}_1 T_1^i\right) + 2\text{Fo}\frac{\Delta x}{k}q_{\text{abs}}^i \qquad (38.41)$$

Here we have defined the grid Biot number:

$$\text{Bi}_1 \equiv \frac{h_1 \Delta x}{k} \qquad (38.42)$$

As before, the coefficient on T_M^i must be positive for numerical stability, further restricting our choice of Δt.

Similarly, for the rear boundary condition (node 0), we have

$$T_0^{i+1} = (1 - 2\text{Fo} - 2\text{Fo}\text{Bi}_2)T_0^i + 2\text{Fo}\left(T_{m=1}^i + \text{Bi}_2 T_2^i\right) \qquad (38.43)$$

where we have written $T_{m=1}^i$ for the node 1 temperature to distinguish it from the T_1^i external air temperature, and Bi_2 is defined in the same way as is Bi_1, replacing h_1 with h_2.

Our numerical solution domain is sketched in Fig. 38.4. Spatial position is plotted along the horizontal axis, and time marches out upward. The left- and rightmost nodes are the rear and front boundaries, respectively, and the bottommost row of nodes represents the initial state T_i. The temperature of any inner node is calculated using its preceding

38.14   Thermal Analysis and Design

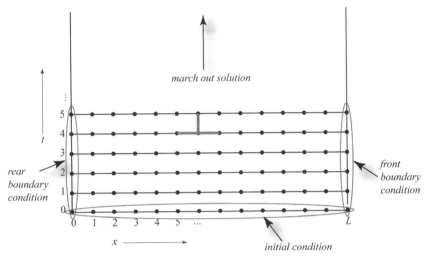

Figure 38.4 Schematic of numerical solution domain. The thick-lined inverted T shows that the nodal temperature value at a particular time step is calculated from three nodal temperatures at the preceding time step.

value, along with the preceding value of the nodes on either side. The solution procedure is a two-step process:

1. Calculate boundary nodes at step $i + 1$ according to Eqs. (38.41) and (38.43).
2. Calculate interior nodes at step $i + 1$ according to Eq. (38.35).

The counter i is then incremented, and the process is repeated until the desired simulation duration has been achieved.

Although there are more complex numerical schemes, this simple explicit approach allows the problem to be solved on a spreadsheet or in a simple Matlab program. Each node in the numerical model corresponds to a cell in the spreadsheet (or an array element in Matlab). Arbitrarily changing boundary conditions can simply be input as columns of data. This allows the use of measured weather data in a mirror simulation. For long simulation durations with high spatial resolution, the stability requirement [Eq. (38.39)] may require a very large number of time steps to be used. In this case, an implicit numerical scheme may be used to reduce the number of required time steps [see, e.g., Jaluria (1988)].

Some Examples Using 0D and 1D Models

Repeat lumped-capacity model test case

First of all, let us repeat the test case presented in "Lumped-Capacity Model Test Case" above, using the 1D,t model. Using the same input curves and properties as for the lumped-capacity (0D) run, we obtain the solution shown in the bottom panel of Fig. 38.3. The temperatures of the front and rear surfaces are shown by a solid curve and a dotted curve, respectively. Because of the symmetry of the applied boundary conditions, these two curves overlay. Temperature T rises and falls in qualitatively the same manner as does the ambient air temperature. However, there is a significant time lag between T and T_1, T_2. The peak T lags the maximum T_1 by over 3 h. In addition, the sharp slope changes in T_1 at 0 and 12 h have been smeared out. Temperature T does not return to its initial value until more than 24 h has elapsed. The 1D solution is quite similar to the 0D solution. Recall the condition of validity of the lumped-capacity solution: the Biot number $\text{Bi} \ll 1$. In this test case, $\text{Bi} = 0.2$ and the 0D approach works well.

Validation of 1D,t model

Again, consider a meniscus mirror 100 mm thick, made of ULE with $\rho = 2300$ kg/m^3, $c_p = 766$ J/kg-K, and $k = 1.3$ W/m-K. The mirror is initially at 0°C. The frontside of the mirror is exposed to calm, ambient air ($h_1 = 5$ W/m^2-K). The ambient air temperature $T_1(t)$ varies sinusoidally from 0 to 15°C at 6 h, and back down to 0°C at 12 h, after which the temperature remains constant at 0°C. Solar radiation impinges on the frontside of the mirror, and approximately 10 percent is absorbed by the mirror coating. The absorbed solar heat flux varies as $\dot{q}''_{\text{abs}} = 75(1 + \sin(\pi t/12))$ W/m^2 from 0 to 12 h (t in h), after which it is zero. The mirror is cooled from the rear side by air at $T_2(t)$ with a heat-transfer coefficient $h_2 = 9.49$ W/m^2-K. The cooling air temperature varies as $T_2(t) = 3.17\sin(\pi t/12) - 4.73$°C from 0 to 12 h, after which it is zero. The input curves are shown in the upper panel of Fig. 38.5.

To check the validity of our 1D,t model, we set up a fully three-dimensional, time-dependent model in a commercial finite-element analysis package (NASTRAN). Using the identical boundary conditions, we ran our 1D,t and 0D,t models and compared the results, which are shown in the bottom panel of Fig. 38.5. The solid and dotted curves show the front and rear mirror surface temperatures from the 1D model. The stars are the result of the NASTRAN 3D model, and the dot-dashed curve is the result of our lumped-capacity treatment. During the first 8 h, the front mirror surface temperature rises steeply as the mirror absorbs the solar heat flux, to a maximum temperature of approximately 17°C. The surface temperature then falls, gradually at first as both the environment temperature and solar irradiance decrease, then more rapidly at nightfall when the solar heat input is removed entirely. As in the previous example, there is a large time lag between the peak mirror surface temperature and the peak T_1. During the daytime, the

38.16 Thermal Analysis and Design

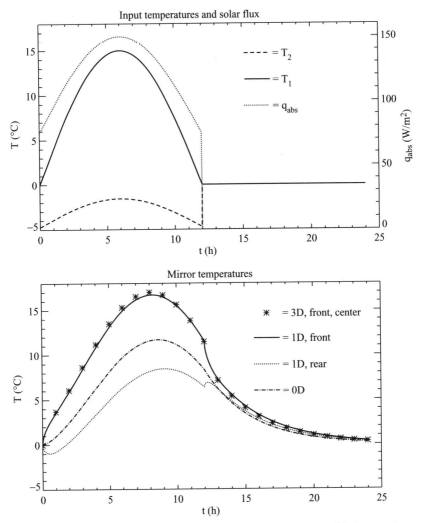

Figure 38.5 Validation of 1D,*t* model. Upper panel: input curves, with temperatures (dashed curve and solid curve) on the left axis and heat flux (dotted curve) on the right axis. Lower panel: result of the calculations. Lumped-capacity results are shown by a dot-dashed curve. Results from the 1D,*t* model are shown by a solid curve and a dotted curve. Three-dimensional results are shown as stars.

absorbed flux on the front surface causes a large through-thickness thermal gradient, so that the rear surface is almost 9°C cooler than the front surface. At nightfall, however, the front and rear surface temperatures quickly converge. Throughout the simulation, the 1D,*t* model gives results that are very close to those for the 3D model. In fact, the root-mean-square difference is only 0.19°C. The lumped-capacity model gives a result that lies between the front and rear surface temperatures during the daytime, and at nighttime the result is quite close to those for the 1D and 3D models. The effective Bi is large during daytime, when solar flux drives the large temperature gradient in the mirror, but at night Bi decreases, allowing the 0D treatment to be surprisingly effective.

An example using measured weather data

In this example, the mirror properties are identical to those in the previous examples, but instead of being a hypothetical curve, $T_1(t)$ is measured air temperature at Sac Peak, New Mexico, over a 60-h period during June 23–25, 2001. The solar flux input curve is modeled using a simple atmospheric transmittance model, accounting for the geographic position of Sac Peak and the elevation above sea level [see, e.g., Kreider and Kreith (1977)]. The coolant temperature $T_2(t)$ is held constant at 15°C. On the front surface, the air is calm, with $h_1 = 5$ W/m^2-K. The mirror is actively cooled by blowing airjets on the rear surface with $h_2 = 56$ W/m^2-K. Input curves are shown in the upper panel of Fig. 38.6, with T_1 and T_2 shown by a solid curve and a dashed curve, respectively, and $\dot{q}''_{abs}$ by a dotted curve. The results of the 1D and 0D models are shown in the bottom panel. Again, the front and rear surface temperatures from the 1D model are shown by a solid curve and a dotted curve, respectively, and the 0D model result is shown by a dot-dashed curve. As in previous examples, there is a lag between the applied boundary conditions and the mirror surface response. During the daytime, the mirror experiences a large through-thickness thermal gradient of approximately 8°C through 100 mm. At night, however, the front and rear surface temperatures converge and the 0D model becomes a much better approximation. Figure 38.7 shows an enlarged version of the first night, showing that the 0D model does very well except during periods of rapid change (e.g., near t of 7 or 8 h).

Optimizing coolant temperature

If the ambient temperature curve $T_1(t)$ is known a priori, then it is possible to optimize the coolant temperature $T_2(t)$ so that the mirror front surface temperature follows $T_1(t)$ within some small tolerance. In this way, the deleterious effects of buoyant plumes on image quality can be avoided. Figure 38.8 shows the result of optimizing $T_2(t)$ for the particular timespan of the previous example. The upper panel shows the input curves, as before, and the lower panel overlays the resulting mirror front surface temperature (solid curve) and the ambient air temperature (dot-dashed curve). After initial transient effects die away, the two curves track one another nicely.

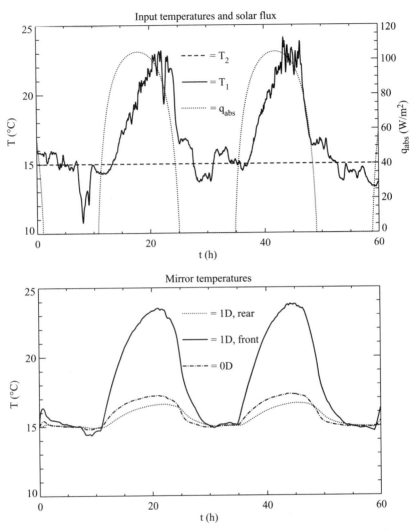

Figure 38.6 A model run using measured air temperature at Sac Peak, New Mexico, over a 60-h period during June 23–25, 2001. The solar flux input curve is modeled using a simple atmospheric transmittance model, accounting for the geographic position of Sac Peak and the elevation above sea level. The coolant temperature $T_2(t)$ is held constant at 15°C. Upper panel: input curves, with T_1 and T_2 shown by a solid curve and a dashed curve, respectively, and $\dot{q}''_{abs}$ by a dotted curve. Bottom panel: results. The front and rear surface temperatures from the 1D model are shown by a solid curve and a dotted curve, respectively, and the 0D model result is shown by a dot-dashed curve.

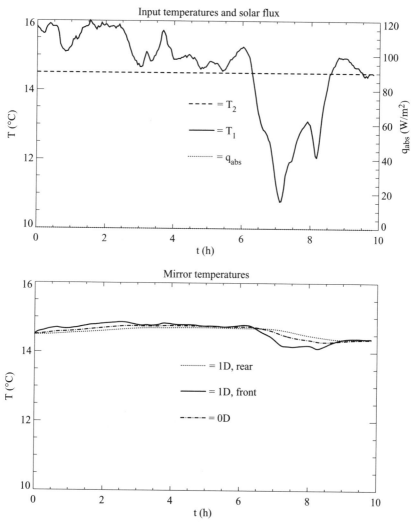

Figure 38.7 An enlarged version of the first night of conditions shown in Fig. 38.6. Upper panel: input curves, with T_1 and T_2 shown by a solid curve and a dashed curve, respectively, and $\dot{q}''_{abs}$ by a dotted curve. Bottom panel: results. The front and rear surface temperatures from the 1D model are shown by a solid curve and a dotted curve, respectively, and the 0D model result is shown by a dot-dashed curve. The 0D model does very well except during periods of rapid change (e.g., near t of 7 or 8 h).

Closure

Two mirror thermal models were presented: a lumped-capacity (0D) model that considered the mirror as a single thermal mass and a one-dimensional (1D) model that considered unsteady conduction in the mirror substrate. The 0D case has a simple analytical solution that is

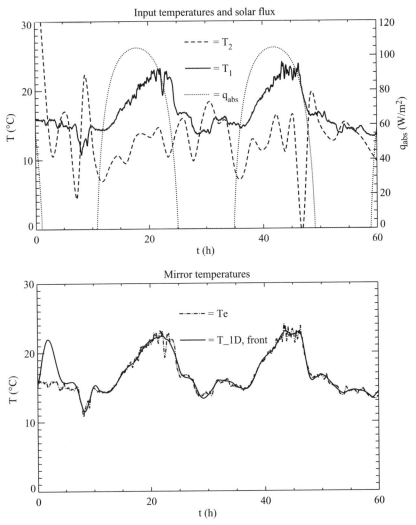

Figure 38.8 Optimized coolant temperature profile [$T_2(t)$] drives the mirror front surface temperature very close to the ambient air temperature [$T_1(t)$]. Upper panel: input curves, with T_1 and T_2 shown by a solid curve and a dashed curve, respectively, and $\dot{q}''_{abs}$ by a dotted curve. Lower panel: the resulting mirror front surface temperature (solid curve) and the ambient air temperature (dot-dashed curve).

easily evaluated for a variety of boundary conditions. However, the 0D model is valid only when the mirror sees no solar radiation and when the convective heat-transfer coefficient on the mirror surfaces is not too large. Under these conditions, the 0D model performs surprisingly well, producing results that are comparable to more complex 1D and 3D treatments.

Our 1D solution was an explicit finite-difference numerical scheme. The 1D model gave results that were within 0.19°C (rms) of a fully three-dimensional finite-element model, and ran in fractions of a second on a desktop computer. Because of its simplicity and speed, our 1D solution is well suited for iterative optimization processes or other studies in which a large number of numerical experiments must be performed quickly.

This work was supported by the Air Force Office of Scientific Research.

References

Greenberg, M. D., *Methods of Applied Mathematics*, Prentice-Hall, Englewood Cliffs, N.J., 1978.
Jaluria, Y., *Computer Methods for Engineering*, Allyn and Bacon, Boston, 1988.
Kreider, J. F., and Kreith, F., *Solar Heating and Cooling*, McGraw-Hill, New York, 1977.
Mills, A. F., *Heat Transfer*, Irwin, Boston, 1992.

Chapter 39

Thermal Design and Operation of a Portable PCM Cooler

R. Letan and G. Ziskind
Heat Transfer Laboratory
Mechanical Engineering Department
Ben-Gurion University of the Negev
Beer-Sheva, Israel

Abstract

A case study is presented in which a real-size room at daytime is cooled by a latent-heat-storage unit. The unit is designed as a shell-and-tube heat exchanger. The tubes are vertical and filled with a phase-change material which melts at 23°C, while the ambient temperature is 35°C.

Three cases are considered: an insulated room, a room heated by the ambient, and a room heated by both the ambient and a heat source.

Calculations of the room cooling by the PCM unit were carried out in three ways: analytically, by Matlab computation, and by a numerical simulation using Fluent software.

Introduction

Herein are presented calculations of a latent-heat-storage unit utilized for temperature moderation of an enclosed space. The specific case examined relates to a real-size room, and a phase-change-material (PCM)–based unit, which absorbs heat from the room air circulated

39.2 Thermal Analysis and Design

through the unit. The PCM melts and the room air is cooled down to a comfortable level, which lasts as long as the PCM is melting.

The design of the unit and its mode of operation determine the rate of air cooling and the temperature evolution in the room. The larger is the PCM mass in the unit, the longer is the cooling provided.

Thermal storage units that utilize latent heat of PCMs are used for domestic heating or cooling in climates characterized by high temperatures in the daytime and cool nights. During the day the PCM absorbs heat in the building and melts, while at night it releases its latent heat to the ambient and solidifies. Thus, a PCM-based unit replaces an air conditioner, which requires considerable electrical energy for its operation.

For the case analyzed here, we have chosen a storage unit designed as a shell-and-tube heat exchanger. The tubes are vertical and contain the melting/solidifying material. The room air is fan-driven through the shell.

Units of similar structure have been investigated in the literature. Farid and Kanzawa [1] theoretically investigated a thermal storage unit designed as a shell-and-tube heat exchanger. The tubes were filled with PCM, and air was flowing in the shell across the tubes. Their simulations showed that the performance of the unit improves by using PCMs with different melting temperatures in the same unit.

Lacroix [2] investigated a shell-and-tube unit in which the PCM was stored in the shell and the heat-transfer fluid circulated inside the vertical tubes. The unit was numerically simulated, and experimental tests were conducted to validate the model. Turnpenny et al. [3,4] studied a unit in which a heat pipe was embedded in a phase-change material. The unit was utilized as a ventilation cooling system in buildings. Mozhevelov [5] performed three-dimensional transient simulations in a real-size room and a portable storage unit with cooling elements of various shapes in a shell: vertical plates, horizontal plates, horizontal square tubes in in-line and staggered configurations, and vertical square tubes in an in-line configuration.

Mozhevelov et al. [6] have also studied thin vertical storage units installed parallel to walls in a room. In daytime heat was free-convected from the room air and the PCM melted. At night, by both free and forced convection, heat was released from the unit into the ambient, and the PCM solidified.

Arye and Guedj [7] experimented with a shell-and-tube unit, in which the tubes were vertical and filled with PCM. The room air was force-convected by fans through the shell. The PCM used was a paraffin wax melting at 22 to 24°C, while the ambient temperature at daytime was 30 to 35°C and at night, 18 to 19°C. Thus, the paraffin that melted at daytime solidified at night, and thermal comfort was preserved in the room. Our calculations, that follow relate to a case study of a real-size room cooled at daytime by a unit in which the PCM melts at 23°C, while the ambient temperature is 35°C. The calculations are conducted analytically, by Matlab computation, and by numerical simulation using the Fluent software.

Thermal Design

Prescribed geometries and operating conditions

The cooler is a shell-and-tube heat exchanger operated in crossflow: PCM in vertical tubes is unmixed, airflow in shell is mixed. The external dimensions of the cooler are height H, 0.80 m; width W, 0.21 m, length L, 1.35 m.

The fans are positioned vertically; with length 0.20 m; height and width are the same as in the cooler. The overall length of the fans is $L_{tot} = 1.55$ m. In the fans the free-stream velocity of air, outside the tube array, is $u_\infty = 1.4$ m/s.

The external size of the cooler approximates the size of a conventional, wall air conditioner. The cooler is portable. Figure 39.1 depicts the structure of the cooler.

The room has the following dimensions: height 2.5 m, length 4.0 m, and width 4.0 m. The initial temperature of air in the room is $T_{r,0} = 35°C$. The room temperature at any instant is uniform (the room air is mixed). Radiation inside the room is ignored.

The phase-change material is a paraffin wax (RT-25 by Rubitherm). The thermophysical properties of the PCM are as follows: melting temperature T_m, 23°C (22 to 24°C); specific enthalpy of melting Δh_m, 206 kJ/kg; density in liquid state ρ_l, 750 kg/m^3; density in solid state ρ_s, 800 kg/m^3 (not used); thermal conductivity k, 0.2 W/m K; specific-heat capacity c_p, 2500 J/kg K.

For simplification of calculations, all the thermophysical properties are assumed independent of temperature. The PCM is homogeneous and isotropic as both a liquid and a solid.

Configuration of tubes

The tubes are vertical, thin, and circular and are made of aluminum, with the following dimensions: diameter d, 0.01 m; height H, 0.80 m.

The in-line configuration of the tubes is $S_p = S_n = 1.5 \times d$, as shown in Fig. 39.2, where S_n is the tube pitch normal to flow and S_p is the tube pitch parallel to flow.

39.4 Thermal Analysis and Design

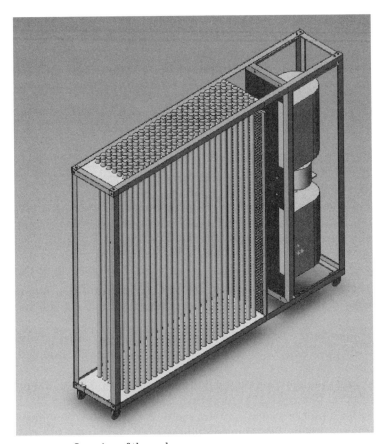

Figure 39.1 Overview of the cooler.

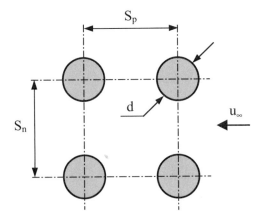

Figure 39.2 Square in-line configuration of tubes in the cooler.

Calculations

Number of tubes. The number of tubes comprises the following:

N_n Number of tubes in a row normal to flow
N_p Number of tubes in a row parallel to flow
N Total number of tubes

The width of the configuration is $W = N_n \cdot S_n = N_n \cdot (1.5 \times d) = N_n \cdot (1.5 \times 0.01) = 0.21$ m:

$$N_n = 14$$

The length of the configuration is $L = N_p \cdot S_p = N_p \cdot (1.5 \times d) = N_p \cdot (1.5 \times 0.01) = 1.35$ m:

$$N_p = 90$$

The total number of tubes in the cooler is $N = N_n \cdot N_p = 14 \times 90 = 1260$:

$$N = 1260$$

PCM mass in tubes M_{PCM}. The liquefied PCM fills each tube up to its top. The liquid density ($\rho_l = 750$ kg/m^3) is to be used in PCM mass calculations:

$$M_{PCM} = N \cdot V_{tube} \cdot \rho_l = N \cdot \left(\frac{\pi \cdot d^2}{4} \cdot H \right) \cdot \rho_l \qquad (39.1)$$

$$= 1260 \times \left[\frac{\pi \times (0.01)^2 \times 0.80}{4} \right] \times 750 = 59.37 \text{ kg} \approx 60 \text{ kg}$$

Heat-transfer area. The *heat-transfer area* A is the surface of the tubes. Because of the thinness of the tube wall, the inside and outside surfaces are the same:

$$A = N \cdot \pi \cdot d \cdot H$$

$$= 1260 \times \pi \times 0.01 \times 0.80 = 31.667 \text{ m}^2 \approx 31.67 \text{ m}^2 \qquad (39.2)$$

Convection heat-transfer coefficient outside the tubes h_o. For the crossflow of air, the Grimson correlation [8] is applied:

$$\frac{h_o \cdot d}{k_a} = C \cdot \left(\frac{u_{max} \cdot d}{\nu_a} \right)^n \cdot \text{Pr}^{1/3} \qquad (39.3)$$

All the properties in Eq. (39.3) are of air at atmospheric pressure and room temperature of 27°C (300 K): air density $\rho_a = 1.177$ kg/m^3, specific-heat capacity $c_{p,a} = 1006$ J/kg K, thermal conductivity $k_a = 0.026$ W/m K, kinematic viscosity $v_a = 15.7 \times 10^{-6}$ m²/s, and Prandtl number $\text{Pr}_a = 0.708$; $u_{\max}$ is the air velocity at the minimum frontal area. In an in-line arrangement, this is

$$u_{\max} = u_\infty \cdot \left(\frac{S_n}{S_n - d}\right) \quad (39.4)$$

Presently, for $u_\infty = 1.4$ m/s and $S_n/d = 1.5$, we have

$$u_{\max} = 1.4 \times \left(\frac{1.5}{1.5 - 1.0}\right) = 4.2 \text{ m/s}$$

The Grimson correlation [8] provides the constants of Eq. (39.3).

For an in-line configuration and the pitch-to-diameter ratio $S_n/d = S_p/d = 1.5$, we obtain $C = 0.278$ and $n = 0.620$. Substituting these values in Eq. (39.3) yields

$$\frac{h_o \times 0.01}{0.026} = 0.278 \times \left[\frac{4.2 \times 0.01}{15.7 \times 10^{-6}}\right]^{0.620} \times (0.708)^{1/3}$$

$$h_o \approx 87 \text{ W/m}^2 \text{ K}$$

Air heat capacity rate. Our PCM cooler is a shell-and-tube heat exchanger. The shell dimensions are $W \cdot H \cdot L = 0.21 \times 0.8 \times 1.35$ m³. The free cross-sectional area A_c for flow of air from the fans into the tube array is

$$A_c = W \cdot H = 0.21 \times 0.8 = 0.168 \text{ m}^2$$

The mass flow rate of air across this area is

$$\dot{m}_a = u_\infty \cdot \rho_a \cdot A_c$$

$$= 1.4 \times 1.177 \times 0.168 \approx 0.277 \text{ kg/s} \quad (39.5)$$

Thus, the air heat capacity rate is

$$C_a = \dot{m}_a \cdot c_{p,a} = 0.277 \times 1006 = 278.5 \text{ W/K}$$

Mass of air in the room. As above, the air density is $\rho_a = 1.177$ kg/m³:

$$M_a = V_{\text{room}} \cdot \rho_a = (4 \times 4 \times 2.5) \times 1.177 = 47.08 \text{ kg}$$

Operation of the PCM Cooler

The operation of the PCM cooler is analyzed for three conditions of the room in which the cooler operates:

- An insulated room
- A room heated by the ambient
- A room heated by the ambient and a heat source

Each condition is to be analyzed in two ways:

- An ideal performance of the cooler, where the wall temperature of all tubes is preserved at the constant temperature of melting T_m. This is an extreme condition.
- The actual performance of the cooler, where the temperature of a tube wall varies with the melt fraction accumulated in the tube.

Both performances are compared with respect to the temperature achievable in the room and the range of operation of the cooler.

The cooler discussed here has 90 rows of tubes normal to the airflow. The melting of the PCM advances gradually from the first row to the last one. At a given time, we may find that in the first row all the PCM has liquefied, and the wall temperature has risen above the melting point, while the tubes downstream are full of solid PCM and their wall is preserved at the melting temperature.

Room perfectly insulated

Ideal performance. The melting-point temperature is preserved throughout the whole operation at the wall of all the tubes in the cooler: $T_w = T_m$.

The lowest temperature achievable in the room could be $T_{r,\min} = T_m$. In such a case, the heat absorbed from the room by the cooler would have to be

$$Q_a = (M \cdot c_p)_a \cdot (T_{r,0} - T_m)$$
$$= 47.08 \times 1006 \times (35 - 23) = 568{,}350 \text{ J} \quad (39.6)$$

The PCM mass to be melted would have to be

$$\Delta Q_{\text{PCM}} = Q_a = (\Delta M \cdot \Delta h)_{\text{PCM}} = (\Delta M)_{\text{PCM}} \times 206{,}000 = 568{,}350 \text{ J}$$

$$(\Delta M)_{\text{PCM}} \approx 2.76 \text{ kg}$$

and the melt fraction f_m in the tubes would rise up to

$$f_m = \left(\frac{\Delta M}{M}\right)_{PCM}$$

$$= \frac{2.76}{60} = 0.046 \tag{39.7}$$

namely, 4.6 percent. At such a low melt fraction most of the tubes remain fully solid and at a temperature T_m.

The time required to reach a comfortable temperature of, say, 25°C is assessed from the heat balance over the air:

$$-(M \cdot c_p)_a \cdot \frac{dT}{dt} = (\dot{m} \cdot c_p)_a \cdot (T_r - T_m) \tag{39.8}$$

$$\Delta t = -\left(\frac{M}{\dot{m}}\right)_a \cdot \ln\left(\frac{T_r - T_m}{T_{r,0} - T_m}\right) \tag{39.9}$$

Substituting the previously obtained figures yields

$$\Delta t = -\frac{47.08}{0.277} \times \ln\left(\frac{25 - 23}{35 - 23}\right) = 304 \text{ s}$$

$$\approx 5.1 \text{ min}$$

We could expect the same figures in an actual performance.

Actual performance. The full analysis of an actual performance is presented in the "Room Heated by the Ambient" section. However, if the equations of the case described in that section are adopted to the insulated room (without external heat fluxes), we would obtain the evolution of the room temperature with time and the melt fraction in tubes. The results are presented in Fig. 39.3 [9].

The room temperature T_r versus time is plotted in Fig. 39.3a. As estimated above, the room temperature decreases from 35 to 25°C in 5 min. Thus, the ideal and actual performances are the same. Figure 39.3a also shows the room temperature reaching 23°C in about 15 to 20 min. In the insulated room, this temperature remains at that level.

Figure 39.3b depicts the melt fraction in four tube rows. In the first row the melt fraction reaches 38 percent of the tube volume. As we proceed to the fifth row, the melt fraction decreases to 28 percent, and in the fifteenth row to 10 percent. The average melt fraction in the unit is about 4.5 percent, as predicted analytically.

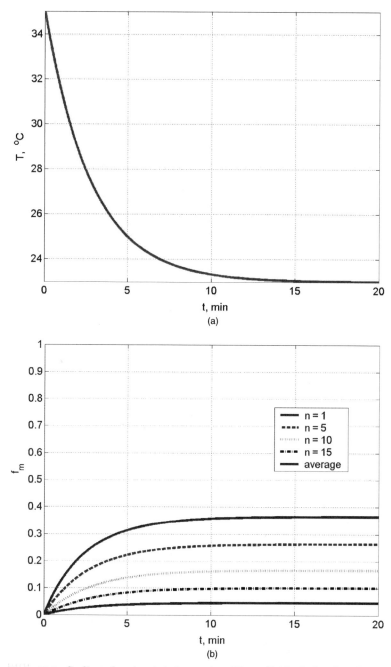

Figure 39.3 Cooling of an insulated room (see "Room Perfectly Insulated" section): (*a*) room temperature versus time; (*b*) melt fraction versus time.

39.10 Thermal Analysis and Design

Room heated by the ambient

Heat is transferred from the surroundings to the room through the walls and ceiling. The floor is adiabatic.

An overall heat-transfer coefficient U_r is assigned to both the walls and the ceiling: $U_r = 1.2$ W/m^2 K.

The ambient temperature during the whole operation of the cooler is $T_\infty = 35°$C.

Ideal performance. The heat balance over the room is

$$(M \cdot c_p)_a \cdot \frac{dT}{dt} = (A \cdot U)_r \cdot (T_\infty - T_r) - (\dot{m} \cdot c_p)_a \cdot (T_r - T_m) \qquad (39.10)$$

where the inlet temperature of air to the cooler is the room temperature, $T_{\text{in}} = T_r$ and its outlet temperature is $T_{\text{out}} = T_m$ in the very long heat exchanger:

$$\text{NTU} = \frac{A \cdot h_o}{C_a} = 10$$

At steady state, we obtain

$$(A \cdot U)_r \cdot (T_\infty - T_r)_{ss} = (\dot{m} \cdot c_p)_a \cdot (T_r - T_m)_{ss} \qquad (39.11)$$

$$T_{r,\,ss} = \frac{[(A \cdot U)_r / (\dot{m} \cdot c_p)_a] \cdot T_\infty + T_m}{1 + (A \cdot U)_r / (\dot{m} \cdot c_p)_a} \qquad (39.12)$$

where

$$A_r = 4 \times (4 \times 2.5) + 4 \times 4 = 56 \text{ m}^2$$

$$(A \cdot U)_r = 56 \times 1.2 = 67.2 \text{ W/K}$$

$$T_{r,\,ss} = \frac{(67.2/278.5) \times 35 + 23}{1 + 67.2/278.5}$$

$$\approx 25.3°\text{C}$$

At this temperature the rate of heat gains through the walls and ceiling amounts to

$$q_{\text{amb}} = (A \cdot U)_r \cdot (T_\infty - T_{r,\text{ss}})$$

$$= 67.2 \times (35 - 25.3) \approx 650 \text{ W} \qquad (39.13)$$

The duration of the cooler operation in this mode can be estimated from the heat capacity of the PCM mass:

$$Q_{\text{PCM}} = (M \cdot \Delta h_m)_{\text{PCM}}$$
$$= 60 \times 206{,}000 = 12.36 \times 10^6 \text{ J} \qquad (39.14)$$

$$\Delta t = \frac{Q_{\text{PCM}}}{q_{\text{amb}}} = \frac{12.36 \times 10^6}{650} = 19{,}015 \text{ s} = 5.28 \text{ h}$$

In this calculation the time duration of the operation is estimated as 5.3 h, ignoring the initial cooling down to the steady state, and final heating back to the initial temperature of the room. All the figures given above relate to an ideal operation in which the tubes are maintained at the melting temperature.

Actual performance. Let us consider again the same cooler and the same room as described above. The in-line array of tubes in shell consists of N_p rows normal to the air stream, where N_n tubes are arranged in each row.

Those rows of tubes filled with solid PCM gradually absorb heat from the circulated air stream. The PCM melts, and the liquid forms a concentric envelope around the central cylindrical solid in the vertical tube, due to the uniform heating and radial heat conduction. Temperature gradients along the tubes are not expected, and convective effects in the liquid PCM are assumed negligible. The liquid PCM envelope forms an axisymmetric thermal resistance between the tube wall and the interface, melting at the constant temperature T_m. As the heat absorption from the room circulating air continues, the liquid envelope grows in thickness toward the center, until all the solid PCM is liquefied. The mass fraction of liquid in the tube is the melt fraction, denoted as f_m. The largest melt fraction is formed in the first row of tubes, and it gradually decreases in the rows downstream. Phase-change material in the last row is the last material to melt. The circulating air temperature decreases downstream, as does the wall temperature T_w of the tubes. The cooled air flows back into the room and mixes uniformly with the room air. A schematic flow diagram is presented in Fig. 39.4.

Our calculation proceeds from row to row, with each row denoted by n; the last row is N_p. The heat-transfer area of a row of tubes is $A_n = A/N_p$.

The temperature of air stream flowing into row n is T_n, while at its outlet from the row it is T_{n+1}; q_n is the rate of heat absorbed in row n. The melt fraction in row n is $f_{m,n}$. The calculations are conducted at time intervals, Δt_j. The following procedure is adopted:

39.12 Thermal Analysis and Design

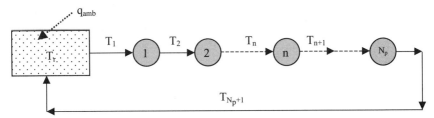

Figure 39.4 Schematic flow diagram of the "Room Heated by the Ambient" section.

1. The heat-transfer rate from the flowing air to tubes in row n, at time t is

$$q_n = (\dot{m} \cdot c_p)_a \cdot (T_n - T_{n+1}) = (A_n \cdot h_o) \cdot [T_n - T_{w,n}] \quad (39.15)$$

where $T_{w,n}$ is the tube wall temperature in row n and $q_n(t=0) = 0$.

2. The melt fraction in row n at time t is obtained as follows

$$\Delta f_{m,n} = \frac{q_n \cdot \Delta t}{Q_{\text{PCM},n}} \quad (39.16)$$

where

$$Q_{\text{PCM},n} = \frac{Q_{\text{PCM}}}{N_p} \quad (39.17)$$

and

$$f_{m,n}(t + \Delta t) = f_{m,n}(t) + \frac{q_n}{Q_{\text{PCM},n}} \cdot \Delta t \quad (39.18)$$

where $f_{m,n}(t=0) = 0$

3. The heat-transfer rate inside the PCM tube in row n at time t is

$$q_{\text{tube}} = \frac{q_n}{N_n} = -(2 \times \pi \cdot L \cdot k_{\text{PCM}}) \cdot r \cdot \frac{dT}{dr} \quad (39.19)$$

where r, which is a radial coordinate, is expressed as $r_i \leq r \leq r_o$.

Integration with the boundary conditions $T_{w,n}(r_o)$ and $T_m(r_i)$, yields

$$\frac{q_n}{N_n} = \frac{2 \times \pi \cdot L \cdot k_{\text{PCM}}}{\ln(r_o/r_i)} \cdot (T_{w,n} - T_m) \quad (39.20)$$

For a constant density of the PCM, the volumetric melt fraction is also the mass fraction. Therefore

$$f_m = 1 - \left(\frac{r_i}{r_o}\right)^2 \quad (39.21)$$

$$\frac{r_i}{r_o} = (1 - f_m)^{1/2} \quad (39.22)$$

$$r_i = r_o \quad \text{at} \quad f_{m,n}(t=0) = 0$$

Let us define the thermal resistance R_n of the liquid PCM inside a tube in row n as

$$R_n = -\frac{d}{2 \times k_{\text{PCM}}} \cdot \ln\left(\frac{1}{\sqrt{1 - f_{m,n}}}\right) \quad (39.23)$$

where $R_n(t = 0) = 0$, and the surface area is

$$A_{\text{tube}} = \frac{A_n}{N_n} = \pi \cdot d \cdot L \quad (39.24)$$

Combining Eqs. (39.15), (39.20), (39.23), and (39.24) leads to

$$q_n = A_n \cdot \frac{T_{w,n} - T_m}{R_n} = A_n \cdot h_o \cdot (T_n - T_{w,n}) \quad (39.25)$$

From Eq. (39.25) the wall temperature of tubes in row n at time t is obtained:

$$T_{w,n} = \frac{h_o \cdot R_n \cdot T_n + T_m}{1 + h_o \cdot R_n} \quad (39.26)$$

where $T_{w,n}(t = 0) = T_m$. Now, $T_{w,n}$ is substituted into Eq. (39.15), yielding finally

$$q_n = \frac{A_n \cdot h_o}{1 + h_o \cdot R_n} \cdot (T_n - T_m) = B_n \cdot (T_n - T_m) \quad (39.27)$$

where

$$B_n = \frac{A_n \cdot h_o}{1 + h_o \cdot R_n}$$

4. The heat balance over the room includes the input and output rates

$$(M \cdot c_p)_a \cdot \frac{dT_r}{dt} = (A \cdot U)_r \cdot (T_\infty - T_r) - (\dot{m} \cdot c_p)_a \cdot (T_r - T_{N_p+1}) \quad (39.28)$$

where $T_{(N_p+1)} = T_{a,\text{out}}$, which is the outlet temperature of air from the PCM cooler.

Thus

$$\Delta T_r = \frac{\Delta t}{(M \cdot c_p)_a} \cdot [(A \cdot U)_r \cdot (T_\infty - T_r) - (\dot{m} \cdot c_p)_a \cdot (T_r - T_{N_p+1})] \quad (39.29)$$

and the room temperature is

$$T_r(t + \Delta t) = T_r(t) + \Delta T_r \quad (39.30)$$

where $T_r(t = 0) = T_\infty$.

The calculations are carried out in the following sequence of steps:

1. $f_{m,n}(t=0) = 0$, $T_r(t=0) = T_{r,0} = T_1(t=0)$, $R_n(t=0) = 0$, $B_n(t=0) = A_n \cdot h_o$. At $t = t_1$ by Eq. (39.27)

$$q_1 = B_1 \cdot (T_1 - T_m) = (\dot{m} \cdot c_p)_a \cdot (T_1 - T_2) \quad \text{results in } T_2$$
$$q_2 = B_2 \cdot (T_2 - T_m) = (\dot{m} \cdot c_p)_a \cdot (T_2 - T_3) \quad \text{results in } T_3$$
$$\vdots$$
$$q_n = B_n \cdot (T_n - T_m) = (\dot{m} \cdot c_p)_a \cdot (T_n - T_{n+1}) \quad \text{results in } T_{(n+1)}$$
$$\vdots$$
$$q_{N_p} = B_{N_p} \cdot (T_{N_p} - T_m) = (\dot{m} \cdot c_p)_a \cdot (T_{N_p} - T_{N_p+1})$$

yields $T_{(N_p+1)}$, which is the outlet temperature of the circulating air from the cooler.

2. The melt fraction is calculated by Eqs. (39.16) and (39.18), and used in Eq. (39.23) for $R_n(t_1)$ and in Eq. (39.27) for $B_n(t_1)$.
3. The room temperature is obtained from Eqs. (39.29) and (39.30), and applied back in the sequence of q_n.

In our present case, a program was run in Matlab at time intervals $\Delta t = 1$ s. The results of $T_r(t)$ and $f_{m,n}(t)$ are illustrated in Fig. 39.5a and b, respectively [9].

We see in Fig. 39.5a that the room temperature of 35°C falls to 25.3°C in a few minutes, and is maintained in the room below 26°C for 5 h. The temperature of the room returns to its initial value in about 10 h. In the ideal performance the steady-state temperature of 25.3°C was predicted to prevail in the room for 5.3 h. Thus, the analytical, simplified calculations provide a good prediction of the room temperature.

Figure 39.5b illustrates the complete melting in the first row of tubes ($n = 1$) in approximately 50 min. In the rows downstream the melting starts much later and is completed in the last row in approximately 340 min.

Room heated by the ambient and a heat source

Again, only the floor is adiabatic. Heat is transferred at the walls and the ceiling. The overall heat-transfer coefficient U_r is the same as in the preceding section, and so are the ambient and initial temperatures.

The heat source releases heat into the room at a constant rate $q_{hs} = $ const.

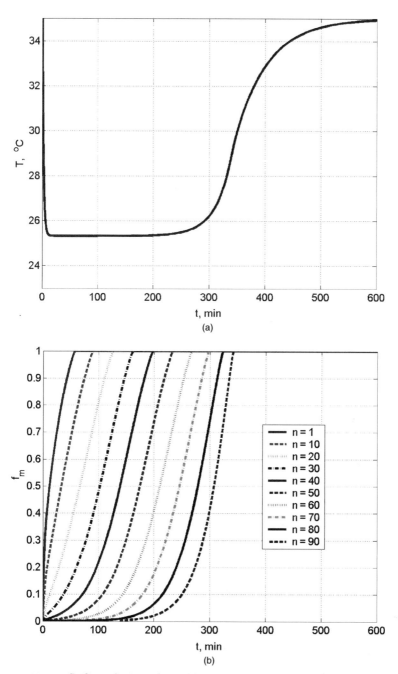

Figure 39.5 Cooling of a room heated by the ambient (see the "Room Heated by the Ambient" section): (a) room temperature versus time; (b) melt fraction versus time.

Now the heat balance over the room is

$$(M \cdot c_p)_a \cdot \frac{dT_r}{dt} = (A \cdot U)_r \cdot (T_\infty - T_r) + q_{\text{hs}} - (\dot{m} \cdot c_p)_a \cdot (T_r - T_{a,0}) \quad (39.31)$$

In an ideal performance of the PCM cooler, the outlet temperature should be $T_{a,0} = T_m$.

For such case, Eq. (39.31) is integrated, using the initial condition $T_r(t = 0) = T_{r,0}$. We obtain the explicit room temperature

$$T_r(t) = a + (T_{r,0} - a) \cdot \exp\left(-\frac{1}{b} \cdot t\right) \quad (39.32)$$

where

$$a = \frac{(A \cdot U)_r \cdot T_\infty + q_{\text{hs}} + (\dot{m} \cdot c_p)_a \cdot T_m}{(A \cdot U)_r + (\dot{m} \cdot c_p)_a} \qquad b = \frac{(M \cdot c_p)_r}{(A \cdot U)_r + (\dot{m} \cdot c_p)_a}$$

For a heat source of $q_{\text{hs}} = 500$ W, a room temperature of 27°C is reached in approximately 8 min. The operation of the cooler at this room temperature is estimated to last for approximately 3.3 h.

In an actual performance of the cooler, all the previously presented equations in the "Room Heated by the Ambient" section apply in this case as well, except for Eq. (39.30), which is being replaced by Eq. (39.31), where $T_{a,0} = T_{(N_p+1)}$.

For a heat source of $q_{\text{hs}} = 500$ W, the Matlab program yields the room temperature $T_r(t)$ and the melt fraction $f_{m,n}(t)$ presented in Fig. 39.6a and b, respectively [9].

Figure 39.6a illustrates the room temperature falling from 35°C to approximately 27°C. This temperature lasts for about 150 min (2.5 h) and then rises up in 400 min or so to 42°C, due to the internal heat source. In this particular case, after approximately 220 min heat is being transferred from the hot room to the ambient of 35°C.

The melt fraction–time plot illustrated in Fig. 39.6b again shows the same pattern of melting as in the "Room Heated by the Ambient" section, however, it is faster and completed within 220 min. These figures illustrate that the ideal performance prediction has provided a good estimate of the room temperature.

Numerical simulation of operation

The numerical simulation was conducted in a two-dimensional PCM cooler. The system was defined similarly to that described in the "Room

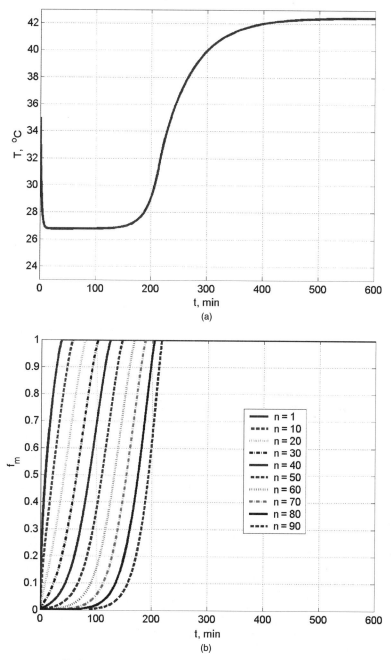

Figure 39.6 Cooling of a room heated by the ambient and a heat source (see the "Room Heated by the Ambient and a Heat Source" section): (*a*) room temperature versus time; (*b*) melt fraction versus time.

Heated by the Ambient" section, except for the tube shape and size, and the mass flow rate of air:

1. The room was of the same size. Its temperature was uniform, and it was heated in the same way, with the same heat flux and the same heat-transfer coefficient. Air properties were the same, as specified in the "Thermal Design" section.
2. Tubes were of a square cross section: 8.0 × 8.0 mm. Tube height was $H = 1.0$ m. Configuration was in-line, $S_n = S_p = 1.5$ cm, $N_n = 14$, $N_p = 90$, and $N = 1260$. Surface area of tubes was $A = 40.32$ m^2. Cooler dimensions were $H = 1.0$ m, $W = 0.21$ m, $L = 1.35$ m.
3. Phase-change material mass was $M_{\text{PCM}} = 60$ kg, as in the "Room Heated by the Ambient" section. Melting occurred at $T_m = 22$ to $24°C$. The initial temperature of the PCM was $22°C$. Properties of PCM were as specified in the "Thermal Design" section.
4. Air velocity was: $u_\infty = 1.4$ m/s, as in the "Room Heated by the Ambient" section. The mass flow rate of air was $\dot{m}_a = 0.346$ kg/s, while in the "Room Heated by the Ambient" section it was $\dot{m}_a = 0.277$ kg/s.

The convection heat-transfer coefficient around the tubes does not have to be calculated, as it is inherent in the software employed.

Although the numerically simulated system is close to that described in the "Room Heated by the Ambient" section, it is not identical, because of the enormous number of cells and computation time required to simulate a three-dimensional system of 1260 cylindrical tubes, in which change of phase takes place.

The numerical solution was obtained using Fluent 6.1 software [10]. The computational grid comprised $1403 \times 115 = 161{,}345$ cells, each cell measuring 1×1 mm in area. The time step in the calculations was 1.0 s for the first 10 min, and was increased to 5.0 s for the remaining time.

The results of the numerical simulation are presented in Fig. 39.7. Figure 39.7a illustrates the room-temperature evolution. A comfortable temperature of 24.3 to $25.5°C$ is maintained in the room for about 250 min. Comfortable temperature below $27°C$ lasts up to 300 min (5 h). These figures are close to the Matlab results of the "Room Heated by the Ambient" section. Figure 37.7b shows the process of melting at 10, 60, 120, 180, and 240 min. The solid PCM (dark area) retreats in time, while the melt (bright area) advances downstream as time passes. This process can be visualized tube by tube, and row by row.

The numerical simulation provides all the data for any comparison with the Matlab calculations. However, as the systems are not identical, the approximate comparison is sufficient.

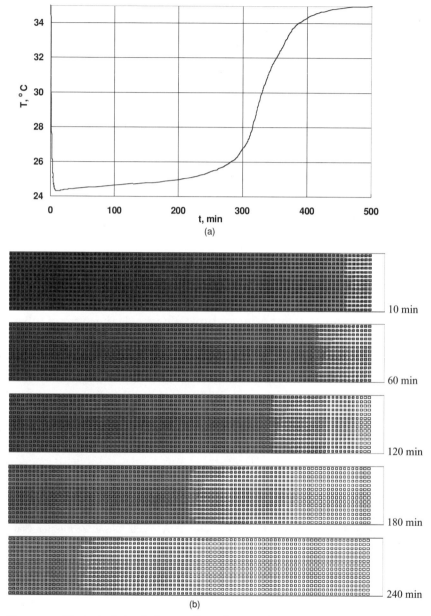

Figure 39.7 Numerical simulation—cooling of a room heated by the ambient: (*a*) room temperature versus time; (*b*) visualization of the melting process.

Closure

In all the calculations presented herein a few simplifying assumptions were involved. The density of liquid PCM was used for both liquid and solid. The room temperature was assumed uniform, as if the air were perfectly mixed. Air properties were considered constant. Radiation in the room was ignored. These simplifications affect the closeness of the results to a real situation, but enable the comparison of solutions. The analytical solution is the simplest, while the numerical simulation is a heavily time-consuming procedure.

Three cases were analyzed: cooling of an insulated room, a room heated by the ambient, and a room heated by both the ambient and a heat source. All three cases were solved analytically, for an "ideal" performance, and by Matlab for their actual performance. In all cases the analytical results were close to the computed results and could have been used for estimates of the room temperature.

The analytical and Matlab calculations related to exactly the same system. The numerical simulation was based on tubes of different shapes and sizes. It was carried out by Fluent software for the room heated by the ambient. The advantage of this solution is that we do not intervene into the flow and temperature fields inside the array of tubes. The results obtained by this method could have been the most accurate. However, in the present study the numerical simulation provides only the visualization of the melting process.

References

1. Farid, M. M., and Kanzawa, A., 1989, "Thermal Performance of a Heat Storage Module Using PCMs with Different Melting Temperatures: Mathematical Modeling," *Journal of Solar Energy Engineering*, **111**, 152–157.
2. Lacroix, M., 1993, "Numerical Solution of a Shell-and-Tube Latent Heat Thermal Energy Storage Unit," *Solar Energy*, **50**(4), 357–367.
3. Turnpenny, J. R., Etheridge, D. W., and Reay, D. A., 2000, "Novel Ventilation Cooling System for Reducing Air Conditioning in Buildings. I. Testing and Theoretical Modeling," *Applied Thermal Engineering*, **20**, 1019–1038.
4. Turnpenny, J. R., Etheridge, D. W., and Reay, D. A., 2000, "Novel Ventilation Cooling System for Reducing Air Conditioning in Buildings. II. Testing of a Prototype," *Applied Thermal Engineering*, **21**, 1203–1217.
5. Mozhevelov, S., 2004, *Cooling of Structures by a Phase-Change Material (PCM) in Natural and Forced Convection*, M.Sc. thesis, Mechanical Engineering Department, Ben-Gurion University of the Negev, Beer-Sheva, Israel.
6. Mozhevelov, S., Ziskind, G., and Letan, R., 2005, *Numerical Study of Temperature Moderation in a Real Size Room by PCM-Based Units, Heat-SET 2005: Heat Transfer in Components and Systems for Sustainable Energy Technologies*, Grenoble, France, April 5–7, 2005.
7. Arye, G., and Guedj, R., 2004, *A PCM-Based Conditioner*, final report 13-04, Heat Transfer Laboratory, Mechanical Engineering Department, Ben-Gurion University of the Negev, Beer-Sheva, Israel.
8. Holman, J. P., *Heat Transfer*, 8th ed., McGraw-Hill, New York, 1997.

9. Dubovsky, V., 2004, "Matlab Calculations," in HTL report 04-12, Heat Transfer Laboratory, Mechanical Engineering Department, Ben-Gurion University of the Negev, Beer-Sheva, Israel.
10. Mozhevelov, S., 2004, "Numerical Simulation of a PCM Unit," in HTL report 04-12, Heat Transfer Laboratory, Mechanical Engineering Department, Ben-Gurion University of the Negev, Beer-Sheva, Israel.

Chapter 40

A First-Order Thermal Analysis of Balloon-Borne Air-Cooled Electronics

Angela Minichiello
Space Dynamics Laboratory
Utah State University
North Logan, Utah

Background

The heat-transfer problem presented here is a *real* engineering problem that came to light less than 3 months prior to the final engineering design review for a multiorganizational balloon-borne science experiment. The goal of this "first-order" thermal analysis is to assert the mission viability of a proposed engineering solution. To meet the requirements of the design review and proceed with the project, adequate analytical detail is required to show that the engineering solution has a high probability of success, despite the rather uncertain and variable thermal environment in which it operates.

In addition to heat-transfer calculations, the work presented outlines the thought process used to arrive at the proposed "solution." It should be noted that several design alternatives were investigated. First-order calculations, such as those presented here, were performed for each alternative. The solution proposed at the engineering design review was identified using such calculations. Once the solution was chosen and

40.2 Thermal Analysis and Design

ultimately accepted by the review panel, detailed calculations, namely, finite-element analysis and flow network modeling, were performed to optimize and further refine the design. The detailed calculations are not presented here.

Statement of the Problem

A commercial, air-cooled computer used for data acquisition is to be part of a science instrument payload for an earth atmospheric experiment. The payload will ride aboard the gondola of a helium balloon to an altitude above sea level (ASL) of 100,000 ft, where the science instrument will collect data on the earth's atmosphere. During the experiment, the computer will process the data, record the data to its hard drive for post-float analysis, and send portions of the data to the telemetry downlink for real-time evaluation of the experiment.

Clearly, the commercial computer plays a vital role in the success of the science experiment. Should it fail, the entire payload would be rendered incapable of performing its mission. As engineers, our concern is how we can ensure the operational functionality of the computer during its mission.

Atmospheric environment and computer vendor specifications

To ensure operational functionality, it is critical to verify that the environmental conditions to which the computer will be exposed fall within the computer vendor's specifications. The operating specifications for the computer were determined from the vendor to be as follows:

Operating inlet air temperature: 0 to 50°C

Operating altitude: sea level to 10,000 ft

Operating relative humidity (RH): 10 to 90 percent noncondensing

Heat energy to be dissipated: 200 W

For comparison purposes, we determine the atmospheric air conditions at the computer's maximum allowable operating altitude of 10,000 ft from the U.S. Standard Atmosphere [1]:

Air temperature: −4.8°C (23.36°F)

Air pressure: 10.1 psia

Acceleration due to gravity: 9.80 m/s^2 (32.143 ft/s^2)

Air density: 0.905 kg/m^3 (1.756 × 10^{-3} slugs/ft^3)

Air dynamic viscosity: 1.692 × 10^{-5} N · s/m^2 (3.534 × 10^{-7} lb · s/ft^2)

These data indicate the minimum air pressure and density in which the computer can operate. Finally, we determine the atmospheric air conditions at the mission float altitude of 100,000 ft from the U.S. Standard Atmosphere [1]:

Air temperature: −46.2°C (−51.10°F)

Air pressure: 0.162 psia

Acceleration due to gravity: 9.71 m/s^2 (31.868 ft/s^2)

Air density: 0.017 kg/m^3 (3.318 × 10^{-5} slugs/ft^3)

Air dynamic viscosity: 1.478 × 10^{-5} N · s/m^2 (3.087 × 10^{-7} lb · s/ft^2)

As one would suspect, a comparison of the computer operating specifications and the atmospheric environment at 100,000 ft shows that the air-cooled computer cannot operate on its own at this altitude. While the air temperature at 100,000 ft is well below the computer's specification for inlet air temperature, the more disconcerting fact is the lack of air for cooling. This lack of air mass is seen through extremely low air pressure and density values (less than 2 percent of the 10,000 ft values) at 100,000 ft. Despite the extremely low inlet air temperature, the computer would most probably overheat because of a lack of available air for cooling!

One solution to this unique thermal management problem is to "bring some sea level atmosphere along for the ride" with the computer. To accomplish this, we require that the computer be operated within a pressure vessel during the mission. The vessel, which we call the *electronics enclosure* or *enclosure* for short, is sealed and pressurized (backfilled with dry nitrogen gas if sealed at altitudes above sea level) to a sea level pressure of 14.696 psia. It is assumed that this vessel will be structurally capable of withstanding, with an adequate margin of safety, the near one atmosphere (14.696 psia) differential pressure that it will encounter at float altitude. When operated within this enclosure, the computer operates at sea-level conditions even though the enclosure is actually floating at 100,000 ft altitude.

While this design requirement fulfills the computer's operating altitude specifications, the inlet air temperature and humidity specifications must be examined in detail; the enclosure must be designed to maintain the air within the enclosure within the vendor acceptable inlet air temperature and relative humidity requirements. Thus, the transfer of heat and its effect on air temperature and humidity must be analyzed to ensure the feasibility of this design concept.

40.4 Thermal Analysis and Design

First-Order Analytical Approach

A schematic of the proposed electronics enclosure located on the balloon gondola is shown in Fig. 40.1. From a mass/volume perspective, it is desired to make the electronics enclosure as small as possible. In this way, the balloon and gondola can rise to altitude more quickly, remain at altitude longer, and/or more ancillary equipment can be flown on the gondola. This goal, however, is directly opposed to thermal considerations; as the enclosure size (and mass) is reduced, the equilibrium internal air temperature of the enclosure will increase. Depending on the thermal boundary conditions of the enclosure during the mission, it may be difficult to maintain the air temperature within the enclosure within the allowable limits (0 to 50°C). Thus, our goal is to determine the *least massive* enclosure size that can maintain the internal air temperature and relative humidity values within required limits. In analyzing this thermal system, we first focus on determining the required enclosure size on the basis of its internal air temperature. Later, we address the relative humidity specification.

The purpose of the calculations that follow is to estimate the required size of the electronics enclosure using first-order techniques—calculations that can be performed rather quickly by hand or with the aid of a simple calculator or common computer equation solver. The equations and correlations used are readily available in the open literature and/or in standard undergraduate engineering texts. This approach allows us to quickly run through basic design parameters and reduce the set of possible solutions. It also serves to answer the question as to whether this is possible while keeping the resource investment low.

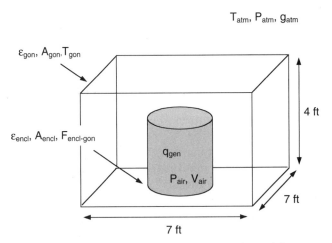

Figure 40.1 Electronics enclosure schematic inside gondola.

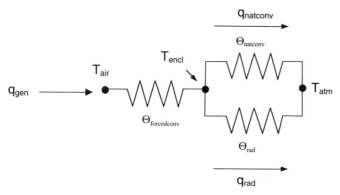

Figure 40.2 One-dimensional thermal network.

Most simply, the electronics enclosure can be modeled using one-dimensional thermal network as shown in Fig. 40.2. The heat energy dissipated by the electronic components within the computer is considered to be energy generated within the enclosure. This heat is transferred to the air within the enclosure, as it passes through the computer, via forced convection driven by fans integral to the computer. (This heat-transfer process will not be modeled, since we are concerned primarily with the temperature of the inlet air and not that of the circuit boards within the computer. Thus we treat the computer as a "black box" that generates a known amount of heat, and concern ourselves only with providing it the required air supply.) The heat is then transferred to the walls of the enclosure by the appropriate mechanism: either conduction through still air, natural convection (if the internal enclosure geometry supports the development of buoyancy-driven air currents), or forced convection (if a pressure differential is "forced" on the air inside the enclosure). Finally, heat is removed from the enclosure and transferred to its surroundings via natural convection to the atmospheric air and radiation to the gondola sidewalls. We assume that conduction between the enclosure and the gondola surface it rests on is negligible, since thermal contact resistances will be large at float altitude.

Analytical Assumptions

To perform these calculations, we make some important, simplifying assumptions about the thermal boundary conditions of the electronics enclosure and the computer itself. It may appear, at first glance, that these assumptions oversimplify the problem. They are, however, necessary to complete the calculations and suitable for their intent.

We start by assuming that the electronics enclosure is isothermal and will reach thermal equilibrium during the mission. Thus, we

40.6 Thermal Analysis and Design

perform a steady-state analysis to ultimately determine the equilibrium enclosure air temperature. The thermal environment of the enclosure consists of the atmospheric air at altitude and the gondola structure, since heat is transferred from the enclosure via natural convection to atmospheric air and radiation to the gondola structure. Atmospheric air properties are determined from the U.S. Standard Atmosphere [1] and property tables available in standard engineering textbooks [1–3]. We assume that the radiation boundary condition, the gondola structure, is isothermal and of constant temperature. While we know that, in reality, the temperature of the gondola structure varies continually according to environmental heat loads (solar, earth albedo), we attempt to span the range of variable temperatures by defining two constant boundary temperatures: a maximum and a minimum (hot-case and cold-case, respectively) gondola structure temperature. Each calculation set is run for both the hot- and cold-case gondola structure temperatures, which are identified from temperature data from previous missions. Thus, the resultant steady-state enclosure air temperatures should serve as a conservative estimate by bounding actual enclosure air temperatures. Furthermore, we assume that the radiation exchange is that of a two surface enclosure as discussed in Ref. 3. Both surfaces are considered to be opaque, diffuse, and gray.

Now, consider the thermal environment inside the enclosure. In order to maximize heat transfer, we assume that the enclosure includes a mechanism to create forced convection within in order to maximize heat transfer from the air inside to the enclosure structure. The "mechanism" will be an air mover that provides global circulation of air inside the enclosure and appropriate baffling so as to create a (nearly) circular pattern of airflow within the enclosure, similar to that shown in Fig. 40.3. Thus, heat is transferred from the air to the enclosure via *forced* convection. We assume that, because of this global circulation mechanism, the air inside the enclosure is sufficiently well mixed to be considered isothermal. We also assume that any changes in air pressure (and hence density) of the air inside the enclosure due to a temperature change at constant volume have a negligible effect on the forced-convection heat-transfer coefficient. Finally, we choose to model the enclosure itself as cylindrical in shape, since this is a common pressure vessel geometry and is easily analyzed from both the structural and thermal perspectives.

In summary, the assumptions required by the first-order analysis are:

1. *The mission is sufficiently long that the electronics enclosure will reach thermal equilibrium (steady-state conditions).* For this reason, a steady-state analysis is used to determine conservative estimates of temperature reached by the air inside the enclosure.

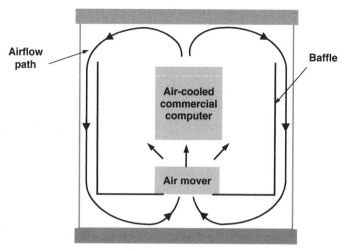

Figure 40.3 Cross section of electronics enclosure with internal airflow.

2. *The electronics enclosure is isothermal.* Temperature gradients within the enclosure walls can be investigated in follow-on finite-element analysis (FEA) studies.

3. *The gondola structure (sidewalls, ceiling, and floor) is isothermal and of known constant temperature.* The hot- and cold-case temperatures of the gondola structure are known from data from previous float missions. Temperature gradients within the gondola structure, as well as temporal variations, due to environmental heat loads can be investigated in follow-on FEA studies.

4. *Enclosure and gondola surfaces are opaque, gray, diffuse surfaces for radiation heat transfer. Atmospheric air is a nonparticipating medium.* These are common assumptions for most engineering surfaces experiencing radiative heat transfer in air. Clearly, the walls of the enclosure and the gondola are opaque. We assume that the surfaces of the enclosure are painted to increase their value of emissivity above that of bare metal, thereby increasing radiative heat transfer. Additionally, the paint will help their surface emission more closely approximate diffuse emission, since surfaces of nonconductors tend to emit more diffusely than do conductors [3]. The "gray" assumption applies since the radiative exchange is considered to be two surface exchanges between the gondola structure and the enclosure. Both surfaces have nearly equivalent spectral regions of irradiation and surface emission (since the temperature difference between them is small), largely in the infrared region of electromagnetic radiation. Thus, it is reasonable to assume that the spectral values of absorptivity and emissivity of the enclosure sidewalls and gondola structure do not vary significantly with wavelength within this region and can be considered gray.

40.8 Thermal Analysis and Design

5. *The air inside the enclosure is well mixed and isothermal.* Forced global air circulation is provided within the enclosure to enhance mixing and increase heat transfer to the enclosure. Temperature gradients within the air inside the enclosure can be investigated in follow-on computational fluid dynamics (CFD) studies, if required.

6. *The flow path of the air inside the enclosure is known.* The flow path of the air inside the enclosure is assumed to be similar to that shown in Fig. 40.3. An air mover and baffles will be required by the system design to perpetuate this flow pattern.

7. *The pressure change (due to the temperature change at constant volume) of the air inside the enclosure has a negligible effect on the enclosure's internal convective heat-transfer coefficient.* The behavior of a real gas can be modeled by the ideal-gas equation when its pressure is very low and its temperature very high compared to critical-state values. From Ref. 2 we know that, although air is a mixture of gases and does not have true critical properties, "pseudo-critical" properties can be defined using Kay's law for mixtures. Pseudo-critical properties determined in this manner for air are

$$T_c = -140.37°C \ (132.8 \ K)$$

$$P_c = 545.4 \ psi$$

where T_c is the critical temperature and P_c is the critical pressure. Thus, the air in the enclosure closely approximates ideal-gas behavior. From the ideal-gas equation $Pv = RT$, where R is the specific-gas constant for air, we can find the change in pressure P due to a change in temperature T of a constant specific volume v air mass:

$$\frac{P_1}{T_1} = \frac{P_2}{T_2} \rightarrow \frac{P_2}{P_1} = \frac{T_2}{T_1}$$

Therefore, the change in pressure due a maximum allowable temperature change of the air in the enclosure of 50°C is

$$\frac{P_2}{P_1} = \frac{50 + 273}{273} = 1.183$$

or about 18 percent over the range of allowable air temperatures. Since the air density ρ is directly proportional to the air pressure P, we can say that the density of air within the enclosure also varies by approximately 18 percent over the range of allowable air temperatures.

Modeling the axial flow over the sidewalls of the enclosure as a flat plate, we see that the average heat-transfer coefficient $\bar{h}$ is directly proportional to the Reynolds number Re_L, taken to a positive power multiplied by the Prandtl number Pr, taken to the one-third ($1/3$) power.

The value of the power of the Reynolds number depends on whether the flow is laminar or turbulent:

$$\overline{h} \propto \text{Re}_L^{1/2} \cdot \text{Pr}^{1/3} \quad \text{for laminar flow}$$

$$\overline{h} \propto \text{Re}_L^{4/5} \cdot \text{Pr}^{1/3} \quad \text{for turbulent flow}$$

where $\text{Re}_L = (\rho V L_c / \mu)$, ρ is the density of air, μ is the absolute viscosity of air, V is the velocity of the airflow, and L_c is the characteristic dimension of the flow.

Assuming the global circulation airflow within the enclosure to be laminar, the value of the heat-transfer coefficient should vary by $18^{1/2}$ percent or 4.2 percent over a maximum 50°C air temperature change (the values of V, L_c, μ, and Pr are assumed to be independent of changes in pressure). This is well within our ability to approximate the heat-transfer coefficient using standard textbook correlations (± 20 percent accuracy). Therefore, we consider the change in the convective heat-transfer coefficient to be negligible with the change in pressure inside the enclosure. Thus, we will assume the air pressure within the enclosure to remain constant at 14.696 psia.

Developing a Thermal System of Equations

Note: All calculations are performed in the SI system of units. All temperatures are in absolute units (kelvins).

Overall energy balance

From an overall energy balance on the enclosure at steady state we can write

$$\dot{E}_{in} - \dot{E}_{out} + \dot{E}_{gen} = \dot{E}_{st}$$

$$0 - \dot{E}_{out} + \dot{E}_{gen} = 0$$

$$\dot{E}_{gen} = \dot{E}_{out}$$

$$q_{gen} = q_{natconv} + q_{rad} \quad (40.1)$$

where the generated energy q_{gen} is the heat energy dissipated by the electronic components in the computer and heat energy is lost from the enclosure via natural convection and radiation.

We know from the computer specifications that

$$q_{gen} = 200 \text{ W} \quad (40.2)$$

40.10 Thermal Analysis and Design

Enclosure external heat transfer via natural convection

Heat transferred via natural convection from the outside surfaces of the enclosure can be broken down into that from two sets of surfaces: the vertical cylindrical surface and the top and bottom horizontal surfaces. The enclosure area is broken down in this manner so as to calculate heat-transfer coefficients based on correlations available in the open literature. The natural convection heat transfer can be written as

$$q_{natconv} = (\bar{h}_{vert, out} \cdot A_{vert, out} + \bar{h}_{horz, out} \cdot A_{horz, out}) \cdot (T_{encl} - T_{atm}) \quad (40.3)$$

where T_{encl} is the isothermal enclosure temperature and T_{atm} is the atmospheric air temperature outside the enclosure, known from the U.S. Standard Atmosphere data at the float altitude of 100,000 ft:

$$T_{atm} = 226.8 \text{ K} \quad (40.4)$$

It should be noted that the atmospheric air trapped inside the gondola structure will probably be warmer than the atmospheric air surrounding the gondola. We assume for these initial calculations that there is some interchange of air so that the temperature of the air inside the gondola is approximately the same as that outside. Once the calculation procedure is set, the value of T_{atm} can easily be varied to see its effect.

The external area of the enclosure's vertical sidewalls $A_{vert,ext}$ and the horizontal top surface $A_{horz,ext}$ are calculated to be

$$A_{vert,ext} = \pi \cdot D_{ext} \cdot H \quad (40.5)$$

$$A_{horz,ext} = \frac{\pi}{4} \cdot D_{ext}^2 \quad (40.6)$$

where the top and bottom covers are assumed to be the same diameter as the cylindrical sidewalls D_{ext} and H is the height of the cylinder. Note that only the top cover surface area participates in natural-convection heat transfer since the enclosure is resting on the bottom horizontal surface.

The vertical and horizontal surface external average heat-transfer coefficients are approximated from correlations published in Ref. 3. The vertical surface external average heat-transfer coefficient $\bar{h}_{vert,ext}$ is approximated using the Churchill-Chu Nusselt number relation valid for all Ra_L for an isothermal vertical surface [3]:

$$\overline{Nu}_{L_{vert, ext}} = \left\{ 0.825 + \frac{0.387 \, Ra_{L_{vert, ext}}^{1/6}}{[1 + (0.492/Pr_{atm})^{9/16}]^{8/27}} \right\}^2 \quad (40.7)$$

knowing that, in general, $\overline{\mathrm{Nu}}_L = (\overline{h} \cdot L_c)/k_{\mathrm{fl}}$, where L_c is the characteristic length of the enclosure sidewalls and k_{fl} is the thermal conductivity of the ambient fluid. As discussed in Ref. 3, this correlation can be applied to vertical cylinders when the boundary-layer thickness is much less than the cylinder diameter:

$$\frac{D_{\mathrm{ext}}}{H} \geq \frac{35}{\mathrm{Gr}_{L_{\mathrm{vert, ext}}}^{1/4}}$$

or

$$\frac{B}{C} \geq 1$$

where

$$B = \frac{D_{\mathrm{ext}}}{H} \tag{40.8}$$

$$C = \frac{35}{\mathrm{Gr}_{L_{\mathrm{vert, ext}}}^{1/4}} \tag{40.9}$$

and the Grashof number based on length for the external vertical surface $\mathrm{Gr}_{L_{\mathrm{vert,ext}}}$ is

$$\mathrm{Gr}_{L_{\mathrm{vert, ext}}} = \frac{\mathrm{Ra}_{L_{\mathrm{vert, ext}}}}{\mathrm{Pr}_{\mathrm{atm}}} \tag{40.10}$$

The Rayleigh number based on length for the external vertical surface $\mathrm{Ra}_{L_{\mathrm{vert,ext}}}$ is defined as

$$\mathrm{Ra}_{L_{\mathrm{vert, ext}}} = \frac{g_{\mathrm{atm}} \beta_{\mathrm{atm}} (T_{\mathrm{encl}} - T_{\mathrm{atm}}) L_{c,\mathrm{vert}}^3}{\alpha_{\mathrm{atm}} \nu_{\mathrm{atm}}} \tag{40.11}$$

where the characteristic length scale for the vertical external surface $L_{c,\mathrm{vert}}$ is equal to the height H of the cylindrical surface:

$$L_{c,\mathrm{vert}} = H \tag{40.12}$$

The vertical surface external average heat-transfer coefficient $\overline{h}_{\mathrm{vert, ext}}$ is finally calculated from:

$$\overline{h}_{\mathrm{vert, ext}} = \frac{\overline{\mathrm{Nu}}_{L_{\mathrm{vert, ext}}} \cdot k_{\mathrm{atm}}}{L_{c,\mathrm{vert}}} \tag{40.13}$$

The horizontal surface external average heat-transfer coefficient $\overline{h}_{\mathrm{horz, ext}}$ is approximated using a common correlation used to model the upper surface of a heated plate, valid for $10^4 \leq \mathrm{Ra}_L \leq 10^7$ [3]:

$$\overline{\mathrm{Nu}}_{L_{\mathrm{horz, ext}}} = 0.54 \, \mathrm{Ra}_{L_{\mathrm{horz, ext}}}^{1/4} \tag{40.14}$$

The Rayleigh number based on length for the external horizontal surface $\text{Ra}_{L_{\text{horz,ext}}}$ is

$$\text{Ra}_{L_{\text{horz,ext}}} = \frac{g_{\text{atm}}\,\beta_{\text{atm}}(T_{\text{encl}} - T_{\text{atm}})L_{c,\text{horz}}^3}{\alpha_{\text{atm}}\,\nu_{\text{atm}}} \quad (40.15)$$

where the characteristic length scale for the horizontal external surfaces $L_{c,\text{horz}}$ is defined:

$$L_{c,\text{horz}} = \frac{A_s}{p} = \frac{\frac{\pi}{4}D_{\text{ext}}^2}{\pi \cdot D_{\text{ext}}} = \frac{D_{\text{ext}}}{4} \quad (40.16)$$

The horizontal surface external average heat-transfer coefficient $\overline{h}_{\text{horz,ext}}$ is calculated from

$$\overline{h}_{\text{horz,ext}} = \frac{\overline{\text{Nu}}_{L_{\text{horz,ext}}} \cdot k_{\text{atm}}}{L_{c,\text{horz}}} \quad (40.17)$$

A final check should be made that the constraint $10^4 \leq \text{Ra}_{L_{\text{horz,ext}}} \leq 10^7$ holds.

To calculate $\overline{h}_{\text{vert,ext}}$ and $\overline{h}_{\text{horz,ext}}$, the gravitation constant g_{atm}, as well as the atmospheric air properties β_{atm}, α_{atm}, ν_{atm}, Pr_{atm}, and k_{atm}, must be known. The gravitation constant g_{atm} is found from the U.S. Standard Atmosphere data for the appropriate altitude of 100,000 ft:

$$g_{\text{atm}} = 9.71 \text{ m/s}^2 \quad (40.18)$$

The calculation of the atmospheric air properties at altitude is more involved since the heat-transfer correlations dictate that the fluid properties be evaluated at the "film" temperature of the air—the mean temperature of the heated surface and the ambient fluid—in order to account for property variation over the temperature range. Thus, the exterior film temperature $T_{\text{film,ext}}$ is defined:

$$T_{\text{film,ext}} = \frac{T_{\text{encl}} + T_{\text{atm}}}{2} \quad (40.19)$$

As shown in Ref. 3, the volumetric expansion coefficient for an ideal gas β can be approximated from its definition as

$$\beta = -\frac{1}{\rho}\left(\frac{\partial \rho}{\partial T}\right)_P = \frac{1}{\rho}\frac{P}{RT^2} = \frac{1}{T}$$

where T is the absolute temperature. Since air at 100,000 ft exists well above its pseudo-critical temperature and below its pseudo-critical pressure, it can be considered to behave as an ideal gas. Thus we can say that

$$\beta_{atm} = \frac{1}{T_{film,ext}} \quad (40.20)$$

Using fluid property tables found in Ref. 3, polynomial curve fits of α_{atm}, ν_{atm}, Pr_{atm}, and k_{atm} over varying temperatures are developed using standard plotting techniques and are shown in Figs. 40.4 to 40.6. The curve fits are used in the system of equations to determine the property values at the appropriate film temperature:

$$\alpha_{atm} = [(-6.5052 \cdot 10^{-19} T^3_{film,ext} + 0.00018286 \cdot T^2_{film,ext}$$
$$+ 0.030286 \cdot T_{film,ext} - 3.0771) \cdot 10^{-6}] \cdot \frac{14.696}{0.162} \quad (40.21)$$

$$\nu_{atm} = [(-9.3333 \cdot 10^{-8} T^3_{film,ext} + 0.00019429 \cdot T^2_{film,ext}$$
$$+ 0.0036619 \cdot T_{film,ext} - 0.16571) \cdot 10^{-6}] \cdot \frac{14.696}{0.162} \quad (40.22)$$

$$k_{atm} = (-3.1429 \cdot 10^{-5} T^2_{film,ext} + 0.097057 \cdot T_{film,ext} - 0.031429) \cdot 10^{-3}$$
$$(40.23)$$

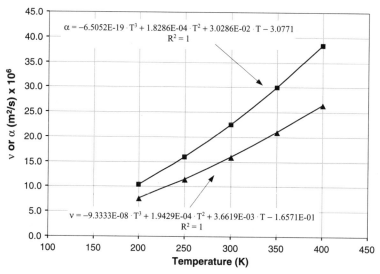

Figure 40.4 Curve fit of thermal diffusivity α and kinematic viscosity ν of air at 1 atm.

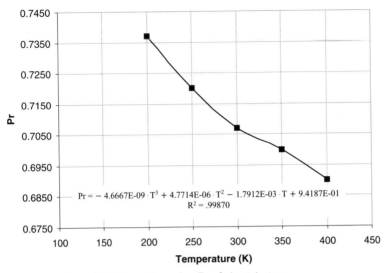

Figure 40.5 Curve fit of Prandtl number Pr of air at 1 atm.

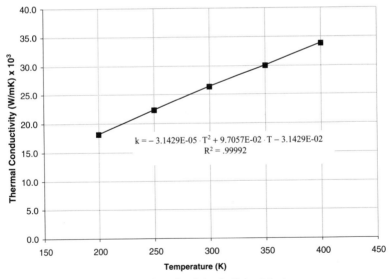

Figure 40.6 Curve fit of thermal conductivity k of air at 1 atm.

$$\text{Pr}_{\text{atm}} = -4.667 \cdot 10^{-9} T_{\text{film,ext}}^3 + 4.7714 \cdot 10^{-6} T_{\text{film,ext}}^2$$
$$-0.0017912 \cdot T_{\text{film,ext}} + 0.94187 \tag{40.24}$$

As discussed in Ref. 3, values for the thermal diffusivity α and the kinematic viscosity ν must be corrected for pressures other than 1 atm (14.696 psi). The last term in Eqs. (40.21) and (40.22) represents this correction. The values of the thermal conductivity k and Prandtl number Pr are assumed to be independent of pressure.

Enclosure external heat transfer via radiation

Referring to Fig. 40.1, the enclosure located within the gondola is modeled as a two-surface diffuse gray enclosure as described in Ref. 3. The inherent assumption exists that the gondola structure—ceiling, floor, and sidewalls—and electronics enclosure are isothermal. Thus, the net radiation leaving the electronics enclosure is

$$q_{\text{rad}} = \sigma \cdot \frac{T_{\text{encl}}^4 - T_{\text{gon}}^4}{\dfrac{1-\varepsilon_{\text{encl}}}{\varepsilon_{\text{encl}} \cdot (A_{\text{vert,ext}} + 2 \cdot A_{\text{horz,ext}})} + \dfrac{1}{(A_{\text{vert,ext}} + 2 \cdot A_{\text{horz,ext}}) F_{\text{encl-gon}}} + \dfrac{1-\varepsilon_{\text{gon}}}{\varepsilon_{\text{gon}} \cdot A_{\text{gon}}}} \tag{40.25}$$

where σ is the Stefan-Boltzmann constant:

$$\sigma = 5.67 \times 10^{-8} \frac{\text{W}}{\text{m}^2\,\text{K}^4} \tag{40.26}$$

It should be noted that both horizontal surfaces (top and bottom) of the enclosure participate in radiation; at high altitude the gap conductance between the bottom enclosure horizontal surface and the gondola floor is very low because of low air pressure and density. Thus, it is safe to assume that radiation will play a role at this interface. The factor of 2 in front of the $A_{\text{horz,out}}$ term ensures that both horizontal surfaces participate in the radiation calculation. The gondola temperature T_{gon} and gondola interior surface area A_{gon} are estimated from data of previous float missions:

$$\left.\begin{array}{l} T_{\text{gon}} = 293\,\text{K} \quad \text{hot case} \\ T_{\text{gon}} = 268\,\text{K} \quad \text{cold case} \end{array}\right\} \tag{40.27}$$

$$A_{\text{gon}} = 210\,\text{ft}^2 = 19.5\,\text{m}^2 \tag{40.28}$$

In reality, the temperature of the gondola structure is nonuniform and changing as the gondola encounters the varying thermal environment of the upper atmosphere. Thus, an attempt to bound the heat-transfer problem is made by obtaining solutions for a hot- and cold-case environment—where the gondola structure is at its highest and lowest average temperatures as indicated by data from previous missions. Because the electronics enclosure is convex, the view factor between the enclosure and the gondola $F_{\text{encl-gon}}$ is unity:

$$F_{\text{encl-gon}} = 1 \tag{40.29}$$

Additionally, the emissivities of the enclosure outer surface $\varepsilon_{\text{encl}}$ and the gondola inner surface ε_{gon} are known from previous missions to be

$$\varepsilon_{\text{encl}} = 0.85 \tag{40.30}$$

$$\varepsilon_{\text{gon}} = 0.20 \tag{40.31}$$

Currently, we have developed all equations required to calculate the effect of enclosure size (exterior diameter and height) on enclosure temperature assuming internal energy generation and external heat transfer via natural convection and radiation.

Enclosure internal air temperature

Ultimately, the goal of our calculations is to determine the required enclosure size to maintain the air internal to the enclosure within its specified temperature limits of 0 to 50°C (273 to 323 K). Therefore, another equation that relates the dissipated heat to the internal air temperature of the enclosure is required. This equation can be obtained from the one-dimensional (1D) thermal network shown in Fig. 40.2:

$$q_{\text{gen}} = \frac{T_{\text{air}} - T_{\text{encl}}}{\Theta_{\text{forcedconv}}}$$

$$\Theta_{\text{forcedconv}} = \frac{1}{\overline{h}_{\text{int}} A_{\text{int}}}$$

$$q_{\text{gen}} = \overline{h}_{\text{int}} A_{\text{int}} \cdot (T_{\text{air}} - T_{\text{encl}}) \tag{40.32}$$

where T_{air} is the "perfectly mixed" temperature of the air inside the enclosure. The internal area of the enclosure A_{int} can be obtained from the equation

$$A_{\text{int}} = [\pi \cdot (D_{\text{ext}} - 2t) \cdot H] + 2 \cdot \left[\frac{\pi}{4}(D_{\text{ext}} - 2t)^2\right] \tag{40.33}$$

where t is the thickness of the enclosure sidewalls.

The internal forced-convection heat-transfer coefficient is approximated in the same manner as are the external natural convection heat-transfer coefficients, except that an appropriate *forced-convection* heat-transfer coefficient correlation is used. For simplicity, the flow inside the enclosure is modeled as flat-plate flow down the sides of the enclosure and across the top and bottom surfaces as shown schematically in Fig. 40.3. It is assumed that the internal airflow contacts all available internal surface area. In order to add some degree of conservatism, the internal flow velocity V_{int} is assumed to remain laminar ($\text{Re}_{L_{\text{int}}} < 5 \cdot 10^5$ for flat-plate flow), where the Reynolds number for the internal flow $\text{Re}_{L_{\text{int}}}$ is defined as

$$\text{Re}_{L_{\text{int}}} = \frac{V_{\text{int}} \cdot L_{c,\text{int}}}{\nu_{\text{air}}} \qquad (40.34)$$

The correlation used to approximate the average heat-transfer coefficient inside the enclosure $\overline{h}_{\text{int}}$ is then [3]

$$\overline{\text{Nu}}_{\text{int}} = 0.664 \, \text{Re}_{L_{\text{int}}}^{1/2} \, \text{Pr}_{\text{air}}^{1/3} \qquad (40.35)$$

where

$$\overline{h}_{\text{int}} = \frac{\overline{\text{Nu}}_{\text{int}} \cdot k_{\text{air}}}{L_{c,\text{int}}} \qquad (40.36)$$

To approximate an internal heat-transfer coefficient, the internal forced velocity of the air V_{int} must be approximated or assumed. For our purposes, we assume a fairly conservative, yet laminar value:

$$V_{\text{int}} = 300 \text{ linear ft/min} = 1.5 \text{ m/s} \qquad (40.37)$$

The resulting Reynolds number should be verified to remain laminar during the calculations. The characteristic length $L_{c,\text{int}}$ of the flat plate will be defined as

$$L_{c,\text{int}} = \frac{H}{2} \qquad (40.38)$$

The internal air property values, ν_{air}, Pr_{air}, k_{air} must be determined at the appropriate film temperature $T_{\text{film,int}}$:

$$T_{\text{film,int}} = \frac{T_{\text{air}} + T_{\text{encl}}}{2} \qquad (40.39)$$

The previous air property curve fits for air at one atmosphere can be used as long as they are evaluated at $T_{\text{film,int}}$:

$$\nu_{\text{air}} = [-9.3333 \cdot 10^{-8} T_{\text{film,int}}^3 + 0.00019429 \cdot T_{\text{film,int}}^2 \\ + 0.0036619 \cdot T_{\text{film,int}} - 0.16571] \cdot 10^{-6} \qquad (40.40)$$

40.18   Thermal Analysis and Design

$$k_{air} = [-3.1429 \cdot 10^{-5} T_{film,int}^2 + 0.097057 \cdot T_{film,int} - 0.031429] \cdot 10^{-3} \tag{40.41}$$

$$\Pr_{air} = -4.667 \cdot 10^{-9} T_{film,int}^3 + 4.7714 \cdot 10^{-6} T_{film,int}^2 \\ - 0.0017912 \cdot T_{film,int} + 0.94187 \tag{40.42}$$

Note that, since the air inside the enclosure is assumed to remain constant at one atmosphere (14.696 psi) throughout the mission, ν_{air} [Eq. (40.40)] does not have to be corrected for variable pressure effects.

Solving the Thermal System of Equations

We have developed the complete set of equations for determining the air temperature within a cylindrical enclosure given the enclosure diameter, height, and sidewall thickness. For clarity, the system of equations is listed below:

$$q_{gen} = q_{natconv} + q_{rad} \tag{40.1}$$

$$* \, q_{gen} = 200 \tag{40.2}$$

$$q_{natconv} = (\bar{h}_{vert,ext} \cdot A_{vert,ext} + \bar{h}_{horz,ext} \cdot A_{horz,ext}) \cdot (T_{encl} - T_{atm}) \tag{40.3}$$

$$* \, T_{atm} = 226.8 \tag{40.4}$$

$$A_{vert,ext} = \pi \cdot D_{ext} \cdot H \tag{40.5}$$

$$A_{horz,ext} = \frac{\pi}{4} \cdot D_{ext}^2 \tag{40.6}$$

$$\overline{Nu}_{L_{vert,ext}} = \left\{ 0.825 + \frac{0.387 \, Ra_{L_{vert,ext}}^{1/6}}{[1 + (0.492/\Pr_{atm})^{9/16}]^{8/27}} \right\}^2 \tag{40.7}$$

$$B = \frac{D_{ext}}{H} \tag{40.8}$$

$$C = \frac{35}{Gr_{L_{vert,ext}}^{1/4}} \tag{40.9}$$

$$Gr_{L_{vert,ext}} = \frac{Ra_{L_{vert,ext}}}{\Pr_{atm}} \tag{40.10}$$

$$Ra_{L_{vert,ext}} = \frac{g_{atm} \beta_{atm} (T_{encl} - T_{atm}) L_{c,vert}^3}{\alpha_{atm} \nu_{atm}} \tag{40.11}$$

$$L_{c,vert} = H \tag{40.12}$$

$$\overline{h}_{\text{vert,ext}} = \frac{\overline{\text{Nu}}_{L_{\text{vert,ext}}} \cdot k_{\text{atm}}}{L_{c,\text{vert}}} \tag{40.13}$$

$$\overline{\text{Nu}}_{L_{\text{horz,ext}}} = 0.54 \text{Ra}_{L_{\text{horz,ext}}}^{1/4} \tag{40.14}$$

$$\text{Ra}_{L_{\text{horz,ext}}} = \frac{g_{\text{atm}} \beta_{\text{atm}} (T_{\text{encl}} - T_{\text{atm}}) L_{c,\text{horz}}^3}{\alpha_{\text{atm}} \nu_{\text{atm}}} \tag{40.15}$$

$$L_{c,\text{horz}} = \frac{D_{\text{ext}}}{4} \tag{40.16}$$

$$\overline{h}_{\text{horz,ext}} = \frac{\overline{\text{Nu}}_{L_{\text{horz,ext}}} \cdot k_{\text{atm}}}{L_{c,\text{horz}}} \tag{40.17}$$

$$* g_{\text{atm}} = 9.71 \tag{40.18}$$

$$T_{\text{film,ext}} = \frac{T_{\text{encl}} + T_{\text{atm}}}{2} \tag{40.19}$$

$$\beta_{\text{atm}} = \frac{1}{T_{\text{film,ext}}} \tag{40.20}$$

$$\alpha_{\text{atm}} = [(-6.5052 \times 10^{-19} T_{\text{film,ext}}^3 + 0.00018286 \cdot T_{\text{film,ext}}^2 \\ + 0.030286 \cdot T_{\text{film,ext}} - 3.0771) \times 10^{-6}] \cdot \frac{14.696}{0.162} \tag{40.21}$$

$$\nu_{\text{atm}} = [(-9.3333 \times 10^{-8} T_{\text{film,ext}}^3 + 0.00019429 \cdot T_{\text{film,ext}}^2 \\ + 0.0036619 \cdot T_{\text{film,ext}} - 0.16571) \times 10^{-6}] \cdot \frac{14.696}{0.162} \tag{40.22}$$

$$k_{\text{atm}} = (-3.1429 \times 10^{-5} T_{\text{film,ext}}^2 \\ + 0.097057 \cdot T_{\text{film,ext}} - 0.031429) \times 10^{-3} \tag{40.23}$$

$$\text{Pr}_{\text{atm}} = -4.667 \times 10^{-9} T_{\text{film,ext}}^3 + 4.7714 \cdot 10^{-6} T_{\text{film,ext}}^2 \\ - 0.0017912 \cdot T_{\text{film,ext}} + 0.94187 \tag{40.24}$$

$$q_{\text{rad}} = \sigma \cdot \frac{T_{\text{encl}}^4 - T_{\text{gon}}^4}{\dfrac{1-\varepsilon_{\text{encl}}}{\varepsilon_{\text{encl}} \cdot (A_{\text{vert,ext}} + 2 \cdot A_{\text{horz,ext}})} + \dfrac{1}{(A_{\text{vert,ext}} + 2 \cdot A_{\text{horz,ext}}) F_{\text{encl-gon}}} + \dfrac{1-\varepsilon_{\text{gon}}}{\varepsilon_{\text{gon}} \cdot A_{\text{gon}}}} \tag{40.25}$$

$$* \sigma = 5.67 \times 10^{-8} \tag{40.26}$$

$$*T_{\text{gon}} = 293 \atop *T_{\text{gon}} = 268 \Bigg\} \tag{40.27}$$

$$*A_{\text{gon}} = 19.5 \tag{40.28}$$

$$*F_{\text{encl-gon}} = 1 \tag{40.29}$$

$$*\varepsilon_{\text{encl}} = 0.85 \tag{40.30}$$

$$*\varepsilon_{\text{gon}} = 0.20 \tag{40.31}$$

$$q_{\text{gen}} = \overline{h}_{\text{int}} A_{\text{int}} \cdot (T_{\text{air}} - T_{\text{encl}}) \tag{40.32}$$

$$A_{\text{int}} = [\pi \cdot (D_{\text{ext}} - 2t) \cdot H] + 2 \cdot \left[\frac{\pi}{4}(D_{\text{ext}} - 2t)^2\right] \tag{40.33}$$

$$\text{Re}_{L\,\text{int}} = \frac{V_{\text{int}} \cdot L_{c,\text{int}}}{\nu_{\text{air}}} \tag{40.34}$$

$$\overline{\text{Nu}}_{\text{int}} = .664 \, \text{Re}_{\text{int}}^{1/2} \, \text{Pr}_{\text{air}}^{1/3} \tag{40.35}$$

$$\overline{h}_{\text{int}} = \frac{\overline{\text{Nu}}_{\text{int}} \cdot k_{\text{air}}}{L_{c,\text{int}}} \tag{40.36}$$

$$*V_{\text{int}} = 1.5 \tag{40.37}$$

$$L_{c,\text{int}} = \frac{H}{2} \tag{40.38}$$

$$T_{\text{film,int}} = \frac{T_{\text{air}} + T_{\text{encl}}}{2} \tag{40.39}$$

$$\nu_{\text{air}} = [-9.3333 \cdot 10^{-8} T_{\text{film,int}}^3 + 0.00019429 \cdot T_{\text{film,int}}^2 \\ + 0.0036619 \cdot T_{\text{film,int}} - 0.16571] \times 10^{-6} \tag{40.40}$$

$$k_{\text{air}} = [-3.1429 \times 10^{-5} T_{\text{film,int}}^2 + 0.097057 \times T_{\text{film,int}} \\ - 0.031429] \times 10^{-3} \tag{40.41}$$

$$\text{Pr}_{\text{air}} = -4.667 \times 10^{-9} T_{\text{film,int}}^3 + 4.7714 \times 10^{-6} T_{\text{film,int}}^2 \\ - 0.0017912 \cdot T_{\text{film,int}} + 0.94187 \tag{40.42}$$

$$*D_{\text{ext}} = \text{user-defined} \tag{40.43}$$

$$*H = \text{user-defined} \tag{40.44}$$

$$*t = \text{user-defined} \tag{40.45}$$

Within this system of equations, there are 13 knowns, each marked with an asterisk, and 32 unknowns and equations. The system can be solved using any standard software package that solves simultaneous equations. For the purposes of this text, the equations were solved using Interactive Heat Transfer (IHT) software [5]. It is important to remember that, for each solution set, the following checks must be made to ensure the validity of the assumptions going into the analysis:

1. Check the validity of the Churchill-Chu Nusselt number relation for natural convection on the vertical surfaces of the enclosure:

$$\frac{D_{ext}}{H} \geq \frac{35}{Gr_{L_{vert, ext}}^{1/4}} \rightarrow \frac{B}{C} \geq 1$$

2. Check that the calculated Rayleigh number used in the horizontal Nusselt number relation for natural convection off the top of the enclosure holds in the required range:

$$10^4 \leq Ra_{L_{horz, ext}} \leq 10^7$$

3. Check that the laminar flow assumption holds for the airflow inside the enclosure:

$$Re_{int} \leq 5 \times 10^5$$

If any of these checks fail for a given run, a more representative heat-transfer correlation should be found for that set of input values. Additionally, it should be noted that while IHT software contains integral subroutines to calculate air property values, the air property curve fits were used in order to present a solution method that can be used with any generic equation solver.

Thermal Calculation Results

The equation set as programmed into IHT software is shown in Fig. 40.7. Solution sets for the inputs of $D_{ext} = 3$ ft, $H = 2.5$ ft, and $t = 3/16$ in. for the hot-case ($T_{gon} = 293$ K) and cold-case ($T_{gon} = 268$ K) gondola temperatures are shown in Figs. 40.8 and 40.10, respectively. The air temperature within the enclosure reached 314.5 K (41.5°C) for the hot case and 295.4 K (22.4°C) for the cold case. Since both of these results are within the 0 to 50°C inlet air temperature specification for the computer, the proposed enclosure dimensions are feasible. Plots of air and enclosure temperatures for varying values of enclosure diameter D_{ext} at $H = 2.5$ ft and $t = 3/16$ in. are shown in Figs. 40.9 and 40.11 for the hot and cold cases, respectively. Plots such as these are valuable

40.22 Thermal Analysis and Design

```
//Overall energy balance

qgen=qnatconv+qrad
qgen=200

//Natural convection on the outside of the enclosure

qnatconv=(hbarvertext*Avertext+hbarhorzext*Ahorzext)*(Tencl-Tatm)
Tatm=226.8
Avertext=3.14*Dext*H
Ahorzext=(3.14/4)*Dext^2
NuLbarvertext=(0.825+(0.387*(RaLvertext^(1/6)))/(1+((0.492/Pratm)^(9/16)))^(8/27))^2
B=Dext/H
C=35/GrL^(1/4)
GrL=RaLvertext/Pratm
RaLvertext=(gatm*Betaatm*(Tencl-Tatm)*Lcvert^3)/(alphaatm*nuatm)
Lcvert=H
hbarvertext=NuLbarvertext*katm/Lcvert
NuLbarhorzext=0.54*RaLhorzext^(1/4)
RaLhorzext=(gatm*Betaatm*(Tencl-Tatm)*Lchorz^3)/(alphaatm*nuatm)
Lchorz=Dext/4
hbarhorzext=NuLbarhorzext*katm/Lchorz
gatm=9.71
Tfilmext=(Tencl+Tatm)/2
Betaatm=1/Tfilmext
alphaatm=(((-6.5052*10^(-19)*Tfilmext^3)+(0.00018286*Tfilmext^2)+(0.030286*Tfilmext)-3.07710)*10^(-6))*(14.696/.162)
nuatm=(((-9.3333*10^(-8)*Tfilmext^3)+(0.00019429*Tfilmext^2)+(0.0036619*Tfilmext)-0.16571)*10^(-6))*(14.696/.162)
katm=((-3.1429*10^(-5)*Tfilmext^2)+(0.097057*Tfilmext)-0.031429)*10^(-3)
Pratm=-4.667*10^(-9)*Tfilmext^3+4.7714*10^(-6)*Tfilmext^2-0.0017912*Tfilmext+0.94187

//Radiation from outside the enclosure

qrad=(sigma*(Tencl^4-Tgon^4))/(((1-
epsilonencl)/(epsilonencl*(Avertext+2*Ahorzext)))+(1/((Avertout+2*Ahorzext)*Fenclgon))+((1-
epsilongon)/(epsilongon*Agon)))

sigma=5.678*10^(-8)
Tgon=293
Agon=19.5
Fenclgon=1
epsilonencl=.85
epsilongon=.2

//Convection inside the enclosure

qgen=hbarint*Aint*(Tair-Tencl)
Aint=(3.14*(Dout-2*t)*H)+2*(3.14/4*(Dout-2*t)^2)
ReLint=Vint*Lcint/nuair
NuLbarint=.664*ReLint^(1/2)*Prair^(1/3)
hbarint=NuLbarint*kair/Lcint
Vint=1.5
Lcint=H/2
Tfilmint=(Tair+Tencl)/2
nuair=(((-9.3333*10^(-8)*Tfilmint^3)+(0.00019429*Tfilmint^2)+(0.0036619*Tfilmint)-0.16571)*10^(-6))
kair=((-3.1429*10^(-5)*Tfilmint^2)+(0.097057*Tfilmint)-0.031429)*10^(-3)
Prair=-4.667*10^(-9)*Tfilmint^3+4.7714*10^(-6)*Tfilmint^2-0.0017912*Tfilmint+0.94187


//Input
Dext=3*.3048
H=2.5*.3048
t=(3/16)*(.3048/12)
```

Figure 40.7 IHT equation set. // denotes a comment statement.

Ahorzext	0.6564	hbarhorzext	0.6392
Aint	3.451	hbarint	4.461
Avertext	2.188	hbarvertext	0.4565
B	1.2	kair	0.02688
Betaatm	0.003786	katm	0.02341
C	1.129	nuair	1.667E-5
GrL	9.233E5	nuatm	0.001147
Lchorz	0.2286	qnatconv	106
Lcint	0.381	qrad	94.01
Lcvert	0.762	Agon	19.5
NuLbarhorzext	6.241	Dext	0.9144
NuLbarint	63.22	Fenclgon	1
NuLbarvertext	14.86	H	0.762
Prair	0.7064	Tatm	226.8
Pratm	0.7156	Tgon	293
RaLhorzext	1.784E4	Vint	0.5
RaLvertext	6.607E5	epsilonencl	0.85
ReLint	1.143E4	epsilongon	0.2
Tair	314.5	gatm	9.71
Tencl	301.5	qgen	200
Tfilmext	264.2	sigma	5.678E-8
Tfilmint	308	t	0.004763
alphaatm	0.001604		

Figure 40.8 IHT solution, hot case for $D_{ext} = 3$ ft, $H = 2.5$ ft, $t = 3/16$ in.

in determining parameter trends and bounds for a feasible design. Similar plots can be made for varying H and t, as well as other parameters.

Developing the Humidity Equations

Now that we have calculated the steady-state internal air temperature for a given volume enclosure, we calculate the resultant percent

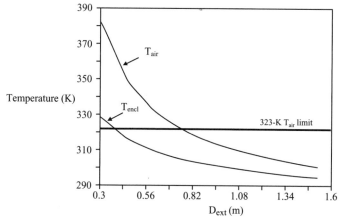

Figure 40.9 IHT solution set plot, hot case for $D_{ext} = 1$ to 5 ft, $H = 2.5$ ft, $t = 3/16$ in.

40.24 Thermal Analysis and Design

Ahorzext	0.6564	hbarhorzext	0.5996
Aint	3.451	hbarint	4.469
Avertext	2.188	hbarvertext	0.4278
B	1.2	kair	0.02538
Betaatm	0.003928	katm	0.02264
C	1.165	nuair	1.486E-5
GrL	8.147E5	nuatm	0.001072
Lchorz	0.2286	qnatconv	73.93
Lcint	0.381	qrad	126.1
Lcvert	0.762	Agon	19.5
NuLbarhorzext	6.054	Dext	0.9144
NuLbarint	67.08	Fenclgon	1
NuLbarvertext	14.4	H	0.762
Prair	0.7101	Tatm	226.8
Pratm	0.7181	Tgon	268
RaLhorzext	1.58E4	Vint	0.5
RaLvertext	5.85E5	epsilonencl	0.85
ReLint	1.282E4	epsilongon	0.2
Tair	295.4	gatm	9.71
Tencl	282.4	qgen	200
Tfilmext	254.6	sigma	5.678E-8
Tfilmint	288.9	t	0.004763
alphaatm	0.001496		

Figure 40.10 IHT solution, cold case for $D_{ext} = 3$ ft, $H = 2.5$ ft, $t = 3/16$ in.

relative humidity at that temperature using a typical psychometric analysis as described in Ref. 6. *Psychometrics* is the study of atmospheric air, considered to be a mixture of two ideal gases: dry air and water vapor. Our goal in this part of the analysis is to check that the relative humidity at the internal air temperatures calculated in the

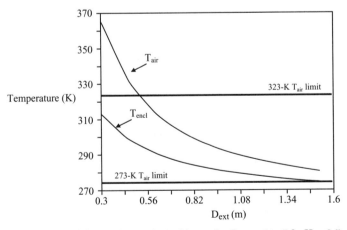

Figure 40.11 IHT solution set plot, cold case for $D_{ext} = 1$ to 5 ft, $H = 2.5$ ft, $t = 3/16$ in.

Figure 40.12 Thermodynamic states of the enclosure.

previous section is within the product specifications (10 to 90 percent, noncondensing).

The thermodynamic initial and final states of the air inside the enclosure are shown in Fig. 40.12. It is assumed that the initial conditions of the air inside the enclosure (as the enclosure is sealed prior to the mission) is known; thus the initial dry-bulb temperature $T_{db,i}$ and the initial percent relative humidity ϕ_i are known. The final air dry-bulb temperature $T_{db,f}$ is known from the calculations of T_{air} of the previous section. Note that dry-bulb temperatures are the temperatures that a "dry" thermometer would read if exposed to atmospheric air and thus, at any instant, the temperatures of the air and water vapor are equal:

$$T_{db,i} = T_{wvap,i} \tag{40.46}$$

$$T_{air} = T_{db,f} = T_{wvap,f} \tag{40.47}$$

where $T_{wvap,i}$ and $T_{wvap,f}$ are the initial and final temperatures of the water vapor in the enclosure. The percent relative humidity ϕ defined as the partial pressure of the water vapor P_{wvap} divided by the water vapor saturation pressure $P_{sat,wvap}$:

$$\phi = \frac{P_{wvap}}{P_{sat,wvap}}$$

The water vapor saturation pressure is read from the steam tables [7] as the pressure corresponding to the dry-bulb temperature of the atmospheric air. The enclosure volume V_{encl} is calculated as

$$V_{encl} = \frac{\pi}{4} \cdot (D_{ext} - 2 \cdot t)^2 \cdot H \tag{40.48}$$

where the values of D_{ext}, H, and t are the inputs of the calculations of the previous section.

Note that, since atmospheric air is assumed to be a homogenous mixture of dry air and water vapor, the initial and final volumes of the water vapor $V_{\text{wvap},i}$ and $V_{\text{wvap},f}$ are both equal to the volume of the enclosure:

$$V_{\text{encl}} = V_{\text{wvap},i} = V_{\text{wvap},f} \tag{40.49}$$

To perform the humidity calculations, the initial water vapor pressure $P_{\text{wvap},i}$ of the air inside the enclosure is first calculated from the initial value of percent relative humidity ϕ_i and the initial temperature $T_{\text{db},i}$ of the air in the enclosure. For the purposes of this text, we assume the initial percent relative humidity and air temperature to be

$$\phi_i = 0.30 \tag{40.50}$$

$$T_{\text{db},i} = 20°C = 293 \text{ K} \tag{40.51}$$

We can calculate the initial partial pressure of the water vapor $P_{\text{wvap},i}$

$$P_{\text{wvap},i} = \phi_i \cdot P_{\text{sat,wvap},i} \tag{40.52}$$

where the water vapor saturation pressure $P_{\text{sat,wvap},i}$ is read from the steam tables as the pressure corresponding to $T_{\text{db},i}$. The initial mass of water within the enclosure $m_{\text{wvap},i}$ is then calculated from the ideal-gas equation

$$m_{\text{wvap},i} = \frac{P_{\text{wvap},i} \cdot V_{\text{wvap},i}}{R_w \cdot T_{\text{wvap},i}} \tag{40.53}$$

where $T_{\text{wvap},i}$ is in absolute units (K) and R_w is the specific-gas constant for water

$$R_w = \frac{R^*}{\text{MW}_w} \tag{40.54}$$

where R^* is the universal gas constant ($R^* = 8314.3$ J/kmol · K) and MW_w is the molecular weight of water ($\text{MW}_w = 18.016$ kg/kmol).

Thermodynamically, to get from the initial state to the final state, heat is added to the system. The mass of the water vapor is unchanged during this process (the enclosure is a sealed vessel) such that

$$m_{\text{wvap},i} = m_{\text{wvap},f} \tag{40.55}$$

Thus, knowing the final dry-bulb temperature of the air $T_{\text{db},f}$, one can calculate the water vapor pressure at the final state:

$$P_{\text{wvap},f} = \frac{m_{\text{wvap},f} \cdot R_w \cdot T_{\text{wvap},f}}{V_{\text{wvap},f}} \tag{40.56}$$

TABLE 40.1 Humidity Calculation Results
($D_{ext} = 3$ ft, $H = 2.5$ ft, $t = 3/16$ in.)

Case	T_{gon}, K	ϕ_i (assumed)	$T_{db,i}$ (assumed), K	$m_{wvap,i} = m_{wvap,f}$, kg	$T_{db,f}$	ϕ_f
Hot	293	0.30	293	0.00234	314.5	0.094
Cold	268	0.30	293	0.00234	295.4	0.260

The percent relative humidity of the air inside the enclosure at its final, steady-state temperature is then:

$$\phi_f = \frac{P_{wvap,f}}{P_{sat,wvap,f}} \tag{40.57}$$

where $P_{sat,wvap,f}$ is found from the steam tables at $T_{db,f}$.

Humidity Calculation Results

Results from humidity calculations for the hot- and cold-case enclosure internal air temperatures at $D_{ext} = 3$ ft, $H = 2.5$ ft, and $t = 3/16$ in. are shown in Table 40.1. These results predict that the percent relative humidity of the air inside the enclosure drops from 30 to 9.4 percent for the hot case and 26 percent for the cold case. Because the hot-case value is slightly below the computer specification for percent relative humidity of 10 to 90 percent, the enclosure design should be modified so as to reduce the hot-case air temperature and increase the resultant hot-case percent relative humidity.

Follow-on Analyses

The first-order calculation procedures presented here provide a means of estimating the effect of the enclosure size on the temperature and relative humidity of the air inside the electronics enclosure. Thus, we have a means for determining the feasibility of the proposed design: the likelihood that the computer will function properly during the mission. It is important to realize that these calculations are not an end, but rather a beginning, in fully solving this engineering problem. They serve to focus the system design to the few most feasible options for the enclosure design. Using this approach, constant-temperature boundary conditions, surface emissivity values, power levels, and geometric parameters can be quickly varied to determine their effects on the final values of interest.

Once an enclosure design that meets all requirements is attained, more detailed analyses should be pursued to optimize and refine the design. Finite-element calculations can be pursued to take into account the temperature variations with the gondola structure and the enclosure walls. Constant-temperature boundary conditions can be replaced by pursuing orbital modeling techniques, which accurately model environmental heat loads, and transient analysis solutions available with many FEA radiation solvers. Flow network modeling or CFD techniques can be used to refine internal airflow patterns and optimize heat transfer within the enclosure. The basic calculations shown here serve as the springboard for more detailed analyses to come.

Nomenclature

A	Area
B	D_{ext}/H
C	$35/\text{Gr}_{L_{\text{vert,out}}}^{1/4}$
D	Diameter
$\dot{E}$	Energy per unit time
F	Geometric view factor
g	Acceleration due to gravity
Gr	Grashof number
$\bar{h}$	Average heat-transfer coefficient
H	Height
k	Thermal conductivity
m	Mass
MW	Molecular weight
$\overline{\text{Nu}}$	Average Nusselt number
p	Perimeter
P	Pressure
Pr	Prandtl number
q	Heat-transfer rate
R	Specific gas constant
R^*	Universal gas constant
Ra	Rayleigh number
Re	Reynolds number
T	Temperature
t	Wall thickness
v	Specific volume
V	Velocity
$\forall$	Volume

Greek

α	Thermal diffusivity
β	Volumetric expansion coefficient
ε	Emissivity
Θ	Thermal resistance
μ	Absolute or dynamic viscosity
ν	Kinematic viscosity
ρ	Density
σ	Stefan-Boltzmann constant
ϕ	Relative humidity

Subscripts

1	State 1
2	State 2
air	Air inside the enclosure
atm	Atmospheric
c	Critical
db	Dry bulb
encl	Enclosure
encl-gon	Enclosure to gondola
ext	Exterior
f	Final
fl	Fluid
film	Film
forcedconv	Forced convection
gen	Generated
gon	Gondola
horz	Horizontal
i	Initial
in	Into the control volume
int	Internal
L	Length
natconv	Natural convection
out	Out of the control volume
rad	Radiation
s	Surface
st	Stored

sat	Saturated
vert	Vertical
w	Water
wvap	Water vapor

References

1. Munson, B., Young, D., and Okiishi, T., *Fundamentals of Fluid Mechanics*, 2d ed., Wiley, New York, 1994.
2. Black, W. Z., and Hartley, J. G., *Thermodynamics*, 2d ed., HarperCollins, New York, 1991.
3. Incropera, F. P., and DeWitt, D. P., *Fundamentals of Heat and Mass Transfer*, 5th ed., Wiley, Hoboken, N. J., 2002.
4. Ellison, G. N., *Thermal Computations for Electronic Equipment*, Van Nostrand-Reinhold, New York, 1984.
5. Interactive Heat Transfer (IHT) Software, version 2.0, to accompany *Fundamentals of Heat and Mass Transfer*, 5th ed., Wiley, New York and Intellipro Inc., 2001.
6. Lindeburg, M. R., *Mechanical Engineering Reference Manual for the PE Exam*, 11th ed., Professional Publications, 2001, pp. 38.1–38.4.
7. Keenan, J. H., Keyes, F. G., Hill, P. G., and Moore, J. G., *Steam Tables (International System of Units—S.I.)*, Krieger, New York, 1992.

Chapter 41

Thermal Analysis of Convectively Cooled Heat-Dissipating Components on Printed-Circuit Boards

Hyunjae Park
Thermofluid Science and Energy Research Center (TSERC)
Department of Mechanical Engineering
Marquette University
Milwaukee, Wisconsin

Problem Statement and Objective

One of the most critical concerns in the electronic packaging industry is the effective removal of the heat generated by surface-mounted components in order to reduce thermal stresses and ensure reliability. The printed-circuit board (PCB) thermal management issue requires that device junction temperatures be maintained below maximum point for reliability considerations. Among the various techniques employed to cool the heating components, forced air cooling is commonly utilized because of its simplicity. Air cooling is common and a low-cost, effective method of thermal management of PCBs.

The physical system evaluated is based on a typical industrial electronic device: PCBs held in housings or PCB racks (see Fig. 41.1). The flow is uniform and unidirectional as it enters the channel. The model

41.2 Thermal Analysis and Design

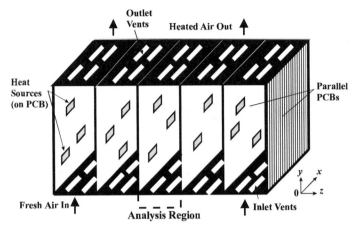

Figure 41.1 Schematic diagram of PCB channel.

considered two parallel plates; the lower plate had arbitrary (in-line) arranged discrete heat sources and the upper plate was adiabatic. This is shown in the side view of Fig. 41.2. The spacing between heat sources is considered adiabatic, and the two-dimensional flow is laminar.

Analytical solutions for forced-convection heat transfer over a flat plate with arbitrary surface boundary conditions can be obtained using the method of superposition [1] because of the linearity of the boundary-layer energy equation. Figliola and Thomas [2] and Park and Tien [3] used the superposition method to develop the approximate solutions for the forced- and natural-convection heat-transfer problem having discrete heat sources along the plate.

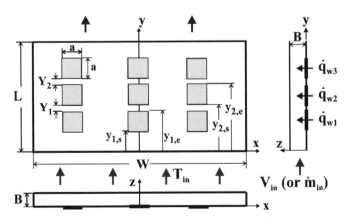

Figure 41.2 Schematic of flow channel and discrete heat-dissipating components.

This work discusses the development and use of an approximate analysis model for predicting the temperature distribution on a PCB containing embedded heat-dissipating electronic components cooled by laminar airflow induced by a fan. The direction of the airflow is parallel to the PCB. Temperature results obtained from the model are presented in dimensionless curves as a function of the PCB design parameters, including the board dimensions, air velocity (or flow rate), and number, position, and heat dissipation of the components. These curves enable the engineer to quickly select the optimal values of the design parameters which meet the design specification imposed or the maximum allowable board wall temperature. Information gained from these curves can be employed to establish a set of guidelines (steps) for the thermal design and layout of the PCB and components.

Problem Formulation and Analysis

An approximate method for the forced-convection heat transfer for flow between parallel plates having (embedded) discrete heat sources is presented in this work. As a simplification, each heat-dissipating component is assumed to have uniform, constant surface heat flux, although the values for heat flux and heat-source length may differ between sources. Important variables included in the formulation are the channel spacing and length, heater spacing and size, the heat flux variation among sources, and airflow rate (or velocity). The analysis is performed to coincide with modern electronic equipment in geometric and operating similarity. Table 41.1 shows the typical input used in this work. These values are used as a base-case condition. Furthermore, this work considers the case where the hydrodynamic and thermal entry lengths (L_v and L_T, respectively) defined by Eq. (41.1) are sufficiently larger than the PCB channel length (L)

$$L \ll L_v \approx 0.05\,\mathrm{Re}_{D_h} D_h$$
$$L \ll L_T \approx 0.05\,\mathrm{Re}_{D_h} D_h\,\mathrm{Pr} \tag{41.1}$$

TABLE 41.1 Typical PCB Dimensions and Analysis Input Used in This Work

PCB channel length L	100 mm
PCB channel width W	200 mm
Heater dimensions $a \times a$	20 × 20 mm
Channel spacing B	24 mm
Inlet air velocity V_{in}	1 m/s
Heaters' heat amount $\dot{Q}_{wi}$	0.5 W

41.4 Thermal Analysis and Design

where D_h is the hydraulic diameter and equivalent to $2B$, in which B is the channel spacing.

Equation (41.1) represents the limiting case of internal channel flow situation that can be treated as external flow over a flat plate. While the pressure drop should be considered in the channel flow model, this limitation excludes the pressure drop in the analysis model, eliminating the required system fan load. Other simplifications and limitations are (1) the elements are flushed with the surface, (2) the laminar flow is two-dimensional and unidirectional, and (3) buoyancy and radiation effects are treated as being negligible.

One-heater solution

Using the approximate boundary-layer velocity and temperature solutions (cubic parabola) for flow along a flat plate with an unheated starting section [1] and the superposition method, the temperature profiles for one heater with constant heat flux can be obtained as

$$\tau = T - T_{\text{in}} = u_{y-y_{1,s}} \tau_{w1,s} f_{1,s} - u_{y-y_{1,e}} \tau_{w1,e} f_{1,e} \qquad (41.2)$$

where the wall temperature excess ($\tau_w = T_w - T$) and corresponding variables are given by

$$\tau_{w1,p} = \frac{C_1}{k} \Pr^{-1/3} \text{Re}_y^{-1/2} y \, \dot{q}_w'' \, \beta_{a_{1,p}} \qquad (41.3)$$

$$f_{1,p} = 1 - \frac{3}{2}\left(\frac{z}{\Delta_{1,p}}\right) + \frac{1}{2}\left(\frac{z}{\Delta_{1,p}}\right)^3 \qquad (41.4)$$

$$\beta_{a_{1,p}} = \int_0^{a_{1,p}} Z^{m-1}(1-Z)^{n-1} dZ \qquad m = 1/3 \text{ and } n = 4/3 \qquad (41.5)$$

$$a_{1,p} = 1 - \left(\frac{y_{1,p}}{y}\right)^{3/4} \qquad (41.6)$$

$$u_{y-y_{1,p}} = 1 \quad \text{for} \quad y \geq y_{1,p} \quad \text{and} \quad 0 \quad \text{for} \quad y < y_{1,p}$$

where $C_1 = 0.8307$, the subscript p stands for s or e ($y_{1,e} = y_{1,s} + a$), β is the incomplete beta function, and $\Delta_{1,p}$ is thermal boundary-layer thickness for an unheated starting length $y_{1,p}$. The heater's start or end position is shown in Fig. 41.2.

Wall temperature can be obtained by substituting $z = 0$ into Eq. (41.2) as

$$\tau_w = \frac{C_1}{k} \Pr^{-1/3} \operatorname{Re}_y^{-1/2} y \, \dot{q}_w'' \, \psi \tag{41.7}$$

where

$$\psi(y, y_{1,s}, y_{1,e}) = u_{y-y_{1,s}} \beta_{a_{1,s}} - u_{y-y_{1,e}} \beta_{a_{1,e}} \tag{41.8}$$

Different dimensionless wall temperatures can be defined in terms of heat flux and channel spacing or channel length as

$$\theta_w = \frac{\tau_w}{\dot{q}_w'' B / k} \tag{41.9}$$

and

$$\phi_w = \frac{\tau_w}{\dot{q}_w'' L / k} \tag{41.10}$$

Depending on the preliminary design parameters, the following expressions can be obtained as

$$\theta_w = C_2 \Pr^{-1/3} \left(\frac{y}{L}\right)^{1/2} \left(\frac{B}{L}\right)^{-1/2} \psi \operatorname{Re}_B^{-1/2} \tag{41.11}$$

or

$$\theta_w = C_1 \Pr^{-1/3} \left(\frac{y}{L}\right)^{1/2} \left(\frac{B}{L}\right)^{-1} \psi \operatorname{Re}_L^{-1/2} \tag{41.12}$$

and

$$\phi_w = C_2 \Pr^{-1/3} \left(\frac{y}{L}\right)^{1/2} \left(\frac{B}{L}\right) \psi \operatorname{Re}_B^{-1/2} \tag{41.13}$$

or

$$\phi_w = C_1 \Pr^{-1/3} \left(\frac{y}{L}\right)^{1/2} \psi \operatorname{Re}_L^{-1/2} \tag{41.14}$$

where $C_2 = \sqrt{2} C_1$ and

$$\operatorname{Re}_B = \frac{\rho V_{in}(2B)}{\mu} = \frac{2 \dot{m}_{in}}{\mu W} \tag{41.15}$$

Since the present analysis model uses the flat-plate solution, channel spacing does not affect the temperature result. Considering the Reynolds number in terms of mass flow rate and choosing the preliminary design parameters as heater size, heater location ($y_{1,s}$ and $y_{1,e}$),

41.6 Thermal Analysis and Design

and mass flow rate per channel ($\dot{m}_{in}$), a new dimensionless wall temperature can be defined as

$$\Theta_w = C_2 \operatorname{Pr}^{-1/3} \left(\frac{y}{L}\right)^{1/2} \psi \operatorname{Re}_B^{-1/2} \tag{41.16}$$

where

$$\Theta_w = \frac{T_w}{\frac{\dot{q}_w''}{k}\sqrt{BL}} \tag{41.17}$$

It can be found that the maximum wall temperature for one heater occurs at $y = y_{1,e}$ (endpoint of the heater). In this work, the following dimensionless maximum wall temperature is used as

$$\Theta_{w,\max} = C_2 \operatorname{Pr}^{-1/3} \left(\frac{y_{1,e}}{L}\right)^{1/2} \beta_{a_{1,s}^*} \operatorname{Re}_B^{-1/2} \tag{41.18}$$

where

$$a_{1,s}^* = 1 - \left(\frac{y_{1,s}}{y_{1,e}}\right)^{3/4} \tag{41.19}$$

Equation (41.18) is used to calculate the maximum wall temperature in terms of design input such as incoming airflow rate and heater location and size. The corresponding wall temperature at a distance y from the channel inlet is calculated by using Eqs. (41.7) and (41.16).

Multiple-heater solution

For multiple heaters having constant heat flux(es), air temperature is obtained by superposing the results of one heater

$$\tau = T - T_{\mathrm{in}} = \sum_{i=1}^{N} \left[u_{y-y_{i,s}} \tau_{wi,s} f_{i,s} - u_{y-y_{i,e}} \tau_{wi,e} f_{i,e} \right] \tag{41.20}$$

and the wall temperature is given by

$$T_w = \frac{C_1}{k} \operatorname{Pr}^{-1/3} \operatorname{Re}_y^{-1/2} y \left[\sum_{i=1}^{N} \dot{q}_{w,i}'' \psi_i \right] \tag{41.21}$$

where N is the total number of heaters

$$\psi_i(y, y_{i,s}, y_{i,e}) = u_{y-y_{i,s}} \beta_{a_{i,s}} - u_{y-y_{i,e}} \beta_{a_{i,e}} \tag{41.22}$$

and for $p = s$ or e, $u_{y-y_{i,p}} = 0$ for $y < y_{i,p}$ and 1 for $y \geq y_{i,p}$.

Analysis of Convectively Cooled Components on PCBs   41.7

Since the maximum wall temperature for each heater occurs at its end position of the heater, the expression for the maximum temperature for kth heater can be obtained as

$$\tau_{wk,\max} = \frac{C_1}{k} \mathrm{Pr}^{-1/3} \mathrm{Re}_{y_{k,e}}^{-1/2} y_{k,e} \left[\sum_{i=1}^{k} \dot{q}''_{w,i} \psi_i^* \right] \quad (41.23)$$

where

$$\psi_i^* (y_{i,s}, y_{i,e}) = \beta_{a_{i,s}^*} - u_{y_{k,e}-y_{i,e}} \beta_{a_{i,e}^*} \quad (41.24)$$

where for $p = s$ (or e), we have

$$a_{i,p}^* = 1 - \left(\frac{y_{i,p}}{y_{k,e}} \right)^{3/4} \quad (41.25)$$

Since the temperature of the heater(s) located downstream is affected by the upstream heater(s), in this work, the dimensionless wall temperature is defined as

$$\Theta_w = C_2 \mathrm{Pr}^{-1/3} \left(\frac{y}{L} \right)^{1/2} \left[\sum_{i=1}^{N} R_{1i} \psi_i \right] \mathrm{Re}_B^{-1/2} \quad (41.26)$$

where the dimensionless wall temperature for multiple heaters is defined as

$$\Theta_w = \frac{\tau_w}{\dfrac{\dot{q}''_w}{k} \sqrt{BL}} \quad (41.27)$$

and R_{ij} is the ratio of wall heat flux:

$$R_{ij} = \frac{\dot{q}''_{w,j}}{\dot{q}''_{w,i}} \quad (41.28)$$

The corresponding dimensionless maximum wall temperature is given by

$$\Theta_{wk,\max} = C_2 \mathrm{Pr}^{-1/3} \left(\frac{y_{k,e}}{L} \right)^{1/2} \left[\sum_{i=1}^{k} R_{1i} \psi_i^* \right] \mathrm{Re}_B^{-1/2} \quad (41.29)$$

where

$$\Theta_{wk,\max} = \frac{\tau_{wk,\max}}{\dot{q}''_{w1} \sqrt{BL}/k} \quad (41.30)$$

41.8 Thermal Analysis and Design

Equation (41.29) can be explicitly used to predict the maximum wall temperature for the given design parameters and input conditions. For $k = 1$, Eq. (41.29) is reduced to Eq. (41.18).

Temperature influence factor

To investigate the effect of thermal behavior of the upstream heater(s) on the downstream heater(s), the *temperature influence factor* (TIF) for maximum wall temperature ($\text{TIF}_{\max}$) is proposed and defined as

$$\text{TIF}_{\max,ij} = \frac{\Theta_{wj,\max}}{\Theta_{wi,\max}} \qquad i < j \tag{41.31}$$

where i and j represent each heater and $\text{TIF}_{\max,ij} = 1$ for $i = j$ and it can be shown that

$$\text{TIF}_{\max,ik} = \text{TIF}_{\max,ij} \cdot \text{TIF}_{\max,jk} \tag{41.32}$$

This factor represents the relative magnitude(s) of the maximum wall temperature in each heater in which the downstream heater temperature can be affected by an upstream heater(s). Furthermore, this factor can be used to properly locate downstream heater(s) on the PCB.

Sample Analysis Results and Thermal Design Procedure

Single-heater analysis results

The effect of the heater location (distance from the channel inlet) and Re_B (or flow rate per channel) on the maximum wall temperature for one heater is shown in Fig. 41.3. For the given heater size, the maximum wall temperature decreases with Re_B and increases with $y_{1,s}/L$. Furthermore, at low Reynolds number (or low flow rate), the changes in maximum wall temperature are perceivable. This is due to the fact that increasing flow rate into the channel enhances the convective heat transfer. Also, the longer unheated length from the inlet leads to a thicker hydrodynamic boundary layer above the heater, resulting in a higher wall temperature.

Figure 41.4 shows the effect of heater size and location on the maximum wall temperature for one heater when a constant value of Re_B is used. It can be seen that the wall temperature increases with heater size and location. Figures 41.3 and 41.4 also show that the heater location is an important parameter affecting the wall temperature. These results can be used to properly locate one heater on the PCB to lower the maximum wall temperature.

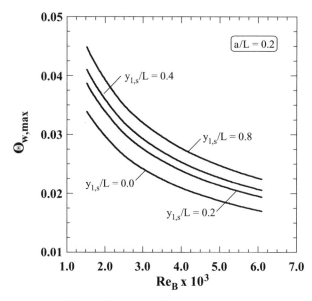

Figure 41.3 Effect of heater location and Re_B (or $\dot{m}_{in}$) on maximum wall temperature (one heater).

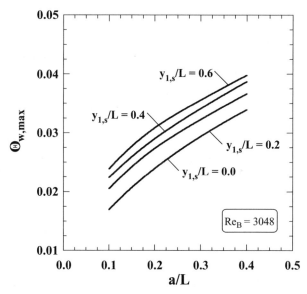

Figure 41.4 Effect of heater location and size on maximum wall temperature (one heater).

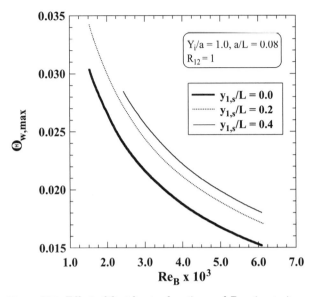

Figure 41.5 Effect of first-heater location and Re_B (or $\dot{m}_{in}$) on maximum wall temperature (two heaters).

Multiple-heater analysis results

As described earlier, the variations of wall temperature for multiple heaters located on the PCBs can be obtained by superposing the results of one heater. Also, the thermal behavior of the heater(s) located in the downstream is affected by the upstream heater(s), especially the first heater from the channel inlet. The effects of the first heater location and Re_B on the maximum wall temperature for two and three heaters on the PCB are shown in Figs. 41.5 and 41.6 when the heater spacing is fixed ($Y_i/a = Y_1/a = Y_2/a = 1.0$) and the same amounts of heat flux in each heater are used ($R_{12} = R_{13} = 1$). The results show that the location of the first heater influences the maximum wall temperature of the downstream heater(s). Since the hydrodynamic boundary layer becomes thicker as the unheated starting length (or the first heater location) increases, resulting in a higher wall temperature of the first heater, the wall temperature of the downstream heater(s) increases as $y_{1,s}$ increases. The effect of heater spacing (Y_i/a) on the wall temperature variation of the downstream heater(s) was relatively small (see Fig. 41.7). This is because the downstream heater(s) faces the thermal boundary layer developed by the upstream heater(s), reducing heat transfer from the downstream heater(s).

The variations of the maximum wall temperature for multiple heaters with Re_B are shown in Fig. 41.8 when the given heater conditions are used. For different numbers of heaters, the maximum wall temperature

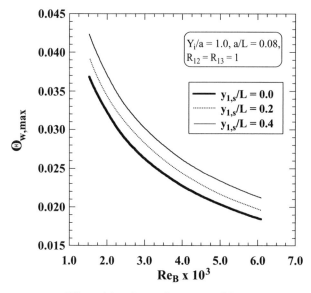

Figure 41.6 Effect of first-heater location and Re_B (or $\dot{m}_{in}$) on maximum wall temperature (three heaters).

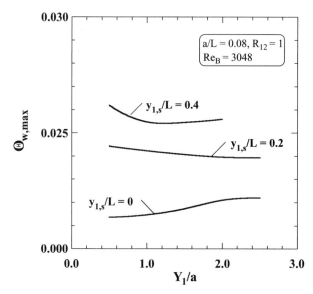

Figure 41.7 Effect of $y_{1,s}$ and Y_1 on maximum wall temperature (two heaters).

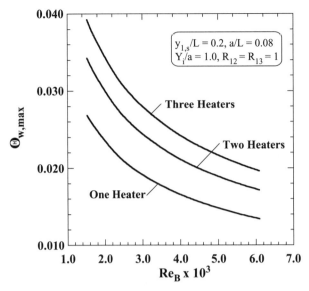

Figure 41.8 Effect of Re_B on $\Theta_{w,\max}$ (three heaters).

occurs at the last downstream heater. To predict the wall temperature of the downstream heater(s), the TIFs defined by Eq. (41.31) are used. Calculation of $TIF_{\max}$ values is based on the results shown in Fig. 41.8. Figure 41.9 shows that the $TIF_{\max}$ values are invariant with Re_B [see Eq. (41.31)]. Also, the effect of the first upstream heater on the third heater is larger than that of the second heater.

The effects of different amounts of heat fluxes in each heater on the TIF values are investigated. Figure 41.10 shows the variation of $TIF_{\max,12}$ with different ratios of heat fluxes. The maximum wall temperature for two heaters occurs at the second heater when R_{12} is about 0.65 to 0.8 for $y_{1,s}/L = 0$ to 0.4. It also shows the maximum wall temperature occurring on the second heater increases rapidly even though the heat flux amount of the second heater is slightly larger than that of the first heater. As mentioned earlier, the effect of the heater spacing (Y_1/a) on the TIF was small.

The effect of three heaters on their TIF values was investigated by fixing the values of $y_{1,s}$ and $Y_i = Y_1 = Y_2$. Figure 41.11 shows the variations of $TIF_{\max,ij}$ for different heat flux ratios of the first and third heaters. Since $TIF_{\max,12}$ is greater than 1, the maximum wall temperature of the second heater is higher than that of the first heater. The variation of $TIF_{\max,23}$ explains that the maximum wall temperature occurs at the third heater if R_{13} is greater than approximately 0.8. Furthermore, the variation of $TIF_{\max,13}$ shows the extent to which the first heater affects the third heater wall temperature.

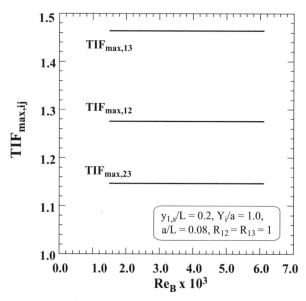

Figure 41.9 Effect of Re_B (or $\dot{m}_{in}$) on $TIF_{max,ij}$ (three heaters).

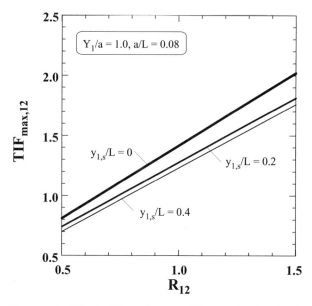

Figure 41.10 Effect of R_{12} and $y_{1,s}$ on $TIF_{max,12}$ (two heaters).

41.14 Thermal Analysis and Design

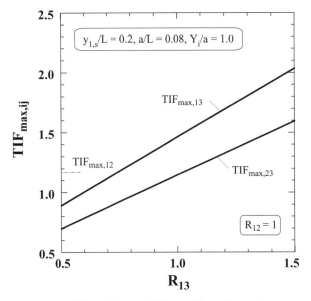

Figure 41.11 Effect of R_{13} on $\text{TIF}_{\text{max},ij}$ (three heaters).

For arbitrarily assigned values of heat flux in each heater, the TIFs can be used to predict the location and the relative magnitude of maximum wall temperature of a downstream heater. Figures 41.12 and 41.13 show the variations of $\text{TIF}_{\text{max},13}$ and $\text{TIF}_{\text{max},23}$ for the given conditions. Figure 41.12 shows $\text{TIF}_{\text{max},13}$ increases monotonically with R_{13} and R_{12}. However, $\text{TIF}_{\text{max},23}$ increases with R_{13} but decreases with R_{12} (see Fig. 41.13) because $\text{TIF}_{\text{max},12}$ increases with R_{12}, decreasing $\text{TIF}_{\text{max},23}$ as shown in Eq. (41.32).

Sample thermal design procedure

As a case study, a sample design guideline is presented in this section. The design constraints considered are (1) maximum allowable heater wall temperature ($T_{\text{allow}} = T_{w1,\text{max}}$), assumed to be occurring at the first heater; (2) component heat fluxes $\dot{q}''_{w,1} > \dot{q}''_{w,2} > \dot{q}''_{w,3}$; and (3) position of the first component, $y_{1,s}$. The design method consists of the following three steps:

1. Allowable dimensionless wall temperature for the first heater ($\Theta_{\text{allow}} = \Theta_{w1,\text{max}}$) is obtained using Eq. (41.17). With this dimensionless maximum wall temperature, Fig. 41.3 is used to determine the minimum Reynolds number ($\text{Re}_{B,\text{min}}$) for the given heater location, $y_{1,s}$. Reynolds number $\text{Re}_{B,\text{min}}$ is used to determine the corresponding airflow rate using Eq. (41.15).

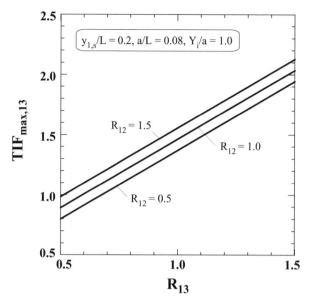

Figure 41.12 Effect of R_{13} and R_{12} on $TIF_{max,13}$ (three heaters).

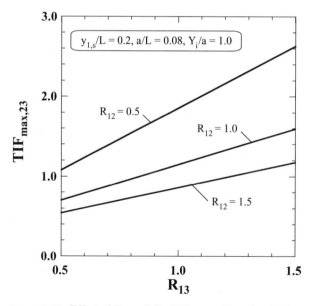

Figure 41.13 Effect of R_{13} and R_{12} $TIF_{max,23}$ (three heaters).

41.16 Thermal Analysis and Design

TABLE 41.2 Sample Thermal Design Procedure

Design constraints	Design solution	Component temperatures
$T_{in} = 50°C$	$\Theta_{w1,max} = 0.027$	$T_{w1,max} = 105°C$
$T_{allow} = 105°C$	$\text{Re}_{B,min} = 3200$	$T_{w2,max} = 89.6°C$
$Q_{w1} = 0.5\text{ W}$	$\dot{m}_{in} = 241.2$ kg/h	$T_{w3,max} = 77.5°C$
$Q_{w2} = 0.25\text{ W}$	Figure 41.10	
$Q_{w3} = 0.125\text{ W}$	$R_{12} = 0.5$	
$a = 0.02$ m	$\text{TIF}_{max,12} = 0.72$	
	$\Theta_{w2,max} = 0.019$	
	Figure 41.12	
	$R_{13} = 0.25$	
	$\text{TIF}_{max,13} = 0.5$	
	$\Theta_{w3,max} = 0.013$	

2. The heat flux ratio R_{12} is calculated using Eq. (41.28) and then Fig. 41.10 is used to determine $\text{TIF}_{max,12}$. This allows the temperature of the second heater to be determined with Eq. (41.31).

3. Similar to the previous step, the heat flux ratio R_{13} along with R_{12} is used in Fig. 41.12 to determine $\text{TIF}_{max,13}$. This allows the third-heater temperature to be determined.

Table 41.2 illustrates this method for a hypothetical design. Examination of Figs. 41.10 and 41.12 show that for large heat flux ratios the $\text{TIF}_{max,ij}$ values are greater than 1. In this situation, the allowable temperature of the first heater must be reduced to prevent the second or third component from exceeding their allowable wall temperature.

Summary of analysis work

- The approximate analysis model proposed predicts the thermal behavior of PCBs with embedded (discrete) heating components by using the boundary-layer theory and the superposition method.
- The variations of maximum (heater) wall temperature were investigated by parametrically changing the heater location and spacing, Reynolds number (or flow rate), and heat flux.
- In order to estimate the relative magnitude and location of the maximum wall temperature of the heater, the temperature influence factor (TIF) is proposed.
- The effect of the first-heater location ($y_{1,s}$) on the maximum wall temperature is significant. However, the effect of the heater spacing is small.
- The maximum wall temperature usually occurs at the last heater even though the amounts of the heat flux of the downstream heater(s) are smaller than those of the upstream heater(s).

- Temperature influence factor values are useful for PCB thermal design during the conceptual layout.
- As an extension of the current work, some additional works have been performed by Park et al. [4–8], in which lateral conduction and radiation heat transfer for both forced- and natural-convection cooling modes are investigated.

Nomenclature

a	Heating component size, m
B	PCB channel spacing, m
C_1, C_2	Constants
k	Thermal conductivity, W/m-K
L	PCB channel length, m
$\dot{m}_{in}$	Mass flow rate per PCB channel, kg/s
N	Total number of heater
Pr	Prandtl number
$\dot{Q}_w$	Heater heat amount, W
$\dot{q}''_w$	Wall heat flux, W/m²
R_{ij}	Wall heat flux ratio $\dot{q}''_{wj}/\dot{q}''_{wi}$
Re	Reynolds number
T	Temperature, K
TIF	Temperature influence factor
u	Unit step function
V	Velocity, m/s
W	PCB wall width, m
x	Lateral distance, m
y	Distance from channel inlet, m
Y_i	Heater spacing, m
z	Distance from PCB wall, m

Greek

β	Beta function
Δ	Thermal boundary-layer thickness, m
θ, ϕ, Θ	Dimensionless temperature
μ	Dynamic viscosity, N-s/m²
ρ	Density, kg/m³
τ	Temperature excess $(T - T_{in})$, K

Subscripts

e	Endpoint of heater
i, j, k	Heaters
in	Inlet
max	Maximum
s	Starting point of heater
w	Wall

References

1. Kays, W. M., and Crawford, M. E., 1993, *Convective Heat and Mass Transfer*, McGraw-Hill, New York.
2. Figliola, R. S., and Thomas, P. G., 1995, "An Approximate Method for Predicting Laminar Heat Transfer between Parallel Plates Having Embedded Heat Sources," *Journal of Electronic Packaging*, vol. 117, pp. 63.
3. Park, S. H., and Tien, C. L., 1990, "An Approximate Analysis for Convective Heat Transfer on Thermally Nonuniform Surfaces," *Journal of Heat Transfer*, vol. 112, p. 192.
4. Park, H., Nigro, N., Gollhardt, N., and Lee, P., 2001, *Development and Integration of a Semi-Analytical PCB Thermal Design Technique with an Infrared Thermal Imaging System*, ASME IMECE 2001/AES-23600.
5. Park, H., Nigro, N., Elkouh, A., Gollhardt, N., and Lee, P., 1999, *Estimation of Effective Radiation Heat Loss from Heaters on Two Parallel PCBs and Its Effect on Heater Wall Temperature*, ASME 99-IMECE/EEP-1.
6. Park, H., Nigro, N., Elkouh, A., Gollhardt, N., and Lee, N., 1998, *Approximate Thermal Analysis of Printed Circuit Boards Cooled by Laminar Natural Convection*, ASME 98-WA/EEP-20.
7. Park, H., Nigro, N., Elkouh, A., Gollhardt, N., and Lee, P., 1998, *Effect of Design Parameters on Thermal Behavior of Printed Circuit Boards Cooled by Uni-Directional Flow*, ASME 98-WA/EEP-21.
8. Park, H., Nigro, N., Elkouh, A., Gollhardt, N., and Lee, P., 1997, *Guidelines for Preliminary Thermal Design of Convectively Cooled Printed Circuit Boards*, ASME HTD, vol. 351, p. 315.

Chapter 42

Design of a Fusion Bonding Process for Fabricating Thermoplastic-Matrix Composites

F. Yang and R. Pitchumani
Advanced Materials and Technologies Laboratory
Department of Mechanical Engineering
University of Connecticut
Storrs, Connecticut

Abstract

Processing of layered thermoplastic-matrix composite products involves applying heat and pressure to contacting thermoplastic surfaces and consolidating the interface. The combined processes of interfacial contact area increase, referred to as *intimate contact*, and polymer interdiffusion (*healing*) across the interfacial areas in contact, are responsible for the development of interlaminar bond strength in the composite products. On the basis of the modeling of the involved transport phenomena, the interfacial bond strength is related to the processing conditions (i.e., temperature, pressure, and time), which, in turn, govern the manufacturing cost. Higher interfacial strength may be obtained under the processing conditions of high temperature and pressure, and long processing time, all of which correspond to higher costs. To address these competing considerations of part quality and fabrication cost, a methodology is presented to allow the determination of the processing conditions necessary to achieve the desired quality with constraints on the manufacturing time and cost. The process design considerations

42.2 Thermal Analysis and Design

are illustrated for an example thermoplastic composite system of AS4 carbon-fiber-reinforced poly(ether-ether-ketone) (PEEK).

Introduction

Fabrication of thermoplastic-matrix composites is achieved by a number of techniques, including tow placement, tape laying, resistant welding, and autoclave forming. The basic approach in all these techniques is that of stacking up and consolidating prepreg layers to the desired final form. A brief description of two example fabrication methods is given, followed by a discussion of the common physical mechanisms that govern all the processing methods.

Autoclave forming involves carefully stacking layers of thermoplastic prepregs to the final thickness and form of the composite product, and placing this stack in an autoclave oven, where the layup is subject to a temperature and pressure schedule [1,2]. The pressure cycle could be either a positive pressure that pushes normally on the stack, or a vacuum that pulls the layers in the stack together. The high temperature causes the material to soften, and the applied pressure, together with the temperature, serves to trigger the mechanisms leading to consolidation of the layers to form a monolithic structure. The consolidated product is cooled before removal from the autoclave. A schematic representation is shown in Fig. 42.1a. *Tape laying* and automated *tow placement* (Fig. 42.1b) are based on incrementally laying down and continuously consolidating prepreg layers to build the composite product. The two processes are particularly suited for fabrication of large structures such as aircraft wing skins and fuselage [3,4] and principally differ in the size of the prepregs used; whereas wider tapes are used in tape laying, the prepreg tows used in tow placement are smaller

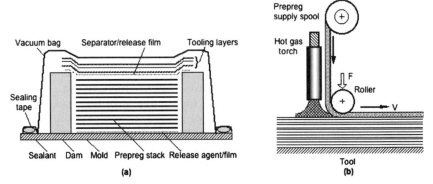

Figure 42.1 Schematic of (a) autoclave and (b) tow placement/tape-laying processes for fabrication of thermoplastic matrix composites.

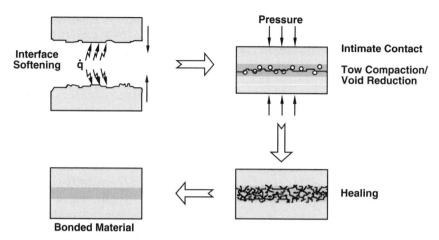

Figure 42.2 Schematic illustration of the steps in a fusion bonding process.

(on the order of approximately 0.25 in. wide). Other fabrication techniques, including filament winding and pultrusion, are discussed in Ref. 2.

Fabrication of composites from prepreg using the foregoing processes is based on the principle of *fusion bonding*, which involves applying heat and pressure to the interface between two contacting material layers (Fig. 42.2). The applied temperature and pressure cause the interfacial contact area to increase by the spreading of the softened surface asperities—a process referred to as *intimate contact*—while the elevated temperatures initiate intermolecular diffusion processes, termed *autohesion* or *healing*, across the areas in intimate contact. The combined transport processes of intimate contact and healing are responsible for the development of interfacial bond strength—a critical quality parameter—in the composite products.

The processes of intimate contact and healing are coupled in that healing can occur only across areas of the interface that are in intimate contact. While intimate contact development is a function of the applied pressure, temperature, and time, healing is governed mainly by the temperature history and time. As a result of the coupling, however, the interlaminar bond strength σ is a function of the processing temperature T, the consolidation pressure p, and the processing time t [5–10]:

$$\sigma = f(T, p, t) \qquad (42.1)$$

A reliable description of the microscale intimate contact and healing phenomena as well as the coupling effect of the two processes is therefore essential for arriving at effective process designs in practice.

42.4 Thermal Analysis and Design

The models for fusion bonding processes are briefly outlined in the following section. These models are used to illustrate fusion bonding process parameter design in the "Results and Discussion" section.

Process Models

Intimate contact process

Modeling of the intimate contact process fundamentally requires a description that captures the intrinsic features of the asperity structures on thermoplastic ply surfaces. The geometric complexity of the surface asperities, however, has posed a challenge to their quantitative description, and in turn, of the intimate contact process. A fundamental view of thermoplastic surfaces (Fig. 42.3a) reveals that the geometric structure is random, and that the roughness features are found at a large number of length scales. Further, Fig. 42.3b shows that the power spectrum of the asperity profile, obtained as a fast Fourier transform (FFT) of the profilometric scan in Fig. 42.3a, exhibits a power-law variation with the spatial frequency (equivalently, the asperity scale). This suggests the existence of a fractal structure formed by the asperity elements in the profile, implying that when a portion of the asperity surface is magnified appropriately, the magnified structure will be statistically similar to the original [11–13].

A fractal geometry–based description, which accounts for the roughness features at multiple length scales, is therefore appropriate for describing the surface asperities and their spreading process. This forms the basis of the intimate contact model used in the present calculation. Yang and Pitchumani [12] developed a fractal Cantor set construction to represent the multiscale asperity structures on the surfaces. A squeeze flow model based on the fractal description was then developed to simulate the intimate contact evolution as a function of the process and surface parameters.

The construction of a Cantor set surface, shown in Fig. 42.4a, is as follows. Starting from a rectangle of length L_0 and an arbitrary height, a small rectangle of height h_0 is removed from the middle, so that the remaining length of the rectangle is $L_1 = L_0/f$, where the scaling factor $f > 1$. This results in two identical rectangles of height h_0 and width $L_1/2$, which are called the *first-generation asperities*. Next, from each of the first-generation asperities, a rectangle of height $h_1 = h_0/f$ is removed, which yields 4 second-generation rectangular asperity elements of height h_1 and width $L_2/4$, where $L_2 = L_1/f$. Continuing the procedure results in the self-affine Cantor set in Fig. 42.4a. Note that in this procedure, one rectangle is removed from each asperity, which results in $s = 2$ smaller asperities.

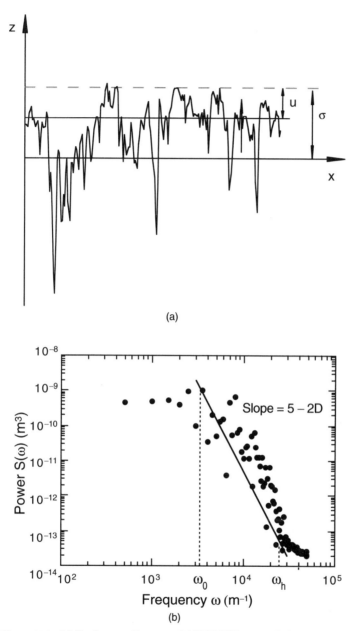

Figure 42.3 (a) Surface profile scan of AS4/PEEK tow and (b) its power spectrum.

42.6 Thermal Analysis and Design

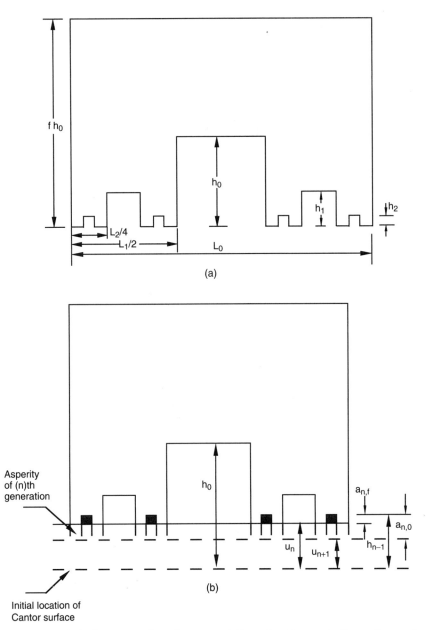

Figure 42.4 (a) A Cantor set fractal representation of a thermoplastic surface and (b) deformation of a Cantor set surface by a contact plane.

Starting from the self-affine scaling of a rough surface and statistical considerations, the fractal dimension of the Cantor set surface D may be related to the scaling factor f and s [12]:

$$s = f^{D/(2-D)} \tag{42.2}$$

The fractal parameters D, L_0, h_0, f, and s in the description above can be determined from the roughness measurements as discussed below:

1. For a given thermoplastic tape, its surface profile $z(x)$ may be obtained using a contact or noncontact profilometer. Using Fourier's theorem, the surface profile function $z(x)$ can be spectrally decomposed into its components of sine and cosine waves, whose amplitudes can be determined from the profile's power spectrum obtained using a fast Fourier transformation. It can be shown that for a fractal surface, its power spectrum $S(\omega)$ follows a power law of the form [14]

$$S(\omega) = \frac{C}{\omega^{5-2D}} \tag{42.3}$$

where C is a constant and ω is the frequency (or the wavenumber), which denotes the reciprocal of the length scale. The power law suggests the self-affine scaling of a fractal surface mentioned above, where the parameter D is the fractal dimension of the profile. The surface profile measurements of AS4/PEEK thermoplastic tapes in Fig. 42.3 reveal the existence of a distinct interval in which the power spectrum obeys a power law, and the fractal dimension D can be determined from the slope of a log-log plot of the power spectrum with respect to the frequency ω within the frequency range $\omega_0 \leq \omega \leq \omega_h$.

2. The upper limit of the frequency interval ω_h corresponds to the smallest length scale in the surface profile while the lower limit ω_0 is associated with the largest repeating unit in the profile. Hence, the lower bound ω_0 of the interval is chosen in this study to determine the length L_0 of the Cantor set block as $L_0 = 1/\omega_0$.

3. Since h_0 denotes the largest asperity height in the Cantor set model, its value may be estimated on the basis of the standard deviation σ of the surface roughness profile. Considering the asperity features within a $\pm \sigma$ range about the mean to be dominant in the intimate contact process, h_0 is evaluated as 2σ.

4. For the contact process between a Cantor set surface and a flat surface, as shown in Fig. 42.4b, the linear contact area of the fractal surface with the plane L_i is related to the displacement u of the plane by following an approach similar to that described in Ref. 15. The resulting

42.8 Thermal Analysis and Design

expression is given by [12]

$$L_i = \frac{\sigma - u}{f(h_0/L_0)} \qquad (42.4)$$

From a real surface profile, L_i may be evaluated as the total intercepted lengths of the profile with horizontal lines at various displacement $\sigma - u$ from the profile mean (see Fig. 42.3b). The parameter f can then be obtained by a statistical regression of the measured intercepted length data over the range $0 \leq u \leq 2\sigma$ to Eq. (42.4).

5. The parameter s can be determined using the measured values of f and D in Eq. (42.2).

Note that in the preceding description, all the fractal parameters are uniquely determined for a surface from its profile measurements. In the next discussion, the fractal description is used to model the interfacial contact area evolution during the intimate contact process, which is modeled as the flattening of the Cantor set surface by a rigid half-plane, as shown in Fig. 42.4b. The contact process starts with the squeeze flow of the highest-generation asperities (theoretically, asperities at generation ∞), and progressively goes to the first-generation asperities. When the asperities from the $(n+1)$th generation are deformed, the nth and lower generations are supposed to retain their shape and all the deformed material is considered to flow into the troughs between the $(n+1)$th-generation asperities. When the troughs are completely filled, the rigid half-plane has a cumulative displacement u_{n+1}, which is measured from the initial location of the Cantor surface as shown in Fig. 42.4b. The $(n+1)$th-generation asperities are then combined into the (n)th generation, and the process proceeds with the deformation of the (n)th-generation asperities.

According to the description of the Cantor set surface deformation process, the *degree of intimate contact* $D_{ic}^{(n)}$ (defined as the fraction of the total surface area in contact at any time) for the (n)th-generation asperities, was given by Yang and Pitchumani [12] as

$$D_{ic}^{(n)}(t) = \frac{1}{f^n}\left[\frac{5}{4}\left(\frac{h_0}{L_0}\right)^2 \frac{f^{(2nD)/(2-D)+n+4}}{(f+1)^2}\int_{t_{n+1}}^{t}\frac{p}{\mu}dt + 1\right]^{1/5}, \quad t_{n+1} \leq t \leq t_n \qquad (42.5)$$

where D is the fractal dimension of the thermoplastic surface, p is the applied gauge pressure, μ is the viscosity of the material, and t_{n+1} and t_n are the start and end times for the consolidation process of the (n)th-generation asperities, respectively. Note that the foregoing equation pertains to the (n)th-generation asperities being in contact with the rigid flat plane. The degree of intimate contact evolution with time is obtained by recording the process from the highest generation (for which $t_{n+1} = 0$) down to the first. It must be pointed out that the geometric parameters in Eq. (42.5) can be completely determined from surface profile measurements [12], which eliminates the need for tuning the model to experimental data, as required in the other models in the literature [8,16].

Nonisothermal healing process

Healing of a polymer interface via molecular interdiffusion is often described using the *reptation theory*, which models the motion of individual linear polymer chains in amorphous bulk [17]. In the model, a polymer chain of length L is considered to be confined in a tube, which represents the steric effects of neighborhood chains (Fig. 42.5). The tube forms a geometric constraint such that the chain can move only along

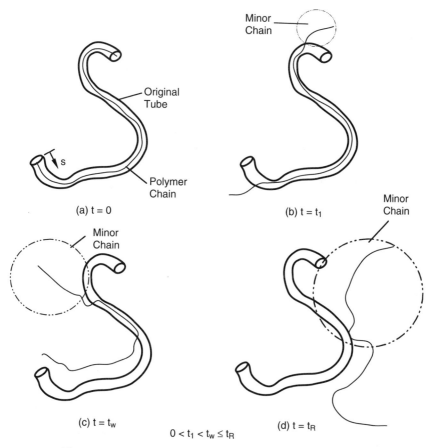

Figure 42.5 The reptation movement of a linear polymer chain in an entangled melt. The chain escapes from its original tube at the reptation time t_R, and the maximum bond strength is achieved at the welding time t_w ($\leq t_R$).

its curvilinear length. At the beginning of the process $t = 0$, the chain, represented by the thin solid line in Fig. 42.5, is totally encompassed by the original tube. The chain moves within the tube in a brownian motion manner, and after a period of time, the chain ends escape from the original tube, forming the "minor chains." The length ℓ of the minor chains increases with time and reaches L at the reptation time t_R (when the entire polymer chain escapes from the tube). Considering an interface between two thermoplastic layers in perfect contact, as time proceeds and as the lengths of the minor chains grow, some of the chains from each layer move across the interface, and the net interpenetration depth χ contributes to the bond strength buildup. After the reptation time, the interpenetration and entanglement of all the polymer chains are fully developed and the molecular configuration at the interface is identical to that of the virginal bulk material.

The interlaminar bond strength σ is proportional to the interpenetration distance χ, which is related to the minor chain length as $\chi \sim \sqrt{\ell}$, by considering the minor chains to amble across the interface via a random-walk motion [6]. The ultimate bond strength σ_∞ is achieved as the interpenetration depth and the minor chain length reach certain critical values. At high molecular weights, the maximum bond strength σ_∞ is achieved at the welding time $t_w < t_R$ as shown in Fig. 42.5c, where χ and ℓ have the values of $\chi_w < \chi_\infty$ and $L_w < L$, respectively. The *degree of healing* may be defined as the ratio of the instantaneous interfacial bond strength to the ultimate bond strength as [17,18]

$$D_h(t) = \frac{\sigma}{\sigma_\infty} = \frac{\chi}{\chi_w} = \left(\frac{\ell}{L_w}\right)^{1/2} \tag{42.6}$$

Considering the description of the reptation motion of a polymer chain encased in a surrounding tube, a *probability density function* $P(s, t)$ is defined as the probability of finding a particular chain segment at some position s at time t, where s is the curvilinear coordinate along the encompassing tube (Fig. 42.5a). The density function $P(s, t)$ was determined through a random-walk analysis of the chain segments in the tube, and was shown to be mathematically governed by a diffusion equation [17–19]

$$\frac{\partial P}{\partial t} = D\frac{\partial^2 P}{\partial s^2} \tag{42.7}$$

where D is the reptation diffusion coefficient. Generally, the diffusivity D depends on the molecular weight M, temperature T, and hydrostatic pressure p. For a typical polymer and its processing, M and p are constants, and the present focus is on the influence of the temperature, which is a function of time. Thus, for isothermal healing, the diffusion coefficient is a constant, whereas in the more general nonisothermal case, the diffusion coefficient is time-variant through its dependence on temperature.

At time $t = 0$, the chain segment considered is at the origin $s = 0$, and the initial condition associated with Eq. (42.7) may be written as

$$P(s, 0) = \delta(0) \tag{42.8}$$

where δ is the Dirac delta function. The diffusion domain is considered to be infinitely large ($|s| \to \infty$), and since the chain segment will never be able to move across an infinite distance, the following natural boundary conditions must be automatically satisfied:

$$P(s, t) = 0 \qquad \frac{\partial P(s, t)}{\partial s} = 0 \qquad \text{as} \qquad |s| \to \infty \tag{42.9}$$

Under isothermal conditions ($T = T_0$), the reptation diffusivity $D_0 = D(T_0)$ is a constant, and the solution of Eqs. (42.7) and (42.8) is given as [18,19]

$$P(s, t) = \frac{1}{[4\pi D_0 t]^{1/2}} \exp \frac{-s^2}{4 D_0 t} \tag{42.10}$$

The mean-square displacement of the polymer chain at time t corresponds to the square of the minor chain length $\ell^2(t)$ and may be evaluated as follows using the isothermal probability density function [Eq. (42.10)]:

$$<s^2> = \ell^2 = \int_{-\infty}^{+\infty} s^2 P(s, t) ds = 2 D_0 t \tag{42.11}$$

Recall that the welding time for a given temperature is defined as the time required for the minor chain length to reach L_w, at which point the maximum bond strength is obtained. From Eq. (42.11), we may write

$$t_w(T_0) = \frac{L_w^2}{2 D_0} \tag{42.12}$$

For an arbitrary temperature history $T(t)$, the reptation diffusivity is a function of time $D(t)$, and the nonisothermal bond strength evolution with time is obtained as a solution of the governing equation [Eq. (42.7)] with a variable coefficient and the associated conditions [Eqs. (42.8) and (42.9)]. The system of equations may be solved using the Fourier transformation or using the separation of variables technique [20]. Omitting the details of the solution steps, which may be found elsewhere [21,22],

42.12 Thermal Analysis and Design

the probability density function can be obtained as

$$P(s,t) = \frac{1}{2\pi}\int_{-\infty}^{+\infty} \exp\left[\int_0^t -D(t)\omega^2 dt\right]\cos(\omega s)\,d\omega \quad (42.13)$$

This equation forms the basis for nonisothermal healing analysis. It may be verified that for isothermal healing, Eq. (42.10) is recovered by setting $D(t) = D_0 = \text{const}$ in Eq. (42.13) and evaluating the resulting integrals.

Equation (42.12), which relates the welding time to the reptation diffusivity at any temperature, T, is used to replace the diffusion coefficient $D(t)$ in terms of the welding time in Eq. (42.13), as $D(t) = L_w^2/2t_w[T(t)]$. Furthermore, defining $f(t)$ for brevity as

$$f(t) = \int_0^t \frac{1}{2t_w}dt \quad (42.14)$$

the probability distribution function $P(s,t)$ [Eq. (42.13)] is evaluated as follows:

$$P(s,t) = \frac{1}{2\pi}\int_{-\infty}^{+\infty} \exp\left[-f(t)L_w^2\omega^2\right]\cos(\omega s)\,d\omega$$

$$= \frac{1}{\sqrt{4\pi f(t)L_w^2}} \exp\frac{-s^2}{4f(t)L_w^2}$$

Following Eq. (42.11), the square of the minor chain length at time t for a nonisothermal temperature history is determined using the probability density function as

$$\ell^2(t) = \int_{-\infty}^{+\infty} s^2 P(s,t)\,ds$$

$$= \int_{-\infty}^{+\infty} s^2 \frac{1}{\sqrt{4\pi f(t)L_w^2}} \exp\frac{-s^2}{4f(t)L_w^2}\,ds$$

$$= 2f(t)L_w^2 \quad (42.15)$$

From Eqs. (42.6), (42.14), and (42.15), the nonisothermal degree of healing evolution with time is given by the following expression:

$$D_h(t) = \left(\frac{\ell}{L_w}\right)^{1/2} = [2f(t)]^{1/4} = \left[\int_0^t \frac{1}{t_w(T)}dt\right]^{1/4} \quad (42.16)$$

The foregoing result is based on the solution of the fundamental equations governing reptation dynamics, and therefore constitutes a rigorous model for nonisothermal healing. For a given temperature history, and the temperature dependence of the welding time, the degree of healing can be evaluated using Eq. (42.16). Note that this equation is valid for the full range of molecular weight because in the low-molecular-weight range, $t_w = t_R$, $\chi_w = \chi_\infty$, $L_w = L$, where t_R is the reptation time, χ_∞ and L are the interpenetration depth and minor chain length at the reptation time, respectively (Fig. 42.5d). For a constant-temperature healing process, $t_w(T)$ is a constant and Eq. (42.16) reduces to the expression for isothermal healing given by de Gennes [17].

Fusion bonding model

Since polymer healing can take place only across interfacial areas in intimate contact, the interfacial bond strength is determined by the coupled effects of the two processes. At any time t, during the fusion bonding process, an incremental intimate contact area $(dD_{ic}/dt)\,dt$ develops at the interface. The degree of healing for this incremental area at a bonding time t_b is denoted as $D_h(t_b - t)$, where $t_b - t$ represents the time available for the area to heal. The effective bond strength can therefore be expressed as an area average of the healing of all the incremental areas, and may be given in a nondimensional form as the degree of bonding D_b [10]:

$$D_b(t_b) = D_{ic}(0) \cdot D_h(t_b) + \int_0^{t_b} D_h(t_b - t) \cdot \frac{dD_{ic}(t)}{dt} \cdot dt \qquad (42.17)$$

The first term on the right-hand side of this equation represents the contribution by the area in intimate contact at the beginning of the bonding process. The initial area contact $D_{ic}(0)$ can be obtained from the fractal parameters of the tow surface, as discussed in Ref. 13.

Results and Discussion

In this section, validation of the nonisothermal healing model and the intimate contact model with experimental data is presented first. Various combinations of processing conditions are adopted to illustrate the physical trends of the physical models. The healing model is coupled with the intimate contact model to predict the required bonding time for specified temperature, pressure schedules and value of D_b. An example cost function is subsequently introduced to quantitatively determine the processing conditions that ensure a target value of D_b at a minimum cost.

Figure 42.6 shows the validation of the intimate contact model for the case of fusion bonding of IM-7/PIXA-M thermoplastic tapes [12]. The isothermal processing temperatures ranged from 260°C (which is

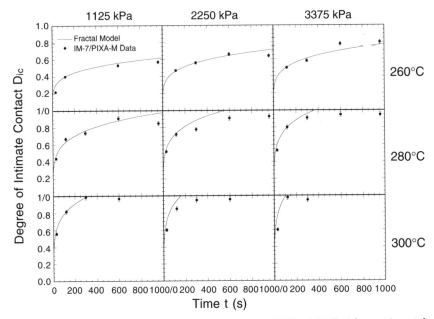

Figure 42.6 Validation of the fractal intimate contact model [Eq. (42.5)] with experimental data on IM-7/PIXA-M prepreg tapes.

slightly above the glass transition point of 250°C) to 300°C, and for each temperature, the consolidation pressure was varied from 1.125 to 3.375 MPa. The degree of intimate contact values for the fabricated specimens were evaluated from the ultrasonic C-Scan images of the interface. The prepreg viscosity at the various temperatures, given in Ref. 12, and the surface parameters, listed in Table 42.1, were used in the model to predict the degree of intimate contact as a function of time for the various process parameters. The experimental measurements and the predicted values of the degree of intimate contact are presented in Fig. 42.6, where the solid lines represent the model prediction and the symbols denote experimental data along with their respective error bars.

TABLE 42.1 Fractal Surface Geometric Parameters for AS4/PEEK and IM-7/PIXA-M Materials

Prepreg material	f	D	h_0/L_0
AS4/PEEK	1.45	1.32	0.050
IM-7/PIXA-M	1.40	1.38	0.018

SOURCE: Ref. 12.

The data comparisons in Fig. 42.6 reveal a good agreement between the model predictions and the experimental measurements. The intimate contact is seen to evolve rapidly as the pressure is increased, for a fixed temperature, and as the temperature is increased, for a fixed pressure. Overall, it may be noted that the model tracks the development of interfacial area contact accurately for all the processing conditions shown in Fig. 42.6. A general trend observed in both the experimental and the theoretical results in Fig. 42.6 is that the intimate contact development is initially rapid, but slows down with time. This trend may be explained using the Cantor set model. The smaller, later-generation asperities in the Cantor set, which undergo consolidation initially, are spread easily leading to a rapid contact area increase at the beginning of the process. As the process continues, however, the smaller asperities fuse with the larger asperities, which require a longer time for consolidation, thereby slowing down the contact area development. It must be mentioned that other material systems and processing conditions were explored in Ref. 12 for validation of the intimate contact model.

Experimental validation of the healing model developed in the study was examined using the strength measurements on PEEK ribbons [23]. Figure 42.7 compares the model prediction with experimental data in terms of the degree of healing development with time under nonisothermal temperature conditions. The temperature schedules used in the experiments have the rampup and hold stages, shown by the dashed lines in Fig. 42.7. The temperature-dependent welding time of PEEK used in the model predictions is given by the expression from Lee and Springer [8]:

$$t_w = \left(\frac{1}{44.1} \exp \frac{3810}{T} \right)^4 \qquad (42.18)$$

Numerical calculations are carried out until D_h reaches a value of unity. Once a degree of healing of unity is achieved, the state of complete healing is preserved regardless of the further temperature variation. The predictions of the current model are denoted by the solid lines, and the experimental data are represented by the symbols with the error bars. As the temperature gradient decreases from that in Fig. 42.7a to that in Fig. 42.7b, the healing development slows down as physically expected. Overall, a good agreement between the healing model and the experimental data is observed for the nonisothermal schedules and through the entire healing process. Validation of the healing model using other materials and processing conditions is discussed further in Refs. 21 and 22.

With validation of the healing and intimate contact models as the basis, the coupled bonding model in Eq. (42.17) is employed to provide

42.16 Thermal Analysis and Design

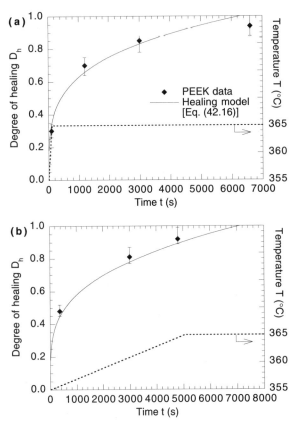

Figure 42.7 Validation of the nonisothermal healing model [Eq. (42.16)] with experimental data on PEEK films.

practical guidelines for the processing of an example thermoplastic composite system, AS4/PEEK [12,21]. In the remainder of this section, temperature (pressure) schedules are considered to be consist of a linear ramp from T_0 (p_0) to T_f (p_f) within t_{ramp} seconds, followed by a hold stage with a constant temperature T_f (pressure p_f), as shown in Fig. 42.8. To simplify the discussion, the initial values are taken to be $T_0 = 340°\mathrm{C}$ and $p_0 = 0$ kPa (gauge), and the ramp times t_{ramp} are 360 and 60 s for the temperature and pressure schedules, respectively. For the AS4/PEEK system, the prepreg viscosity given in Ref. 24, the temperature-dependent welding time in Eq. (42.18) and the surface parameters listed in Table 42.1 are adopted in the numerical integration of Eq. (42.17).

Figure 42.9a presents a contour plot of the bonding time t_b for the AS4/PEEK system to achieve a 90 percent degree of bonding ($D_b = 0.9$)

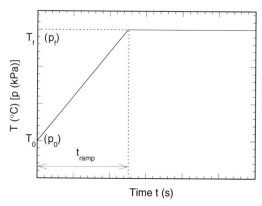

Figure 42.8 Schematic of the profile of the temperature (pressure) variation considered in the illustrative calculations.

as a function of the hold pressure p_f and the hold temperature T_f. It is seen that t_b decreases monotonically with increasing p_f and T_f, owing to faster healing and intimate contact processes. Under the lowest hold temperature and pressure, $T_f = 350°C$ and $p_f = 10$ kPa, a relatively long bonding time, $t_b = 8910$ s, is required; significant decrease to $t_b = 1230$ s is obtained when the hold temperature and pressure increase to $T_f = 410°C$ and $p_f = 100$ kPa. Following the presentation format in Fig. 42.9a, the bonding time required to achieve 95 and 100 percent degrees of bonding is shown in Fig. 42.9b and c, respectively. For a given combination of T_f and p_f, the bonding time increases notably with increasing D_b, as physically expected. When $T_f = 350°C$ and $p_f = 10$, the bonding time is almost doubled from 8910 s in Fig. 42.9a to 16,660 s in Fig. 42.9c. The results in Fig. 42.9 provide a convenient tool to determine the processing pressure, temperature, and time for a desired degree of bonding of the AS4/PEEK system.

Evidently, higher hold temperature and pressure are desirable for shorter processing time; however, the manufacturing cost may also increase with increasing T_f and p_f. To quantitatively address the competing effects, a cost function C is introduced in the study as

$$C = \int_0^{t_b} \left[\frac{T}{T_{\text{ref}}} + \frac{p}{p_{\text{ref}}} \right] dt \qquad (42.19)$$

where the reference temperature and pressure are taken to be $T_{\text{ref}} = 410°C$ and $p_{\text{ref}} = 100$ kPa. It must be mentioned that Eq. (42.19) serves as an example to introduce the methodology of cost analysis for the fusion bonding process, and different forms of cost functions may be adopted for different manufacturing considerations.

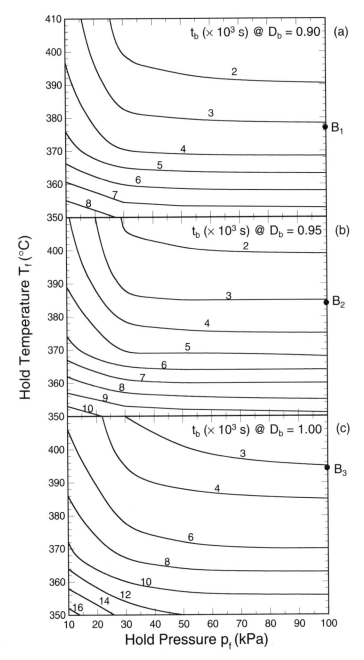

Figure 42.9 Contour plot of the required bonding time t_b as a function of the hold pressure p_f and the hold temperature T_f for the AS4/PEEK system to achieve (a) $D_b = 0.90$, (b) $D_b = 0.95$, and (c) $D_b = 1.00$.

Using the bonding time and the corresponding hold temperature and pressure obtained in Fig. 42.9, the cost function in Eq. (42.19) is readily evaluated, and the results are presented again as contour plots (with p_f and T_f as coordinate axes) in Fig. 42.10 for the three values of D_b considered. For a fixed hold temperature, it is observed that the cost function C may not be increasing monotonically with increasing hold pressure, since smaller p_f may lead to longer processing time and higher cost, and higher cost is also needed to maintain relatively high pressures. Under the current definition of C in Eq. (42.19), however, the cost decreases monotonically with increasing T_f, owing to significant reduction in the bonding time. To achieve a 90 percent degree of bonding, the combination of $T_f = 410°C$ and $p_f = 50$ kPa, denoted as point A_1 in Fig. 42.10a, yields a minimum cost around 2010 s. Similarly, the processing conditions that correspond to minimum cost for the two cases, $D_b = 0.95$ and $D_b = 1$, are shown as points A_2 and A_3 in Fig. 42.10b and c, respectively, and the cost is seen to increase from approximately 2310 to 3675 s with increasing D_b.

It is interesting to note that Figs. 42.9 and 42.10 may be used simultaneously to determine the processing conditions by considering both the bonding time and the cost function. For example, if a cost function $C = 6000$ s is deemed acceptable for the manufacturing, Figs. 42.9a and 42.10a indicate that a combination of $p_f = 100$ kPa and $T_f = 377°C$ yields the shortest processing time of about 3200 s, as denoted by point B_1. As the expected degree of bonding increases, the processing conditions correspond to a target cost of 6000 s shift to $p_f = 100$ kPa and $T_f = 384°C$ (point B_2 in Fig. 42.10b) and $p_f = 100$ kPa and $T_f = 394°C$ (point B_3 in Fig. 42.10c) with processing times of around 3100 s (point B_2 in Fig. 42.9b) and approximately 3050 s (point B_3 in Fig. 42.9c), respectively. It must be mentioned that the decreasing trend of the shortest processing time with respect to increasing D_b may not have physical significance, since the variations are within the accuracy of interpolation of the contour plots.

Note that the discussion for Figs. 42.9 and 42.10 is limited to the case of the AS4/PEEK system with the specific cost function defined in Eq. (42.19). However, the coupled fusion bonding models in Eqs. (42.5), (42.16), and (42.17) and the methodology for the bonding time and cost analyses are generally valid for other material systems and cost considerations.

Conclusions

A complete set of fusion bonding models is developed from first principles for the processing of thermoplastic-matrix composite materials. The nonisothermal healing model and the intimate contact model are

42.20 Thermal Analysis and Design

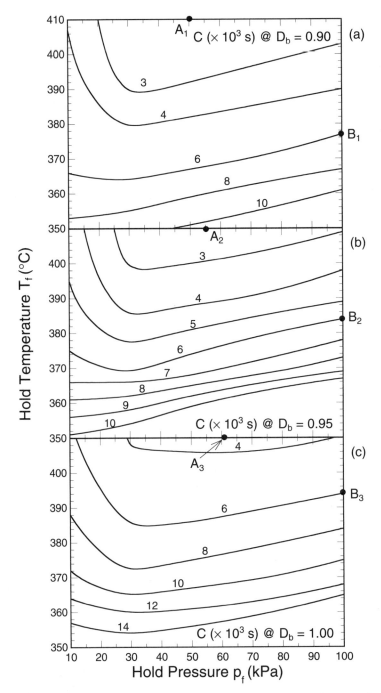

Figure 42.10 Contour plot of the cost function C [Eq. (42.19)] as a function of the hold pressure p_f and the hold temperature T_f for the AS4/PEEK system to achieve (a) $D_b = 0.90$, (b) $D_b = 0.95$, and (c) $D_b = 1.00$.

validated using experimental data from PEEK ribbons and IM-7/PIXA-M thermoplastic tapes, respectively. The healing model is coupled with the intimate contact model to predict the required bonding time and cost for desired degree of bonding for the AS4/PEEK system under a wide range of processing conditions. The bonding time is found to decrease monotonically with increasing hold temperature and hold pressure, owing to accelerated fusion bonding processes. An example cost analysis indicates that the cost function C decreases monotonically with increasing hold temperature as a result of significant saving in the processing time. Since lower pressure corresponds to longer processing time and increased manufacturing cost, and higher cost is required to maintain relatively larger hold pressure, the variation of C may show minimum with increasing hold pressure. The fusion bonding models and the methodology for the bonding time and cost function analyses may be applied to other material systems and cost considerations.

Acknowledgment

The models reported in this chapter were developed as part of a study funded by the National Science Foundation (Grant CTS-9912093).

References

1. B. T. Aström, *Manufacturing of Polymer Composites*, Chapman and Hall, New York, 1997.
2. R. Pitchumani, *Annu. Rev. Heat Transfer*, **12**, 117 (2002).
3. M. A. Lamontia et al., *Proceedings of the Submarine Technology Symposium*, The Johns Hopkins University Applied Physics Laboratory, 1992.
4. M. A. Lamontia, M. B. Gruber, M. A. Smoot, J. Sloan, and J. W. Gillespie, Jr., *J. Thermoplastic Composite Mater.*, **8**, 15 (1995).
5. R. P. Wool, B. L. Yuan, and O. J. McGarel, *Polym. Eng. Sci.*, **29**, 1340 (1989).
6. R. P. Wool and K. M. O'Connor, *J. Appl. Phys.*, **52**, 5953 (1981).
7. Y. H. Kim and R. P. Wool, *Macromolecules*, **16**, 1115 (1983).
8. W. I. Lee and G. S. Springer, *J. Composite Mater.*, **21**, 1017 (1987).
9. R. Pitchumani et al., *Int. J. Heat Mass Transfer*, **39**, 1883 (1996).
10. C. A. Butler, R. L. McCullough, R. Pitchumani, and J. W. Gillespie, Jr., *J. Thermoplastic Composite Mater.*, **11**, 338 (1998).
11. A. Majumdar and B. Bhushan, *J. Tribology*, **112**, 205 (1990).
12. F. Yang and R. Pitchumani, *J. Mater. Sci.*, **36**, 4661 (2001).
13. F. Yang and R. Pitchumani, *Polym. Eng. Sci.*, **42**, 424 (2002).
14. B. B. Mandelbrot, *Phys. Scr.*, **32**, 257 (1985).
15. T. L. Warren, A. Majumdar, and D. Krajcinovic, *J. Appl. Mech.*, **63**, 47 (1996).
16. P. H. Dara and A. C. Loos, *Thermoplastic Matrix Composite Processing Model*, Virginia Polytechnic Institute Report CCMS-85-10, Hampton, Va., 1985.
17. P. G. de Gennes, *J. Chem. Phys.*, **55**, 572 (1971).
18. S. Prager and M. Tirrell, *J. Chem. Phys.*, **75**, 5194 (1981).
19. C. W. Macosko, *Rheology Principles, Measurements, and Applications*, Wiley-VCH, New York, 1994, p. 502.
20. E. Kreyszig, *Advanced Engineering Mathematics*, 6th ed., Wiley, New York, 1988.

21. F. Yang and R. Pitchumani, *Macromolecules*, **35**, 3213 (2002).
22. F. Yang, *Investigations on Interface and Interphase Development in Polymer-Matrix Composite Materials*, Ph.D. thesis, Department of Mechanical Engineering, University of Connecticut, Storrs, 2002.
23. F. Yang and R. Pitchumani, *Polym. Composites*, **24**, 262 (2003).
24. S. C. Mantell and G. S. Springer, *J. Composite Mater.*, **26**, 2348 (1992).

Part

11

Economic Optimization

This concluding part of the handbook goes beyond thermal analysis and design and integrates design-related economic analysis. This is important because designing heat-transfer systems almost always involves tradeoffs, most notably in the form of performance versus cost.

Two chapters follow, both by U.S. academics. The first chapter gives a brief introduction to simple engineering economics and outlines its use in optimizing a heat-transfer system. A specific example is used for illustration—the economics of purchasing a new compressor/water-cooled condenser unit for an existing air-conditioning system. The second chapter presents a set of calculations that deals with a specific problem—which oven should a delicatessen purchase to cook turkeys in?

Chapter 43

Economic Optimization of Heat-Transfer Systems

Thomas M. Adams
Mechanical Engineering Department
Rose-Hulman Institute of Technology
Terre Haute, Indiana

Introduction

Designing heat-transfer systems almost always involves tradeoffs, most notably in the form of performance versus cost. Increasing the surface area of a fin array, for example, will tend to increase the heat-transfer rate from it. Increasing the surface area of the fin array, however, also increases its expense, and as the fin array becomes increasingly large, the resulting expense may become prohibitive. The surface area of a fin array above which any increase in area no longer justifies the increase in cost represents one example of an economic optimum.

Adding a fin array to a surface increases the convective heat transfer from the surface by increasing the effective area for heat transfer. Another option for increasing convective heat transfer is to increase the convective heat-transfer coefficient h. In the case of forced convection this is most easily accomplished by increasing the flow rate of the fluid in contact with the surface. Increasing this flow rate, however, also involves an added expense, in this case in the form of increased power to a fan or pump. We again see the tradeoff between performance and cost, as well as the opportunity to define an economic optimum.

In the previous examples we saw that increased performance often comes at increased cost. However, two different kinds of costs were represented. In the case of the fin array, the cost of a larger surface area is a

43.4 Economic Optimization

one-time, up-front cost, the *capital cost*. In the second example the cost of supplying power to a fan or a pump represents an ongoing expense, the *operating cost*. In some cases only one type of cost is significant and an economically optimum design may be simply defined as the one that minimizes this cost. Often both types of cost are important, however, and an additional tradeoff exists between capital and operating costs.

As an example, consider a fin array used in conjunction with the forced convection of air to achieve a given rate of heat transfer. One possible design consists of a large fin array with a low flow rate of air. This design incurs a large capital cost (due to the large fins) with a low operating cost (due to the low fan power required). Another design may employ relatively small fins with a large flow rate of air. This design has a lower capital cost but a higher operating cost. In this case it is not at all clear that minimizing either capital or operating costs alone constitutes an optimum design. Rather, a more sophisticated economic analysis is required in order to define "optimum."

This chapter gives a brief introduction to simple engineering economics and outlines its use in optimizing a heat-transfer system. Although a specific example is used for illustration, the methods employed here are general and can easily be extended to other applications. The method makes extensive use of computer software as a tool for optimization. Specifically, the software *Engineering Equation Solver*, abbreviated EES (pronounced "ease"), is used in the example. The software has been developed especially for thermal-fluid engineering applications and offers a versatility and ease of solution unavailable with more traditional optimization schemes such as linear programming or the method of lagrangian multipliers.

Simple Engineering Economics

In the previous section we saw that there are two basic types of costs, capital costs and operating costs. Capital costs represent the one-time expenses usually associated with purchasing equipment, facilities, tools, and/or land, whereas operating costs are ongoing expenses that occur repeatedly. From the perspective of designing a specific heat-transfer system, the capital cost is how much the system costs to buy or build; the operating cost is how much it costs to run. (Operating costs are usually reported on a per year basis.) In general, heat-transfer systems with large capital costs are less expensive to operate than systems with small capital costs.

Figure 43.1 shows a graph of possible designs for a given heat-transfer objective, with capital cost on the vertical axis (ordinate) and operating cost on the horizontal axis (abscissa). From Fig. 43.1 we can see that design B is better than design A, as design B has the same operating cost as

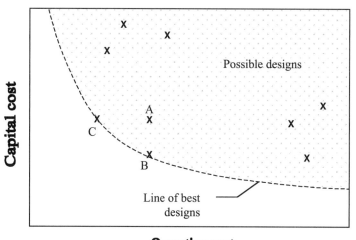

Figure 43.1 Competing effect of capital and operating costs.

design A but has a smaller capital cost. Design C is also a better design than A since it offers less expensive operation than A for the same capital cost. The dashed line bounding the possible designs on the left and the bottom, then, represents a line of best designs. Designs to the left or below this line are impossible because of either physical laws or technological feasibility. The goal of economic optimization is to guarantee that a design falls along this line.

Simple payback period

A number of economic parameters have been suggested to assess the competing nature of capital and operational costs. The most elementary parameter is known as the *simple payback period*, or more succinctly as the *simple payback* (SPB).

Mathematically, it is given as the derivative of capital cost with respect to annual savings

$$\text{SPB} = \frac{d(\text{Cap})}{d(\text{Sav})} \tag{43.1}$$

where Cap is capital cost in dollars and Sav is yearly savings in dollars per year. If yearly savings Sav is given by a decrease in annual operating cost, then

$$\text{SPB} = -\frac{d(\text{Cap})}{d(\text{Oper})} \tag{43.2}$$

where Oper is annual operating cost.

43.6 Economic Optimization

The interpretation of SPB as given in Eq. (43.1) is the amount of time in years required to recoup a capital investment in the form of savings brought about by a decrease in operating cost. Specifically, an *additional* capital dollar invested to improve an existing heat-transfer design will be recovered in SPB years. Geometrically, SPB is given by the negative of the slope of the best design curve in Fig. 43.1. (An SPB for a design not on the best design line is not considered here, as it does not represent an optimum.)

Often the simple payback is cited as the time required to recover an *entire* capital investment in a new design over an existing, base design. Here we will refer to such a payback as the *average* simple payback:

$$\text{SPB}_{\text{avg}} = -\frac{\Delta(\text{Cap})}{\Delta(\text{Oper})} \qquad (43.3)$$

The difference between SPB and SPB_{avg} can be seen in Fig. 43.2. Whereas SPB represents the local slope of the best design curve, SPB_{avg} represents only the slope of a straight line between two discreet designs falling on the curve.

By considering only the simple payback as measured by SPB_{avg} we effectively ignore the design possibilities lying along the best design curve as bases of comparison. For example, design C ostensibly has only a slightly longer simple payback than does design B as measured by SPB_{avg}. It is erroneous, however, to assume that the additional capital spent on design C over design B is recovered in $(\text{SPB}_{\text{avg,C}} - \text{SPB}_{\text{avg,B}})$. Rather, we see that SPB_C is significantly larger than SPB_B, indicating

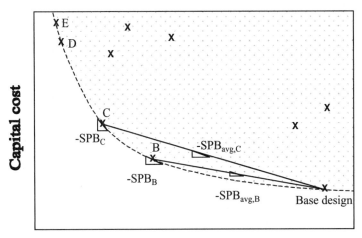

Figure 43.2 Simple payback.

that *additional* capital spent on design C over design B takes a much longer period of time to recover than suggested by the average simple payback concept. This effect is most pronounced at very small operating costs where the best design line becomes quite steep. It is difficult to argue that design E is better than design D, for instance, as additional capital spent on design E over design D will almost certainly never be recovered in the lifetime of the design. This effect may go completely unnoticed when considering only SPB_{avg}.

In short, by defining the simple payback using Eq. (43.2) rather than the average simple payback of Eq. (43.3), we capture an effect not seen in the latter, namely, the diminished returns encountered with increasingly large capital costs. For this reason, in this chapter we will consider only simple payback as defined in Eq. (43.2).

Present discounted value and return on investment

The advantage of using the simple payback period in the economic analysis of heat-transfer systems lies in its simple conceptual and geometric interpretations. An inherent limitation, however, is that SPB does not take into account the time value of money. In other words, SPB does not account for the fact that rather than generating annual savings, one could make annual payments to interest-bearing accounts and generate additional new revenue. Thus the economic question becomes whether to invest capital in new or improved heat-transfer equipment in order to generate future savings, or simply to invest money annually and earn interest. A parameter called the *present discounted value* (PDV) makes such a comparison.

PDV is defined by the following equation:

$$\text{PDV} = \text{Sav} \frac{(1+i)^n - 1}{i(1+i)^n} \tag{43.4}$$

where Sav is yearly savings, i is the annual interest rate, and n is the number of years. PDV represents the *current* value of n years' worth of future savings (decreases in operating costs) taking into account the fact that those savings could be invested at an interest rate of i instead. For example, consider a particular capital purchase of $5000 that is expected to generate $500 of savings per year over the next 10 years. One may be tempted to claim that capital will be recovered in 10 years, as $500 of yearly savings over 10 years gives 10 years × $500/year = $5000. This is not the case, however, as this $5000 of *future* savings generated over 10 years has a *present* value of only

$$\text{PDV} = \$500 \cdot \frac{(1+0.1)^{10} - 1}{0.1(1+0.1)^{10}} = \$3072 \tag{43.5}$$

43.8   Economic Optimization

where we have assumed an annual interest rate of $i = 10$ percent. In other words, one could invest $3072 today at $i = 10$ percent and earn just as much money at the end of 10 years as if one had used $3072 to purchase equipment that generated a $500/year decrease in operating costs. This makes the prospect of investing $5000 of capital in equipment to generate that same annual savings less attractive indeed. (Even if the time value of money were *not* considered here, we see that 10 years of $500 annual savings represents only SPB$_{avg}$ and not the more meaningful SPB.)

Another way to account for the time value of money is to compute an effective interest rate for a capital cost Cap expected to produce future annual savings Sav over n years. This effective interest rate is known as the *return on investment* (ROI), and is found by setting Cap equal to PDV and solving the resulting equation for the interest rate:

$$\text{Cap} = \text{Sav}\frac{(1 + \text{ROI})^n - 1}{\text{ROI}(1 + \text{ROI})^n} \tag{43.6}$$

Although this equation must be solved numerically, the ROI gives a simple comparison of potential future savings to investment returns.

In our previous example, in which a capital purchase of $5000 is expected to generate $500 of savings per year over the next 10 years, the ROI for that 10-year period is 0 percent. Should the capital investment expected to generate a $500/year savings over the next 10 years be only $4000, however, Eq. (43.6) gives

$$\$4000 = \$500 \cdot \frac{(1 + \text{ROI})^{10} - 1}{\text{ROI}(1 + \text{ROI})^{10}} \tag{43.7}$$

$$\text{ROI} = 0.043 = 4.3\% \tag{43.8}$$

If an interest-bearing account can be found with $i > 4.3$ percent, then our $4000 might be better utilized by generating interest rather than purchasing new equipment. The simplicity of this type of comparison has made the return on investment perhaps the most popular economic parameter. (It should also be noted that ROI can take on negative values, indicating that we are losing money over time for a particular capital expenditure.)

Numerous other economic parameters exist, but those given here represent three of the most commonly used parameters in engineering economics. The question of which parameter to use is largely a matter of taste. Luckily, the advent of modern software such as EES has made computing these quantities rather painless, and several different economic parameters can be calculated from the same general analysis.

Case Study: Optimization of a Water-Cooled Condenser Refrigeration Unit

As an example of the economic optimization of heat-transfer systems, let us consider a case study in which a client of an engineering consulting firm is considering upgrading an existing air-conditioning system. In particular, the client is interested in the economics of purchasing a new compressor–water-cooled condenser unit for an existing air-conditioning system. The client has little technical expertise in the area, but is familiar with economic terminology. Thus, the deliverable would most likely be a summary of economic predictions for the purchase of a new unit. This represents an ideal opportunity to perform an economic analysis with the goal of recommending an optimized system.

One of the first tasks should be to define what constitutes an economic optimum. If only capital costs *or* operating costs, but not both, are important, then the optimum design is the one which minimizes that cost. If both costs are important, however, then there are several ways to define an economic optimum, just as there is more than one economic parameter for comparing the tradeoffs between capital and operating costs.

One way to optimize a design is to specify a simple payback period and find the best design meeting that criterion. Here the optimization process amounts to finding a design that lies along the line of best designs in Fig. 43.1, where the local slope of the best design curve represents the negative of the specified SPB. Different specified SPBs result in different optimums. For example, the best design would be different for a client willing to accept a simple payback period of 5 years than for one willing to accept a simple payback of only 3 years. A design that maximizes return on investment represents another economic optimum. In the case study considered here, both of these approaches will be considered.

In order to perform an analysis, the problem definition must first be made more rigorous, and a number of modeling assumptions must be made as well. For our analysis, we will assume the following:

System requirements
- The client requires 1 ton (12,000 Btu/h) of cooling.
- The system uses R134a as a refrigerant.
- The R134a operates at an evaporator temperature of 45°F.

System reliability assumptions
- Evaporator exit superheating should be kept at 10°F.
- Condenser exit subcooling should be kept at 10°F.
- The hot-to-cold fluid temperature difference in any heat exchanger should be no less than 5°F.

43.10 Economic Optimization

Compressor modeling assumptions
- Isentropic efficiency is 65 percent.
- Purchase cost is approximately $800/hp.
- Operating cost is based on local utility rates at approximately $0.07/kWh.

Condenser modeling assumptions
- The condenser is water-cooled with a purchase cost of approximately $0.70/UA, where UA is in W/°C.
- Operating costs for condensers are based on providing city water at a cost of approximately $3.25/100 ft^3. This includes sewer fee.
- City water is available at a temperature of 65°F.
- Fluid streams experience negligible pressure losses.

Other information
- Evaporator fluid streams experience negligible pressure losses.
- The existing system has an annual operating cost of $2380.

This set of assumptions physically constrains some but not all of the system parameters. Choosing the remaining system parameter(s) in such a way that the additional constraint of a specified SPB or a maximized ROI is achieved is the heart of the optimization process.

The preceding assumptions regarding the capital costs and operating costs of the equipment are subject to frequent change and therefore possibly the most spurious of the modeling assumptions. It should be kept in mind that any model is better than none, however; furthermore, the versatility of a software model allows these figures to be easily changed should more reliable data become available.

Let us begin our analysis by analyzing the refrigerant side of the system only. A schematic of the basic vapor refrigeration cycle is shown in Fig. 43.3.

Conservation of energy applied to the evaporator yields

$$\dot{Q}_{\text{evap}} = \dot{m}_{\text{r134a}}(h_1 - h_4) \tag{43.9}$$

The heat-transfer rate into the evaporator is fixed at one ton via the problem statement. The exit enthalpy h_1 is fixed by the refrigerant exit pressure and temperature, which is 10°F higher than the evaporation temperature of 45°F. Both the flow rate of refrigerant and the inlet enthalpy are unknown as of yet.

Conservation of energy for an isentropic compressor yields

$$\dot{W}_s = \dot{m}_{\text{r134a}}(h_{2s} - h_1) \tag{43.10}$$

where h_{2s} is the enthalpy of the refrigerant corresponding to condenser pressure and the compressor inlet entropy. Conservation of energy for

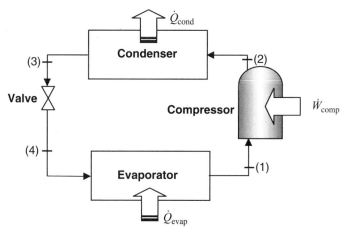

Figure 43.3 Schematic diagram of the refrigerant side of the system.

the actual compressor is given by

$$\dot{W} = \dot{m}_{r134a}(h_2 - h_1) \qquad (43.11)$$

The isentropic compressor efficiency relates the isentropic compressor power to the actual power:

$$\eta_c = \frac{\dot{W}_s}{\dot{W}} \qquad (43.12)$$

Conservation of energy applied to the condenser and the throttling valve give, respectively

$$\dot{Q}_{cond} = \dot{m}_{r134a}(h_2 - h_3) \qquad (43.13)$$

$$h_4 = h_3 \qquad (43.14)$$

The enthalpy h_3 is calculated at the condenser exit pressure and temperature, which is 10°F cooler than the condensing temperature.

Examination of these equations will show that once a condensing temperature or pressure is chosen, all other property information can be found and the resulting equations solved. This represents one degree of freedom in the cycle. Equivalently, assigning a value to the compression ratio, the ratio of the exit compressor pressure to the inlet compressor pressure, completes the equation set:

$$r_C = \frac{P_{cond}}{P_{evap}} \qquad (43.15)$$

43.12 Economic Optimization

At this point let us examine the use of the Engineering Equation Solver (EES) software in the solution of the equation set presented above. It is not the intent here to give a detailed tutorial of EES, but rather to outline its major features and its utility in performing economic optimization. The two main features of EES include the ability to solve large numbers of simultaneous nonlinear algebraic equations and the ability to include thermodynamic and/or transport property lookups as part of the equation set. The interface is very intuitive, and once the syntax for writing a property look up in equation form is learned, a novice can literally start using the software in minutes.

Equations are entered into EES by typing them in the Equations window. A property function call has the following general form:

Variable = Property(Substance, Prop1 = [value],

Prop2 = [value])

where Property is the desired property, Substance is the substance name, and Prop1 and Prop2 are independent properties needed to fix the state for the desired property. For example, in order to calculate the value of h_1, we would type

h_1 = enthalpy(R134a, T = 55, P = 54.8)

The values of the supplied independent properties may be unknowns themselves, as is the case with the enthalpy at state point 3:

h_3 = enthalpy(R134a, T = T_3, P = P_cond)

This ability to couple unknown property information with the governing equations gives EES its versatility and power.

Figure 43.4 shows a screen shot of the preceding equation set typed into the Equations window of EES. The reader will notice that the compression ration r_C has been arbitrarily set to 3.

Once a complete equation set has been entered into the Equations window, choosing the Solve command from the Calculate menu will solve the equation set and list the results in the Solutions window. Figure 43.5 shows the solution to the above equation set.

Parametric trends are easily examined by changing values in the Equations window and recalculating. In our case, different values of the compression ratio can be set in the Equations window and its effect on the rest of the cycle assessed. The process can be streamlined by making use of the Parametric Table feature of EES. In this process, equations assigning values to parameters of interest are removed from the Equations window and their values assigned to individual runs in a Parametric Table. The user then chooses Solve Table from

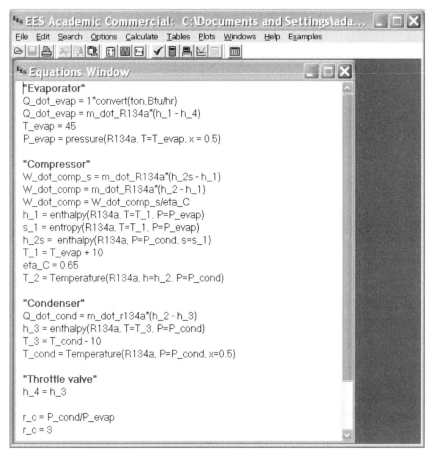

Figure 43.4 The EES Equations window.

the Calculate menu, at which point EES solves all the equations in the Equations window, using the assigned values of the parameter in the table as the remaining equation for each run. Figure 43.6 shows the results of this process for our examples with compression ratios ranging from 2 to 7. Any number of parameters can be assigned in the Parametric Table in this fashion.

Examination of the table in Fig. 43.6 shows that larger compression ratios always mean larger compressor sizes. This translates into both larger capital and operating costs for the compressor, which may lead one to believe that the optimum system is the one that utilizes the smallest possible compressor. However, we have yet to add the more detailed analysis of the water-cooled condenser. Larger compression ratios lead to larger condenser inlet refrigerant temperatures (T_2).

43.14 Economic Optimization

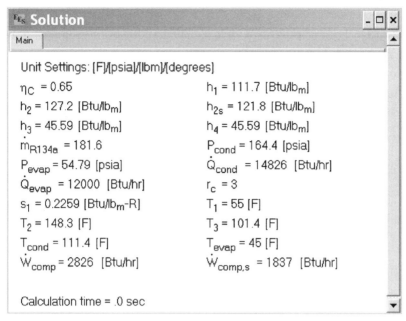

Figure 43.5 The EES Solution window.

As the inlet water temperature of the condenser is set at 65°F, larger refrigerant-to-water fluid temperature differences result within the condenser; therefore, a smaller required heat-transfer surface area is required for the condenser heat exchanger. Thus, larger compressors mean smaller condensers. In economic terms, more capital invested in a compressor results in less capital invested in the condenser, and vice versa.

It is unclear as to how changing the compression ratio affects water flow rate within the condenser. Condensers with smaller surface areas may require larger flow rates of water to accomplish the same heat-transfer rate, thereby increasing operating costs. Since the rate of heat transfer within the condenser increases with compression ratio, this effect may be exacerbated. On the other hand, the larger refrigerant-to-water temperature difference at large r_C may require smaller water flow rates. We see, then, that the tradeoffs between capital and operating costs for the system become quite difficult to intuitively determine. The use of software such as EES for modeling the system thus becomes indispensable in discerning these trends.

Figure 43.7 shows a schematic drawing of the condenser heat exchanger, modeled here as a simple counterflow type. Figure 43.8 gives a temperature-area diagram of the same heat exchanger. Three distinct

Economic Optimization of Heat-Transfer Systems

Figure 43.6 The EES `Parametric Table` window.

	r_c	$\dot{W}_{comp}$ [Btu/hr]	T_2 [F]	$\dot{Q}_{cond}$ [Btu/hr]	P_{cond} [psia]
Run 1	2.00	1560	113	13560	110
Run 2	2.56	2281	134	14281	140
Run 3	3.11	2960	152	14960	170
Run 4	3.67	3621	166	15621	201
Run 5	4.22	4280	179	16280	231
Run 6	4.78	4948	191	16948	262
Run 7	5.33	5637	201	17637	292
Run 8	5.89	6357	210	18357	323
Run 9	6.44	7119	219	19119	353

regions can be inferred from Fig. 43.8, corresponding to those regions of the heat exchanger in which the refrigerant is desuperheated, condensed, and then subcooled.

An energy balance on the superheated region of the condenser gives

$$\dot{Q}_{sh} = \dot{m}_{r134a}(h_2 - h_g) = \dot{m}_{wat} c_{p,wat}(T_{wat,out} - T_{wat,sh}) \quad (43.16)$$

The UA for the superheated section is related to the heat-transfer rate by

$$\dot{Q}_{sh} = UA_{sh} \frac{(T_2 - T_{wat,out}) - (T_{cond} - T_{wat,out})}{\ln \dfrac{T_2 - T_{wat,out}}{T_{cond} - T_{wat,out}}} \quad (43.17)$$

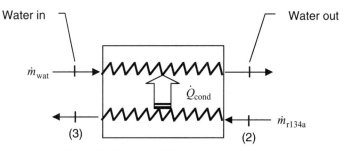

Figure 43.7 Schematic diagram of the water-cooled condenser.

43.16 Economic Optimization

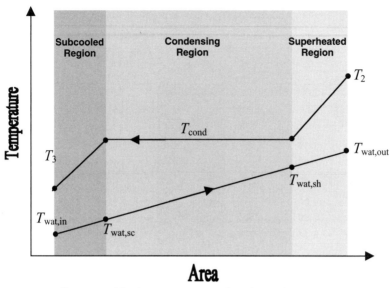

Figure 43.8 Variation of fluid temperatures within the condenser heat exchanger.

The analogous equations for the condensing and subcooled regions, respectively, are

$$\dot{Q}_{\text{sat}} = \dot{m}_{\text{r134a}}(h_g - h_f) = \dot{m}_{\text{wat}} c_{p,\text{wat}}(T_{\text{wat,sh}} - T_{\text{wat,sc}}) \quad (43.18)$$

$$\dot{Q}_{\text{sat}} = UA_{\text{sat}} \frac{(T_{\text{cond}} - T_{\text{wat,sh}}) - (T_{\text{cond}} - T_{\text{wat,sc}})}{\ln \dfrac{T_{\text{cond}} - T_{\text{wat,sh}}}{T_{\text{cond}} - T_{\text{wat,sc}}}} \quad (43.19)$$

$$\dot{Q}_{\text{sc}} = \dot{m}_{\text{r134a}}(h_f - h_3) = \dot{m}_{\text{wat}} c_{p,\text{wat}}(T_{\text{wat,sc}} - T_{\text{wat,in}}) \quad (43.20)$$

$$\dot{Q}_{\text{sc}} = UA_{\text{sc}} \frac{(T_{\text{cond}} - T_{\text{wat,sc}}) - (T_3 - T_{\text{wat,in}})}{\ln \dfrac{T_{\text{cond}} - T_{\text{wat,sc}}}{T_3 - T_{\text{wat,in}}}} \quad (43.21)$$

The total UA for the entire heat exchanger is simply the sum of the UAs for the separate regions:

$$UA = UA_{\text{sh}} + UA_{\text{sat}} + UA_{\text{sc}} \quad (43.22)$$

Figure 43.8 shows that the minimum temperature difference between the refrigerant and the water most likely occurs as the refrigerant enters the condensing region of the heat exchanger. The problem statement requires that this difference by no less than 5°F. Thus

$$T_{\text{wat,sh}} = T_{\text{cond}} - 5°\text{F} \quad (43.23)$$

The operating cost of the condenser is governed by the required volumetric flow rate of water. This is easily found from knowledge of the water mass flow rate

$$\dot{m}_{\text{wat}} = \rho \dot{V} \qquad (43.24)$$

where ρ and $\dot{V}$ are the density and volumetric flow rate of water, respectively.

Once a compression ratio is chosen, the previous refrigerant side analysis fixes T_2. Equations (43.16) to (43.24) can then be solved to size the condenser heat exchanger. In other words, Eqs. (43.16) to (43.24) can be added to our previous analysis with the resulting degrees of freedom remaining at one, specifically, the compression ratio r_C.

To find the economic optimum, the cost data must also be included. Per our modeling assumptions, the capital cost of the system is governed by the purchase of a compressor and condenser, and is given by the equations

$$\text{Cap}_{\text{comp}} = \dot{W}_{\text{comp}} \cdot 800 \frac{\$}{\text{hp}} \qquad (43.25)$$

$$\text{Cap}_{\text{cond}} = UA \cdot 0.70 \frac{\$}{\text{W}/^\circ\text{C}} \qquad (43.26)$$

where $\dot{W}_{\text{comp}}$ is in horsepower and UA is in W/°C. The operating cost of the system is governed by the same two pieces of equipment, and is given by

$$\text{Oper}_{\text{comp}} = \dot{W}_{\text{comp}} \cdot 210 \frac{\text{days}}{\text{year}} \cdot 0.07 \frac{\$}{\text{kWh}} \qquad (43.27)$$

$$\text{Oper}_{\text{cond}} = \dot{V} \cdot 210 \frac{\text{days}}{\text{year}} \cdot 0.0325 \frac{\$}{\text{ft}^3} \qquad (43.28)$$

where we have assumed operation of the unit for 210 days per year. (Care must be taken in handling the units in these two equations. EES facilitates this task as well by inclusion of a unit conversion function call.)

Figures 43.9 and 43.10 show the Equations window for the complete equation set outlined above. By using the Parametric Table feature for various values of r_C, the overall system capital and operating costs are calculated as a function of r_C. Figures 43.11 and 43.12 show the results of this analysis. (The figures were created using the graphing capabilities in EES.)

Figures 43.11 and 43.12 show that increased compression ratio indeed has the opposite effect on the costs of the compressor and the

43.18 Economic Optimization

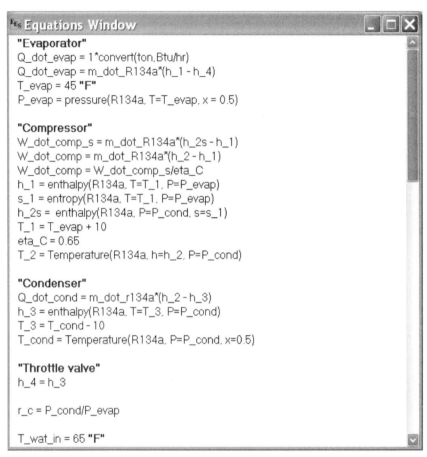

Figure 43.9 Case study equation set within EES Equations window.

condenser. As compression ratio increases, both the capital and operating costs of the compressor steadily rise, whereas both costs decrease for the condenser. Figures 43.11 and 43.12 also show that the total capital cost and the total operating cost for the system occur at very different compression ratios of approximately 2 and 4.5, respectively. As previously stated, in a case where only capital or operating costs, but not both, are important, this analysis alone is sufficient to determine the best design. In our example, if the salvage value of the existing system is significant, capital costs may not be important. Thus the best design would be the one that minimizes total operating cost at a compression ratio of 4.5.

For the case in which both of these costs are important, let us first determine the optimum system by optimizing for a specified SPB. To

Economic Optimization of Heat-Transfer Systems 43.19

Figure 43.10 (*Continued*)

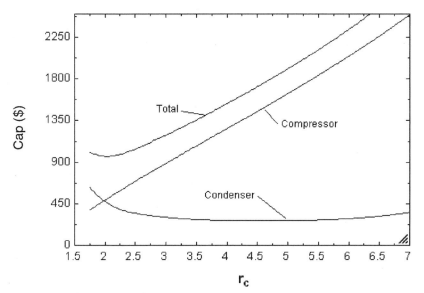

Figure 43.11 System capital costs.

43.20 Economic Optimization

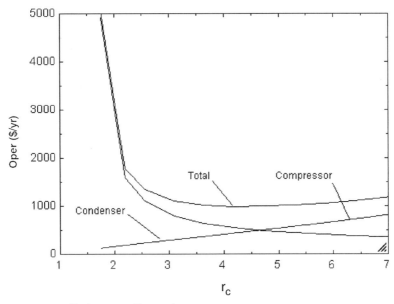

Figure 43.12 System operating costs.

find this optimum, the following figure of merit (FOM) is minimized:

$$\text{FOM} = \text{Cap} + \text{SPB} \cdot \text{Oper} \quad (43.29)$$

The rationale for minimizing this term is seen by taking the first derivative of FOM with respect to Oper and setting it equal to zero:

$$\frac{d(\text{FOM})}{d(\text{Oper})} = \frac{d(\text{Cap})}{d(\text{Oper})} + \text{SPB} = 0 \quad (43.30)$$

Solving for SPB, we find

$$\text{SPB} = -\frac{d(\text{Cap})}{d(\text{Oper})} \quad (43.31)$$

identically, as it should be. The interpretation of FOM is the total cost of a design for a period of time equal to the simple payback, including both capital and operating costs. By minimizing this cost, we ensure that our design falls along the line of best designs in Fig. 43.1.

Adding Eq. (43.29) for FOM to our equation set and setting SPB = 4 years generated the numbers used to create Fig. 43.13, where we see that the best design for a simple payback of 4 years corresponds to a compression ratio of approximately 3.65. This process can be repeated for a range of simple paybacks, which would allow clients to choose the optimum design for the SPB that they are willing to accept. Such an analysis is presented in Fig. 43.14. For our case, it is interesting to note

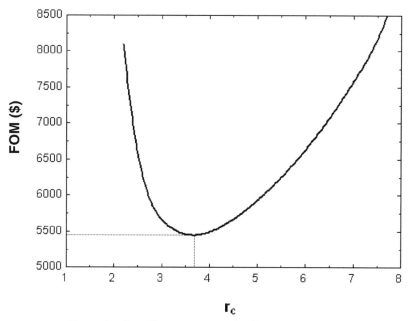

Figure 43.13 Determination of best system size for SPB = 4 years.

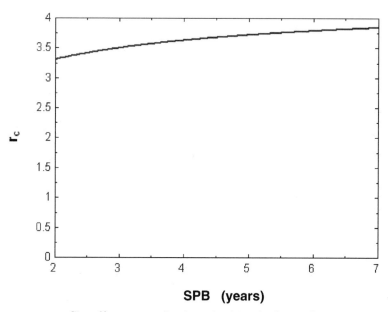

Figure 43.14 Size of best system for given simple payback period.

43.22 Economic Optimization

that the compression ratio for the best system varies only slightly for a wide range of SPB.

Optimizing for a specified SPB allows for the competing effects of capital and operating costs to be taken into account, but neglects the time value of money. By defining the optimum design as the one that maximizes the ROI, both effects are considered. In our example this is accomplished by including the defining equation for ROI [Eq. (43.6)] in our equation set. Here the savings is given by the decrease in annual operating costs for the proposed design over the existing system:

$$\text{Sav} = \$2083 - \text{Oper} \qquad (43.32)$$

Figure 43.15 shows the ROI over a 4-year period as a function of r_C. The optimum system using this criterion occurs at a compression ratio of just under 3, with a ROI reaching close to 100 percent. As interest-bearing accounts do not have interest rates anywhere close to this value, this represents a very attractive design, indeed.

At this point in our case study, we can either remain satisfied with our analysis or add more detail as needed. For example, rather than treating the condenser heat exchanger as a "black box" for which a required UA value was calculated but no physical dimensions, the individual

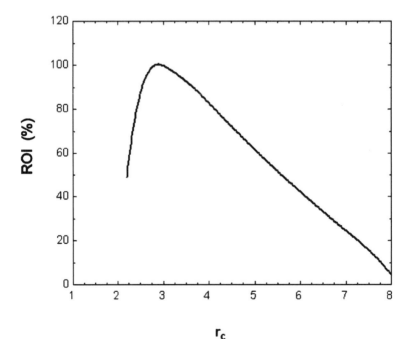

Figure 43.15 Return-on-investment analysis results.

convective heat-transfer coefficients (the h values) could be calculated for specific heat-exchanger types and geometries. This would undoubtedly give us a more accurate estimate of both the capital cost and operating cost of the condenser. Pressure drops within the fluids may also be considered. Whether we should spend the extra time and effort in this more detailed analysis, of course, should be dictated by the time and budget constraints of the project itself.

Even without an extremely detailed model of the system, however, we can easily distinguish the major trends involved. It is clear, for example, that the economically optimum system, however it is defined, should have a compression ratio of 3 to 4. Furthermore, we see from Fig. 43.11 that the capital cost of the compressor dominates the other costs in this range of compression ratios. Therefore, we may want to spend more time in a future analysis exploring the effect of compressor isentropic efficiency on system performance rather than the more tedious heat-exchanger analysis of the condenser.

Conclusion

We have introduced simple engineering economic parameters as a way of assessing the competing costs associated with the design of heat-transfer systems. Through a case study, we have coupled the physical governing equations of a particular heat-transfer system with these parameters in order to assess the economic trends, defining an economic optimum as the best design for a specified simple payback, or one that maximizes return on investment. In the process, we made extensive use of the Engineering Equation Solver (EES) software, as its versatility and functionality make it especially suited for such applications. The methods outlined in this example are easily extended to a very large variety of heat-transfer systems.

Nomenclature

Cap	Capital cost, $
c_P	Specific heat, Btu/lb$_m$·°F
h	Enthalpy, Btu/lb$_m$
i	Interest rate
n	Number of years
Oper	Operating cost, $/year
PDV	Present discounted value, $
$\dot{Q}$	Rate of heat transfer, Btu/h
P	Pressure, psia (lb/in.² absolute)
r_C	Compression ratio

43.24   Economic Optimization

ROI	Return on investment
Sav	Annual savings, $/year
SPB	Simple payback, years
UA	Heat-exchanger combined heat-transfer coefficient, Btu/h·°F
$\dot{V}$	Volumetric flow rate, gal/min (gpm)
$\dot{W}$	Power, Btu/h

Greek

ρ	Density, lb_m/ft^3
η	Efficiency

Subscripts

cond	Condenser
evap	Evaporator
f	Saturated liquid
g	Saturated vapor
r134a	Refrigerant R134a
s	Isentropic
sat	Saturated
sc	Subcooled
sh	Superheated
wat	Water

Chapter 44

Turkey Oven Design Problem

Jason M. Keith
Department of Chemical Engineering
Michigan Technological University
Houghton, Michigan

Introduction and Problem Statement

This problem integrates the analysis of combined conduction and convection heat transfer with the design-related issues of economic analysis and parametric sensitivity.

"Papa G's Deli" has decided to sell oven-roasted turkey. They have asked you to help them decide which oven to purchase to cook the turkeys in. The following information is available:

- The local grocery store sells both 10- and 20-lb turkeys at $0.50 per pound.
- Each turkey has 4 lb of bones which cannot be sold. The remainder of the turkey is sold to customers at $7.50 per pound.
- The deli employs a worker for 14 h per day, 365 days per year, at a rate of $10 per hour. You can be cooking turkeys for a maximum of only 14 h; then everyone has to go home.
- The turkey is initially at 32°F when it is placed into the oven. The oven temperature is 350°F. The turkey must be cooked until its center reaches 160°F.
- Material properties for food can be found in Geankoplis (1993). Turkey has a 75 percent water content, so the density can be approximated as that of water, such that $\rho = 1000$ kg/m^3. Furthermore, the heat

44.2 Economic Optimization

capacity $C_p = 3.31$ kJ/kg-K, and the thermal conductivity $k = 0.50$ W/m-K.

- For simplicity in obtaining an analytical solution, it should be assumed that the turkey is a sphere with uniform material properties. Although it will not be discussed here, one could validate this assumption by using a finite-element solver to compare solutions for ellipsoids and spheres.

- A "convection" oven uses a fan to circulate hot air throughout the oven. With increasing air circulation rates, the heat-transfer coefficient h will increase, leading to reduced cooking times at a constant oven temperature. The ability of the convection oven to heat a sphere of radius R can be characterized by a dimensionless group called the *Biot number*, $\text{Bi} = hR/k$, which is a ratio of the convection $(1/h)$ and conduction (R/k) heat-transfer resistances. At a minimum, one would want to choose an oven with $\text{Bi} = 1$ such that the conduction and convection resistances are equal. For this problem, assume that five "convection" ovens are available which have Biot numbers of 5.0, 2.5, 1.666, 1.25, and 1.0.

Assume that the convection oven has a fixed cost (for the walls, heating elements, etc.) and a variable cost (for a circulating fan). In general, for flow past an object, the Biot number is proportional to the square root of the gas velocity (and thus the airflow rate Q), such that $\text{Bi} \approx \sqrt{Q}$. Assuming that the variable costs are proportional to the airflow rate, they will scale as Bi^2. Thus, it is reasonable that the oven cost may be given by the formula $3000 + 200\,\text{Bi}^2$. It is further assumed that the oven lasts for 3 years and then needs to be replaced.

The following procedure is helpful in solving this problem and determining which parameters have the biggest impact on the optimum oven choice for the delicatessen:

1. Determine the thermal diffusivity of turkey α in units of m²/s.
2. Determine the radius R of a 10-lb turkey and a 20-lb turkey in meters.
3. Given the Biot number values, estimate the heat-transfer coefficient h in W/m²-K for each oven and the 10- and 20-lb turkeys.
4. Use the exact solution to the transient, one-dimensional heat equation in spherical coordinates with convection to predict the time to cook a 10-lb turkey and a 20-lb turkey in each oven.
5. Analyze the exact solution from part 4 and determine the minimum dimensionless time $\alpha t/R^2$ (and corresponding dimensionless temperature θ at the center of the turkey) when the error between the full series solution and a one-term solution is less than 1 percent for all five ovens.

6. Find the number of 10- and 20-lb turkeys that you can cook per day in each oven. (*Note:* You cannot cook two and a half turkeys; you must round down to two.)

7. Find the overall 3-year profit (or loss) for each oven, using only 10-lb and only 20-lb turkeys.

8. Which is the best oven choice and turkey selection?

9. Is there a way to cook 10- and 20-lb turkeys on the same day to increase the profit above the "best choice" scenario from part 8? How would you do this? Explain your reasoning.

10. Economic conditions may affect the price of labor, turkey, or ovens. By performing a parametric sensitivity analysis over a range of ± 50 percent around the base-case costs given in the problem statement (above), determine the most sensitive variable with respect to the 3-year profit. For simplicity, assume that the delicatessen cooks only 10-lb turkeys in an oven with Bi = 5.0.

Solution

1. The thermal diffusivity can be calculated from the following relationship:

$$\alpha = \frac{k}{\rho C_P} = \frac{0.50 \text{ W}}{\text{m-K}} \frac{\text{m}^3}{1000 \text{ kg}} \frac{\text{kg-K}}{3.31 \text{ kJ}} \frac{\text{kJ}}{1000 \text{ J}} \frac{1 \text{ J/s}}{1 \text{ W}} = 1.51 \times 10^{-7} \text{m}^2/\text{s} \tag{44.1}$$

2. Assume a spherical turkey with uniform material properties. The mass of a sphere m and radius R are given by the formula

$$m = \frac{4}{3}\pi R^3 \rho \tag{44.2}$$

For the 10-lb turkey, the radius can be determined as follows:

$$R = \left(\frac{3m}{4\pi\rho}\right)^{1/3} = \left(\frac{3}{4}\frac{10 \text{ lb}}{\pi}\frac{1 \text{kg}}{2.2 \text{ lb}}\frac{\text{m}^3}{1000 \text{ kg}}\right)^{1/3} = 0.10 \text{ m} \tag{44.3}$$

Similarly, $R = 0.13$ m for the 20-lb turkey.

3. The Biot number is a dimensionless group given by the formula

$$\text{Bi} = \frac{hR}{k} \tag{44.4}$$

which relates the convection ($1/h$) and conduction (R/k) heat-transfer resistances. For the case of Bi = 5.0 and the 10-lb turkey, the heat-transfer coefficient can be calculated as follows:

$$h = \frac{k \cdot \text{Bi}}{R} = \frac{0.50 \text{ W}}{\text{m-K}} \frac{5.0}{0.10 \text{ m}} = 25.0 \text{ W/m}^2\text{-K} \tag{44.5}$$

44.4 Economic Optimization

TABLE 44.1 Heat-Transfer Coefficients for Various Turkey and Oven Combinations

Oven type (Biot number)	Heat-transfer coefficient for a 10-lb turkey, W/m²-K	Heat-transfer coefficient for a 20-lb turkey, W/m²-K
1.0	5.0	3.8
1.25	6.3	4.8
1.666	8.3	6.4
2.5	12.5	9.6
5.0	25.0	19.2

The heat-transfer coefficients for all the oven types are given in Table 44.1.

4. The unsteady-state conservation of energy for a symmetric sphere is given as

$$\frac{\partial T}{\partial t} = \alpha \frac{1}{r^2}\left[\frac{\partial}{\partial r}\left(r^2 \frac{\partial T}{\partial r}\right)\right] \quad (44.6)$$

with the initial condition $T = T_0$ at $t = 0$ and boundary conditions: at $r = 0$, $r\, dT/dr = 0$ and at $r = R$, $-k\, dT/dr = h(T - T_1)$, where T_1 is the oven temperature. Incropera and DeWitt (1996) give the solution as a function of position and time:

$$\theta = \frac{T_1 - T}{T_1 - T_0} = \sum_{n=1}^{\infty} C_n \exp\left(-\frac{\zeta_n^2 \alpha t}{R^2}\right) \frac{R}{\zeta_n r} \sin\left(\frac{\zeta_n r}{R}\right) \quad (44.7)$$

where ζ_n is the nth positive root of the transcendental equation

$$1 - \zeta_n \cot \zeta_n = \text{Bi} \quad (44.8)$$

and C_n is the Fourier coefficient, given by

$$C_n = \frac{4\left[\sin \zeta_n - \zeta_n \cos \zeta_n\right]}{2\zeta_n - \sin(2\zeta_n)} \quad (44.9)$$

At the center of the sphere, the formula given by Eq. (44.7) can be simplified using L'Hôpital's rule:

$$\theta = \frac{T_1 - T}{T_1 - T_0} = \sum_{n=1}^{\infty} C_n \exp\left(-\frac{\zeta_n^2 \alpha t}{R^2}\right) \quad (44.10)$$

Since the oven temperature $T_1 = 350°\text{F}$, the initial turkey temperature $T_0 = 32°\text{F}$, and the desired temperature $T = 160°\text{F}$, the value of θ at the center is $(350 - 160)/(350 - 32) = 0.60$. Thus, one must find the time t such that the dimensionless temperature $\theta = 0.60$. In this case, the series expression given in Eq. (44.10) will quickly converge. To be safe,

TABLE 44.2 First Five Roots of Transcendental Equation $1 - \zeta_n \cot \zeta_n = Bi$

Bi = 1.0	Bi = 1.25	Bi = 1.666	Bi = 2.5	Bi = 5.0
1.571	1.715	1.907	2.174	2.571
4.712	4.765	4.849	5.004	5.354
7.855	7.886	7.938	8.038	8.303
10.996	11.018	11.056	11.130	11.335
14.137	14.155	14.184	14.242	14.408

the first five eigenvalues are determined for each Biot number using Eq. (44.8), and are listed in Table 44.2. A plot of the dimensionless temperature θ as a function of the dimensionless time $t^* = \alpha t/R^2$ is shown in Fig. 44.1 for the case of Bi = 2.5.

5. It must be noted that a reasonable approximation can often be made using the first term in the series of Eq. (44.10)

$$\theta_1 = \frac{T_1 - T}{T_1 - T_0} = C_1 \exp\left(-\frac{\zeta_1^2 \alpha t}{R^2}\right) \qquad (44.11)$$

where θ_1 is the solution using just one term. The error between the full solution and the one term solution is given by

$$E(\%) = \left|\frac{\theta - \theta_1}{\theta}\right| \times 100 \qquad (44.12)$$

A trial-and-error solution can be used to determine the critical dimensionless time $\alpha t/R^2$ beyond which only one term is needed in the series. The results for $E = 1$ percent are listed in Table 44.3. As the Biot number is increased, Eq. (44.8) and Table 44.2 predict a larger eigenvalue

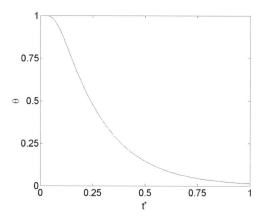

Figure 44.1 Turkey center temperature as a function of time for Bi = 2.5.

44.6 Economic Optimization

TABLE 44.3 Dimensionless Time (and Corresponding Temperature) When a One-Term Solution Is Accurate within 1 Percent

Bi	$t^* = \alpha t / R^2$	θ
1.0	0.178	0.81
1.25	0.186	0.76
1.666	0.194	0.69
2.5	0.201	0.60
5.0	0.197	0.48

ζ_1 and a more rapid transient response of the turkey temperature due to increased external convection heat transfer. In general, as $\theta \to 1$, more terms will be needed for the series to converge.

6. The ovens can operate for only 14 h per day, such that the number of turkeys per day can be calculated by dividing 14 h by the cooking time. Results for the cooking time and number of turkeys per day for the 10-lb turkeys are summarized in Table 44.4 and for the 20-lb turkeys, in Table 44.5. In both tables, the calculated number of turkeys is given with the rounded value in parentheses.

7. This point is usually where the analysis stops. However, we now perform a simple economic analysis for each oven–turkey size combination to maximize the delicatessen's profits.

As an example, the profit for the Bi = 5.0 oven with 10-lb turkeys will now be demonstrated. From Table 44.4 it can be seen that one can cook four turkeys during the course of a day. The costs to the delicatessen are the price of the turkey at

$$\frac{4 \text{ turkeys}}{\text{day}} \frac{10 \text{ lb}}{\text{turkey}} \frac{\$0.50}{\text{lb}} = \frac{\$20}{\text{day}}$$

and the employee salary at

$$\frac{14 \text{ h}}{\text{day}} \frac{\$10}{\text{h}} = \frac{\$140}{\text{day}}$$

for a total cost of $160/day.

TABLE 44.4 Cooking Times and Turkeys per Day for 10-lb Turkeys

Oven type (Biot number)	Cooking time, h	Turkeys per day (rounded values in parentheses)
1.0	5.6	2.5 (2)
1.25	5.0	2.8 (2)
1.666	4.3	3.3 (3)
2.5	3.7	3.8 (3)
5.0	3.0	4.7 (4)

TABLE 44.5 Cooking Times and Turkeys per Day for 20-lb Turkeys

Oven type (Biot number)	Cooking time, h	Turkeys per day (rounded values in parentheses)
1.0	9.5	1.5 (1)
1.25	8.4	1.7 (1)
1.666	7.3	1.9 (1)
2.5	6.2	2.3 (2)
5.0	5.0	2.8 (2)

The amount of meat on the turkey that can be sold is equal to the initial weight (10 lb) minus the weight of the bones (4 lb), such that there is 6 lb of meat. The daily income is

$$\frac{4 \text{ turkeys}}{\text{day}} \frac{6 \text{ lb meat}}{\text{turkey}} \frac{\$7.50}{\text{lb meat}} = \frac{\$180}{\text{day}}$$

The daily profit is obtained from the difference of the income and costs, or $20/day. Over the course of 3 years, the total daily profit is

$$3 \text{ years} \frac{365 \text{ days}}{\text{year}} \frac{\$20}{\text{day}} = \$21,900$$

The oven cost is given by the formula

$$\text{Oven} = 3000 + 200 \, \text{Bi}^{2.0} \tag{44.13}$$

For $Bi = 5.0$, the oven cost is $8000. Thus, the overall 3-year profit is $13,900. Table 44.6 lists the overall deli profit for each oven–turkey combination. Losses are reported in parentheses.

8. From Table 44.6, the best choice for the delicatessen is to cook 20-lb turkeys in an oven with $Bi = 2.5$ and collect a total profit of $83,350 over 3 years.

9. It is noted from Table 44.6 that there is a net profit for the $Bi = 5.0$ oven for both 10- and 20-lb turkeys. Thus, there may be a way to cook

TABLE 44.6 Overall Delicatessen Profit for Various Turkey and Oven Combinations

Oven type (Biot number)	Overall profit (loss) for 10-lb turkeys only, $	Overall profit (loss) for 20-lb turkeys only, $
1.0	(68,900)	(36,050)
1.25	(69,013)	(36,163)
1.666	(25,456)	(36,406)
2.5	(26,150)	83,350
5.0	13,900	79,600

44.8 Economic Optimization

TABLE 44.7 Overall Delicatessen Profit for Various Turkey Combinations with Bi = 5.0

Number of 20-lb turkeys	Number of 10-lb turkeys	Total cooking time, h (must be less than 14 h)	Delicatessen profit (loss), $
0	1	3	(117,500)
0	2	6	(73,700)
0	3	9	(29,900)
0	4	12	13,900
1	0	5	(40,850)
1	1	8	2,950
1	2	11	46,750
1	3	14	90,550
2	0	10	79,600
2	1	13	123,400

both 10- and 20-lb turkeys each day in order to increase profits even further. Table 44.7 shows the total cooking time and profit for combinations of 10- and 20-lb turkeys. It can be seen that the optimum choice is two 20-lb turkeys and one 10-lb turkey with a profit of $123,400. This represents about a 50 percent improvement over the result from part 8.

10. To perform the sensitivity analysis, a base case is developed with the costs given in the problem statement ($10/h for labor, $0.50/lb for turkey, and an exponent of 2.0 in the cost). It is assumed that the Biot number is 5.0 and the delicatessen cooks four 10-lb turkeys per day.

The results are shown as a plot of dimensionless profit (profit for the parameters chosen divided by the base-case profit of $13,900 obtained from Table 44.6) as a function of a dimensionless parameter (parameter value divided by the base value of that parameter) in Fig. 44.2. The solid line with circles shows that the overall profit is most sensitive to

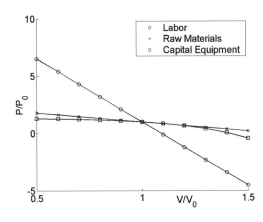

Figure 44.2 Turkey profit sensitivity analysis for 10-lb turkeys with Bi = 5.0.

the labor cost. This plot varies the labor cost while holding the raw-materials cost and oven cost parameter constant. When the labor cost is $15/h, the value of the dimensionless parameter V/V_0 is 1.5, and the yearly profit P/P_0 can be read off the figure as nearly −5, corresponding to a 3-year loss of approximately $70,000. On the other hand, the delicatessen can still make a profit if the turkey cost or oven cost parameter were to change because the P/P_0-versus- V/V_0 graph is relatively flat.

Additional sensitivity analyses can be performed for the amount of time that the delicatessen is open, oven Biot number, and other variables. However, certain discontinuities that arise in the profit-parameter graphs as there are step changes from three to four turkeys, and so on. Nevertheless, this subproblem demonstrates the power of sensitivity analysis.

Discussion

This problem integrates heat transfer with economic analysis. The problem can be made more complex by adding the concept of the time value of money and varying the interest rate or by having other oven and turkey combinations. The heat-transfer problem can also be solved using Web-based JAVA applets developed by Zheng and Keith (2003, 2004) which feature added visualization capabilities.

Acknowledgments

The author would like to thank MTU students Rebecca Hauser, Robert Sandoval, Joshua Enszer, and Thomas Horenziak for their feedback on this problem.

References

Geankoplis, C., *Transport Processes and Unit Operations*, 3d ed., Prentice-Hall, Englewood Cliffs, N.J., 1993, pp. 889–891.
Incropera, F., and DeWitt, D., *Fundamentals of Heat and Mass Transfer*, 4th ed., Wiley, New York, 1996, p. 229.
Zheng, H., and Keith, J. M., "Web-Based Instructional Tools for Heat and Mass Transfer," *Proceedings of the 2003 American Society for Engineering Education Annual Conference and Exhibition*, 2003.
Zheng, H., and Keith, J. M., "JAVA-Based Heat Transfer Visualization Tools," *Chemical Engineering Education*, **38**, 282–285 (2004).

Index

Abraham, John P., 3.3
Absolute viscosity, 7.5
Accounting (*see* Economic optimization)
Activation losses, 6.4
Actuator-controlled surface figure, 38.3
Adams, Scott, 14.1
Adams, Thomas M., 43.3
Adiabatic film effectiveness, 4.9
Adiabatic wall temperature, 4.9
Advanced Technology Solar Telescope (ATST), 38.3–38.4
 (*See also* Large telescope mirror, thermal analysis)
Aerogel insulation, 19.1, 19.2, 20.3
Aeropropulsion tunnels, 26.1
 (*See also* Combustor, hypersonic vehicle, thermal load)
Aerospace engineering (*see* Low Temperature, Low Energy Carrier)
Air, thermophysical properties, 23.14
Air-conditioning system upgrade (*see* Condenser unit, air-conditioning system)
Air-cooled electronics (*see* First-order analysis of balloon-borne electronics)
Air-cooled heat exchanger, 30.1–30.2, 30.11–30.21
Air cooling, 41.1
Air cooling of high-voltage power supply:
 analysis assumptions, 27.5
 calculations, 27.6–27.15
 Colborn *j* factor, 27.9, 27.10
 coldplate fin configuration, 27.6
 duct aspect ratio, 27.6–27.7
 energy loss, 27.13
 fin efficiency, 27.10–27.11
 folded fin, 27.6
 friction factor, 27.9, 27.10
 heat-exchanger surface efficiency, 27.11
 heat-transfer surface-area-to-volume ratio, 27.15
 hydrodynamic entry length, 27.9
 operating point, 27.14
 Reynolds number, 27.9
 statement of problem, 27.4–27.5
 thermal entry length, 27.9
 thermal properties, 27.6
Air heat capacity rate, 39.6
Alumina-pellet interface, 37.12
Aluminum, 38.5
Aluminum hub, tape pack, 12.1, 12.2, 12.13
Ammonia, 18.6–18.15
Archimedes number (Ar):
 bubbling fluidized bed, 33.7, 33.8
 CFB, 33.14
Argon, double-glazed window, 8.9
Arrhenius equation, 11.4
AS4/PEEK system, 42.2
 (*See also* Fabrication of thermoplastic, matrix composites)
ASME steam tables, 31.6
Aspect ratio, 6.8, 27.6–27.7
Asperity profile, 42.4
Atmospheric air conditions, 40.2, 40.3
Atmospheric transmittance model, 38.17
ATST, 38.3–38.4
 (*See also* Large telescope mirror, thermal analysis)
ATST M1 cell, 38.44
Authors (*see* Contributors)
Autoclave forming, 42.2
Autohesion (healing), 42.3, 42.9–42.13, 42.15
Automobile assembly plant, painting process:
 airflow per unit panel length, 14.7
 assumptions/design conditions, 14.5
 Biot number, 14.13
 blackbody, 14.12
 convection coefficient, 14.8–14.9
 convection heat transfer to painted part, 14.12
 heat transfer on panel inside wall surface, 14.6–14.8
 heat transfer through panel wall, 14.9
 nomenclature, 14.16
 Nusselt number, 14.8

Automobile assembly plant, painting process (*Cont.*):
  panel heat-transfer breakdown, 14.15
  predicted painted panel temperatures, 14.14
  problem statement, 14.1–14.4
  radiant panel outer wall surface temperature, 14.9–14.10
  radiation heat transfer, vertical painted panel, 14.10–14.11
  radiation heat-transfer calculations, 14.11–14.12
  Reynolds number, 14.7, 14.8
  spare parts carrier, 14.3
  thermal resistance, 14.9
  time steps, 14.13
  transient panel temperature analysis, 14.12
Axial temperature gradient, 31.5

Backup storage devices for computer data, 12.1
  (*See also* Tape pack cooling)
Balloon-borne science experiment (*see* First-order analysis of balloon-borne electronics)
Basic heat-transfer equations, 2.2–2.4
Benefit-cost analysis (*see* Economic optimization)
Bernoulli's principle, 7.2, 7.6
Bidmus, Hamid O., 25.1
Binary encoding, 32.2
Biot number (Bi), 44.3
  automobile assembly plant, painting process, 14.13
  convectively cooled solids, 38.6
  freezing time of food, 36.3
  graphite sheet substrates, 16.17
  lumped capacitance, 26.11
  printed-circuit board, 16.11
  SS 440 material test piece, 26.11
  tape pack cooling, 12.6–12.7, 12.9
  telescope mirror, 38.6, 38.11, 38.13, 38.17
  temperature distribution, 14.13
  turkey oven design problem, 44.3, 44.6
Blackbody, 14.12
Blasius form of smooth pipe friction factor, 24.9
Blasius smooth pipe correlation, 24.7
Bonding process, 42.3

(*See also* Fabrication of thermoplastic, matrix composites)
Boundary estimation (*see* Parameter and boundary estimation)
Boundary-layer theory, 41.4, 41.16
Boussinesq approximation, 35.9
Boxes housing heat-generating components (*see* Upper bounds of heat transfer from boxes)
Bradley, Royce, 10.1
Brazed component (liquid fuel injector), 26.8–26.10
Bubbling fluidized bed, 33.4–33.5, 33.7–33.11
  Archimedes number, 33.7, 33.8
  bubble phase, 33.4
  dense "emulsion" phase, 33.4
  Ergyn equation, 33.7
  nomenclature, 33.17–33.18
  Nusselt number, 33.8–33.10
  particles (groups A–D), 33.4
  phase map, 33.4
  Prandtl number, 33.10
  Reynolds number, 33.7, 33.8
  U, 33.8
Bunker, Ronald S., 4.1
Buoyant force, 30.3, 30.6
Byrd, Larry W., 9.3, 10.1

Cantor set surface, 42.4, 42.7, 42.8
Capacitances, 15.13–15.16
Capital budgeting (*see* Economic optimization)
Capital costs, 43.4, 43.17, 43.19
CEA code, 26.3
Cell phone, cooling mechanism (*see* Upper bounds of heat transfer from boxes)
CFB (*see* Circulating fluidized bed)
CFD, 40.28
Characteristic curve (fan), 35.12, 35.13
Chart solution method, 12.7, 12.8
Chemical Equilibrium Analysis (CEA) code, 26.3
Chemical processing technology, liquid film coating, 1.3–1.6
Chilton-Colburn analogy, 26.7, 28.4
Churchill-Bernstein equation, 25.7
Churchill-Chu Nusselt number, 40.10, 40.21
Circular duct, heat transfer (*see* Combustor, hypersonic vehicle, thermal load)

Circulating fluidized bed (CFB),
  33.5–33.7, 33.11–33.16
  Archimedes number, 33.14
  core-annulus model, 33.6
  gas convection term, 33.14
  heat transport from bed to wall,
    33.6–33.7
  nomenclature, 33.17–33.18
  Nusselt number, 33.14
  particle cluster convective term,
    33.12–33.13
  Prandtl number, 33.14
  radiation term, 33.14–33.16
  total bed-to-wall heat-transfer
    coefficient, 33.16
Closed-loop thermosyphon, 1.6–1.9
Clothes dryer (*see* Energy recovery from
  industrial clothes dryer)
Cloud-point temperature (CPT), 25.2
Coal-fired power plant, 2.6–2.7
Coefficient of performance (COP), 18.3,
  18.4, 18.9–18.12
Coefficient of thermal expansion (α), 30.6
Coil coke deposition, 11.5–11.6
Coke deposits, 11.1
  (*See also* Direct-fired cracking heaters,
    decoking intervals)
Colborn j factor, 27.9, 27.10
Cold-wall convective heat transfer (duct),
  26.6
Combustion tube of prototype PDE (*see*
  Lumped-capacitance model of PDE
  exhaust tube)
Combustor, hypersonic vehicle, thermal
  load:
  Biot number, 26.11
  brazed component (liquid fuel injector),
    26.8–26.10
  cold-wall convective heat transfer, 26.6
  conductive heat transfer inside duct
    wall, 26.10–26.15
  convective/conductive heat transfer,
    circular duct, 26.7–26.10
  convective heat-transfer coefficient,
    26.5–26.6
  convective/radiative heat transfer,
    circular duct, 26.2–26.7
  hot-wall convective heat transfer,
    26.6–26.7
  hydrodynamic length, 26.3–26.4
  JP-10 fuel, 26.8
  liquid fuel injector, 26.7–26.10
  Nusselt number, 26.4–26.5
  Poco Graphite, 26.8
  properties of combustor, 26.2
  recovery temperature inside combustor,
    26.2–26.3
  Reynolds number, laminar/turbulent
    flow, 26.2–26.3
  shear stress, 26.7
  SS heat flux (test specimen),
    26.13–26.14
  stainless steel 440 (SS 440), 26.8
  test specimen, 26.10–26.15
  theoretical temperature distribution
    (test specimen), 26.11–26.13
  thermal entry length, 26.4
  thermocouples (test specimen), 26.13
  wall temperature (test specimen),
    26.14–26.15
Commercial equation solving software, 28.9
  (*See also* Engineering Equation Solver)
Commercial numerical solvers, 38.6
Compact heat exchangers:
  applications, 27.3
  calculating core dimensions (*see* Sizing
    of crossflow compact heat
    exchanger)
  coldplate, 27.3–27.4
  (*See also* Air cooling of high-voltage
    power supply)
  defined, 27.3, 27.15
Component design, levels of analysis,
  4.5–4.6
Computational fluid dynamics (CFD),
  40.28
Concentric tube heat exchangers, 2.4, 2.5
Condensation heat transfer, vertical tube
  heat exchanger:
  ASME steam tables, 31.6
  assumptions, 31.2
  axial temperature gradient, 31.5
  condensation heat-transfer coefficient,
    31.4, 31.9
  final results, 31.9
  given information, 31.2
  iterations, 31.6, 31.8
  local condensation heat-transfer
    coefficient, 31.9
  local heat flux, 31.5
  local stream flow rate, 31.6
  local stream partial pressure, 31.6, 31.8
  modified heat of vaporization, 31.6, 31.7
  nomenclature, 31.2–31.3

Condensation heat transfer, vertical tube heat exchanger (*Cont.*):
 number of tubes needed for heat removal, 31.9–31.10
 steam mole fraction, 31.8, 31.9
 thermal resistances, 31.3–31.4
 tube wall inner surface temperature, 31.7
Condensation heat-transfer coefficient, 31.4, 31.9
Condenser, large steam power plant, 2.9–2.10
Condenser unit, air-conditioning system, 43.9–43.24
 assumptions/system requirements, 43.9
 capital cost, 43.17, 43.19
 compression ratio, 43.11, 43.14, 43.17–43.18, 43.20
 condensing region, 43.16
 conservation of energy, 43.10, 43.11
 economic optimization, 43.17–43.23
 EES (*see* Engineering Equation Solver)
 figure of merit (FOM), 43.20
 nomenclature, 43.23–43.24
 operating cost, 43.17, 43.20
 return on investment (ROI), 43.22
 schematic of vapor refrigeration cycle, 43.11
 simple payback period, 43.20–43.22
 subcooled region, 43.16
 superheated section, 43.15
Condensing heat exchanger (*see* Energy recovery from industrial clothes dryer)
Conduction:
 circular duct, 26.7–26.15
 cooler/freezer, 17.5
 corrosion detection, 9.5–9.6
 double-glazed window, 8.3, 8.4, 8.9
 Fourier heat conduction equation, 17.5, 38.11
 Fourier's law of heat conduction, 23.4, 38.11
 heat-transfer rate, 2.2
 telescope mirror, 38.6
 TEM, 37.6, 37.11
Conductive mirror materials, 38.5
Conservation of energy, 43.10, 43.11
Conservation-of-energy principle, 3.7
Constriction resistance, 37.11–37.12
Continuity of heat flux, 38.11
Contributors:
 Abraham, John P., 3.3

Adams, Scott, 14.1
Adams, Thomas M., 43.3
Bidmus, Hamid O., 25.1
Bradley, Royce, 10.1
Bunker, Ronald S., 4.1
Byrd, Larry W., 10.1
Cooper, Scott A., 33.3
Coronella, Charles J., 33.3
Cross, Alan, 11.1
Dalrymple, Nathan E., 38.3
Dearth, David R., 5.1, 12.1
Ekkad, Srinath V., 21.3
Emery, Ashley F., 34.3
Ghajar, Afshin J., 23.3
Hagen, Kirk D., 27.3
Hall, George, 15.1
Harris, S. Dyer, 24.1
Hodes, Marc, 37.3
Hoke, John L., 10.1
Jae-yong Kim, 23.3
Keith, Jason M., 6.1, 44.1
Kleinfeld, Jack M., 13.1
Lawrence, Thomas M., 14.1
Letan, R., 39.1
Mago, Pedro J., 17.1
Marthinuss, James E., Jr., 15.1
McLeskey, James T., Jr., 7.1
Mehrotra, Anil K., 25.1
Minichiello, Angela, 40.1
Nakayama, Wataru, 16.1, 35.1
Naterer, Greg F., 1.3
Naylor, David, 8.1, 28.1
Oosthuizen, P. H., 8.1, 28.1
Park, Hyunjae, 41.1
Rasul, Mohammad G., 2.1
Schauer, Fred, 10.1
Sekulic, Dusan P., 29.1
Sherif, S. A., 17.1, 18.1
Sparrow, Ephraim M., 3.3
Speyer, Daniel M., 30.1
Swain, James J., 20.1
Tiedeman, J. S., 18.1
Tiliakos, Nicholas, 26.1
Vierow, Karen, 31.1
Wessling, Francis C., 19.1, 20.1
Willey, Ronald J., 22.1
Woodbury, Keith A., 32.1
Ziskind, G., 39.1
Control point (cp), 37.7
Control volume momentum analysis of fluid, 24.5
Controlled-side interface (csi), 37.4, 37.5

Convection:
  automobile assembly plant, painting process, 14.8–14.9, 14.12
  balloon-borne air-cooled electronics, 40.10–40.15
  CFB, 33.12–33.14
  circular duct, 26.2–26.10
  cooling of fuel cell, 6.5–6.9
  cooling printed-circuit boards, 41.1–41.18
  double-glazed window, 8.3, 8.7
  forced (*see* Forced convection)
  fuel cell, 6.1, 6.2
  heat-transfer coefficient, 21.3–21.15
  heat-transfer rate, 2.2
  lumped-capacitance model, PDE, 10.7–10.10
  plate coolers and freezers, 17.2
  portable PCM cooler, 39.5–39.6
  roof moisture surveys, 13.4, 13.6, 13.7, 13.11
  surface-air temperature difference, 38.4
  telescope mirror, 38.4
Convection oven (*see* Turkey oven design problem)
Convective mass-transfer conductance, 28.5
Convectively cooled heat, dissipating component, PCB:
  boundary-layer theory, 41.4, 41.16
  heater location/heater spacing, 41.16
  multiple-heater solution, 41.6–41.8, 41.10–41.14
  nomenclature, 41.17, 41.18
  one-heater solution, 41.4–41.6, 41.8–41.9
  problem formulation and analysis, 41.3–41.8
  problem statement, 41.1–41.3
  Reynolds number, 41.4, 41.8–41.14
  sample analysis results, 41.8–41.14
  sample thermal design procedure, 41.14–41.16
  schematic diagrams, 41.2
  simplification/limitations, 41.4
  summary, 41.16–41.17
  superposition method, 41.2, 41.4
  TIF, 41.8
Convergence, 20.12, 28.9
Coolant exit temperature, 4.9
Coolant temperature rise, 4.13
Cooled gas turbine airfoil:
  adiabatic film effectiveness, 4.9
  component design, 4.5–4.6
  coolant temperature rise, 4.13
  external boundary conditions, 4.7–4.10
  external flow Reynolds numbers/HTCs, 4.11
  external heat-transfer coefficients, 4.7–4.8
  film cooling, 4.8–4.9, 4.14
  film temperature, 4.8
  gas temperature, 4.8
  impingement cooling, 4.11–4.13
  internal cooling, 4.11–4.13
  one-dimensional thermal model, 4.6
  optimization of solution, 4.13–4.15
  problem geometry and flow simplification, 4.4
  Reynolds number, 4.8, 4.9, 4.11
  roughness Reynolds number, 4.9
  solution outline, 4.5–4.7
  surface roughness effect, 4.9–4.10
  TBC, 4.3
  wall temperature, 4.8
Coolers (*see* Heat and mass transfer with phase change)
Cooling efficiency, 37.15
Cooling fin:
  dual-fin concept, 5.2, 5.9–5.11
  FEA methods, 5.6–5.11
  (*See also* FEA methods)
  Grashof number, 5.4–5.5
  Nusselt number, 5.5
  Rayleigh number, 5.5
  sample problem, 5.3–5.6
  single fin, 5.2, 5.10
  temperature distribution along length of fin, 5.6
  T_{film}, 5.4
Cooling of fuel cell:
  aspect ratio, 6.8
  convection cooling, 6.5–6.9
  cooling channels, 6.5–6.6
  E_{cell}, 6.4
  efficiency, 6.5
  fuel usage, 6.4–6.5
  heat generation, 6.5
  laminar flow, 6.9
  M_{fuel}, 6.3, 6.4
  molar hydrogen flow rate, 6.3, 6.4
  Nusselt number, 6.7, 6.8
  power, 6.3, 6.4, 6.9
  Q_{cell}, 6.5, 6.6

Cooling of fuel cell (*Cont.*):
   Reynolds number, 6.7, 6.9
   workings of fuel cell, 6.1–6.4
Cooper, Scott A., 33.3
COP, 18.3, 18.4, 18.9–18.12
Copper windings, turbogenerator rotor cooling, 7.9–7.11
Core-annulus model, 33.6
Core dimensions of heat exchanger (*see* Sizing of crossflow compact heat exchanger)
Coronella, Charles J., 33.3
Corrosion, 22.1
Corrosion detection (*see* Locating corrosion under paint)
Cosmos/M, 13.2
COSMOSWorks, 38.6
Cost-benefit analysis (*see* Economic optimization)
Counterflow concentric heat exchanger, 2.5
Coupled fusion bonding models, 42.8, 42.12, 42.13, 42.19
cp, 37.7
CPT, 25.2
Cracking heaters (*see* Direct-fired cracking heaters, decoking intervals)
Creep, 32.5
Cross, Alan, 11.1
Crossflow compact heat exchanger (*see* Sizing of crossflow compact heat exchanger)
Crossflow effect, 4.12
Crossflow heat exchangers, 2.5
Crude oil (*see* Solids deposition, oil pipelines)
csi, 37.4, 37.5
Cubic polynomial, roots, 30.10
Cyclic calculations (*see* Transient and cyclic calculations)

Dalrymple, Nathan E., 38.3
Dearth, David R., 5.1, 12.1
Decoking intervals (*see* Direct-fired cracking heaters, decoking intervals)
Deflagrative combustion, 10.2
Degree of healing, 42.10
Delicatessen, turkey oven (*see* Turkey oven design problem)
Dendrite fragmentation, 1.23
Density, 7.5
Density-versus-elevation plot, 30.3
Deposit aging, 25.18
Derivatives, 20.4–20.10
Desktop computer, cooling mechanism (*see* Upper bounds of heat transfer from boxes)
Detonative combustion, 10.2
Differential equations:
   corrosion detection, 9.6–9.7
   LoTEC, 19.4
   lumped-capacitance model, PDE, 10.4, 10.9
   moving sheet, moving fluid, 3.16–3.17
   similarity solutions for velocity problem, 3.21
   tape pack cooling, 12.6
   telescope mirror, 38.8
   thermal system transient response, 15.17
Diffusion coefficient, 42.10, 42.11
Dimensionless quantities:
   Archimedes number (Ar), 33.7, 33.8, 33.14
   Bi (*see* Biot number)
   Churchill-Chu Nusselt number, 40.10, 40.21
   Fourier number (Fo), 12.7, 38.12, 38.13
   Gr (*see* Grashof number)
   Grashof and Prandtl numbers (Gr Pr), 30.20–30.21
   Lewis number, 17.5
   Nu (*see* Nusselt number)
   Pr (*see* Prandtl number)
   Ra (*see* Rayleigh number)
   Re (*see* Reynolds number)
   Schmidt number (Sc), 28.4
   Sherwood number (Sh), 28.4
   Stanton number (St), 26.7, 27.9
Dirac delta function, 42.11
Direct-cooled rotors (*see* Turbogenerator rotor cooling)
Direct-fired cracking heaters, decoking intervals:
   coke deposition rates, 11.5–11.6
   hydrocarbon and petroleum fraction decomposition rates, 11.3–11.5
   maximum allowable coke thickness, 11.6
   maximum heat flux, 11.2–11.3
   nomenclature, 11.8–11.9
   pressure drag, 11.2
   reaction velocity constant equations, 11.5
   run length, 11.6–11.7

Dittus-Boelter correlation, 22.3, 25.6
Dittus-Boelter equation, 14.8
Donohue equation, 22.2
Double-door technique, 19.2
Double-glazed window:
   air vs. argon, 8.9
   aspect ratio, 8.6
   combined radiation/convection, 8.3
   conduction, 8.3, 8.4, 8.9
   convective heat-transfer coefficient, 8.7
   iterative solution procedure, 8.7–8.8
   kinematic viscosity, 8.7
   nomenclature, 8.2–8.3
   Nusselt number, 8.7
   problem statement, 8.1
   Rayleigh number, 8.6
   thermal conductivity, 8.7
   thermal resistance, 8.9
   vertical high-aspect-ratio enclosure, 8.6
Drew's equation, 30.8
Dry-bulb temperature, 40.25, 40.26
Duct aspect ratio, 27.6–27.7

E_{cell}, 6.4
EA, 32.2
   (*See also* Heat-exchanger design)
Earth atmospheric experiment (*see* First-order analysis of balloon-borne electronics)
Economic optimization:
   capital costs, 43.4
   example (*see* Condenser unit, air-conditioning system)
   fabrication of thermoplastic composites, 42.17, 42.20, 42.21
   operating costs, 43.4
   present discounted value (PDV), 43.7–43.8
   return on investment (ROI), 43.8
   simple payback period, 43.5–43.7
   turkey oven design problem, 44.1–44.9
EES (*see* Engineering Equation Solver)
Effectiveness:
   adiabatic film, 4.9
   film cooling, 21.5
   NTU method, 2.5–2.6
Effectiveness-NTU method:
   design of heat exchanger with specific specifications, 2.13–2.16
   effectiveness, 2.5–2.6
   execution time, 32.13
Efficiency:
   cooling, 37.15
   energy loss (*see* Energy loss)
   extended surface, 29.28
   fin, 27.10–27.11, 29.27
   fuel cell, 6.2–6.3, 6.5
   heat-exchanger surface, 27.11, 29.28
Eigenfunction, 9.7, 9.9
Eigenvalue, 9.6, 9.10, 9.11
Ekkad, Srinath V., 21.3
Electric transformer air-cooled heat exchanger, 30.11–30.21
Electrical dryer (*see* Energy recovery from industrial clothes dryer)
Electrically heated circular tube, inside-wall heat-transfer parameters:
   bulk Prandtl numbers, 23.20, 23.22
   bulk Reynolds numbers, 23.20, 23.22
   calculated Reynolds number (one-phase), 23.19
   experimental setup, 23.7–23.10
   finite-difference formulations, 23.4–23.6, 23.7–23.13
   kth thermocouple station, 23.12–23.13
   nomenclature, 23.24–23.27
   nonuniform thermal conductivity, 23.5
   Nusselt number, 23.20, 23.22, 23.23
   output, single-phase flow heat-transfer test run, 23.19–23.20
   output, two-phase flow heat-transfer test run, 23.21–23.22
   superficial Reynolds number (two-phase), 23.21
   thermophysical properties, 23.13–23.20
   zero-thickness control volume layer, 23.11
Electronic boxes (*see* Upper bounds of heat transfer from boxes)
Electronic packing industry, 41.1
   (*See also* Convectively cooled heat, dissipating component, PCB)
Electronics cooling, 15.2
   (*See also* Thermal system transient response)
Elemental tubular heat-exchanger design equations, 22.2
Elements, 5.9
Elitism, 32.5–32.6
Emery, Ashley F., 34.3
Energy balance:
   intercooler-flash tank, 18.5
   outlet temperature of hot fluid, 2.12
   solids deposition, oil pipeline, 25.2–25.4

Energy balance (*Cont.*):
  superheated region of condenser, 43.15
  TEM, 37.7–37.8
  thermal systems, 15.4
  transient heat and mass transfer, 17.3
Energy loss:
  activation losses, 6.4
  air cooling of high-voltage power supply, 27.3
  entrance loss, 27.13
  gradual enlargement loss (fan transition), 27.13
  I^2R losses, 7.9
  load loss, 30.11
  mass-transfer losses, 6.4
  ohmic losses, 6.4
  pressure drop (*see* Pressure drop/loss)
  single-phase natural circulation loops, 30.4
Energy recovery from industrial clothes dryer:
  coils, addition of, 28.9–28.10
  condensation heat transfer, 28.6–28.7
  conditions after coil, 28.9–28.10
  convective mass-transfer conductance, 28.5
  final result underestimated, 28.10–28.11
  Gnielinski equation, 28.7
  Hilpert's correlation, 28.4, 28.5
  internal pipe flow (forced convection), 28.7
  iterations, 28.4–28.9
  LMTD analysis, 28.8–28.9
  mass-transfer resistance, 28.4–28.6
  noncondensable vapors, 28.11
  Nusselt number, 28.6–28.7
  overall heat-transfer rate, 28.8–28.9
  problem statement, 28.1–28.3
  Reynolds number, 28.5, 28.7
  sensible-heat transfer, 28.11
  specific/relative humidity, 28.10
  thermal resistance diagram, 28.3
  vapor shear, 28.11
Engineering economics (*see* Economic optimization)
Engineering Equation Solver (EES), 28.9, 43.4, 43.12
  Equations window, 43.12, 43.13
  Parametric Table, 43.12–43.13, 43.15
  property function cell, 43.12
  Solutions window, 43.12, 43.14

Engineers/academics (*see* Contributors)
Enhancement factor, 17.4
Ensemble design, 32.9 32.11
Enszer, Joshua, 44.9
Enthalpy:
  two-phase pressure drop in pipes, 24.8
  vent ports, 35.7–35.15
Entrance loss, 27.13
Envelope of box, 35.2
ε_{mf}, 33.4
Equation solving software, 28.9
  (*See also* Engineering Equation Solver)
Ergyn equation, 33.7
`Estimate_kappa`, 34.7
Estimation of parameters in models:
  estimation, 34.7
  minimizing least-squares error, 34.4
  noise, 34.7
  pseudocode/Matlab programs, 34.7
  sensitivity, 34.4–34.7
Ethane cracking, 11.5
Ethylene glycol/water mixture, thermophysical properties, 23.15–23.16
Evolutionary algorithm (EA), 32.2
  (*See also* Heat-exchanger design)
Excel, 8.8, 31.8, 38.12
Exergetic efficiency, 18.5, 18.6, 18.13–18.14, 18.15
Exergy destruction, 18.4
Exergy gained, 18.5
Exergy supplied, 18.5
Explicit finite-difference numerical scheme, 38.11–38.14, 38.21
Extended surface efficiency, 29.28

Fabrication of thermoplastic, matrix composites:
  bonding time, 42.21
  cost function, 42.17, 42.20, 42.21
  fabrication methods, 42.2–42.3
  fusion bonding model, 42.13
  interfacial bond strength, 42.3, 42.13
  interlaminar bond strength, 42.3, 42.10
  intimate contact, 42.4–42.9, 42.13–42.15
  nonisothermal healing process, 42.9–42.13, 42.15
  reptation theory, 42.9
  results and discussion, 42.13–42.19
  schematic of fusion bonding process, 42.3
Fan, cooling, computer case, 35.11–35.15

Fan as cooling agent for box, 35.11–35.15
Fanning friction factor (*see* Friction)
Faraday's constant, 6.3
Fast Fourier transform (FFT), 42.4, 42.7
FDA/FEM, 15.1
FEA methods:
   analysis models, 5.9
   balloon borne air-cooled electronics, 40.27–40.28
   cooling fin, 5.6–5.11
   elements, 5.9
   historical overview, 5.8–5.9
   mesh, 5.9
   NASTRAN, 38.15
   nodes, 5.9
   radiation solvers, 40.28
   roof moisture surveys, 13.2–13.4
   tape pack cooling, 12.9–12.14
   what is it, 5.9
FEA radiation solvers, 40.28
FEMLab, 38.6
FFT, 42.4, 42.7
Figure of merit (FOM), 43.20
Filament winding, 42.3
Fill-burn-exhaust cycle, 10.2
Fill-burn-purge cycle, 10.4
Film coefficients, 22.1–22.9
Film cooling, 4.8–4.9, 4.14, 21.5
Film cooling effectiveness, 21.5
Fin efficiency, 27.10–27.11, 29.27
Fin pitch, 27.6
Fin problem (*see* Cooling fin)
Finite-difference/finite-element method, 16.1
Finite-difference analysis/finite-element analysis (FDA/FEA), 15.2
Finite-element analysis (*see* FEA methods)
Fired heaters (*see* Direct-fired cracking heaters, decoking intervals)
First-generation asperities, 42.4
First-law efficiency, 18.1–18.3
First law of thermodynamics, 15.3, 30.4
First-order analysis of balloon-borne electronics:
   assumptions, 40.5–40.8
   atmospheric air conditions, 40.2, 40.3
   background, 40.1–40.2
   Churchill-Chu Nusselt number, 40.10, 40.21
   enclosure (electronics enclosure), 40.3–40.5

   enclosure external transfer (convection), 40.10–40.15
   enclosure external transfer (radiation), 40.15–40.16
   enclosure internal air temperature, 40.16
   follow-on analyses, 40.27–40.28
   gondola, 40.4, 40.6, 40.7
   Grashof number, 40.11
   humidity equations, 40.23–40.27
   IHT software, 40.21
   nomenclature, 40.28–40.30
   operating specifications, 40.2
   overall energy balance, 40.9
   polynomial curve fits, plotting techniques, 40.13, 40.14
   Prandtl number, 40.8, 40.14
   problem statement, 40.2–40.3
   Rayleigh number, 40.11, 40.12, 40.21
   Reynolds number, 40.8–40.9, 40.17
   thermal calculation results, 40.21–40.23
   thermal system of equations, 40.9–40.18, 40.18–40.20
First-order forward difference, 38.12
Fitness function, 32.3, 32.8, 32.9
Flash curve data, 11.7
Flat-crest folded fin, 27.6
Flow network modeling, 40.28
Flow resistance (*see* Resistance)
Fluent 6.1 software, 39.18
Fluidization hydrodynamics, 33.3–33.4
Fluidized bed:
   applications, 33.3
   basic principles, 33.3–33.4
   bubbling regime (*see* Bubbling fluidized bed)
   CFB (*see* Circulating fluidized bed)
   freezer, 36.1–36.5
   nomenclature, 33.17–33.18
Fluidized-bed freezer, 36.1–36.5
Fo, 12.7, 38.12, 38.13
Folded fin, 27.6
FOM, 43.20
Food:
   freezing time, 36.1–36.5
   thermal properties, 36.4
Forced air cooling, 41.1
Forced convection:
   balloon-borne air-cooled electronics, 40.16–40.17
   dimensionless groups, 2.2
   flat plates, 17.2

Forced convection (*Cont.*):
  internal pipe flow, 28.7
  turbogenerator rotor cooling, 7.6
  two-phase flow patterns (pipe), 24.4–24.5
  (*See also* Convection)
Fouling, 22.1
Fourier coefficient, 44.4
Fourier heat conduction equation, 17.5, 38.11
Fourier modulus (Fo), 12.7
Fourier number (Fo), 38.12, 38.13
Fourier series, 10.3
Fourier transformation, 42.11
Fourier's law, 3.17
Fourier's law of heat conduction, 23.4, 38.11
Fourier's theorem, 42.7
Fractal intimate contact model, 42.4–42.9, 42.13–42.15
Fractional thermal resistance (θ_d), 25.15–25.17
Freezers (*see* Heat and mass transfer with phase change)
Freezing time of foods, 36.1–36.5
Friction:
  air-cooled heat exchanger, 30.17, 30.18
  air cooling of high-voltage power supply, 27.9, 27.10, 27.13
  crossflow compact heat exchanger, 29.10–29.17, 29.21, 29.25, 29.26
  laminar flow, 30.8
  Prandtl's universal law, 30.8
  skin friction coefficient, 26.7
  turbogenerator rotor cooling, 7.7
  turbulent flow, 30.17
  two-phase pressure drops in pipes, 24.6, 24.7, 24.9
Frontside convective cooling, 38.44
Frost density, 17.5
Frost deposition rate per unit area, 17.6
Frost surface temperature, 17.6, 17.9
Frost thermal conductivity, 17.6
Frost thickness, 17.6, 17.8–17.10
Fuel cell, 6.1–6.4
Fuel cell cooling (*see* Cooling of fuel cell)
Fuel cell stack, 6.2, 6.3
Fully rough surface heat-transfer coefficient, 4.10
Furnace wall, coal-fired power plant, 2.6–2.7
Fusion bonding process, 42.3
(*See also* Fabrication of thermoplastic, matrix composites)

GA, 32.2
Gas-fluidized bed (*see* Fluidized bed)
Gas oil cracking, 11.5
Gas-solid contacting methods, 33.4
Gas turbine cooling (*see* Cooled gas turbine airfoil)
Gating and risering system, 1.9–1.11
Gauss algorithm, 20.4, 20.11
Gemini telescopes, 38.3
(*See also* Large telescope mirror, thermal analysis)
General jet array impingement geometry, 4.12
Genetic algorithm (GA), 32.2
Geometric mean method, 18.3, 18.4, 18.10, 18.15
Ghajar, Afshin J., 23.3
Gnielinski equation, 28.7
Gradual enlargement loss (fan transition), 27.13
Graphic sheet substrates, 16.15–16.19
Grashof number (Gr):
  balloon-borne air-cooled electronics, 40.11
  cooling fin, 5.4–5.5
  tape pack cooling, 12.5
Grashof and Prandtl numbers (Gr Pr), 30.20–30.21
Gray-body radiation, 30.20
Green's function, 9.7–9.9
Grimson correlation, 39.5, 39.6
Group A particles, 33.4
Group B particles, 33.4
Group C particles, 33.4
Group D particles, 33.4

Hagen, Kirk D., 27.3
Hall, George, 15.1
Harris, S. Dyer, 24.1
Hauser, Rebecca, 44.9
Healing, 42.9–42.13, 42.15
Heat and mass transfer with phase change:
  air velocity, 17.8–17.9
  applications, 17.1
  calculation procedure, 17.2–17.7
  energy balance, 17.3
  enhancement factor, 17.4
  example, 17.7

frost density, 17.5
frost deposition rate per unit area, 17.6
frost surface temperature, 17.6, 17.9
frost thermal conductivity, 17.6
frost thickness, 17.6, 17.8–17.10
heat-transfer rate (conduction), 17.5
heat-transfer rate (mass diffusion), 17.3
humidity ratio, 17.4
latent heat of sublimation, 17.5
local mass-transfer coefficient, 17.5
nomenclature, 17.10–17.11
published works, 17.2
quasi-steady-state analysis, 17.3
Reynolds number, 17.9, 17.10
saturation pressure, 17.4
Heat exchanger:
  air-cooled, 30.1–30.2, 30.11–30.21
  air cooling of high-voltage power supply, 27.3–27.16
  compact (see Compact heat exchangers)
  condensing, 28.1–28.11
  corrosion, 22.1
  defined, 2.4
  design, 2.13–2.16
  design using evolutionary algorithms, 32.1–32.16
  effectiveness-NTU method, 2.5–2.6, 2.13–2.16
  elemental tubular design equation, 22.2
  energy recovery from industrial dryer, 28.1–28.11
  example, 2.11–2.12, 2.13–2.16
  fouling, 22.1
  LMTD method, 2.5
  methodology, 27.1
  optimization methods, 32.1, 32.2
  passive condenser, 31.1–31.10
  room cooling (see Portable PCM cooler)
  Shah's design procedure, 32.2
  single-phase natural circulation loops, 30.1–30.21
  sizing of crossflow compact heat exchanger, 29.1–29.32
  SPHE, 32.6–32.14
  step-by-step methodology, 29.1–29.32
  types, 2.4–2.5
  vertical tube, 31.1–31.10
  Wilson analysis, 22.1–22.9
Heat-exchanger design:
  "asking for too much," 32.11
  convergence history for EA SPHE design, 32.9, 32.10

creep, 32.5
crossover, 32.4–32.5
EAez-m, 32.9, 32.11, 32.14
elitism, 32.5–32.6
ensemble design, 32.9–32.11
evolutionary algorithm (EA), 32.2–32.5
execution times, 32.13
fitness function, 32.8, 32.9
heat-transfer coefficient, 32.6
inequality constraints, 32.11–32.13
initial population, 32.3
LMTD, 32.6
mating the parents, 32.4
mutation, 32.5
nomenclature, 32.14–32.15
Nusselt number, 32.6
overview, 32.1–32.2
pressure drop, 32.7
properties and design data, 32.7–32.8
reproduction, 32.4
results from 10 EA designs, 32.9–32.11
Reynolds number, 32.6
selection of parents, 32.3–32.4
Shah's procedure, 32.2
SPHE, 32.6–32.14
Heat-exchanger surface efficiency, 27.11
Heat/mass-transfer analogy, 28.4
Heat recovery system (see Energy recovery from industrial clothes dryer)
Heat-sink resistance, 37.15–37.16
Heat transfer:
  basic equations, 2.2–2.4
  modes, 2.1
  when it happens, 2.1
"Heat Transfer by Natural Convection across Vertical and Inclined Air Layers" (Sherbiny et al.), 8.7
Heat-transfer coefficient determination:
  convective heat-transfer coefficient, 21.3–21.5
  Dittus-Boelter relation, 22.3
  Donohue equation, 22.2
  elemental tubular heat-exchanger design equations, 22.2
  film cooling, 21.5
  heat-transfer film coefficients, 22.1–22.9
  Newton's law of cooling, 22.3
  Nusselt number, 22.3, 22.8
  semi-infinite solid assumption, 21.3–21.5
  Wilson analysis, 22.1–22.9
  Wilson plot, 22.4, 22.8

Heat-transfer film coefficients, 22.1–22.9
Heat-transfer rate:
  concentric tube heat exchanger, 2.5
  conduction, 2.2
  convection, 2.2
  defined, 2.2
  high rate required, 2.1
Heisler chart, 12.8
Helium, 38.5
Helium balloon-science experiment (see First-order analysis of balloon-borne electronics)
Helmholtz function, 19.3
HGP components, 4.1
High-pressure turbine (HPT), 4.1, 4.2
High-temperature TEM, 37.18
Hilpert's correlation, 28.4, 28.5
Hodes, Marc, 37.3
Hoke, John L., 10.1
Homogeneous algebraic equations, 9.10
Horenziak, Thomas, 44.9
Hot-gas-path (HGP) components, 4.1
Hot-wall convective heat transfer (duct), 26.6
HPT, 4.1, 4.2
Humidity, 40.23–40.27
Humidity ratio, 17.4
Hydrocarbon fraction decomposition, 11.3–11.5
Hydrodynamic entry length, 26.4, 27.9
Hydrogen:
  fuel cell, 6.1, 6.3, 6.4
  molar flow rate, 6.3, 6.4
  molecular weight, 6.4
  turbogenerator rotor cooling, 7.3, 7.9–7.11
Hydrogen molar flow rate, 6.3, 6.4
Hydrostatic head, 30.4
Hypersonic vehicle (see Combustor, hypersonic vehicle, thermal load)

I^2R losses, 7.9
Ideal-gas law, 7.5
IHT software, 40.21
IM-7/PIXA-M thermoplastic tapes, 42.13, 42.14
Impingement cooling, 4.11–4.13
Infrared thermography, 13.1, 21.4
  (See also Transient response of roof to diurnal variations)
Integration, 3.15–3.16, 3.18
  corrosion detection, 9.7, 9.11
  moving sheet, moving fluid, 3.16–3.17
  normalization integral, 9.11
  telescope mirror, 38.8, 38.9
  trapezoidal rule (numerical integration), 24.10
  two-phase pressure drop in pipes, 24.6
Interactive Heat Transfer (IHT) software, 40.21
Intercooler–flash tank, 18.2–18.5
Interfacial bond strength, 42.3, 42.13
Interferometers, 37.9
Interlaminar bond strength, 42.3, 42.10
Intermolecular diffusion process (healing), 42.9–42.13, 42.15
Interpenetration depth, 42.10
Intimate contact, 42.4–42.9, 42.13–42.15
Introductory calculations:
  chemical processing technology, liquid film coating, 1.3–1.6
  closed-loop thermosyphon, 1.6–1.9
  concentric tube heat exchanger, specified conditions, 2.11–2.12
  condenser, large steam power plant, 2.9–2.10
  design of heat exchanger, 2.13–2.16
  duct, air conditioning plant, 2.10–2.11
  furnace wall, coal-fired power plant, 2.6–2.7
  gating and risering system, 1.9–1.11
  refractory lining, smokestack, 2.7–2.9
IR thermography, 13.1
  (See also Transient response of roof to diurnal variations)
Isothermal probability density function, 42.11
Iterations:
  condensing heat exchanger, 28.4–28.9
  double-glazed window, 8.7–8.8
  parameter estimation, 20.12
  passive condenser, nuclear reactor, 31.6, 31.8
  single-phase saturated circulation loops, 30.9, 30.19
  step-by-step procedure, 29.4, 29.32
  thermal system transient response, 15.20
  two-phase pressure drop in pipes, 24.6
Iterative optimization processes, 38.21

Jae-yong Kim, 23.3
Jet array impingement correlations, 4.12
JP-10 fuel, 26.8

Kay's law for mixtures, 40.8
Keith, Jason M., 6.1, 44.1
Kinematic viscosity:
   balloon-borne air-cooled electronics, 40.13, 40.15
   double-glazed window, 8.7
   turbogenerator rotor cooling, 7.6
Kleinfeld, Jack M., 13.1

LabVIEW Virtual Instrument, 23.7
Laminar flow, 6.9
Laminar heat-transfer coefficient, 4.7
Laminar vs. turbulent flow (*see* Reynolds number)
Laminate structures (*see* Thermal response of laminate structures)
Laplace transform, 19.4, 19.5
Laplace transform inversion theorem, 19.7
Laptop computer, cooling mechanisms (*see* Upper bounds of heat transfer from boxes)
Large flow resistance, 35.14
Large telescope mirror, thermal analysis:
   Biot number, 38.6, 38.11, 38.13, 38.17
   differential equation, 38.8
   explicit scheme, 38.12
   Fourier number (Fo), 38.12, 38.13
   integration, 38.8, 38.9
   lumped-capacity model (OD), 38.6–38.10, 38.19–38.20
   M1 thermal environment, 38.5
   measured weather data (Sac Peak, New Mexico), 38.17, 38.18, 38.19
   mirror seeing, 38.5
   1D model, 38.11–38.14, 38.21
   optimizing coolant temperature, 38.17, 38.20
   problem statement, 38.5–38.6
   repeat lumped-capacity model test case, 38.14–38.15
   solar telescopes, 38.4, 38.7, 38.11
   validation of 1D model, 38.15–38.17
Latent heat of sublimation, 17.5
Latent-heat-storage unit, room cooling (*see* Portable PCM cooler)
Lawrence, Thomas M., 14.1
Layered thermoplastic composites (*see* Fabrication of thermoplastic, matrix composites)
Leak detection (*see* Transient response of roof to diurnal variations)

Least-squares error, 34.4
Letan, R., 39.1
Lewis analogy, 17.5
Lewis number, 17.5
L'Hôpital's rule, 44.4
Linear models, 5.9
Liquid crystal coating, 21.4
Liquid film coating, 1.3–1.6
Liquid fuel injector, 26.7–26.10
Liquid reaction velocity constants, 11.5
LMTO method:
   condensing heat exchanger, 28.8–28.9
   heat-exchanger design, 32.6
   heat-transfer rate, concentric tube heat exchanger, 2.5
ln() function, 32.13
Load loss, 30.11
Local gas static temperature, 4.9
Locating corrosion under paint:
   detection concept, 9.3
   differential equations, 9.6–9.7
   eigenfunction, 9.7, 9.9
   eigenvalue, 9.6, 9.10, 9.11
   Green's function, 9.7–9.9
   heat conduction equation, 9.5–9.6
   homogeneous algebraic equations, 9.10
   incident flux, 9.3, 9.4
   integration, 9.7, 9.11
   normalization integral, 9.11
   paint layer thickness, 9.14
   periodic component of temperature, 9.12
   phase lag, 9.3, 9.13–9.14
   problem geometry, 9.5
   separation-of-variables technique, 9.6
   superimposed ripple, 9.13
   surface temperature, 9.12–9.15
   trigonometric calculations, 9.12
Log-log plot of power spectrum, 42.7
Logarithmic mean-temperature-difference method (*see* LMTO method)
Loss (*see* Energy loss)
Low Temperature, Low Energy Carrier (LoTEC):
   aerogel insulation, 19.1, 19.2, 20.3
   derivatives, 20.4–20.10
   differential equation, 19.4, 19.5
   double-door technique, 19.2
   Gauss algorithm, 20.4, 20.11
   Laplace transform, 19.4, 19.5
   Laplace transform inversion theorem, 19.7

Low Temperature, Low Energy Carrier (LoTEC) (*Cont.*):
  nonlinear least-squares parameter estimation technique, 20.3, 20.4, 20.11–20.13
  one-dimensional representation, 19.3
  parameter estimation, 20.1–20.13
  phase-change materials (PCMs), 19.1
  photograph of LoTEC, 19.2
  power, 19.1
  residual error, 20.11
  sensitivity coefficients, 20.11
  sensitivity matrix, 20.4, 20.11
  standard deviation, 20.12
  sum of squared residuals, 20.12
  sum-of-squares error, 19.8
  transient analysis, 19.1–19.9
Lumped capacitance:
  Biot number, 26.11
  PDE exhaust tube (*see* Lumped-capacitance model of PDE exhaust tube)
  SS 440 material test piece, 26.11
  telescope mirror, 38.6–38.10, 38.19–38.20
  thermal system transient response, 15.11–15.13, 15.20–15.30
Lumped-capacitance model of PDE exhaust tube:
  differential equation, 10.4, 10.9
  nomenclature, 10.10–10.11
  prototype PDE, 10.2
  sinusoidal function, 10.8
  steel tube, 10.7–10.10
  superimposed ripple, 10.7
  water-cooled aluminum tube, 10.2–10.7

M_{fuel}, 6.3, 6.4
M1 mirror, 38.4–38.5
  (*See also* Large telescope mirror, thermal analysis)
Magnesium alloy substrate (graphic sheet), 16.16–16.18
Magnetic tape pack (*see* Tape pack cooling)
Mago, Pedro J., 17.1
Marthinuss, James E., Jr., 15.1
Mass-transfer losses, 6.4
Mass-transfer resistance, 28.4–28.6
MathCad, 7.8
Matlab, 38.12, 38.14, 39.14
  (*See also* individual functions)

McLeskey, James T., Jr., 7.1
Mean-square displacement (polymer chain), 42.11
Mechanical draft ventilation, computer case, 35.11–35.15
Mehrotra, Anil K., 25.1
Mesh, 5.9
Method of superposition, 41.2, 41.4
Microsale intimate contact and healing, 42.3
  (*See also* Fabrication of thermoplastic, matrix composites)
Miniaturization of electronics, 24.13
Minichiello, Angela, 40.1
Mirror, telescope (*see* Large telescope mirror, thermal analysis)
Mirror seeing, 38.5
Modified heat of vaporization, 31.6, 31.7
Moist air heat and mass transfer (*see* Heat and mass transfer with phase change)
Molar hydrogen flow rate, 6.3, 6.4
Molecular interdiffusion (healing), 42.9–42.13, 42.15
Momentum analysis of fluid, 24.5
Momentum transfer, 25.18
Moving sheet, moving fluid:
  analysis of heat-transfer problem, 3.7–3.10
  boundary conditions, 3.9–3.10
  dimensionless temperature, 3.8
  Fourier's law, 3.17
  heat-transfer coefficients, 3.11–3.12
  illustrative example, 3.12
  integration, 3.15–3.16, 3.18
  linear differential equation, 3.16–3.17
  local heat flux, 3.17
  materials, 3.4
  nomenclature, 3.19–3.20
  numerical solution scheme, 3.10
  Nusselt number, 3.11
  Prandtl number, 3.11
  processing configurations, 3.4
  Reynolds number, 3.14
  similarity solutions for velocity problem, 3.5–3.7, 3.20–3.23
  streamwise temperature variations of moving sheet, 3.5, 3.12–3.19
  temperature decrease (sheet), 3.18–3.19
MSC/Nastran, 5.6, 12.11, 12.12
Multicomponent paraffin waxes (*see* Solids deposition, oil pipelines)
Multicomponent two-phase flow, 24.1

Multiorganizational balloon-home science experiment (*see* First-order analysis of balloon-borne electronics)
Mutation, 32.5

n-type pellets, 37.4
Nakayama, Wataru, 16.1, 35.1
Naphtha, 11.5
NASTRAN, 38.6, 38.15
Naterer, Greg F., 1.3
National Instruments data acquisition system, 23.7
Natural circulation, 30.1
  (*See also* Single-phase natural circulation loops)
Natural draft ventilation, computer case, 35.7–35.11
Natural selection, 32.1
Naylor, David, 8.1, 28.1
Newton's law of cooling, 22.3, 36.2, 38.11
Ni-based superalloy, 4.5
Nighttime telescopes, 38.6
Nodes, 5.9
Noise, 34.7
Noncondensable vapors, 28.11
Nonisothermal healing process, 42.9–42.13, 42.15
Nonlinear least-squares parameter estimation technique, 20.3, 20.4, 20.11–20.13
Nonlinear models, 5.9
Nonuniform thermal conductivity, 23.5
Normalization integral, 9.11
NTU method, 2.5–2.6
  (*See also* Effectiveness-NTU method)
Nuclear reactor, passive condenser (*see* Condensation heat transfer, vertical tube heat exchanger)
Numerical analysis, 16.1
Numerical solvers (NASTRAN, etc.), 38.6
Nusselt number (Nu), 8.7
  automobile assembly plant, painting process, 14.8
  bubbling fluidized bed, 33.8–33.10
  CFB, 33.14
  circular duct, 26.4–26.5
  condensing heat exchanger, 28.6–28.7
  cooling fin, 5.5
  Dittus-Boelter relation, 22.3
  double-glazed window, 8.6
  electronically heated circular tube, 23.20, 23.22, 23.23
  forced convection, 2.2
  fuel cell cooling, 6.7, 6.8
  heat-exchanger design, 32.6
  heat-transfer film coefficients, 22.3, 22.8
  liquid film coating, 1.6
  moving sheet, moving fluid, 3.11
  reference (Holman), 30.8
  sealed box, 35.3
  sensitivity, 34.5
  tape pack cooling, 12.5
  thermal system transient response, 15.9

Ohmic heating, 37.13, 37.15, 37.17
Ohmic losses, 6.4
Ohmic resistance, 37.7
Oil pipeline (*see* Solids deposition, oil pipelines)
One-dimensional (1D) analysis, 13.2
Oosthuizen, P. H., 8.1, 28.1
Open-circuit voltage, 6.4
Operating costs, 43.4, 43.17, 43.20
Operating point, 27.14
Operational functionality of air-cooled computer (*see* First-order analysis of balloon-borne electronics)
Optical router (ROADM), 37.8–37.19
Optimization methods, 32.2
  (*See also* Heat-exchanger design)
Optimization of refrigeration cycles:
  ammonia, 18.6–18.15
  calculation procedure, 18.4–18.6
  COP, 18.3, 18.4, 18.9–18.12
  examples, 18.6–18.14
  exergetic efficiency, 18.5, 18.6, 18.13–18.14, 18.15
  first-law efficiency, 18.1–18.3
  geometric mean method, 18.3, 18.4, 18.10, 18.15
  intercooler-flash tank, 18.2–18.5
  nomenclature, 18.16–18.17
  optimum interstage pressure, 18.7, 18.9, 18.12–18.13, 18.15
  polynomial expression, 18.10, 18.13
  R22, 18.6–18.15
  second-law efficiency, 18.3–18.4
Optimizing coolant temperature, 38.17, 38.20
Orbital modeling techniques, 40.28
Orifices, 30.7, 30.16
Orthogonality, 9.7
Oven-roasted turkey (*see* Turkey oven design problem)

Overall heat-transfer coefficient (U):
  composite materials, combined
    conduction/convection, 2.2
  condensing heat exchanger, 28.8–28.9
  heat exchangers, 22.2
  inverse, 22.2
  thin-walled pipe, 28.3

p-type pellets, 37.4
Paint drying, curing ovens (*see* Automobile
  assembly plant, painting process)
Paint layer thickness, 9.14
Paraffin wax, 39.3
  (*See also* Solids deposition, oil pipelines)
Parallel-flow concentric heat exchanger,
  2.5
Parameter and boundary estimation:
  estimating freezing time of foods,
    36.1–36.5
  estimation of parameters, 34.3–34.10
  LoTEC, 20.1–20.13
    (*See also* Low Temperature, Low
      Energy Carrier)
  upper bounds of heat transfer of boxes,
    35.1–35.5
Park, Hyunjae, 41.1
Partial derivatives, 20.9
  (*See also* Derivatives)
Partial differential equations (*see*
  Differential equations)
Particle cluster (CFB), 33.12–33.13
Passive condenser, 31.1
  (*See also* Condensation heat transfer,
    vertical tube heat exchanger)
Passive cooling for sealed box, 35.2–35.7
PCB (*see* Convectively cooled heat,
  dissipating component, PCB)
PCM, 19.1
PCM cooler (*see* Portable PCM cooler)
PDE, 10.2
  (*See also* Lumped-capacitance model of
    PDE exhaust tube)
PDV, 43.7–43.8
PEEK ribbons, 42.15
Peltier cooling, 37.13, 37.15
Peltier heating, 37.17
Penetration distance, 16.4, 16.10–16.11
Petroleum fraction decomposition rates,
  11.3–11.5
Petroleum solid deposits (*see* Solids
  deposition, oil pipelines)
Phase-change material (PCM), 19.1

Phase-change-material (PCM)–based
  cooler (*see* Portable PCM cooler)
Phase lag, 9.3, 9.13–9.14
Phase map, 33.4
Pin-banks, 4.3
Pin-fins, 4.3
Pitchumani, R., 42.1
Plastic substrate (graphic sheet),
  16.16–16.18
Plate coolers, 17.2
Platinum catalyst, 6.1
Plexiglass, 21.4
Poco Graphite, 26.8
Polycarbonate, 21.4
Polymer interdiffusion (healing),
  42.9–42.13, 42.15
Polynomial curve fits, plotting techniques,
  40.13, 40.14
Polynomial expression, 18.10, 18.13
Polynomial/power-law regressions, 24.7
Portable PCM cooler:
  air heat capacity rate, 39.6
  applications, 39.2
  calculations, 39.5
  conclusion, 39.20
  configuration of tubes, 39.3, 39.4
  convection heat-transfer coefficient,
    39.5–39.6
  Grimson correlation, 39.5, 39.6
  heat-transfer area, 39.5
  insulated room, 39.7–39.9
  literature review, 39.2–39.3
  mass of air in room, 39.6
  number of tubes, 39.5
  numerical solution of operation,
    39.16–39.19
  operation of PCM cooler, 39.7–39.16
  PCM mass in tubes (M_{PCM}), 39.5
  phase-change material, 39.3
  room heated by ambient, 39.10–39.14
  room heated by ambient and heat
    source, 39.14–39.16
  thermal design, 39.3
Porter, Richard, 15.35
Power:
  fuel cell, 6.3, 6.4, 6.9
  LoTEC, 19.1
Power-law regressions, 24.7
Prandtl number (Pr):
  balloon-borne air-cooled electronics,
    40.8, 40.14
  bubbling fluidized bed, 33.10

CFB, 33.14
Colborn j factor, 27.9
electrically heated circular tube, 23.20, 23.22
forced convection, 2.2
inside heat-transfer coefficient, 25.6
moving sheet, moving fluid, 3.11
sealed box, 35.2
thermal system transient response, 15.9
turbogenerator rotor cooling, 7.9
Prandtl's universal law of friction for smooth pipes, 30.8
Precision temperature control:
  background, 37.3–37.5
  constriction resistance, 37.11–37.12
  control point (cp), 37.7
  cooling efficiency, 37.15
  energy balance, 37.7–37.8
  governing equations, 37.6–37.8
  heat equation, conduction, pellets, 37.6
  heat-sink resistance, 37.15–37.16
  nomenclature, 37.18–39.19
  Ohmic heating, 37.13, 37.15, 37.17
  parabolic temperature distribution, pellets, 37.6
  Peltier cooling, 37.13, 37.15
  Peltier heating, 37.17
  problem solution, 37.10–37.18
  problem statement, 37.8–37.10
  surface energy balances, 37.6
  TEM (*see* Thermoelectric module)
  voltage (V_{TEM}), 37.7
  work (W_{TEM}), 37.7
Preliminary design, 4.6
Prepreg viscosity, 42.14
Present discounted value (PDV), 43.7–43.8
Pressure drop/loss:
  heat exchanger, 32.6
  pipes (*see* Two-phase pressure drop in pipes)
  turbogenerator rotor cooling, 7.7–7.8
Printed-circuit board, 16.11–16.14
  (*See also* Convectively cooled heat, dissipating component, PCB)
Probability density function, 42.10–42.12
Professional engineers/professors (*see* Contributors)
Profilometer, 42.7
Proton-exchange membrane fuel cell, 6.1–6.4
  (*See also* Cooling of fuel cell)

Pseudo-critical properties, 40.8
Psychometrics, 40.24
Pulsed detonation engine (PDE), 10.2
  (*See also* Lumped-capacitance model of PDE exhaust tube)
Pultrusion, 42.3

Q_{cell}, 6.5, 6.6
Quality, classical thermodynamics definition, 24.8
Quasi-steady-state analysis, 17.3

R22, 18.6–18.15
Radiation:
  air-cooled heat exchanger, 30.20
  automobile assembly plant, painting process, 14.10–14.12
  balloon-borne air-cooled electronics, 40.15–40.16
  CFB, 33.14–33.16
  circular duct, 26.2–26.7
  double-glazed window, 8.3
  equivalent heat-transfer coefficient, 2.4
  FEA radiation solvers, 40.28
  lumped-capacitance model, PDE, 10.7–10.10
  roof moisture surveys, 13.4–13.6
  sealed box, 35.3
  Stefan-Boltzmann law, 2.4
  telescope mirror, 38.6
Ramjet performance analysis (RJPA) chemical equilibrium code, 26.2
Random-walk motion, 42.10
Rasul, Mohammad G., 2.1
Rayleigh number (Ra):
  balloon-borne air-cooled electronics, 40.11, 40.12, 40.21
  cooling fin, 5.5
  double-glazed window, 8.6
  sealed box, 35.5, 35.6
  tape pack cooling, 12.5
Reaction velocity constant equations, 11.5
Real-number encoding algorithm, 32.3
Reconfigurable optical add/drop module (ROADM), 37.8–37.19
Refractory lining, smokestack, 2.7–2.9
Refrigeration cycles (*see* Optimization of refrigeration cycles)
Relative humidity, 28.10, 40.23–40.27
Reptation diffusivity, 42.11
Reptation theory, 42.9
Residua, 11.5

Residual error, 20.11
Residuals, 34.4
Resistance:
   constriction, 37.11–37.12
   cylinder, 2.3
   heat-sink, 37.15–37.16
   mass-transfer, 22.6
   mechanical draft ventilation, 35.14
   Ohmic, 37.7
   parallel, 2.3
   series, 2.3
   sphere, 2.4
   spreading, 37.12
   spreading-resistance calculator, 37.12
   thermal (see Thermal resistance)
   thermal system transient response, 15.5–15.10
   Wilson analysis, 22.6
Retrofit older steam turbines, 7.1
Return on investment (ROI), 43.8, 43.22
Reynolds number (Re):
   air-cooled heat exchanger, 30.16, 30.18
   automobile assembly plant, painting process, 14.7, 14.8
   balloon-borne air-cooled electronics, 40.8–40.9, 40.17
   bubbling fluidized bed, 33.7, 33.8
   circular duct, 26.2–26.3
   closed-loop thermosyphon, 1.9
   condenser, steam power plant, 2.9
   condensing heat exchanger, 28.5, 28.7
   cooled gas turbine airfoil, 4.8, 4.9, 4.11
   cooling printed-circuit boards, 41.4, 41.8–41.14
   crossflow compact heat exchanger, 29.12, 29.13, 29.15, 29.24
   crude oil in pipeline, 25.6
   electrically heat circular tube, 23.19–23.22
   forced convection, 2.2
   friction (pressure drop, pipes), 24.7, 24.9
   friction factor (laminar flow), 30.8
   frost thickness, 17.9, 17.10
   fuel cell cooling, 6.7, 6.9
   heat-exhanger design, 32.17
   laminar/turbulent flow, 14.8, 15.9, 26.2–26.3
   moving sheet, moving fluid, 3.14
   roughness, 4.9
   sealed box, 35.2

single-phase natural circulation loops, 30.7, 30.16, 30.18
   thermal system transient response, 15.8–15.9
   turbogenerator rotor cooling, 7.9
   two-phase pressure drop in pipes, 24.7, 24.9
Rheological (thixotropic) behavior, 25.18
Rib rougheners, 4.3
Rinick, Richard, 15.35
Ripple (see Superimposed ripple)
Ritz method of numerical analysis, 5.7
RJPA chemical equilibrium code, 26.2
ROADM, 37.8–37.19
ROI, 43.8, 43.22
Roof moisture surveys (see Transient response of roof to diurnal variations)
Room cooling (see Portable PCM cooler)
Roots of cubic polynomial, 30.10
Rotor cooling (see Turbogenerator rotor cooling)
Roughness features, 42.4
RT-25, 39.3
Run length, 11.6–11.7

Sandgrain roughness, 4.9
Sandoval, Robert, 44.9
Saturation pressure, 17.4
Schauer, Fred, 10.1
Schmidt number (Sc), 28.4
Sealed box, 35.2–35.7
Sealed entrance windows, 38.5
Second-law efficiency, 18.3–18.4
Second-order central difference, 38.12
Seebeck coefficient, 37.4, 37.9
Sekulic, Dusan P., 29.1
Self-affine scaling, 42.7
Semi-infinite solid, 38.6
Semi-infinite solid assumption, 21.3–21.5
Sensible-heat transfer, 28.11
Sensitivity, 34.4–34.7
Sensitivity analysis, 44.8–44.9
Sensitivity coefficients, 20.11
Sensitivity matrix, 20.4, 20.11
Separation of variables technique:
   heat conduction in slab, 9.6
   intimate contact, 42.11
   partial differential equation, 12.6
   use, 9.6
Shah's design procedure, 32.2
Shear stress (duct wall), 26.7
Shell-and-tube heat exchangers, 2.5

Sherif, S. A., 17.1, 18.1
Sherwood number (Sh), 28.4
Silicon carbide, 38.5
Similarity-based, relative-velocity model, 3.22
Similarity solutions for velocity problem, 3.5–3.7, 3.20–3.23
Similarity variable, 3.6
Simple payback (SPB), 43.5
Simple payback period, 43.5–43.7, 43.20–43.22
Simultaneous algebraic equations, 16.9
Singh, R. Paul, 36.1
Single-component, two-phase flow, 24.1
Single-phase natural circulation loops:
  air-cooled heat exchanger, 30.1–30.2, 30.11–30.21
  assumptions, 30.3
  buoyant force, 30.3, 30.6
  coefficient of thermal expansion (α), 30.6
  downcomer, 30.4
  Grashof and Prandtl numbers, 30.20–30.21
  hydrostatic head, 30.4
  iterations, 30.9, 30.19
  local minor losses, 30.4
  local Reynolds number, 30.7
  radiation heat transfer, 30.20
  Reynolds number, 30.7, 30.16, 30.18
  riser, 30.4
  surface roughness, 30.8, 30.17
  viscous shear, 30.4
Single-thermocouple TEM, 37.5
Single-time-constant fallacy, 15.17
Sizing of crossflow compact heat exchanger:
  abstract, 29.1–29.2
  assumptions, 29.3
  Colborn j factor, 29.10–29.17, 29.25–29.27
  computer routine, 29.32
  core mass velocities, 29.14
  determination of core dimensions (MGC procedure), 29.4, 29.9–29.31
  determination of thermal size (TDG procedure), 29.4, 29.5–29.8
  Fanning friction factor, 29.10–29.17, 29.21, 29.25, 29.26
  iterative calculation procedure, 29.4, 29.32
  pressure drop, 29.21–29.23, 29.30, 29.31
  problem formulation, 29.2–29.4
  Reynolds number, 29.12, 29.13, 29.15, 29.24
  step-by-step procedure, 29.5–29.31
  total extended surface efficiency, 29.28
Skin friction coefficient, 26.7
Slab_Sensitivity, 34.5
Smokestack, 2.7–2.9
Solar flux, 13.5, 13.6
Solar flux input curve, 38.17
Solar irradiance, 38.44
Solar telescopes, 38.4, 38.7, 38.11
  (*See also* Large telescope mirror, thermal analysis)
Solidification of liquid metal, 1.23
Solids deposition, oil pipelines:
  application of θ_d, 26.17
  Churchill-Bernstein equation, 25.7
  crude-oil composition (WAT), 25.8–25.10
  crude-oil flow rate, 25.10–25.11
  crude-oil/seawater temperatures, 25.12–25.14
  crude-oil temperature to prevent solids deposition, 25.14–25.17
  deposit thermal conductivity, 25.11–25.12
  determination of deposit thickness, 25.5–25.6
  Dittus-Boelter correlation, 25.6
  energy-balance/heat-transfer equations, 25.2–25.4
  fractional thermal resistance (θ_d), 25.15–25.17
  insulated pipeline, 25.7–25.8
  Prandtl number, 25.6
  Reynolds number, 25.6
  thermal resistance, 25.4, 25.15–25.17
  WAT, 25.2
Space shuttle, 19.1
  (*See also* Low Temperature, Low Energy Carrier)
Sparrow, Ephraim M., 3.3
Spatial derivative, 38.12
SPB, 43.5
Specific heat, 7.6
Specific humidity, 28.10
Speyer, Daniel M., 30.1
Spiral plate heat exchanger (SPHE), 32.6–32.14
  (*See also* Heat-exchanger design)
Spreading resistance, 37.12

Spreading-resistance calculator, 37.12
Square jet arrays, 4.12
Squeeze flow model, 42.4
SS 316, 23.14
SS 440, 26.8
Staggered square triangular pitch, 30.7
Stainless Steel 316 (SS 316), 23.14
Stainless steel 440 (SS 440), 26.8
Standard deviation, 20.12
Stanton number (St):
  Chilton-Colburn analogy, 26.7
  Colborn j factor, 27.9
Steady-state calculations:
  cooled gas turbine airfoil, 4.1–4.15
  cooling fin, 5.1–5.11
  double-glazed window, 8.1–8.9
  fuel cell cooling system, 6.1–6.10
  moving sheet, moving fluid, 3.3–3.23
  turbogenerator rotor cooling, 7.1–7.12
Steam mole fraction, 31.8, 31.9
Steam tables, 31.6
Stefan-Boltzmann constant, 26.6, 35.4, 40.15
Step-by-step methodology (*see* Sizing of crossflow compact heat exchanger)
Strawberries, freezing time, 36.1–36.5
Sub-sea pipelines, 25.2
  (*See also* Solids deposition, oil pipelines)
Sublimation, 17.5
Sum of squared residuals, 20.12
Sum-of-squares error, 19.8
Superimposed ripple:
  corrosion detection, 9.13
  lumped-capacitance model, PDE, 10.7
Superposition method, 41.2, 41.4
Surface asperities, 42.4
Surface roughness, 30.8, 30.17
Surface roughness effect, 4.9–4.10
Swain, James J., 20.1

T_{film}, 12.4
Tangent orifices, 30.7
Tangential slot film cooling, 4.8
Tape laying/automated tow placement, 42.2–42.3
Tape pack cooling:
  aluminum hub, 12.1, 12.2, 12.13
  Biot number, 12.6–12.7, 12.9
  chart solution method, 12.7, 12.8
  FEA method, 12.9–12.14
  Grashof number, 12.5
  Heisler chart, 12.8
  Nusselt number, 12.5
  partial differential equation, 12.6
  problem statement, 12.1
  Rayleigh number, 12.5
  T_{film}, 12.4
  time required for cooldown, 12.9, 12.12, 12.15
TBC, 4.3
Telescope mirror (*see* Large telescope mirror, thermal analysis)
TEM (*see* Thermoelectric module)
Temperature control (*see* Precision temperature control)
Temperature influence factor (TIF), 41.8
Thermal analysis and design:
  balloon-borne air-cooled electronics, 40.1–40.30
  boxes, computer cases, 35.1–35.15
    (*See also* Upper bounds of heat transfer from boxes)
  cooling printed-circuit boards, 41.1–41.18
  fabrication of thermoplastic composites, 42.1–42.22
  portable PCM cooler, 39.1–39.21
  room cooling (latent-heat-storage unit), 39.1–39.21
  telescope mirror, 38.3–38.21
Thermal barrier coating (TBC), 4.3
Thermal capacitance, 15.11
Thermal conductivity:
  balloon-borne air-filled electronics, 40.14
  double-glazed window, 8.7
  frost, 17.6
  nonuniform (circular tube), 23.5
  particle cluster (CFB), 33.13
  solids deposition, oil pipeline, 25.11–25.12
  turbogenerator rotor cooling, 7.6
Thermal decomposition (hydrocarbon/petroleum fractions), 11.3–11.5
Thermal diffusivity, 44.3
Thermal entry length, 26.4, 27.9
Thermal epoxy-alumina interface, 37.12
Thermal mess, 15.11
Thermal modeling, 15.19
Thermal resistance:
  automobile assembly plant, painting process, 14.9
  basic equation, 2.3

condensing heat exchanger, 28.3
corrosion, 9.3
double-glazed window, 8.9
graphite sheet saturates, 16.18
mechanical draft ventilation, 35.14, 35.15
passive condenser, nuclear reactor, 31.3–31.4
sealed box, 35.7
solids deposition, oil pipeline, 25.4, 25.15–25.17
TEM, 37.7, 37.11
(*See also* Resistance)
Thermal response of laminate structures:
  amplitude of temperature swing at heat source, 16.6
  case studies, 16.11–16.19
  disparity relations, 16.6
  graphic sheet substrates, 16.15–16.19
  magnesium alloy substrate (graphic sheet), 16.16–16.18
  numerical analysis, 16.1
  penetration distance, 16.4, 16.10–16.11
  plastic substrate (graphic sheet), 16.16–16.18
  printed-circuit board, 16.11–16.14
  problem definition (formal solutions), 16.2–16.11
  simultaneous algebraic equations, 16.9
Thermal storage units, 39.2
Thermal system transient response:
  approximating transient response, 15.22–15.30
  capacitances, 15.13–15.16
  differential equation, 15.17
  energy balance, 15.4
  first law of thermodynamics, 15.3
  governing equations, 15.3–15.4
  heat path description, 15.2
  iterative solution, 15.20
  long transients, 15.32–15.34
  lump masses where possible, 15.11–15.13
  lumped-capacitance approach, 15.11
  lumped-mass approximation, 15.20–15.30
  multiple-transient response, 15.17–15.19
  nomenclature, 15.35–15.36
  Nusselt number, 15.9
  Prandtl number, 15.9
  problem definition, 15.2–15.3
  proximity to thermal event, 15.34–15.35
  resistances, 15.5–15.10
  resistor network, 15.5
  Reynolds number, 15.8–15.9
  short transients, 15.31–15.32
  single-time-constant fallacy, 15.17
  steady-state temperatures, 15.10–15.11
  system response, 15.30
  thermal mess, 15.11
  thermal modeling, 15.19
  time constant, 15.20–15.22
  transient characteristics, 15.3–15.4
Thermal warpage, 38.3
Thermocouples, 26.10, 26.13, 37.4
Thermoelectric effects, 37.3
Thermoelectric module (TEM):
  advantages/disadvantages, 37.5
  controlled-side interface (csi), 37.4, 37.5
  cutaway view, 37.4
  high-temperature TEM, 37.18
  n-type pellets, 37.4
  p-type pellets, 37.4
  schematic representation, 37.5
  single-stage, 37.4–37.5
  thermocouple, 37.4
  uncontrolled-side interface (usi), 37.4, 37.5
  (*See also* Precision temperature control)
Thermooptic phase shifters, 37.9
Thermophysical properties:
  air, 23.14
  mixture of ethylene glycol and water, 23.15–23.16
  Stainless Steel 316, 23.14
  water, 23.14
Thermoplastic composites (*see* Fabrication of thermoplastic, matrix composites)
Thermosyphon, 1.6–1.9
Thin meniscus mirror, 38.3
  (*See also* Large telescope mirror, thermal analysis)
Thixotropic behavior, 25.18
Three-dimensional (3D) model:
  CFD and heat-transfer modeling, 4.6
  detailed final design, including edge effects, 38.6
  interaction on roof on wet/dry areas, 13.3
  3D calculations, 38.6
Three-temperature problem, 4.8, 21.5
Tiedeman, J. S., 18.1
Tiliakos, Nicholas, 26.1

Time constant, 15.20–15.22
Time derivative, 38.12
Time steps:
　automobile assembly plant, painting process, 14.13
　roof moisture surveys, 13.8–13.9
Total extended surface efficiency, 29.28
Tournament selection, 32.4
Tow placement/tape-laying processes, 42.2–42.3
Transcendental equation, 44.4, 44.5
Transient and cyclic calculations:
　automobile assembly plant, painting process, 14.1–14.16
　combustion tube of prototype PDE, 10.1–10.11
　corrosion detection, 9.3–9.15
　fired cracking heater, decoking intervals, 11.1–11.9
　heat and mass transfer with phase change, 17.1–17.13
　LoTEC, 19.1–20.13
　optimization of refrigeration cycles, 18.1–18.17
　tape pack cooling, 12.1–12.15
　thermal response of laminate structures, 16.1–16.19
　thermal system transient response, 15.1–15.36
　transient response of roof to diurnal variations, 13.1–13.13
Transient heat-transfer problem (*see* Tape pack cooling)
Transient response of roof to diurnal variations:
　assumptions, 13.4
　boundary conditions, 13.4–13.8
　calculation results, 13.9
　convection, 13.4, 13.6, 13.7, 13.11
　Cosmos/M software, 13.2
　FEA methods, 13.2–13.4
　IR thermography, 13.1
　radiation, 13.4–13.6
　solar flux, 13.5, 13.6
　temperature difference, wet/dry roof, 13.1
　temperature histories, 13.9–13.12
　time steps, 13.8–13.9
　varying the conditions, 13.11–13.12
　wet material property, 13.8
Transient thermal events (*see* Thermal system transient response)

Trapezoidal rule for numerical integration, 24.10
Trigonometric calculations:
　corrosion detection, 9.12
　lumped-capacitance model, PDE, 10.5
Tubes, pipes, ducts:
　circular duct, heat transfer, 26.1–26.15
　combustor, hypersonic, thermal load, 26.1–26.15
　duct, air conditioning plant, 2.10–2.11
　electronically heated circular tube, inside-wall temperatures, 23.3–23.27
　solids deposition, oil pipeline, 25.1–25.18
　two-phase pressure drop in pipes, 24.1–24.14
Turbine airfoils, 4.3
Turbine vane cooling designs (*see* Cooled gas turbine airfoil)
Turbogenerator rotor cooling:
　absolute viscosity, 7.5
　cooling principle, 7.2
　copper temperature rise, 7.9–7.11
　density, 7.5
　forced convection, 7.6
　friction, 7.7
　gas properties, 7.3–7.6
　hydrogen, 7.3, 7.9–7.11
　I^2R losses, 7.9
　kinematic viscosity, 7.6
　maximum copper winding temperature, 7.11
　operating conditions, 7.5
　Prandtl number, 7.9
　pressure loss, 7.7–7.8
　Reynolds number, 7.9
　self-generated pressure, 7.6
　specific heat, 7.6
　thermal conductivity, 7.6
　volume flow rate, 7.6–7.8
Turbulators, 4.3
Turbulent convective flows, 4.3
Turbulent heat-transfer coefficient, 4.7
Turbulent vs. laminar flow (*see* Reynolds number)
Turkey oven design problem:
　Biot number, 44.3, 44.6
　problem statement, 44.1–44.2
　results, 44.8–44.9
　sensitivity analysis, 44.8–44.9

solution, 44.3–44.8
steps in procedure, 44.2–44.3
Two-dimensional design, 4.6
Two-dimensional (2D) model, 13.2
Two-phase forced-convection flow patterns, 24.4
Two-phase pressure drop in pipes:
  acceleration term, 24.8
  adiabatic flow, 24.13
  annular models, 24.14
  assumptions, 24.2
  Blasius form of smooth pipe friction factor, 24.9
  Blasius smooth pipe correlation, 24.7
  dividing pipe into segments, 24.7
  elevation term, 24.8
  enthalpy, 24.8
  flow instability, 24.3
  friction, 24.6, 24.7, 24.9
  fundamental principles, 24.3–24.4
  given information, 24.2
  homogeneous model, 24.2, 24.13–24.14
  increasing system pressure, 24.11, 24.12
  increasing tube diameter, 24.11, 24.12
  increasing tube power, 24.12
  limiting flows, 24.12–24.13
  miniaturization of electronics, 24.13
  momentum equation, 24.5–24.6
  nomenclature, 24.2–24.3
  polynomial/power-law regressions, 24.7
  pressure profile, 24.10, 24.11
  procedural steps, 24.6–24.7
  quality, classical thermodynamics definition, 24.8
  results, 24.10–24.13
  Reynolds number, 24.7, 24.9
  solution details, 24.7–24.10
  trapezoidal rule for numerical integration, 24.10
  two-phase forced-convection flow patterns, 24.4
Two-stage vapor compression refrigeration cycles (*see* Optimization of refrigeration cycles)

U (*see* Overall heat-transfer coefficient)
U.S. Standard Atmosphere, 40.2, 40.3, 40.6, 40.10
Ultra-low-expansion (ULE) glasses, 38.3
Uncontrolled-side interface (usi), 37.4, 37.5

Unit step function, 19.3
Upper bounds of heat transfer from boxes:
  cooling mode (three choices), 35.1
  exterior envelope of box, 35.2
  fan cooling, 35.11–35.15
  goal, prevent components from overheating, 35.1
  mechanical draft ventilation, 35.11–35.15
  natural draft ventilation, 35.7–35.11
  passive cooling (sealed box), 35.2–35.7
  sealed box, 35.2–35.7
  vent ports, 35.7–35.15
Useful exergy gained, 18.5
usi, 37.4, 37.5

Vapor compression refrigeration cycles (*see* Optimization of refrigeration cycles)
Vapor shear, 28.11
Variational calculus, 12.11
Velocity problem, 3.5–3.7, 3.20–3.23
Vent ports, computer case, 35.7–35.15
Vertical high-aspect-ratio enclosure, 8.6
Vertical tube heat exchanger (*see* Condensation heat transfer, vertical tube heat exchanger)
Vierow, Karen, 31.1
Viscous shear, 30.4

WAT, 25.2
Water, thermophysical properties, 23.14
Water vapor saturation pressure, 40.25
Wax appearance temperature (WAT), 25.2
Waxy crude oil (*see* Solids deposition, oil pipelines)
Wessling, Francis C., 19.1, 20.1
Willey, Ronald J., 22.1
Wilson analysis, 22.1–22.9
Wilson plot, 22.4, 22.8
Window (*see* Double-glazed window)
Woodbury, Keith A., 32.1

Yang, F., 42.1

Zero-dimensional (0D) model, 38.6–38.10, 38.19–38.20
Zero-thickness control volume layer, 23.11
Zerodur, 38.3
Ziskind, G., 39.1

ABOUT THE EDITOR

Myer Kutz has been President of Myer Kutz Associates, Inc., a publishing and information services consulting firm, since 1990. He was vice president in charge of professional and sci-tech publishing at John Wiley & Sons for 5 years. He was a member of the board of the Online Computer Library Center and is a former chair of the publications committee of the American Society of Mechanical Engineers (ASME). He has a BS in mechanical engineering from MIT and a master's from RPI. Mr. Kutz is the editor of *Standard Handbook of Biomedical Engineering & Design* and *Handbook of Transportation Engineering*, both from McGraw-Hill, and other books on engineering.